Thank god For Eve

Why Religion Exists And Why It Must Go

To my new friend and fellow
freedom fighter Pedro.
Best of luck in all you do!
—Max

Max Humana

Thank god For Eve; Why Religion Exists And Why It Must Go.

1st Edition

Published by Reality Publishing, LLC.

For more information, contact Reality Publishing, LLC at realitypublishing2100@gmail.com.

Cover design by Mrs. Humana and Max Humana.

ISBN: 978-1-7332857-1-1

Other Books By The Author

Four Degrees of Separation

America As A Case Study In The Harmful Effects of Religion (Chapter Six from the book *Thank god For Eve* as a separate pamphlet)

An Opening Salvo On Coexistence; Plus Everybody's An Atheist And I Can Prove It (Chapter Eleven from the book *Thank god For Eve* as a separate pamphlet)

Thou Shalt Not Think; How Psychology Subverts Reason And Opens Pandora's Box of Faith (Chapter Twelve from the book *Thank god For Eve* as a separate pamphlet)

Religions Are Fruitful, They Multiply, And They Are Very Bad For You (Chapter Fifteen from the book *Thank god For Eve* as a separate pamphlet)

Why Is Any Moral Person Still A Catholic? (The Church of Evil, Abnormality and Hypocrisy) (Chapter Twenty-One from the book *Thank god For Eve* as a separate pamphlet)

America's Completely Secular Founding Documents (More Christian Attempts At Rewriting History) (Chapter Twenty-Seven from the book *Thank god For Eve* as a separate pamphlet)

Religious Freedom Laws, Also Known As "My god Says You're A Piece of Shit" (Chapter Twenty-Eight from the book *Thank god For Eve* as a separate pamphlet)

Dedication

This book is dedicated to all brave, courageous and happily godless freethinkers, humanists, secularists, scientists and supporters of science and reason in all times and in all places. It is dedicated to every person who has ever advanced the causes of science and reason at the expense of dogma. All these wonderful fellow humans deserve heartfelt credit and our collective thanks for their expansion of human knowledge.

I give a special dedication to all the strong-willed women throughout history who have challenged the establishment and demanded that their equality be recognized. Brave and courageous women have pushed humanity to many of its moral awakenings and the expansion of its freedoms.

From the first evidence-minded freethinking *Homo heidelbergensis* who raised a furry skeptical eyebrow at his fellow troopmate's obeisance to an imaginary entity, to our likeminded godless kin in the far distant future who will quite likely have to continue today's fight to keep religion and state safely separate, I salute you all.

Your courage in standing up for reason in the face of persecution and ignorance, your battles against mindless tradition and those who believe their god is on their side, and your never-ending pursuit of a better life for all creatures on this planet all demonstrate that imaginary gods only get in the way of doing what is right and what is good.

Religion's long war on reason and the reasonable has left long trails of tears and blood across our world. The image below depicts the kind of Inquisitional treatment which the deeply and sincerely faithful have so frequently and so eagerly inflicted upon fellow human beings who chose to think for themselves.

Spanish Inquisition.

The faithful have tortured and killed more people, and damaged more minds, bodies and lives, than we will ever be able to count. I would be inconceivably remiss if I did not sincerely dedicate this book to all of the men, women, children and the unborn who have lost their freedom and their lives at the willing hands of the devout.

Acknowledgements

All humans owe their very existence to their ancestors, and this existence is made flesh in the form of our bodies and brains. Our interactions with our parents, families and societies strongly shape our attitudes and perceptions. The yin and yang of nature and nurture make us who we are.

Thus in the true spirit of genesis, I must first and foremost thank my parents for passing on to me the blend of genes which made me an empiricist and a realist. Fortunately, my inherited DNA orients me away from fear, superstition and its unjustified dogmas and toward courage, reason and an empirical reality.

I must also thank them, and primarily my mother, for providing an environment conducive to learning, thinking and advancement in the pursuit of education and intellectual self-development. The exclusion from our home of any requirements for dogmatic thinking and blind obedience to authority, combined with the freedom we had to explore ideas and information as a pure and unconstrained pursuit, set the conditions for many years of delightful reading, talking and thinking.

I also cannot thank my mother enough for the numerous instances of her selfless parenthood which I witnessed from my earliest years on this planet. My memories of these acts will remain with me until my very last conscious moment. Without ever calling attention to them, she set the example of what it means to actually do things for other people without any expectation of reciprocation, recognition or reward.

Sometimes it really is the smallest and most unconscious acts of selflessness and kindness that say the most about people and human nature. My mother was also instrumental in keeping me in the best schools she could find even when we could not afford them, and has been focused on education and critical thinking for as long as I can remember. I owe her, quite literally, everything I am, and I will always be grateful for it.

I must also thank my wife for her endless patience as I perused references, pored over transcripts, pounded away at the keyboard and spent time in dialogue with the faithful. I very gratefully thank her for her reviews of various parts of this book as it was being written and for her helpful insights and comments. I also need to thank and congratulate her for her creation of the Eve and garden artwork on the cover of this book, which does such a good job of translating the vision of the book and Chapter 24 in particular into visual form. And most of all, I want to thank her for her love and companionship, for which I will be eternally grateful.

Preface

If the world were simple, we would not need to expend much time or effort to learn about it. But the world is far from simple. Life and the universe are extremely complicated. Gaining even a rudimentary understanding of life and the universe requires deliberate study, an open mind and often a quite considerable amount of time and effort. A typical high school education in America will not get you there; it will barely scratch the surface.

Millions of Americans thus don't reach even a basic level of understanding on these subjects. Achieving a moderate level of understanding requires further study of a wide range of subjects and even more time and effort. Few people reach this level of understanding. Achieving a truly deep understanding of either of these subjects requires an equally deep intellect and far more time than most people can dedicate to the pursuit, even if they wanted to. Therefore only a very few people in the world become truly deep experts in more than a single subject or field of study.

If life and the universe were simple, we would not need experts to explain them. As people are not born knowing economics, history, mathematics, grammar, politics and religion, people are not born with any kind of understanding of biology, genetics, psychology, chemistry, paleontology, physics or astronomy. Each and every bit of information in our heads enters our brains through a very long process of learning. Learning even a single one of these extremely complicated subjects in depth gives insight into our world, and learning multiple disciplines enables a person to gain both broad and deep insight into the world.

The farther one climbs up the ladder of knowledge, the more one can see and understand. Just as a person can see far more of the world from an airplane than from a car, having a broad and deep understanding of multiple fields of study enables a person to see and understand things from a far different and far more informed perspective. But climbing the ladder of knowledge is a long process and it requires time and effort.

Far too many people want things to be simple and easy. People prefer to read summaries of very complicated subjects and to listen to five-second sound bites which can only present the tiniest bit of an issue. These people take in a very small amount of information but believe that after being exposed to only the tiniest microscopic bits of a subject they are somehow adequately informed. They do not let their own personal lack of knowledge or their lack of understanding stop them from forming opinions.

They often do not want to take the time or the effort to learn enough about issues to make truly informed decisions. This contemptible anti-intellectual and anti-scientific mindset is now mainstream in America, and it is causing extreme and long-term damage to our country. Uninformed and ignorant people do and say uninformed and ignorant things, and we see more of this every day.

With the above in mind, this is not a hit-the-highlights or a sound-bite type of book. It covers a wide range of topics and often in detail, but is much less dense and much more readable than a scientific paper. Therefore it is not a short book, but it is easy to understand. I discuss topics that are rarely, if ever, found in the same place, and I needed a longer than average book to do them justice. I also included some humorous satire to break up the book and poke fun at mindless superstition, so there is a mix of science, history and satire, but the reader should never forget that the overall point of this book is deadly serious.

I aim to inform, inflame and motivate, and my goal is to rid the world of the mindless and destructive poison of religion and the lack of critical thinking that enables it.

Every reader of the book is likely to learn new things and will be more informed on a wide range of subjects. This is always a very good thing. Many readers will be angry or appalled at something in the book. This is good as well, because some of the facts are quite appalling. The faithful will be angry at my direct and uncompromising attacks on and destruction of their opinions and superstitions. I aim to make as many of these believers so angry or appalled at the actual immorality, lies and evil within religion that they leave it immediately and bring as many people with them as they can on the way out.

Nonbelievers will be angry as they discover that the problem with religion is far worse than even they thought, just how much damage religion has done to our world, how much evil it is doing around us today and how much damage it will do if it is not eradicated as soon as possible. I aim to motivate every single nonbeliever to prepare for and fight a lifelong battle against the armies of ignorance and superstition that are trying desperately to return our world to the Dark Ages.

With those understandings up front, please dive in.

Contents

Part One
Setting the Stage

Part Two
Seeing More Clearly

Part Three
A Deeper Understanding

Part Four
Accepting Unpleasant Facts

Part Five
The Women's Perspective

Part Six
America's Secular Politics

Part Seven
Miracles Are Imaginary

Part Eight
The Proper Path

Part Nine
In Conclusion, Plus Quotable Quotes

Part Ten
References and Further Reading

Part One

Setting the Stage

Chapter One

Putting Things Into Perspective

"I the LORD thy God am a jealous God, visiting the iniquity of the fathers upon the children unto the third and fourth generation of them that hate me."
- Jehovah

"That the Roman pontiff alone can with right be called universal. That for him alone is it lawful, according to the needs of the time, to make new laws... That of the pope alone all princes shall kiss the feet.
That it may be permitted to him to depose emperors. That a sentence passed by him may be retracted by no one; and that he himself, alone of all, may retract it. That he himself may be judged by no one.
That the Roman church has never erred; nor will it err to all eternity, the Scripture bearing witness."
- Dictates of Pope Gregory VII

"Wives, submit yourselves unto your own husbands, as unto the Lord. For the husband is the head of the wife..."
- Paul the Apostle

"I would rather be a rebel than a slave."
- Emmeline Pankhurst

"*Thank god for Eve*". Anyone reading those words will surely wonder what they mean. They mean, quite simply, a celebration of courage, and I do not mean a mere quiet independence of thought. I mean true fearless courage and the kind of strong backbone and iron will that has enabled humanity to break through formerly impossible barriers, to face up to desperate situations, to stand up to tremendous dangers and to accept the most unpleasant truths.

Courage enables a person to look the universe straight in the eye, take its measure, truly understand how fleetingly ephemeral a single life is, and then accept this fact and to resolve to make the most of life with no childish wishful thinking about enduring after death. Having real courage indicates the true strength of a person's mind, although it should never be confused with psychological or ideological rigidity. Having real courage means one is willing to face and deal with reality as it is, not as we wish it would be, and this has always been and will always be the best path forward for humanity as individuals, as societies and as a species.

The extraordinarily weird mythology of Christianity contains many parables, fables and stories from the ridiculous to the detestable. All religions and cults have the same problem, but Christianity and its two archaic dusty desert monotheistic siblings are absolutely overflowing with them. Christianity's central text is almost completely ignorant and evil but is nevertheless advertised as a so-called "good book" and is celebrated as the so-called "good news" of an ancient human torture and sacrifice.

Those who exalt at that cruel and utterly inhuman yet continually celebrated "good news" have used the contents of their book throughout its history as justification for the most horrible tortures and murders of perfectly innocent human beings which it is possible to imagine.

Burning people alive was only one example among a vast collection of cruel and painful tortures inflicted upon the innocent by those with "deeply and sincerely held beliefs", to use the phrase so widely used by aggressive theofascists in America today.

Jews have endured almost two thousand years of individual and collective persecution at the hands of devout Christians for the imaginary crime of denying, betraying and murdering a god, or at least one of its offspring. That an allegedly all-powerful god would allow itself or its "only son" to be tortured then brutally killed by a band of extremely ignorant and weak semi-evolved primates would seem to negate any fault on the part of those alleged to have perpetrated the act, particularly when it was all supposed to have happened according to the all-powerful god's plan, which he had been allegedly waiting to put in motion for untold trillions of years.

As the murdered individual was supposed to be able to perform miracles from resurrections to healings to casting out demons to walking on liquid water, the only way this individual could have been killed by crucifixion or any other act was if he deliberately let it happen.

Which he obviously did if you believe the fairy tale, but admitting this would ruin the story and also remove the justification used by Christian faithful to persecute a group of people they have always despised.

Thus this is a prime example of how the faithful ignore logic, leap to and then cling desperately to the irrational and erroneous conclusions which form the quicksand upon which all religions are constructed.

Divine fathers are not the only ones who let their children be tortured or killed. Human fathers in the Christian bible are held up as paragons for agreeing to murder their deeply loved children in cold blood. One need only take Abraham in Genesis and Jephthah in the book of Judges as proof.

When mere filicide isn't enough to satisfy the bloodthirsty god, male warriors and chieftains are given divine orders to commit genocide and they gleefully make the rivers run red with the blood of the innocent, including that of women, children, babies and huge numbers of the unborn. The more evil the biblical male, and the more mindlessly and slavishly and cruelly he follows orders, the more praiseworthy he is in this cult of death and sadism.

Such psychopathy is, according to their god's own book, exactly how and why we are told the rest of us need to praise their big daddy in the sky. This god is quite literally the universal source of evil and cruelty in not just Christianity but also in the other two Mesopotamian monotheistic cults. It is nearly impossible to find a more ignorant, evil and antihuman book than the Christian bible.

Women in that bible, when they are not being recommended for death by stoning or being shunned by men for any of an astronomically large number of offenses or 'unclean' states such as menstruation and childbirth, are normally held in extremely low regard, at or perhaps sometimes slightly below the level of sheep and goats.

This is because they aren't as mindlessly and robotically obedient to godly commands as their male owners. In Genesis 19:26 (King James Version (KJV)), Lot's wife was turned into a pillar of salt simply because she disobeyed the order not to look back at her god's "shock and awe" divine carpet-bombing of Sodom and Gomorrah during her family's hasty departure from Ground Zero.

Genesis also describes the so-called 'loving creator' of the universe's total and eternally unrelenting ire at Eve and all her many billions of innocent descendants for her disregard of the order from the voice in the sky not to partake of the fruit of the tree of knowledge of good and evil (Genesis 3, KJV).

And many billions of women ever since the story of Eve have suffered cruel oppression and much, much worse, merely for the genetic circumstance of being born with a vagina, or for the apparently infinitely more horrific crime of daring to use it.

Hundreds of millions of women even in the 21st century still suffer from oppression and misogyny on a daily basis, from being told to shut up and subordinate themselves to their husbands, being forced to stuff themselves into cloth bags, being made to cover whichever parts of their bodies their menfolk can't bear to look at, to suffering brutal beatings to rape to acid attacks to murder, all fully justifiable in the name of the biblical god or his Islamic doppelganger.

The biblical messages are clear; if you want what's good for you, do what the Big Guy, both the imaginary one in the sky and the real one threatening you in this world, says without question, no matter how disgustingly cruel the order, and *don't* be born female.

Quite contrary to how the Eden story was intended to impact its audience, I personally find the fable of Eve's defiance to be thoroughly inspirational and uplifting, on par with those of other brave women throughout the history of humanity who have persevered[1] and persisted[2] in the face of repression, threats, danger and the unknown. It is the story of that mythology's first scientist and freethinker, and how she was treated is thoroughly representative of how monotheisms in general have treated science, scientists, freethinkers and indeed women of any enhanced intellectual persuasion and independence even unto this day.

I see Eve as an inquisitive and restless self-made woman and independent thinker, not content to live out eternity in a tiny fishbowl with every day a mind-numbingly boring Groundhog Day repetition of the previous one. She, unlike her gutless male companion, had a brain, a spine and free will and she used them. How she used her vagina has not been well documented.

The biblical Eve was not content to live her entire life like a mindless vegetable, again unlike her spineless and pathetic sidekick. She had no intention of merely existing, lost in a fog of intellectual darkness. I picture her looking up at the sun, the moon and the stars not with slack-jawed stupefaction and spineless weak-kneed unquestioning masochistic worship, but a driving and intense curiosity and a burning need to know the underlying realities in those celestial bodies.

I imagine her looking around her tiny world and wanting to know the "why" in everything she saw. Rather than being a cautionary tale about the need for blind and unquestioning obedience to the ruling male authority, as intended by its writers and its purveyors even today, I see her story as exemplifying the need for rebellion against oppression, as a pioneer's journey into the unknown, and as an inspiration for the bold, the thoughtful and the truly moral in every age.

The Eden story itself is quite simple. It is a fable much less Aesop and much more Brothers Grimm, designed to make the reader or listener think that blind and unquestioning obedience to authority and orders is the right way to live, and that defiance of the psychopathic authoritarian in the sky has truly eternal consequences, both for the offender and for every single one of his or her descendants.

Those of us with a skeptical, a curious and a scientific mindset who also have spines will never cravenly knuckle down to authority. We refuse to live like that. We will question everything, *especially* authority, because we know how badly power and influence corrupt the minds of the weak, the vain and the power-hungry.

We freethinkers know that people lie and they seek to manipulate others for personal gain. We are keenly aware of how unbelievably hard a slog it can be to discover the amazing facts about our universe. We live for the joy of discovery, and for the full-body rush of pleasure which a scientific or intellectual insight can bring. We know that we must keep pushing the boundaries of the known to discover the unknown, and that it will sometimes be extremely difficult to push through social norms, conventional wisdom and common sense in those pursuits.

We know this and yet we still push on. We will never give in to the mindset of the sheeple and their compulsion to flock together. We will never be like those herd animals in human clothing who hide safely in the middle of the pack, eyes downcast and spines turned to jelly, afraid to question the leader and afraid to take bold steps in the pursuit of independence, knowledge and enlightenment.

That strong and independent mindset, that spine, that 'spirit' if you will, is what this book's title and contents seek to convey. Yes, indeed, thank god for Eve! I wish this world had several billion women and girls just like her.

As for the totality of this book, it is neither a debate nor a carefully balanced consideration of the arguments for and against certain religions or metaphysical propositions. It is an antitheistic, unapologetic, fact-based and very direct polemic. It is a full frontal attack on the foolishness and evil of religion and the faith mindset. It describes in clear and direct detail just how much completely preventable damage religion has done and continues to do to individual people and to our societies. It contains quite a lot of epistemology, a bit of exegesis and much political incorrectness. It states facts and truth which need to be stated quite plainly, and is a combination of those facts, with analysis, commentary and pointed satire on this very contentious subject liberally mixed in.

I have no interest in allowing these pages to become contaminated by apologist argumentation for 'equal time', or in allowing wishful thinking as the justification for including illogical apologetic special pleading. To customize Exodus 22:18 (KJV), I shall not suffer evil, ignorance or stupidity to live. Religion needs to be shelved in the library next to astrology and alchemy, right behind alien mind control, werewolves, leprechauns and witchcraft, as all of these faith positions are equally imaginary and equally indefensible in an informed world or brain.

I have heard a great many apologetic arguments covering the entire range of possible appeals for religion, and they all fail catastrophically. To cite just one example, I do not wish to listen for the thousandth time to some religious person trot out the names of some scientists who happened to be highly religious and made some positive contribution to one field or another despite this serious handicap. We secularists and freethinkers respond with three points to this misinformed epic fail.

One is that we freely admit there have been many scientists throughout human history who were religious. This is for an extremely good reason which sharply boomerangs and jabs the religious apologist straight in the brain. You can pithily sum it up by saying "we believed such things because we didn't know any better". For most of human history the religious perspective was the one which seemed most logical given the extreme paucity of actual scientific knowledge.

While it was the only reasonable or apparently reasonable explanation, religion could be and was part of the dialogue. But that is not the case today. In the last 150 years or so, science has moved from a crawl to a walk to an Olympic-level sprint and in many fields it is even accelerating. During that time the most intelligent among us have had many truly good and properly informed scientific ideas about life, the real world and the universe.

It is only in recent decades that supporting data have become available to us to prove many of these ideas true, and zero gods were required for this. This is fortunate, since none of humanity's gods have ever sent even a single syllable of data. It is only since 1859 that the mortal blow to religion was inflicted by the once-religious but later enlightened and nontheistic Charles Darwin in his absolutely masterful *On the Origin of Species*.

The aftershocks from that blow are still reverberating around our world, while various forms of social and religious inertia, along with massive amounts of ignorance and fear, prevent the acceptance of scientific findings that are damaging to religious beliefs. The hurdles one must cross today to be both a religious person and a scientist require multiple simultaneous catastrophic breakdowns of reason and also leaps of faith long enough to hurdle the Grand Canyon in a single evidence-free bound.

The second point is that the number of classically religious scientists has plummeted since Darwin's time, and the decline would be even greater except for the fact that scientists are people too. No one wants to die and go out like a burned-out light bulb. No one wants to lose their families forever. Most people would love to actually transition to a truly wonderful afterlife, and this includes me and I am sure the vast majority of godless freethinkers and scientists as well. There just isn't even a molecule of evidence that there is one.

But like drugs to the addict, this fantasy is irresistible to far too many people. Thus the powerful allure of an imagined afterlife becomes the central nexus around which these people form their worldviews. Thus rationalizations, cognitive dissonance and mountains of special pleading enable some people to employ reason and logic to do science while at work and to turn those rational modules off when they leave the lab in order to enable their fact-free and faith-driven anesthetic religious modules.

So there are some religious scientists today but it is a fraction of what it once was, that number is shrinking all the time, and the only justification for their faith today is…faith.

And finally, for every single religious scientist which apologists cite, we secularists can point to the graves or the ashes of millions of men, women and children who were displaced, imprisoned, tortured and/or killed during the Crusades, the Inquisitions, less widely known crimes against humanity like the Cathar persecutions, any of the myriad other religious wars and "ethnic cleansing" campaigns from practically every country on earth throughout the entire history of religion, and of course the Holocaust.

That particular permanent disfigurement of human history was a genocidal mass murder for which Christianity in general and the Catholic sect in particular had a large hand in causing and against which the Vatican didn't have too much bad to say before, during or after the carnage. I'll take science over religion every day of the week and a billion times on any believer's most holy day.

To address another line of pleading, arguments about the so-called "fine tuning" of the universe for we now-clothed apes are merely arguments from ignorance and what I call arguments from gullibility and wishful thinking. Applying the apologists' twisted illogic to science and reason results in unbelievably jaw-droppingly vacuous non sequiturs.

Among an extremely well-populated field of this sort of logic-free lunacy, we find the creationist Ray Comfort's hilariously idiotic description of a banana[3] as proof of not merely the existence of some random and unknowable god somewhere, but *his own personal specific god's* existence. It is almost impossible to appropriately describe such a display of terminally unrecoverable uncritical thinking. It is equivalent to seeing an apple or a frozen waterfall and claiming that to be all the proof one needs that Zeus, Marduk, Manitou or Osiris is actually the One True God.

The tropes and attempts at argumentation offered by William Lane Craig and other current apologists follow this same well-worn path far away from both science and logic and straight to the dark depths of Christian mythology. What passes for their reasoning "tree" has one branch. They think all knowledge and all reality come from their sole, single specific particular pet god and all things unknown also come from their sole, single specific particular pet god. The coin of their realm is that of Janus, in which heads Christianity wins, tails Christianity still wins, and nothing you say or demonstrate to them will ever change or open their closed minds.

This so-called "fine tuning" argument is merely the well-documented and apparently permanent religious stupefaction at very large and very small numbers combined with the fact that we very likely aren't asking the right questions. Two plus two always equals four, and we didn't need a god to tell us that four was the correct answer. There is no magic or divine intervention needed to form snowflakes, rocks, fossils, organic compounds, the original life forms, oceans, planets and galaxies.

Our universe does of course have certain characteristics and its matter and energy have certain properties, but they need not be any more deliberately fine-tuned than the dropping of a stone into a pond requires deliberate divine intervention or fine tuning to keep circular ripples moving away from the center.

We will soon know more about physics than we do today, and the fine tuning special pleading will wilt and evaporate just like all of the other thousands of apologetic arguments have, like lurking vampires exposed to a full Mesopotamian noon sun. We don't need gods to do math or science.

The same goes for biology versus the deliberately deployed Trojan Horse of the wholly disingenuous "intelligent design" attempt at Constitutional subversion. The creationist asks "How can it possibly be that such marvels like eyes, wings, bananas, the immune response, our emotions and our consciousness not be proof of the **Christian** god?" He is convinced by his personal biases and worldview that not only do these observations lead to the conclusion that some god somewhere can be the only possible reason for their existence, but also and again that the creationist's *own personal pet god* is the one responsible.

Thus fallacy is piled upon error, then both are draped atop ignorance and underlying biases to construct a flawed and skewed but often unalterable structure of thought. This is the creationist and apologist model, and is a self-fulfilling Ouroboros[4]-like escapism. It's a quite damning indictment of human psychology that so many people fall victim to such massive and obvious non sequiturs and logical fallacies. Wishful thinking is armor plating around one's brain that logic and reason simply cannot penetrate.

As a scientist and a realist, I freely admit science may run into walls it cannot cross and understanding it cannot gain. But there are two extremely important points to note about that statement.

The first is that even with the rapid pace of technological advancement, some of the problems facing us are quite difficult and may take some considerable time to work out. An intelligent and honest person would concur that it would be far wiser to wait until problems or limits to our knowledge have stood for several hundred or even several thousand years from now before foolishly proclaiming certain information as "beyond science", because even in the extremely short current Age of Science, the world is absolutely littered with the dried corpses of failed predictions that something was unknowable, unattainable or impossible.

The second point is that at no time is it ever anything but special pleading and a total abdication of reason to try to insert a god or a supernatural cause or explanation into any current spot of uncertainty. The infantile 'god of the gaps' approach is a complete failure of everything that makes us human, from our opposable thumbs to our cooperation on intellectual endeavors to our relatively highly developed brains with their problem-solving and critical-thinking capacities. Invoking any kind of god at any time to try to explain any gap in our knowledge is a weak and spineless crashing to one's knees, an abandonment of reason and a contemptible abdication of our responsibility to grow up and think for ourselves. Such foolishness should be strongly challenged and ridiculed whenever it is encountered.

The pictures in Figure 1 below were taken of a small Midwestern town's church marquee in 2018. This is the mindset of far too many of the faithful, that being 'don't you worry yourself none about thinkin' and them ole' so-called facts, just do what we tell you to do because *we* know what our god wants you to do and think.'

This is what I mean by abdicating responsibility for thoughts and actions. This is what I mean when I say religion seeks to suffocate critical thinking and to make the mindless follow the herd like blind and fearful sheep. This isn't just my opinion; they openly advertise and sell it as their business model. This is no way for a self-respecting person to live.

Speaking of thinking, every person of faith should take the following challenge, which lies at the core of the faith vs science discussion. It is very simple. All you need do is list as many as you can of the questions humans have ever had, and name which ones have been answered by science and which ones have been answered by religion. The current answer to this challenge is zero for religion and in the millions and rapidly growing for science.

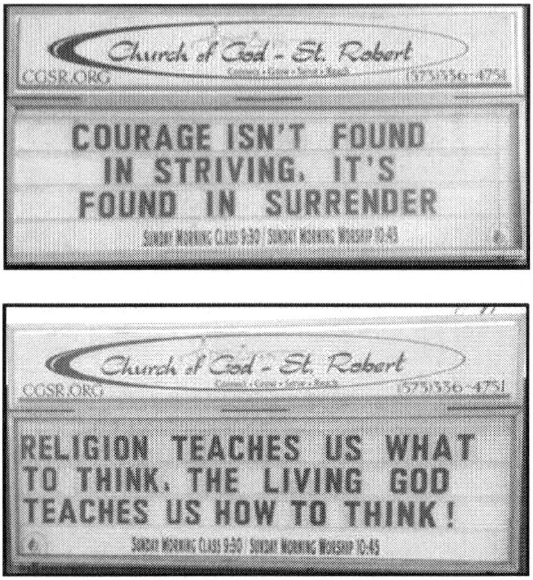

Figure 1. Pictures of a church marquee in small-town USA, 2018.

And the more we know, and the more dangerous the world becomes, the less tolerance we can have for denials of reality. Weapons of mass destruction and the ability to wreak worldwide havoc on the food, water, commerce, energy and communications systems are now much more accessible to people with insane ideas and ideologies, many of which center on the more evil manifestations of how to bring specific eschatological fantasy/cataclysms to life.

Worldwide caliphates, bringing about the savior's next return from wherever he's been hiding for so long, rapturing souls into the skies, the utterly unhinged End Times eroto-schadenfraudic fetish obsession so many evangelicals in America have; none of these and many other fanciful imaginings are peaceful propositions when religiously deluded people think they can do things in the real world to bring them about.

One extremely important fact which the faithful assiduously avoid admitting is how often the religious majority persecutes and/or kills religious dissenters, with the Catholic sect being the most obvious offender for more than a thousand years. Those making any claim about how widely held a specific religion or belief is now held also completely fail to mention that in many times and places nonbelievers were subject to loss of job, loss of freedom, loss of family or even loss of life, thus artificially inflating the reported rates of belief in the so-called Truth mandated by the local religious enforcers.

Openly expressing an opinion that will get you jailed or killed by the religious thought police and holy murder squads will always suppress reported rates of nonbelief and will make it seem like few actually doubt the official Pravda. Such is the way of religion and the faithful to always see the world from the wrong perspective, always ask the wrong questions and always arrive at the wrong conclusions.

Fortunately for all of us, those times are finally waning. Religion, like the life-draining and blood-sucking vampire it is, continues to have the darkness and ignorance in which its explanations and gods hide sundered and shrunken by ever brighter and wider pools of knowledge.

Even so, there are millions if not billions of people of faith who even today have the gall to say that 'yes, indeed, religions and the religious have done and said very bad things over time, but *we* are different today. *We* are good people who don't go on Crusades and Inquisitions. *We* don't believe in an old man in the sky. *We* don't persecute and kill heretics. Other people might be doing and saying bad things but *we* don't do that.' This is deluded, disingenuous and dangerous. For one thing, any person anywhere who believes in anything without proof, which is to say every metaphysical and most every other claim from the faithful, is equal in point of objective fact.

To put it bluntly, you may be a faith-based socialist soup kitchen worker, community volunteer and philanthropist who deliberately lives below the poverty line, which means you are very much like the Jesus depicted in the Christian bible and *extremely* unlike Christians in America today, but if you actually believe in any supernatural claims at all about Jesus or his father, you are equally as beholden to unsupported beliefs as were the 9/11 faith-based mass murderers.

You are all exactly the same in that you feel you are free to hold unprovable metaphysical beliefs without evidence. You are also all equally unable to determine objective reality and morality. You can't prove that the 9/11 murderers and their supporters have it wrong about their god and how to get a carnal paradise in the afterlife, or that they were metaphysically wrong by committing their mass murder, just as their supporters cannot prove you and your specific beliefs are wrong. You are all collectively fed by the same inhuman, illogical and poisonous umbilical of unjustified beliefs. Your embrace of your beliefs without proof gives everyone else, anywhere, an equal ability to assert their beliefs without proof.

What does that mean? You believers, by allowing yourselves and other people to believe things without evidence and demanding your unjustified right to do so, deliberately keep the door wide open for any and every unsupported insane and dangerous belief to fly out of

Pandora's Box into our world. You sit at home and in your churches and needlepoint groups thinking how great and wonderful you are, and you may very well actually be doing some good things, but the door to evil is nevertheless wide open.

Your failure to require proof for your unsupportable beliefs means that every past, present and future faith-based murderer need only cite his own personal unsupportable but deeply and sincerely held beliefs as justification as he pulls the trigger or presses the detonation switch.

Every single belief in the natural or the supernatural must be proven to be supportable using the tools of science. There is no other empirical method to do this, and your belief in your faith is just one opinion among billions.

With the above, did I just blame soup-kitchen and needlepointing believers for faith-based terrorism? Partially, yes I did, and you should too. Read it again, because it's extremely important and far too many people don't understand how much of a problem this is. To make it perfectly clear, I will state it again.

As long as you say it's acceptable for you to believe things without evidence, you can't tell anyone else they can't believe what they believe if they have no evidence. Your personal religious beliefs are 100% exactly the same as theirs and none of you can prove any of them.

Therefore no matter how evil or inhuman a religious belief may actually be, you by your own reasoning can't say it is wrong or evil or inhuman, because all religious beliefs are equally contrived and unprovable, and you all point back to your evil holy books for justification.

As long as you say it's acceptable or praiseworthy for anyone to believe things without evidence, you enable lies and evil to flourish. If you are religious, you are therefore an enabler of lies and evil, even if you haven't realized it before now.

What I am saying here is a direct challenge to all religious believers everywhere. This is likely to be the first time the faithful will have heard of this, and they will likely reject this fact out of hand with a supercilious sniff of arrogance and no small amount of anger at my perceived presumption. I respond with three simple challenges to the self-professed devout Christian:

1. Prove that the Quran, Book of Mormon and all the other holy books and their beliefs are wrong.

2. Prove that your holy book and your beliefs are right.

3. Justify to your fellow believers and to the rest of the world why you do not follow, as Jesus said in Matthew 5:17 (New Revised Standard Edition), every stroke of the letter of the old law, thus every bit of guidance from ***both*** the Old and New Testaments. You ignore almost 100% of what your holy book says so that you can live an actually moral life in a civil society. If you followed every jot and tittle of that book, you'd be performing Inquisitions, going on Crusades, killing disobedient women and children, killing heretics, committing genocide and flying planes into buildings too.

To put it bluntly, it is a binary decision. You are either completely in or completely out of faith. You are out of it if you are a secular freethinker and humanist who is guided in word and deed by the desire to maximize the wellness of humans, animals and the environment. You are in it if you are any person of faith whatsoever. Therefore, the disciples who are alleged to have followed Jesus around and got left holding the bag when he checked out get lumped in with the 9/11 murderers and everyone else who holds supernatural beliefs of any kind. Again, it is a binary choice, whether you like it or not.

And there is no such thing as "beyond science". That other infantile 'get god out of jail free' tactic is as obvious and childish an attempt as obfuscation as is a two-year-old trying to lie about stealing cookies while standing there with one in each hand and chocolate on his nose. It is just an attempt to take the god of the gaps argument and apply it wherever and whenever the believer thinks his god needs an alibi, an excuse or a safe space. This assertion is, more accurately, "beyond childish".

So this book is not a dialogue or a debate. That sort of back-and-forth is well known to those familiar with the arguments for and against any of the absolutely endless parade of religions and gods which humans have dreamed up and will continue to dream up in the future. There are a great many books written by a great many persons on these subjects. If you would like to step into the foul tar pits of illogic known as creationist and apologetic writings, feel free to stop reading this book and go to the nearest bookstore or take a virtual ride on your favorite search engine for an exhaustive and alarmingly simple-minded, ignorant and misinformed collection of, shall we say, extremely dogmatic and strident attempts to either comfort or threaten in their fact-free and circularly illogical passages. Try not to get lost, try not to lose hope in humanity, and we'll see you when you come back. But be careful, a mind is a terrible thing to waste!

So why write this book? I wrote it to satirize religion, to expose its lies, to explain its basal fraud and illogic and to show how it is merely an imaginary appeasement to the inner child seeking comfort and an eternal clasping by strong hands to a protective daddy's chest. I wrote it to describe how religion formed, why it endures, and why it will likely remain a part of our species for a very long time, despite the relentless destruction of its claims by every aspect of scientific and moral research and reality.

I wrote it to show that some people will never, indeed can never, be talked out of their religion, for extremely unfortunate psychological, social and political reasons, and that the rest of us have to do our part to prevent humanity from sliding into theocratic and/or fascistic rule and war, or into a faith-based global apocalypse despite the continuing efforts of the faithful to make that happen.

I wrote it to show that there is a path to a brighter future and that both that path and that future are totally devoid of any religion.

I wrote it to do my part in the near-Sisyphean struggle to help the human race damn well grow up.

I wrote it because if people like me don't fight this fight, my daughters and their daughters will be forced to live under the evil and oppressive thumb of an Islamic State, a Catholic State or a Christian State in which women will be reduced to a biblically-inspired level of chattel and breeding stock, will be accused of imaginary sins such as blasphemy and witchcraft and will be made to suffer through lives made miserable and sorrowful by the deeply and sincerely devout.

I also wrote this book to inform, to show a different perspective, to offer what I hope are enjoyable humor and satire, and sometimes to inflame. If you are a nonbeliever, I hope you will enjoy reading it as much as I enjoyed writing it and that you will also learn something from it. I hope you come up with ideas on how to make your corner of the world less religious and more moral, and that you get motivated to get out there and fight this evil in our midst. Please pass on the facts and the bits you liked to those who need to hear it, and please do all you can in your corner of the world to shore up both your local freethought community and that badly needed Constitutional wall between church and state.

If you are a believer, I hope you use the experience of reading this book as an opportunity to see the universe, your religion and indeed all religions and mythical creatures populating every level of the divine hierarchy from a perspective that is opposite from your own. It will be difficult, for I pull no punches, particularly with regard to Catholicism and its terribly immoral, cruel and criminal history which is unbelievably still poisoning our world. Know that we nonbelievers fully understand that plenty of the faithful across the globe are at their core good people, though if they are truly good and decent and have actually read their holy books all the way through and have actual human decency, they suffer from the cognitive dissonance of what their faith says versus what actual humanistic and positive morality really is.

Know that when a person holds a bad belief but doesn't act on it, we try to "hate the belief, not the believer". It has always been the job of the nonbeliever to show the believer the path to true humanistic morality, preferably by word and deed but by stronger measures if need be, because we know that the road from bad religious beliefs to bad real-world actions is extremely short.

Believers will also read extremely upsetting challenges to the very core of their beliefs about the afterlife, which I ask that you work through, either now or in the future. Trust me, and if you don't trust me, trust in the billions of nonbelievers around the world who get along perfectly fine every single day without believing in that unprovable fantasy and all the others like it.

And for every reader, in the eternal words of the incomparable Douglas Adams and his "Hitchhiker's" universe, I invite all of you to "Share and Enjoy!"

Note that we have a clear and present danger to address at this point. We have only very recently been able to learn tremendous amounts about the brain and human psychology from the inside, and why people believe what they believe. If you think people believe things because they precisely define and thoroughly understand the proposed belief, then rigorously assess all the facets of that potential belief in the light of the most current and relevant empirical evidence and science, then make a rational decision, you are not talking about 99% of *Homo sapiens* at any time in history.

If you think people believe things because they were raised with that belief, because other people believe the same thing, because the belief makes them feel comforted or because their brains aren't working properly, then you are indeed talking about we clothed apes, particularly in places like America and Afghanistan. Logic and reason simply will not work to change many people's minds regardless of how wrong they are. You can show them all the evidence in the world that their belief is wrong, but the human brain has a disappointingly powerful tendency to give in to bias and wishful thinking, and for many people literally nothing whatsoever will change their minds.

The brain does not have some kind of sump pump which turns on and sucks out the garbage when the level of sewage and misinformation has risen above some threshold. For many people, garbage in is always garbage in, and it is indeed garbage out in terms of what the person does and says based on that garbage.

For instance, I personally know many people who are quite intelligent and educated who hold professional jobs which require critical thinking and analysis. Many of these people have succumbed to and hold a religious belief due to fear or weakness. They have an overpowering and quite understandable desire to see dead family members once again, and they fear being ostracized by friends and family and co-workers. They have invested so much of their lives and social capital in a faith lifestyle, and the faithful are so dangerously tribal, that going rogue would incur unacceptable social impacts and loss of status.

These people put their critical thinking skills far out of reach when considering questions of afterlives and seeing the departed forever in some kind of heaven ruled by an overbearing and massively cruel but somehow supposedly also loving god. This is why smart people believe unbelievably dumb things. The fear of death and permanent separation from one's loved ones in this life or the next, the fear of a hell or the fear of permanent cessation of existence are often too strong for the psyche of many people to withstand. Thus this unprovable fantasy endures and it will continue to endure for quite some time in theologically stultified countries like the aforementioned America and Afghanistan.

I am a realist, an empiricist and a student of the human condition. I am thus under no illusions, due to the conditions noted above, that this book or any other will cause billions of people to abandon the pews and the prayer mats, discard their current faith systems and embrace secular humanism. This is indeed my ultimate goal, as it would lead to an immediate improvement in global quality of life.

But without using the instruments of torture and murder which the faithful used for so many centuries to instill and reinforce belief and punish or eliminate disbelief, and using only the instruments of logic and reason, the struggle to remove the deep-rooted parasite of religion will be long and arduous.

Just as some people are born tall and others short, some people are born with an innate weakness and/or predilection for the comforts and mysteries of the magical, the spiritual and the imaginary. They will gravitate toward any ideas they come across which enable them to combine their bias for the unseen and mystical with their ignorance and fears into a comforting or exciting package.

To put it rather more pithily, most people will believe whatever makes them feel good. And freethinkers know how true it is that trying to use logic on the faithful is like pissing on an electric fence. You'll be quite shocked at the result. The faithful didn't arrive at their faith through logic and reason, even if they convinced themselves that they did. They arrived at their faith because they were born into it and/or it makes them feel good, so logic and reason aren't going to get them to change their minds.

Many of them are content to believe solely on the basis of their feelings, and only they will be able to free themselves from their self-imposed mental imprisonment. Many believers feel a need to defend their faith, or to try to convert nonbelievers using words instead of instruments of torture, nooses and burning stakes. In doing so they have harvested the rich hauls of scientific discoveries made by other people and in a frustrating mélange of illogic, misinterpretation and misunderstanding of what they are handling, egregiously confuse and misuse this information to try to prove their specific faith claims true.

You may have agreed with everything in the above paragraphs, but still wonder why I felt this book needed to be written. Aren't there enough religion-bashing books? I say *absolutely not*. We antitheists should not even begin to think about slowing down our assault on the evils and stupidities of religion until there are at least as many of our books on shelves and nightstands in homes and libraries and hotel rooms across the globe as there are copies of the Bible, Quran, Torah, Watchtower, Book of Mormon and all the rest of the mythologies and misconceptions of misguided men put together. And until the number of houses of worship in every city on the planet is in the single digits or is less than the number of brick and mortar public libraries in the city, whichever is smaller.

Humanity would be massively better off if we eradicated this embarrassingly long chapter of ignorance from our history and converted every single place of worship into activities of real good, such as secular schools, shelters, food banks, medical clinics, volunteer centers, libraries and community centers. Much like the Nazi swastika and the Confederate flag, religious symbols such as the cross, the star and crescent and the star of David should be an embarrassment and a reminder of the disastrous history of ignorance and evil our species has perpetrated on itself, and should be relegated to the dustbins of history.

The devout adherents of the world's religions have had their pious knees on humanity's throats for far too many centuries. Every single act of evil committed in the name of these religions, every act of bigotry and persecution perpetrated in accordance with their teachings, each and every one of them, has been another cruel and inhuman lash across the bare and bleeding back of our collective humanity. For far too long the clergy have had the power to imprison, torture and kill those who questioned or challenged them, and they took full advantage of this power. In far too many places today they still hold this power and do not hesitate to use it.

In the West, the first significant rumblings of dissent against this institutional oppression began in the Enlightenment, were supported and strengthened by a growing avalanche of rapid advances in scientific and moral inquiry, and exploded into the spotlight following a single faith-based mass murder in the United States in September 2001.

Since 9/11 the surge in antireligious writings and discussions has taken on a breadth and depth one would have never imagined possible in the mid-1950s, when America became consumed with fear over the threat of a looming nonreligious Soviet Union. It dropped to its embarrassingly weak knees, turned its spineless back on its intentionally wholly secular Constitution and added the words "under God" into the pledge of allegiance, adopted "In God We Trust" as the national motto, thereby replacing the infinitely superior E Pluribus Unum ("from many, one") which had been in use since 1782, and added the same foolishly mindless phrase "In God We Trust" to US currency.

These affronts to common sense and direct attacks on our founding documents and ethos by religiously deranged leaders of our constitutionally secular nation were very bad in their own right, but were mere annoyances compared to the global evil which has been perpetrated by the religious 'superpowers' Christianity and Islam and their devotees.

So while this is not a holy book, nor would most believers ever likely call it a 'good' book in the sense they love to use, it is another brick in the badly needed wall between church and state, another clarion call to reason-based action, another concrete millstone to be hung around the neck of evil and unsupportable dogma, and another strong hand helping to push religion and all the Troubles it causes permanently back into the Pandora's Box from whence it escaped.

Read it with an open mind, and feel free to question everything in it, but do so objectively, and by using the pure lights of reason and evidence as your guide.

I trust you will see things far more clearly that way.

Chapter Two

The Ignorance and Illogic That Enable Faith

"The Universe is not only queerer than we suppose, but queerer than we can suppose."
- J.B.S. Haldane

"All logical arguments can be defeated by the simple refusal to reason logically."
- Steven Weinberg

"Christianity is the belief that a cosmic Jewish zombie who was his own father can make you live forever if you symbolically eat his flesh and telepathically tell him you accept him as your master, so he can remove an evil force from your soul that is present in all humanity because a woman made from a bone was convinced by a talking snake to eat from a magical tree."
- Richard Carrier, others

If you are at or above average intelligence and have interacted with large numbers of people, you are aware that the above quote from J.B.S. Haldane can be reworded as "People are not only more ignorant than we suppose, but more ignorant than we *can* suppose." It is impossible to overestimate how ignorant and illogical most people are. Ignorance and illogic are extremely prevalent in American politics and most people's love lives, but they are most highly concentrated in religion.

Only in religion do billons of people somehow feel it is acceptable for a person to believe quite literally anything without evidence. I will show in this book just how ignorant and illogical people can be, and how religion both exemplifies and dangerously amplifies this significant problem.

If you are an American, you are a member of a highly religious society, even if you yourself are not a believer in or worshiper of any supernatural entity. According to a Pew Research Center study[1] published in 2015, three out of four Americans are members of a faith group, with about 22.8% of the population identifying as nonreligious. The United States is in fact one of the most religious countries in the developed world. Its highways and byways, its suburbs and its rural areas, its gated communities and its poorest neighborhoods, are saturated with religion. And 78.4% of it is some version of Christianity.

Such widely held faith gives a reasonable expectation that all those believers have some level of actual knowledge of the subject. As a quick test of this, I have created a very brief list of questions which you the readers of this book may use to quickly gauge your personal level of religious knowledge, and you are free to try it out on anyone you know. The questions are below, followed by the answers.

Please note that none of the questions are trick questions. Most are based on actual religious beliefs held by large numbers of people, in some cases by over a billion people. In the spirit of everything from the signs of the Zodiac to the number of disciples Jesus is said to have had, there are twelve questions on the test. Feel free to take as much time as you like to answer the questions, but please do not refer to any human or inorganic data sources for help, and please do not skip ahead to the answers. This is a test, after all, and you wouldn't want to cheat, especially if your religion demands honesty, right?

Religious Knowledge Test

1. **Where does this story come from?** God created man from dirt, and the man lived among the animals, one of whom was a sneaky snake. A woman tempted the man with food; the man then covered his nudity and was exiled.

2. **Which belief system has this as a core tenet?** Properly devout men can become gods in the afterlife.

3. **Which belief system is this?** The best thing that ever happened on the planet was that the creator of the universe sent his kid to be tortured and killed by ignorant savages.

4. **Which belief system is this?** The prophet flew to heaven on a winged horse.

5. **Who is this describing?** This boy was hidden in Egyptian marshes for his own safety.

6. **Which belief system is this?** Seventy-five million years ago, the galactic despot Xenu fixed his galaxy's overpopulation problem by sending 13 trillion individuals to Earth, put them into volcanoes and blew them up with atom bombs. The souls of these beings now attach to human beings.

7. **Who is this describing?** The male lead in the story was born of a virgin.

8. **Which belief system has this as a core tenet?** No matter how defensible the act may be, it is forbidden to kill any being, and non-violence is a person's highest religious duty.

9. **Where does this story come from?** God told the male lead in the story to build a boat, giving very precise measurements and directions, in order to survive the great flood the god will be sending. Male lead takes his family and a huge assortment of animals aboard the boat and they survive the great flood, finally using birds to find land.

10. **Who is this describing?** The male lead in the story was a savior who died and was resurrected.

11. **Which belief system holds these tenets?**
    a. There is one all-powerful god who created the world.
    b. This god is omniscient, omnipotent and omnipresent, and the source of all that is good.
    c. There is a holy spirit.
    d. Man has free will granted by the god.
    e. Good people go to heaven, bad people go to hell.
    f. The prophet began his ministry at about age 30.
    g. There is an evil spirit which fights against the powers of good, and it lives in a hell.

12. **Who is this person?**[2]" *"Even before he was born, it was known that he would be someone special. A supernatural being informed his mother the child she was to conceive would not be a mere mortal but would be divine. He was born miraculously, and he became an unusually precocious young man. As an adult he left home and went on an itinerant preaching ministry, urging his listeners to live, not for the material things of this world, but for what is spiritual. He gathered a number of disciples around him, who became convinced that his teachings were divinely inspired, in no small part because he himself was divine. He proved it to them by doing many miracles, healing the sick, casting out demons, and raising the dead. But at the end of his life he roused opposition, and his enemies delivered him over to the Roman authorities for judgment. Still, after he left this world, he returned to meet his followers in order to convince them that he was not really dead but lived on in the heavenly realm. Later some of his followers wrote books about him."*

That's it! You're all done. How do you think you did? Did some of the questions look too easy? Now take a look at the answers below and calculate your score. You get one point per correct answer.

Religious Knowledge Test Answers

1. This story comes from the Epic of Gilgamesh, which was written hundreds of years before the Old Testament was written. The authors of the much later Hebrew bible, which subsequently became the bulk of the Christian bible, included the story in their writings without acknowledging its earlier origin.

2. Mormonism.

3. Christianity.

4. Islam.

5. The Egyptian god Horus, in a tale told about a thousand years before the extremely similar story of Moses in the Old Testament story.

6. Scientology, which has approximately 200,000 believers. If you think it odd that so many people would believe such wacky things, take a close look at Christianity, Islam and Judaism.

7. Zoroaster, Perseus, Huitzilopochtli, Quetzalcoatl, Mars. Also the Jesus character in the Christian biblical story. Virgin births are very common in mythology, as are resurrections. And apparently quite a few American women even today experience pregnancy while they are still virgins, according to a recent study[3]. Unsurprisingly, societal pressure to lie about virginity and overheated male anxiety about sex leads to claims of sex-free impregnation. One need look no further than Catholicism and American evangelicalism for prototypical examples of how to both lie about sex and to have extremely debilitating anxiety about it.

8. Jainism. If you really want to be part of a religion that actually turns the other cheek and which could never be used to justify any act of deliberate misconduct or evil, leave your current religion and join the Jains. Jainism actually is what Christianity claims to be.

9. This tale also comes from the Epic of Gilgamesh, which was written hundreds of years, possibly a thousand years, before the Old Testament was written. This same story was copied in the Hebrew and Christian bibles much later, just as many such tales were told and retold, repackaged and rewritten for new audiences all across the ancient Mediterranean area. Copycatting and plagiarism have been around since the formation of language.

10. Osiris, Attis and Dionysus, who was also the offspring of a male god and human woman. Also the Jesus of Christianity. Resurrections are an absolute dime a dozen in mythologies and religions, including the New Testament and over and above the tiresome claim of Jesus having rising from the dead. It was quite a zombie show in Matthew 27 (KJV), which quite breezily notes that "*{27:52} And the graves were opened; and many bodies of the saints which slept arose, {27:53} And came out of the graves after his resurrection, and went into the holy city, and appeared unto many.*" Too bad none of them, any of their families or indeed anyone else at all in the entire city deemed such a miracle worthy of mentioning in any written records. Either such things were unbelievably common and thus not important enough to write down, or they are all the same flavor of made-up / borrowed ancient desert science fiction / fantasy.

11. Zoroastrianism, which started about 1500 BCE, or far more than a thousand years before the alleged life of Jesus. As with the other stories noted above and many other facets of itself, Christianity stole from other myths to accrete its own story.

12. Apollonius of Tyana.

Total correct answers: _______

So how did you score? If you didn't do so well, don't feel too bad. You have a lot of company. It is a 100% guarantee that the vast majority of people in America, and almost all of those who would call themselves Christian, will get no more than four of these questions correct. Most Americans will get #3 correct because I made it a deliberate gimme question, many will get #4 right, and both current and former Mormons should get #2 correct. Those who connected #6 with Scientology will be a current or former member, will have heard of this group's beliefs or guessed well.

If you scored even a solid six or above, meaning you only got half of the questions right, you will have scored higher than most Americans would, and indeed higher than would most of the faithful anywhere on the planet. Good job.

If you scored an honest, no-cheating score of ten or above, you are exceptionally well-versed in history, mythology and religion. The difference between mythology and religion is, of course, that all religion is mythology except to the true believer, who thinks that only other people's religions are mythology. Great job!

If you scored a full 100% without looking at any references, you are in the microscopic .00001% or so of living humans who actually has done extensive research into the vast ocean of religious beliefs and their literary and mythical origins, rather than restraining your examination or interest to the primary talking points of one particular system. Congratulations on the superior score and the inquisitive mindset!

Think I'm exaggerating? You need not take my estimates at face value; there is factual precedent and unlike religion I have actual evidence behind my statement. Based on that precedent and evidence I can also predict to some extent how well people will score based on their particular religion. Using actual empirical evidence as a guide, the higher you scored on this test, the more likely it is that you are a stone cold atheist or agnostic.

If you are a Catholic, you are likely to have the lowest scores. How do I know this and what is my evidence? In 2010, the Pew Research Center tested various groups of Americans for their religious knowledge. Their study[4] found, quite unsurprisingly, that the religious know much less about religion than do nonbelievers. In some cases, specifically some of the Catholics, they only knew about half as much as did the nonbelievers.

To many people, this score disparity is an apparent violation of common sense. Why would people who don't believe in religion score so much higher on a test of religion than people who actually do believe in it?

This is because nonbelievers as a general rule are more inquisitive, skeptical and investigative of claims or assertions, they have a far higher standard for what they will accept as evidence and they do not accept dogma.

Thus these freethinkers do far more research into the total body of what religion is, what religions proclaim and where they came from than do the faithful. This personal investigation into the facts makes them as an aggregate far more informed on the subject than are the faithful, who are content to passively and unquestioningly absorb platitudes and illogical nonsense like a sponge and to unthinkingly follow the religious leader like sheep follow a herder.

The Pew study results are shown below in Figure Two below, listed by group and by how many and what percentage of the 32 total questions they got correct. The color bubbles graphically represent how well they did, with green being the best and red being the worst.

Group	**Correct Answers**	
Atheist or Agnostic	20.9	(65.3%)
Jewish	20.5	(64.1%)
Mormon	20.3	(63.4%)
White Evangelical Protestant	17.6	(55.0%)
White Catholic	16.0	(50.0%)
White Mainline Protestant	15.8	(49.4%)
Nothing special	15.2	(47.5%)
Black Protestant	13.4	(41.9%)
Hispanic Catholic	11.6	(36.3%)

Figure 2. Results of 2010 Pew Survey on U.S. Religious Knowledge.

This general lack of knowledge of their own and of other faiths is why the devotees of religion are called "believers" and not "knowers". They *believe* quite a lot, but don't actually *know* very much. Even if they do happen to know about the current teachings of their own belief system, they usually have absolutely no idea about the facts of the origin, development and evolution of their faith into its current form.

They accept its teachings dogmatically, and they thoughtlessly and uncritically absorb and retain them like a sponge mindlessly absorbs and retains water.

It is one thing to accept assertions and statements based on massive amounts of rigorously collected, evaluated and replicated processes and information. It is quite another to accept assertions and statements based on accepting what other people or ancient fairy tales tell you. The first path is called science. The second is called faith. The first method leads to well-informed individuals. The second leads to poorly informed or uninformed individuals. Thus the Pew scores you saw above.

On my 12-question test, I obviously constructed some of the questions to specifically demonstrate how much of Christianity is so flagrantly borrowed from earlier religions, beliefs, myths or works of literature. This is why the stories themselves were familiar but the answers were not. Had the relevant copyright and intellectual property laws been on the books when the Hebrew and Christian texts were being written, the heirs of the usually long-dead original story authors would have had plenty of justification for hauling the storytellers of the Bible and Quran into court for much-deserved lawsuits. Christianity, Islam and their parent Judaism were shameless thieves, stealing story ideas from all over the ancient Mediterranean and Middle Eastern civilizations.

It is shameful that so many of the faithful do not know this and even more shameful that they neither care nor want to know it. I have actually had otherwise intelligent people indignantly tell me "so what" when I inform them of this fatal dagger in the heart of their faith. They do not have nor do they care to have any actual understanding of how their religion actually came to be, nor do they question the dogmas of their faith. Their faith is all that matters to them. This book seeks to address that significant problem.

Throughout the book you will find numerous challenges to the "believing" approach to life, and numerous fact-based dismissals of religious thinking and religious statements. You are likely to discover much that you did not know about the history of religion, particularly Christianity and Catholicism, that will enable you to move past dogma to see the actual facts and the actual history of these faiths and also of many relevant and important issues discussed in America today. One of the primary goals of this book is to inform, either through a traditional recitation of facts and descriptions of the unfamiliar and often unknown realities of religion, or through other methods designed to stimulate critical thought such as my short religious knowledge test.

In order to help that process of critical thinking, we must continue to ensure we are all aware of some of the information and terms used in this book and in the real world. This information is necessary for the reader to get a more complete picture of the average American mind, including what it does and does not contain and how it works or does not work, and how this contributes to the prevalence of religious belief in America. We have already discussed the national lack of religious knowledge. You may or may not be aware that Americans are depressingly even less knowledgeable about science than they are about religion, but to set the conditions for this discussion I have listed some of that data below to show how bad the problem really is. You will need to keep this in mind as you read the remainder of the book.

1. **America's Lack of Scientific Knowledge.** I included my religious knowledge test and the Pew study results as part of the message I am conveying about how vitally important it is to be informed on issues as a citizen in the 21st century. That is, to be a "knower" and not a "believer". We have already established that Americans generally don't know much about religion, even their own. A National Science Foundation / National Science Board study[5] on Science and Engineering Indicators

published in 2018 showed that many Americans are extremely ignorant about science. Americans are thus highly religious while being highly ignorant about both religion and science, and this is an extremely toxic combination. It will not be lost on the reader that one would expect high religiosity to be present where there is low knowledge about or low acceptance of science, and this is indeed the case.

So how scientifically ignorant are Americans?

- *"Americans correctly answered an average of 5.6 of 9 factual knowledge questions in 2016…"* (**62%**).
- Only **52%** think humans evolved. This is far below the levels achieved in other advanced nations. Japan is at 78%, Canada is at 74% and South Korea is at 64%. Even *Israel* is at 63%.
- Only **43%** gave the right answers to a set of questions testing "Understanding of scientific inquiry." Those with a HS diploma scored about half as well as those with a bachelor's degree. That is not a typo; they scored **half** as well, thus supporting the fundamentalist fears that higher education does destroy dogmas through the application of actual facts and critical thinking.
- Only **39%** think the universe began with a massive explosion. Canada was almost twice as correct, at 68%. South Korea is at 67% and again even Israel far outshines the US at 64%.
- Somehow **27%** of Americans think the sun goes around the earth. Given the above ignorance, this pathetic figure is unfortunately not too surprising.
- Only **15%** know that very close to 100% of climate scientists "have concluded human-caused global warming is happening."

Relevant Fact. American ignorance extends to many subjects beyond just religion and science, and we nevertheless allow even the most ignorant people to use their religious dogmas as the basis for voting for those who will guide the current top superpower on the planet. These elected officials will share the ignorance and dogmas of the population, and will make decisions based on ignorance and dogma. This is an extremely dangerous situation which is causing huge damage to America, with no apparent end in sight.

Relevant Quote. "*An educated citizenry is a vital requisite for our survival as a free people.*" This quote is attributed to

Thomas Jefferson[6] although he may never have actually said it. Whether he said it or not, it does reflect his views and is absolutely correct. Flocks of the ignorant are the most easily fleeced, and religious Americans are a massively ignorant flock being happily fleeced on a daily basis.

2. **The Non Sequitur, With Examples.** I will discuss the brain and how it works in some detail in Chapter 12, but an explanation of this term with some examples will set the stage for later reading.

The term "non sequitur", which means "does not follow" in Latin, refers to a statement or conclusion that does not make logical sense based on the available evidence. Merriam-Webster defines[7] the term as "*an inference that does not follow from the premises*" and *"a statement (such as a response) that does not follow logically from or is not clearly related to anything previously said*". A simple example would be the statement "My bunny is white. My bunny is an animal. Therefore all animals are white."

The first two parts of that statement may be 100% correct, but the conclusion "does not follow" at all, and is 100% wrong. Non sequitur traps are extremely easy to fall into, especially when one is emotionally wedded to a predetermined or preferred conclusion or position. Religion is built upon and relies on countless non sequiturs, such as shown in the following examples.

a. **Non Sequitur #1.** "The earth is a complicated thing. Only people make complicated things and people could not have made the earth. Therefore the **Christian** god made the earth and **Christianity** is the one true faith."

Explanation. This is actually a two-layered double-fail non sequitur. Fail #1 is the assumption that a god had to create the earth. Fail #2 is that out of many billions of possible divinities both already invented and those not yet invented or imagined, it is somehow always the one worshiped by the speaker of the non sequitur that is said to be the one responsible for whatever is being discussed. For some reason, Christians never credit Zeus, Marduk, Osiris, Ahura Mazda or Mbombo with creating the earth. This is probably the most flagrantly illogical non sequitur seen in religion, and it warrants some further discussion.

Today we need no gods to explain disease or the formation of snowflakes, clouds, lightning, stars and planets. When humans did not know this information, they said that their local god was the responsible agent. They also inserted that specific god or some other local fictional divinities into all of the other unfilled gaps in their scientific knowledge, in the classic "god of the gaps" non sequitur which is still in vigorous practice today in religions across the world. This same mistake has been repeated endlessly since our species learned to speak.

The knowledge gaps in which people have hidden their gods have shriveled up and disappeared on a near-daily basis for thousands of years, with a very dramatic and accelerating loss of these gaps in every scientific field within the last 500 years, when our species embarked on the long-overdue Scientific Revolution.

The core problems are that the faithful suffer from a debilitating lack of imagination, an inability to overcome the constraints of how they have been programmed and what they currently think, and their emotional ties to certain supernatural propositions. They are incapable of truly comprehending the unbelievably long period that two or four or fourteen billion years actually is, or the amount of change that can happen over that length of time, in the way that a science-minded person can. They also often cannot think flexibly. Biological evolution and the origin of the universe provide two illustrative examples.

Biological evolution, both human and otherwise, is a process that has now been understood, explained and documented to an overwhelming degree, with an actual entire planet crammed full of historical evidence to draw on and also the ability to actually observe the process occurring all around us every single day. Despite all this evidence, much of which they can actually hold in their hands and which is present even in their own bodies, the faithful prefer to ignore the mountains of evidence and say they are unconvinced or that it just isn't true.

Regarding the universe, there is no need to think that the lack of gods required for snowflake formation is any different from what would be needed for universe formation. A fully unguided process could have, and until definitely proven otherwise we must assume did, form our universe without the need for any agent whatsoever. But the faithful refuse to see and accept this, or they are unable to do so. They will claim that their so-called "common sense" proves that a god is necessary for universes to form but is not needed for snowflakes or even planets to form. This is because their thinking is constrained by their tie to their religion, their inability to truly understand processes and evidence and their inability to think in a way that does not incorporate their dogmas.

The term "common sense" is undefinable and problematic. Each believer has a personal threshold for what he or she considers common sense, and he places his personal god anywhere he personally thinks something could not naturally occur based on his personal feelings. If you are a believer and you say a god is necessary for universe formation, you should also say a god is needed for snowflake formation. If you don't, you're being inconsistent and you have fallen victim to the trap of a self-serving and special pleading bias where every conclusion derives from what you personally think is possible and if it ends up with your solution incorporating the god you personally like.

Christians in America are doggedly perverse in their compulsion to deliberately pick out a narrative in which a god in the image of a human but with supernatural powers created the universe, and either does or does not intervene thereafter, according to the conclusion they want to reach.

Note that the faithful can assuredly posit a hypothesis that there is a god or a set of them and that they do work in the real world, but to be truly honest, reasonable proof of this hypothetical god's existence absolutely must *precede* any actual belief in it. It is only in religion and politics that our species actually allows its members to hold ridiculous and unprovable beliefs without evidence. Such childish and shameful behavior must be strongly fought wherever it is encountered.

Again, any god or indeed any mythical creature whatsoever can be inserted into the claimed "creator" slot, and not a single one at all can be proven to be real. Jehovah, Allah, Zeus, Thor, the Flying Spaghetti Monster, Luke the Lucky Leprechaun and Mena the Roman goddess of menstruation all have an equal chance at being the One True God.

And don't laugh at someone inventing a goddess of menstruation; there are Catholic patron saints for almost anything, and they unsurprisingly all have their areas of expertise as do the figures in the earlier Greek and Roman pantheons. It's all just like those superhero trading cards kids have today, which show a picture of each hero and describe his or her special superpower(s).

Perhaps advanced beings made of carbon, hydrogen and oxygen and without any supernatural powers whatsoever created the entire universe and the earth, using only highly advanced technology. Perhaps these beings accidentally created the universe and aren't aware of its existence even now.

Our earth formed in our universe via blind and unguided natural processes just as snowflakes, life and fossils have formed. There is no reason other than faith and a desire to protect a specific religion to say any god was required. Selecting any specific god as the causative agent for any of this is pure uninformed, undisciplined and unjustified dogma.

The faithful, almost always the Christian faithful, also permanently and quite pathetically show how unsupportable their own beliefs are when they, as mentioned before, astoundingly, unreasonably and apparently without a trace of shame claim that their god is "beyond science" and "beyond time". This highly immature and disingenuous copout is a cowardly attempt to escape from reality, as the apologists seek to protect their god in a divine "safe room" where he can hide from the advances of empirical critique and reason. Such behavior is the religious version of "because I said so, that's why", and is why religion is always unscientific and childishly unreasonable at its core.

As I will explain later in Chapters 12 and 14, almost all of these objections are both emotional and irrational, and have nothing whatsoever to do with the physical world and quite a lot to do with a needy and immature compulsion to have an afterlife and a protective daddy god.

b. **Non Sequitur #2.** "I think a god created the universe. Therefore the creator of the universe is the ***Christian*** god, who is not only outside of space and time, but is also male, omnipotent, omniscient and omnipresent."

Explanation. The same problems with the first example apply to this one, with yet a third layer of illogic being added on top with the accretion of wholly fabricated characteristics being attributed to the dogmatically chosen divinity. While each additional alleged supernatural ability actually makes the god mathematically far *less* likely to be real, the human brain often perversely comes to the opposite conclusion. Thus the crazier a thing is purported to be, the more people will often believe it. This is the sort of broken thinking and logical breakdown the human brain suffers on a daily basis and why religions, astrology, conspiracy theories and general nuttiness persist. I will discuss the brain and some of these problems in detail in Chapter 12.

c. **Non Sequitur #3.** "There are about 4,000 religions being practiced in the world today. I don't know much about the other 3,999 religions, but I was raised in mine, it makes me feel good, most people I hang around with practice my religion and it's good enough for me. Therefore my faith is the One True Faith."

Explanation. The errors in logic and character in this extremely widely held position are too self-serving and obvious to need much explanation, but again, I will expand on such broken and weak thinking in Chapter 12.

d. **Non Sequitur #4.** "I had a personally moving "religious" experience. I know almost nothing about human psychology, biology, evolutionary biology, evolutionary psychology, biopsychology, anthropology, the history of literary creation and accretion or actually any other relevant field of study. I nevertheless think I am fully qualified to believe that all these highly informed and researched fields of study are wrong and that the ***Christian*** god is the one true god among all the tens of thousands that have been invented and that the worldly and existential teachings of ***Christianity*** are true."

Explanation. This one is obvious. People don't know how or why their brains do what they do, and their lack of knowledge about both religion and science enables their superstitious thinking. Again, see Chapter 12. Also, the voices in the believer's head or the experiences he recounts apparently only ever come from the supernatural source or entity in which the person currently believes or at least knows about. How odd that Zeus and Osiris are no longer responsible for such experiences, and that Jesus only started causing them after about 2000 years ago. How odd that Jews, Christians and Buddhists never get any signs from Allah, and that Jesus never bothers to reach out to Hindus or folk religionists. No god ever sends a sign or does a single damn thing until after he is invented. This is again a default to what little the person knows, which is only a tiny percentage of the easily available information on the subject. Defaulting to Christianity is a pathetic abdication of logic and reason and is a planet-sized non sequitur. If a central African or aboriginal Australian who had truly and demonstrably never heard of Jesus or Allah suddenly began having specific visions and spouting detailed accounts or prophecies which could only have been beamed into his or her head from some deity in the Christian netherworld, we would for the very first time in the history of religion actually have something interesting to look into.

If every human alive either during biblical times or today had all simultaneously received the same mental transmission from the same source, we might begin to consider there might be something to the god stories. Barring any such incidents, we can consider all such religious experiences as non-supernatural brain biology in action, as I will explain in Chapter 12.

e. **Non Sequitur #5.** "I wanted to talk to you. I was able to talk to you. Therefore the god I believe in made it happen."

Explanation. This is actually from a true story that happened to me in 2016. I was in an airport wearing a *Bernie For President* t-shirt and trying desperately to make it through the nauseatingly masochistic book *The Problem of Pain* by the tormented Christian apologist C.S. Lewis. I was reading it to get more insight into the religious mind and mindset and I took no joy from it, believe me. I was taking a Southwest Airlines flight and was in the "C" group, meaning I had to board after all the folks in the "A" and "B" groups boarded, and seat pickings were going to be very slim. This was only going to be a problem if I couldn't get an aisle seat. I don't mind having a nonreclining seat as long as it is an aisle seat. When I got on the plane, I could see an open aisle seat at the very back of the plane so I zipped back there and got it.

A moment later a thirtyish woman and someone traveling with her sat by me. The woman immediately asked me "so what's the deal with Bernie and C.S. Lewis?" in a friendly and inquisitive manner. She quite reasonably had a hard time understanding how a Bernie supporter would masochistically inflict C.S. Lewis on himself. During the flight we had a very good two-hour chat on the subject of religion. Admittedly she asked more questions and listened more than she spoke and gave answers, mostly because she had some extremely mistaken ideas about many things which I had to correct at length and in detail. Our talk was very enjoyable and completely cordial. She wouldn't tell me what her faith group was until our discussion was almost over. I had a strong suspicion what it was from the first minute after we met but I let it ride until she spilled the beans.

She said two things during our talk that I have observed and read to be extremely prevalent among believers. One was the fact that she held her belief because she 'felt it in her heart'. I explained that her feeling was due to confirmation and other biases, a lack of knowledge of human brain processes and psychology, et cetera. She heard my words but as is typical of believers was wholly unconvinced. I don't recall if it was before or after this that she had said she believed in miracles, which is still somehow a very prevalent belief held by the faithful even in the 21st century. Her latest evidence of miracles and her god, in her mind, was, unbelievably, me. She told me that she saw me in the airport and prayed to her as yet un-named deity or subdeity that she would be able to sit by me on the plane, and that since "it came to pass" it was confirmation in her mind that her specific god existed and would respond to prayers offered to it.

When she said this I felt a rush of pity for her due to her extreme susceptibility to logical error and her misattribution of an unremarkable event to the intervention of her imaginary god. Why do I say it was unremarkable? We were on a Southwest flight, which as I mentioned and as many people know, boards by groups and not by a preassigned seat number. I was in the C group, which boards last. The unpopular nonreclining seats at the very back of the plane are often the last ones left open at the end of the boarding process on Southwest flights, and that was the case on this flight. She was in one of the earlier boarding groups but waited to follow me onto the airplane out of her numerical boarding order, then watched me choose one of these undesirable rear seats and then sat next to me. It was as unremarkable as finding an open seat in the front rows of many college classrooms on the first day of school would be. Far less probable things happen millions of times every day.

This whole process was hardly a miracle or even a remarkable coincidence, but she zoomed straight past reason and logic and attributed a divine intervention to a specific god because of something as banal as this situation. And she wasn't joking.

What would have been more believable as a potential miracle would be for us to have been on the same Delta or American Airlines flight, and if our previously purchased tickets had been in adjacent seats, say 12A and 12B, and we ended up sitting next to each other. It would have indeed been a miracle if we had been awaiting a Delta or American Airlines flight and the preassigned seat number had magically changed on one or both of our tickets before our eyes to ensure our next-seat proximity. Since this probably would have also required at least one other person's seat number to change, and it would certainly have required all the various printed paper and electronic ticket and itinerary records to change, the odds of all that happening simultaneously and coincidentally would be sufficient to mark these events as highly worthy of examination for miraculous action.

As we were descending for final approach and landing, my conversation partner admitted she was Mormon, as I had guessed. How did I know? We were flying to the western US and her companion was a thirtyish woman of Pacific Islander descent who stayed very close to her. If you are familiar with Mormonism you will understand these observations.

My first question to her after her admission was "OK, so, how do you feel about your religion being formed on the basis of a guy talking into his hat?" She immediately, and with the speed and precision gained through years of practice, leaped to the defense of the faith on the back of some supposed archeological evidence in the Middle East. She wasn't really sure what it was, but she thought it was "definitely true". This is impossible in reality since no actual archeology confirms Mormonism and all archeological findings quite definitively refute it, as even "Mormon archeologists" have now admitted.

> I then asked her if she got her knowledge of Mormonism from outside or inside the faith; she said she did go to outside sources. I asked her if she knew about the Mormon massacres and she said she did. And then she had some trouble with airsickness. This sickness could have been due to my cordial but quite direct full frontal verbal challenges and debunking of her crazy religion and her misconceptions, or from the fact that the plane was indeed acting like a wild roller coaster. Evidently my companion had neglected to pray to her compliant and miracle-working god for a smooth plane ride.
>
> Upon landing, we parted with a sincere handshake, and I truly wished her well then and I do hope she is doing well now. She seemed like a very nice person who just happened to have been captured by faith's ability to sabotage reason and make people capable of believing and uttering the most ridiculous non sequiturs. Note that Mormonism is far from unique in having massive numbers of devotees who are equally in thrall to nonsense and non sequiturs; this is true of all religions.

All of the above and all other faith-based non sequiturs can be summed up in the short phrase "the end justifies the (il)logic". The thinking is completely backwards. Religious people consciously or unconsciously start with a preferred conclusion that feels good to them, then they build a mental construct which they feel allows them to reach that conclusion by accepting or rejecting facts and evidence as needed.

And since most people are almost wholly lacking in the knowledge and training needed to perform a multidisciplinary logical and empirical analysis of most issues, their feelings, uninformed and incorrect opinions and blatant non sequiturs end up ruling their brains.

So Americans live in one of the most prosperous nations on earth, surrounded by immense physical and virtual libraries, with access to far more information on most subjects than a person can ingest in a lifetime, yet the population is painfully ignorant on some of the basic fundamentals of reality and, as you will see in the rest of this book, is often incapable of thinking and acting logically.

The major takeaway from this chapter is that all people need to understand how much there is to learn and how much they do not know, and to think and act accordingly. It is the duty of every responsible person to first seek out actual true facts and use them to think and act logically and humanely. Conclusions need to *follow* fact-finding and research, not precede it. One cannot use the faith-based process of starting with a conclusion and reverse-engineering everything to reach that preordained conclusion.

Part of this entire process is being aware of how easy it is to commit a non sequitur or acts of broken thinking, and deliberately striving to eliminate these errors in others and especially in yourself. In order to address those problems, the first parts of this book describe standards of evidence, the brain's challenges in thinking properly and the mental traps to avoid when doing actual critical thinking.

Chapter Three

Standards Of Evidence

"Two things are infinite: the universe and human stupidity; and I'm not sure about the universe."
- Albert Einstein

"I even believe in the Devil."
- Antonin Scalia, former United States Supreme Court member

I have not written this book to be an advanced textbook on how to apply logic, probability, statistics, reason, evidence and the burden of proof. There are thousands of such references and I encourage those who are interested in a deeper understanding of those subjects to peruse them. Nevertheless, one of the main purposes of this book is to show that most people you meet, who you know, who are a member of some religion and who vote in political and school board elections are very illogical. They do not truly understand these concepts and do not, cannot or will not apply them. They have no idea about what they should be requiring as a minimum standard of evidence or who has the burden of proof.

To reach conclusions, people must acquire then test and analyze information. This should be done in a straightforward and appropriate manner, and is best done by professionals who have spent many years performing this work and who are also part of a global community of similarly trained professionals who will rigorously review that work. This process cannot include any poisoning of the process by opinions, wishes, beliefs, desired outcomes or preordained conclusions to be supported.

If the conclusion you reach through an objective, rigorous and appropriate process of analysis is undesirable to you, you are free to go back and collect more data or verify if your analysis was properly done, but you are not free to simply reject the conclusion simply because you don't like it. Putting it bluntly, if you emotionally reject appropriately-reached and fact-based conclusions you are a weak and irresponsible fool whose word and judgment cannot be trusted, and you need to go live in a shack by yourself where you won't infect anyone else with your weakness and foolishness.

Neither are you free to twist the analysis or the data so that you reach a conclusion that makes you happy. If you do this, you are a weak, irresponsible and untrustworthy fool and now also a liar, and you need to be kept even farther away from other people. Your desire to escape from reality may result in opinions and actions that only harm you, but history shows that most people who act in this way invariably do and say things that are extremely harmful to the well-being of other people. Religion is the world's number one time-tested path to becoming a weak, irresponsible and untrustworthy fool, and this book is full of examples of how this is the case.

I need to go a bit deeper into the point made above about trained professionals and peer reviews. We rely on doctors and dentists who have been trained and educated in their craft to heal and fix us, not witch doctors and holy books. We rely on pilots to deliver us safely to our destinations, not magic carpet salesmen or enchanted brooms. We rely on engineers and skilled workers, not pastors or theologians, to design and build safe, reliable and properly functioning cars, planes, trains, computers, office buildings, roads, bridges, houses, cell phones and everything else constructed for us.

The standard of evidence we have come to expect in technologically advanced countries is that the people doing technical and highly skilled science, technology and engineering jobs have been appropriately educated and trained to do their jobs properly, due to the certification requirements within each profession and the state, local and federal regulations governing them. The result is billions of properly working cars, planes, buildings, bridges and electronic devices and billions of people safely healed, housed and transported.

There is a tacit understanding that when one gets onto a commercial airplane, buys a commercially produced vehicle or enters an office building, one can trust the pilot and the people who built the plane, the car and the building. One does not need to see the pilot's license and certifications before heading to one's seat. One does not need to check a building's engineering and construction plans and diagrams or the civil engineer's degrees before getting into the elevator to go to work. One can look at the rows of graduation and other certificates on the wall in the medical or dental office if one likes, but it would be unusual to demand a detailed accounting of the practitioner's educational history before being seen. We have a general tacit agreement that the standards of evidence for proper service, design and construction have been met.

For some insane reason, religious people and far too many politicians believe they are entitled to the abject hypocrisy of trusting pilots, doctors, dentists and engineers, but say they need not trust scientists if the findings of those scientists damage or disprove their religion or beliefs. Scientists use the most proven and most powerful fact and truth-finding tool known to humanity to discover facts and truths. They use extremely rigorous standards of evidence in their training, in their education and in the execution of their profession.

These requirements form the foundation of science. Scientists are at least as well trained and educated as their peers in other advanced professions, and it is the findings of science that have given us all those planes, trains, cars, buildings, cell phones and indeed all of medicine itself. If you are over 40 years old and religious, you need to get off your knees and deeply and sincerely thank science and all of the scientists who enable your continued existence, because for most of human history you would have been dead by now due to diseases or injuries which we can now cure or repair.

Science has led to more human well-being and more actual knowledge on every measurable subject than any other human endeavor. Yet the faithful feel that what is preached in their religion and what is in their churches and holy books should somehow, on no justification whatsoever except tribal allegiance, superstition and fear, supersede science. From evolution to climate change to the origins of life and the universe, from the age of the earth to faith healing to morality, the rejection of the only true source of knowledge by the religious is contemptible, illogical and outright dangerous.

Religion and religious beliefs should be held to at least the same standards of evidence and proof as everything else, and to a far higher standard of evidence when the claims become supernatural or metaphysical.

Religions say quite a lot but have proven nothing except that when their claims are examined they are found to be false, that they seek to suppress reason and critical thinking, that they demand adherence to their dogmas while denying the findings of science, and that when practiced by humans they radically diminish rational critical thinking and well-being.

We are long past the time when any religious dogma should get any kind of a free pass or protection from vigorous and critical questioning. From divine creation to miracles to virgin births, every word in every holy book on earth must be put through the same unforgiving crucible as everything else in life. Accepting any of the claims of any religion at face value without vigorously screening each one through this crucible means you are a weak and irresponsible fool and a danger to other people.

Bringing us back to the discussion on standards of evidence, the terms and processes below describe the minimum requirements for understanding and conducting a logical and reasonable analysis of assertions.

- **Logic.**[1]
  - "a science that deals with the principles and criteria of validity of inference[2] and demonstration : the science of the formal principles of reasoning"
  - "a particular mode of reasoning viewed as valid or faulty"
  - "interrelation or sequence of facts or events when seen as inevitable or predictable"

- **Rational.**[3]
  - "having reason or understanding"
  - "relating to, based on, or agreeable to reason"

- **Reason.**[4]
  - "a sufficient ground of explanation or of logical defense"
  - "the power of comprehending, inferring, or thinking especially in orderly rational ways"
  - "intelligence"
  - "proper exercise of the mind"
  - "sanity"

- **Statistics.**[5]
  - "a branch of mathematics dealing with the collection, analysis, interpretation, and presentation of masses of numerical data"
  - "a collection of quantitative data"

- **Probable.**[6]
  - "supported by evidence strong enough to establish presumption but not proof"
  - "establishing a probability"
  - "likely to be or become true or real"

- **Probability.**[7]
  - "the quality or state of being probable"
  - "the ratio of the number of outcomes in an exhaustive set of equally likely outcomes that produce a given event to the total number of possible outcomes"
  - "the chance that a given event will occur"

- **Proof.**[8]
    - "the cogency of evidence that compels acceptance by the mind of a truth or a fact"
    - "the process or an instance of establishing the validity of a statement especially by derivation from other statements in accordance with principles of reasoning"
    - "something that induces certainty or establishes validity"

- **Evidence.**[9]
    - "an outward sign"
    - "something that furnishes proof"

- **Standards of Evidence.** There is one universal path to take in determining if someone has presented sufficient evidence to enable acceptance of a claim. There is another that is far less effective but is nevertheless in wide use in America and has some utility. The first and far more preferred path is through scientific and mathematical means, and it requires extremely solid, reproducible and verifiable empirical evidence. The other path is through using the general levels of evidence used in the American legal system. Science can test current religious or other claims, such as if prayer actually works, if evolution is true or not, or if a modern day medium can actually converse with the dead. Science can determine if a person alive today can or cannot work miracles or bring himself back to life after being dead for days. But just as it cannot prove or disprove that Harry Potter or Merlin were or were not great magicians, it cannot prove or disprove that some ancient real or fictional character could or could not work miracles. In the cases of the alleged magical powers of Harry Potter, Merlin or Jesus, or the existence of any of the gods humans have ever worshiped, we can use the legal levels of evidence.

    - **Science and Math.** As mentioned above, science does not accept claims until they are determined to have met certain standards of evidence and verification. The scientific method is the most powerful and demonstrably proven tool humans have to understand data and reach empirical conclusions. The method has the seven basic steps[10] listed below, although steps can be combined or added as long as they follow the same process.

1. **Make an observation**, perhaps based on an experience, such as you saw a ball bounce.
2. **Ask a question** about your observation, such as "Why did the ball bounce?"
3. **Form a hypothesis / prediction** about your observation which you can test through experimentation. You could hypothesize that the ball bounced because it is inflated with air and it will not bounce anymore if the ball becomes deflated.
4. **Do well-designed, strongly controlled objective and appropriate experiments** to test your hypothesis / predictions, and collect all the appropriate data. You could test the bounciness of the ball at various levels of inflation, for example.
5. **Analyze the data empirically and rigorously** using the appropriate methods. These are statistical[11] and/or other methods which I will not go into in this book.
6. **Make conclusions**, such as if your hypothesis was correct or not, and why or why not. If you like, you can repeat and refine steps 3-6 as many times as you like until you are convinced you've fully tested your original and/or other hypotheses and if they were or were not correct, and why or why not.
7. **Distribute your results**, or not, as you like. If you do, your work will be ruthlessly scrutinized, as it should be.

At the end of the process, if it has been done properly, thoroughly and honestly, the scientific method will provide the best possible empirical and statistical assessment of the assertion. This is an objective process, with feelings and emotions removed, like using a calculator. Calculators don't give different answers based on how they are feeling, but humans do. Science is a rigorous and objective process, and if anyone tries to deviate from that, the peer community will serve as a piranha-like check and balance to eat away subjectivity.

- **The Legal System.** American jurisprudence uses the four general levels shown below to proclaim an assertion as valid or not. They are listed below in order from the highest standard to the lowest standard. Clearly in the case of all religious claims and most others of any significance, we should always require they at a bare minimum easily meet the highest and most stringent level. The problem with this system is that it does not itself contain empirical and statistical assessments and analysis of data. It relies on both the most powerful and the most fallible computing object on earth – the human brain – and many other factors to include the skill of the attorneys, the impact of the judge, the competence of the jury and other factors. It is a **subjective** system, and thus not nearly as rigorous as science.

- Beyond a Reasonable Doubt[12] (Highest). The assertion must be proved such that a person can say with "moral certainty" that an assertion is true. As with much else in the world of law, this is obviously an extraordinarily slippery term and it depends on the person making the judgment to be not only moral, but also of sound mind, honest, intelligent and not beholden to dogma or logical fallacies. This level can be paraphrased as "having so much evidence that only a lunatic, liar or lunkhead would choose not to accept it". Also such that even a dogmatic religionist would accept the truth and evidence and rule in favor of reality. An example of such a failure would be an educated person in the 21st century denying the Big Bang or the theories of relativity, gravity, disease or evolution.

- Clear and Convincing Evidence[13] (Medium High). Defined as being when "...the evidence is highly and substantially more likely to be true than untrue..."

- Preponderance of the Evidence[14] (Medium Low). Defined as "a requirement that more than 50% of the evidence points to something."

- Substantial Evidence[15] (Lowest). Defined as "evidence greater than a scintilla of evidence that a reasonable person would find sufficient to support a conclusion".

- **Burden of Proof.[16]** The person making an assertion accepts 100% responsibility for proving the claim is true. If you claim

> that you are the king of Spain, that you are the monarch of your suburban neighborhood, that you are the latest true prophet or messiah or that the cure for aging that you are selling really works, you must prove it. If you claim to know what the creator of the universe has said and wants me to do, or that he has dictated or inspired a book or three containing this and other guidance, you had better have a huge amount of astronomically rock-solid proof. Without such proof your claim is just irritating hot air and should be treated as such.

The bottom line on all this is that whenever you hear a claim or assertion, you need to suppress your feelings and your opinions, what you've heard and what you personally think, and run the assertion through a very tough screening process. Don't let your emotions run your life or make your decisions. Use reason and logic, think critically and skeptically, and demand that the claimant provide a level of empirical evidence in proportion to the size of the claim. Ordinary claims can be verified with ordinary evidence, but as has been said so often and so well, extraordinary claims require at least equally extraordinary evidence.

Most religions, and the vast majority of their core dogmas, make unbelievably extraordinary claims, yet after century heaped upon century what they still try to offer as embarrassingly pathetic alleged evidence is their holy books and their personal feelings. They still cannot offer even the least bits of ordinary evidence. The reasons why people still believe the fairy tales and ghost stories which religions tell, why they so often abandon reason and critical thinking, and why this is such a danger to the well-being and future of humanity and our planet, will be explained in the rest of this book.

Part Two

Seeing More Clearly

Chapter Four

The Miraculous Becomes The Mundane

"The saddest aspect of life right now is that science gathers knowledge faster than society gathers wisdom."
- Isaac Asimov

While it is distressingly easy to find no end of examples of the ignorance of one's fellow humans on so many subjects, both today and throughout human history, one can never forget just how much progress we have actually made and how truly fortunate we are to live in this time. Our medicine is not perfect but it is immensely superior to that of even 100 years ago. Our lives are not endless worry-free sport and leisure, but for most people they are not endless grueling and painful exertion and misery.

And if you live in a country where you can read a book like this in public and then discuss it at work, at school or with your friends, you enjoy a great deal of personal and social freedom. In far too many places and times your very life would be over or at great risk just by holding such a book in your hands. The author of such a book would be living under a death sentence, applicable upon discovery of the book.

Our knowledge has advanced to such a degree that it would have been impossible for the average human of 200 or even in some cases 100 years ago to even conceive of things which we take completely for granted today. Taken from an extraordinarily long list of advancements, we have clean and cold drinking water and even ice dispensed on demand from refrigerators and freezers which also keep fresh a large supply of food we did not have to grow or hunt. We attach color pictures of our families to these refrigerators with customized flexible magnets which tell the time and temperature. We have hot and cold showers and the indoor plumbing in our air conditioned and heated homes to transport that shower water both in and out. We have readily available and rapidly improving medical care and drugs to treat most diseases and afflictions.

We enjoy instant communications to and from almost anywhere on the globe, electronic point-to-point navigation and the ability to be prepared for the next day's weather due to our satellite-based technologies. We ride on high-speed air, land and sea personal and commercial transportation. We have access to more data than we can absorb in many lifetimes, and we can access that data as well as perform those communications and navigational tasks using electronic devices which can fit in a child's small hand. And with a nod to our evolved nature as highly social creatures, we can use this technology to take videos of our cats or kids and instantly share them with billions of people.

We live in an incredibly fortunate age in comparison to all of our untold billions of human ancestors. Charles Darwin, the man who finally opened the door to the wondrous vistas of evolution and natural selection as recently as 1859, and who was a man of unusually high intellect and perception, was not unusual in his time in that three of his ten children died when they were ten years old or younger.

Pasteurization was only invented in the late 1800s. Infant and maternal mortality rates have plummeted by orders of magnitude. Antibiotics only really came into existence as a widely available technology in the early 1900s. Such developments mean the average human today is infinitely more capable of recovering from illness or injury and living a very long life than all the billions of humans who lived before the early 20th century.

Today if a person in the developed world experiences the same loss of children as did the Darwins, it is rightly considered an unimaginably uncommon though no less tragic experience. As medicine increases its ability to predict and cure, humanity will experience progressively less suffering from the deformities, discomforts or deaths from the diseases that currently afflict us.

Today we can hop on a plane and be anywhere in the world in less time than it used to take to go just a few miles in rough and dangerous country. Today we can look through our most powerful telescopes at objects billions of light years away. We can use our most powerful microscopes to see individual atoms. Think about all the beneficial advancements science and our scientists have discovered for us. Think about what our advances in science have enabled us to understand and to imagine. It is said that each generation makes progress from a position of advantage, by standing on the shoulders of giants, and thus can see farther and reach places those giants could not. We take their work and move it farther and in different directions with amazing results.

We now consider ideas that would quite literally have been impossible to occur to people in generations past. If you could go back and ask even Darwin or Newton for their thoughts on the capability of a smartphone to take megapixel digital images then use wi-fi to instantly post them on the Internet, they would be as confused as if you asked the question in Sanskrit.

Now think about what our science will be able to do in one hundred, five hundred, a thousand or ten thousand years, assuming we and our planet survive that long. Think about the increasingly higher vantage points we will stand on as hundreds of future generations of giants elevate our communal perspective. Think of the ideas we may have.

Consider the following:

The universe we live in is incredibly huge. A rough low estimate of the number of stars in it is in the range of a septillion, which is 1,000,000,000,000,000,000,000,000, the number 1 followed by 24 zeroes. Take a moment to try to imagine that. You can't. It is impossible to actually imagine it with primate brains which evolved to handle and track much smaller numbers of objects in much smaller and closer-in areas, but some of us get far closer to that full and deep understanding than others. In this unfathomably large collection of stars, it is possible that somewhere out there at this very moment that a member of each of a thousand civilizations in a thousand different galaxies are all looking up at the sky, seeing or sensing in our visible spectrum, in ultraviolet, in infrared, or with some unknown sense.

There may be blind batlike or underground creatures which cannot see at all, and which have no conception of the stars. There could be creatures which live underwater or in some other medium. What would their thoughts be like? What would their equivalent be of being afraid of the dark? Would they be curious about their origins or if other life forms existed elsewhere? If these creatures were sentient, would they invent a god or a pantheon of gods? If so, what characteristics would they ascribe to it or them?

Our universe could be the result of the equivalent of a fifth grade science fair experiment being performed in some other universe or plane of existence, and the youngster doing the experiment might be the least intelligent being in its 'class' or 'school'. Our universe could at this moment be sitting in the equivalent of a beaker or a petri dish or a 'container' of matter, energy and space on a long table full of other such experiments and universes. There might indeed be hundreds or even thousands of such universes sitting essentially side by side, with the winner chosen to be kept on an award shelf or in a display case somewhere, and with the losers discarded into some hyperdimensional disposal unit. One day we ourselves may be able to create universes after we discover that it really wasn't that hard after all.

When life on earth began approximately four billion years ago, it was a very rare event. Since that time, billions of species have gone extinct through natural selection or other circumstances. Major mass extinctions have occurred several times in the history of life due to geologically nearly instantaneous causes such as a meteor impact, massive volcanic activity or climate change. The long line of human ancestry includes an unbroken line of predecessors, at least one of whom was alive during every single minute of every single day since the first replicating molecules formed. Each of those organisms died but their descendants now cover the planet and have walked on its moon.

Were this long line to have broken at some point, our world would not now be populated by our species and by all of the plants and animals with which we currently share the biosphere. Earth could very well have ended up being a world inhabited solely by water-dwelling fish, invertebrates, microorganisms and various flora. The land could have remained the domain of reptiles and dinosaurs, with mammals never evolving or growing much above the size of mice.

The human brain is the most complicated organ and machine which humanity has yet encountered. Absolutely microscopic variations in certain chemicals at certain points in fetal and post-natal development contributed to your personality, and they still do. The addition of certain chemicals to your brain can cause you to have an out of body experience, experience visions or feel as if you are one with the universe. Other chemicals can make you more trusting or more loving. Others can make you more aggressive. Others will cause you to feel depressed or ecstatic. Yet others can actually make you more likely to believe the fantasies proposed by the world's religions.

You are an organic robot, part water and part other matter, composed of elements forged in the hearts of long-dead stars. Your body is an organic machine driven by chemicals, chemical processes and electricity. Inside that organic machine lies the brain, which is itself yet another organic machine which controls the body and also functions using chemicals, chemical processes and electricity.

What if one day humanity gains the ability to translate the electric signals in your brain that occur when you dream into a holographic projection or virtual reality image? That would be the most unscripted reality TV ever, and either an enjoyable or horrifying experience to replay for the family over breakfast.

What if one day we can actually come close to replicating the total virtual reality of *The Matrix* movies? If someone committed a crime while in that state, should we prosecute that person in the real world? If not, why not? The intent was clearly to commit a crime which was followed through from the perspective of the perpetrator, so why not face actual prosecution since intent is such a key part of the legal process?

Will we one day enable persons to live their entire lives in an electronic universe, either re-entering the real world or not? Will such persons be able to live in that universe but earn a living through interacting with the real world via tele-operation of machinery or doing other tasks or white collar work? What if we found the key to super-enhancing human intelligence and could grow or alter human brains to take advantage of that knowledge?

One day humans will be able to create life in the laboratory. Would it be ethical to make it intelligent? What if we can one day engineer a cat, dog, elephant, horse, dolphin or chimpanzee which is intelligent enough to function in human society, or at least communicate intelligently with us? What if we could create a wholly new organism tailored to perform specific work? What if we could create customized bodies which could be outfitted with human brains? What rights should these intelligent animals or modified humans have?

Would they be allowed to become citizens of the state or nation in which they reside? Could such creatures work for a living and be compensated for it beyond food and shelter? Could they join sports teams, go to school or do other activities with humans and even move in next door to your great-great grandchildren? Could they go to church with humans? What would a mixed-species Sunday school class be like? What new gods, or variants of current ones, would these creatures invent? What new doctrines and dogmas would be invented? Surely new religions would not be too far behind the formation of a new intelligent organism, if that organism shared certain human mental weaknesses.

What if one day we are able to cure all disease? Even more, what if one day we discover how to maintain the human body and brain in a state of suspended degeneration, with medical care so advanced that death can only occur from very grave and sudden injury? A lucky and careful person could theoretically live for thousands of years in a body that remains at a biological 25 or 30 years old.

Those are all wonderful and interesting things to think about. Truly, a long and vigorous life without the possibility of suffering the ravages of cancer, arthritis, general degeneration and dementia is a goal we should all work toward. But as humans continue to rise along the arc of ever-increasing scientific progress and its benefits for humanity, we must never forget one thing. We must always be vigilant, but never for the dark side of the technology, rather for the dark side of humanity.

What if one day humanity decided that everything, including science and law, should be as dogmatic and as unsupported as its religions? The cores of the major monotheisms have not changed since their inception many centuries ago, and they can never change. They can never change because their books cannot change. Their books cannot change because they form the core of the faiths and they are now in paper and electronic form all over the globe, in libraries, museums, homes, hotels and places of worship. The books are a sunk cost which cannot be clawed back, and are permanently out there for everyone to see. They are embarrassing fossils which the religious cannot wish away, reject or even modify no matter how much they may secretly wish to do so.

Religious beliefs are largely unchanged and in many cases religions continue to accrue bad ideas. These religions and their books come from a time when humans did not know the earth was round or that germs cause disease. The people who wrote these books did not know that sacrificing large numbers of animals and humans would not in fact bring rain, appease the gods, allow the sun to rise, absolve the tribe's sins, ward off pestilence or make the crops grow. Rather than saying anything useful on these subjects, or on anything which would have improved human well-being, the gods of the books only reinforced the authors' abject ignorance and stupidity.

The lack of scientific knowledge in the people who wrote the monotheistic holy books is such that an appropriately educated nine year old today would be the highly advanced instructor in a conversation with such abjectly ignorant adults. An academic standard from the state of Pennsylvania notes that by the end of the fourth grade, a student should be able to[1] "distinguish between a scientific fact and a belief" and "provide clear explanations that account for observations and results." These soon-to-be-fifth graders will know about electric circuits and conductors, the basics of life and the fundamentals of the scientific method, all of which were as wholly unknown to the biblical authors as were exoplanets, computers, penicillin and thousands of other people's gods.

Unless today's children are kept out of actual schools and deliberately raised in a fully biblically-informed isolation chamber of a home, they will also know that animal sacrifice does nothing but result in a dead animal and the deadening of the killer's morality.

The biblical authors were severely limited by their almost total lack of knowledge on nearly every subject except local and regional myths, legends and religions. Thus in their thinking and writing, rather than standing on the shoulders of giants, they were groveling in ignorant superstition at the feet of thoroughly evil imaginary gods. They were kneeling in the dirt in ignorant prostration and supplication, faces and heads down in fear, beholden to the magical tales of their predecessors and contemporaries and to the gods which haunted their imaginations.

Why is it that in the 21st century in the most technologically capable nations on earth, billions of people use iPhones, medicines, cars, computers, electricity and air conditioning, all products of science unheard of and undreamed of by biblical authors, but these same people happily adopt and faithfully practice whatever nonsense these incomparably ignorant ancients said about their gods, our world and indeed our universe? Why do these people think that knowledge about everything else in the world can improve and grow, but this cannot happen on the subject of religion? Or on morality?

Such deliberate schizophrenia is as contemptible as when a person knows all of the benefits of modern medicine, but only takes the decision whether to use them or not after tossing animal bones to decide if he should go to the doctor or just pray longer and harder. Such madness is no different from the person who wakes up, showers and dresses and cooks breakfast using the products of modern technology, drives his car to synagogue, church or mosque while listening to the radio, fully accepting the fruits of the science that gave him these benefits, then takes an hour or more of his life to worship some imaginary god, and in Catholic services actually eating and drinking resurrected bodyparts of that god.

That is the dark side of humanity. This is the kind of thinking that results in Crusades, in Inquisitions, in heresy hunts, in holy wars, in Holocausts, and in persecution of whoever the faithful don't like. This is the lack of thinking that leads millions of Americans to elevate their religion and its ancient dogmas above all other subjects. This is the lack of thinking that leads the faithful to deny human rights to other people who are just as human as themselves, and to deny fully valid science whenever it does not conform with or cannot be twisted to match their religion. This is why religion is a metastatic cancer on humanity and why every scientific advance and every reality-accepting person must ever be on guard for the twisted skeletal claws of faith which seek to suppress science or squeeze the life out of it.

This is also why if I ask any moral person to name the theocracy in which he would most like to live, then to name his reasons, that person will not be able to give a truly moral response. There are plenty of choices, but for some reason the places where religions hold total power are not places where anyone of sound and moral mind actually wants to live. How many devout Christians, and especially any devout Christian women, would like to move to Saudi Arabia and be subject to its rules? Or Iran? How about any of the fundamentalist religious enclaves, whose major crimes are actually the direct result of their patriarchs most faithfully recreating the biblical world in their communities?

Without exception, such places are intolerant, bigoted, immoral and often very dangerous for people to live in. And yes this includes the Vatican, which is a fully theocratic dictatorship ruled by a person who is elected but who is nonetheless a theocratic Maximum Leader whose word is not just law, but treated as infallible whenever he feels like having an Immaculate Communication. This contemptible claim of infallibility is disgusting beyond words and the disgust grows exponentially with the revelation of each child sexually abused by a priest and each revelation of the coverup of those crimes by the perpetrators all the way to the very top.

Many Christians foolishly think turning America into a Christian State theocracy would be a wonderful thing. These people prove that they have absolutely no idea what they are talking about, that they care nothing for human decency, and that they share the unmitigated arrogance of their fellow believers in wanting their fairy tales to be the law of the land over everyone. A Christian State theocracy would be no better than the Islamic State which at the time of writing is killing, enslaving and raping its way through lands which have already seen far too much bloodshed.

In case you have forgotten, the world has already seen what actually happens in a real Christian State. Medieval Europe was the last time Christianity had full control of the reins of power, and no sane or moral person would ever want to return to that sorry time. Make no mistake, any attempt at theocratizing a state would turn it into a quite horrific Dominionist or Christian State.

The medieval Catholic State terrorized the people of Europe and beyond for centuries, with a length and depth of evil that the Islamic State only dreams about. It cannot be said often enough, that when a person thinks his god is on his side, nothing is sacred. And this applies a thousandfold to the monotheistic gods.

I can also ask anyone to name a civilization or culture which became better off and more moral as a result of becoming more religious. As the person struggles to come up with even a single example I can spend the next two days describing in detail all the places which have massively increased overall human well-being and freedom in direct proportion to how much they have *eliminated* religion from their lives and societies while elevating science, reason and humanism above all else.

There are several fundamental reasons why people adhere to religion, and in our modern and well-informed world they all derive from the damnable weakness and fragility of the human mind. I will discuss these reasons throughout the book, and have devoted an entire section to specifically listing and explaining these reasons. All of them are impediments to escaping the insanity of belief in the unseen, the undetectable and the unprovable.

So as you read the rest of this book, please keep the following points in perspective.

- Religion and dogma give us stories of burning bushes, flying horses and magic stones. Science and reason give us medicine, safe food and water, electricity, heated homes, instant communication, clothing, airplanes and computers.

- Religion and dogma give us suicide bombings and witch trials. Science and reason give us secular laws, the knowledge that witchcraft is imaginary, and a means of peacefully solving problems and resolving conflicts.

- Religion and dogma give us sexual slavery, human slavery, misogyny and bigotry. Science and reason give us emancipation from oppression and justify equal rights and freedoms for all people.

- Religion and dogma give us Crusades, Inquisitions, holy wars and no end of justification for doing pure evil to other human beings. Science and reason give us the Enlightenment, the Scientific Revolution, a true understanding of morality and endless reasons to live together in mutually beneficial harmony.

- Religion and dogma persecute and kill. Science and reason help and heal.

- Religion and dogma give us fairy tales, scary tales and pie in the sky when we die. Science and reason give us the enlightening wonders of true reality and a drive to be and do the best we can for ourselves and others in this world while we live the only lives we will ever have.

And if you are a religious person or just a person with a soft spot for religion, and the things you read in this book make you angry or defensive, please keep these other points in perspective:

- There is no need to get angry, all you need to do is prove the contents of this book wrong, prove that your god exists and prove that your religion is true and all the others false. If you can do that according to the appropriate standards of evidence, I will be an honest empiricist. I will follow the data, will change my mind and will proclaim it publicly. If you are a truly honest person you will do the same in reverse.

- Throughout the history of religion, it is the faithful and not the secular who can be counted on to threaten and actually carry out death to those who disagree with them. This is a book full of information and ideas, and they should be discussed in the calming light of reason, not the ghastly light of burning bodies and books.

- If your jealous or vengeful god or one of his minions or avatars is going to take out some kind of hateful and immoral revenge on me for writing this book, I happily look forward to engaging with him, her or it on this after I close my eyes for the last time. Any actions whatsoever which you may take to threaten me or others like me before that time again prove the evil in your religion and in yourself and further demonstrate the truth in and the need for this book.

Chapter Five

Hate the Belief, Not the Believer

"Father, forgive them; for they know not what they do."
- Jesus

One of the social and personal challenges which result from being a strong opponent of religion, particularly its monotheistic strains, is that many religious people are actually very caring, kind and giving people who can be trusted to do the truly moral thing most of the time. They can be very pleasant, friendly and helpful to friends and strangers even when it inconveniences them. They can be honest, hard-working, dedicated and ethical people who do the hard work required to keep a civilization going, while also working to help the less fortunate even when they themselves may be of meager means.

There really are many wonderful people who happen to be religious who have cherry-picked out what actually happen to be some of the very few decent parts of their faith. They try to live what are often actually empirically moral and decent lives based on these small nuggets extracted from what are dumpsters overflowing with bloody stinking garbage. They are often actually good to others even when their behavior is measured by a truly and fully secular and humanistic ethos and system of morals.

Not to sound condescending or trite, but it is a true fact that some of the best people I know are religious, and some of my best friends are religious. Some of them come from backgrounds almost diametrically different from mine, and some of us are on the exact same moral wavelength on almost every issue. Does this mean religion makes people moral? Not at all, or at least no more than their hair or skin color makes them moral, or that their gender or their age or ethnicity make them moral. These people would be good and moral if they were not religious, because being moral and good is who they are. Their religion is a social accretion and they absolutely did not need it to be good.

Religion, like alcohol, often serves as a personality amplifier. It can inspire a moral person do acts of extreme good for other people, and it can inspire an immoral person to do absolutely horrible things to other people. The savage and extremely unfortunate root problem is that both types of person can use the same holy book, sometimes even the same passages, to justify their actions. Particularly for the billions of people who are Mesopotamian monotheists, the references they proclaim as their references for ultimate morality are appallingly hateful books, and the gods they proclaim as the sources of their ultimate morality are actually the epitome of evil and immorality.

Religion is a nearly infinitely variable and powerful instrument of division, tribalism, hatred and bigotry. Humans are highly social primates and many have an intense tribalism. Religion, being tied to an allegedly all-powerful god in the sky, is the most powerful reinforcer of this tribalism and has and will always be used as justification for the most horrible and antihuman policies and practices a person can commit. Yet despite the massive amount of evil in their holy books, some monotheists from each of the three strains manage to actually do good things regardless of the bad things their books tell them to do.

It is important to understand that the extremely small number of religious people who have actually read their holy books all the way through and are nevertheless still moral in their actions are already a very long way down the road to freethinking and secularism. By going through that process of cherry-picking from their holy books they have demonstrated the ability to think somewhat independently, to at least partially reject nonsense when they choose to, and to run the teachings and books through a more or less usable moral filter.

These people mainly remain in the faith for social reasons, because their fear of a permanent death is overpowering, because they do not understand science and probabilities, and/or because they retain a childlike desire for strong hands to take care of or assist them, even if these hands are imaginary.

It is worth restating that the innate and inborn morality or goodness of a person in no way serves as proof of or justification for any beliefs he or she holds regarding the supernatural. There were good people among us tens and hundreds of thousands of years before humanity knew anything about the natural world and how things actually work, and before any religions were invented.

There have almost certainly been good people in every single system of faith ever invented. There are almost certainly good people today in every religious system known to humanity, and there most certainly are huge numbers of very good people who shun all religions and disbelieve in all supernatural imaginings.

The unfortunate fact that separates the faithful from the rest of us is that religious people have been born, wandered, frightened, tricked, trapped, threatened, coerced, encouraged or forced into a religious system or belief when they could have escaped all that. There is nothing whatsoever truly separating the goodness and kindness of a totally secular humanist from the goodness and kindness of a devout religionist except for the trappings, rituals, rewards and punishments within the believer's religion, and the money which the faithful give to their clergy as opposed to a charity or other wholly useful purpose.

Indeed, it is usually the case that the nonbeliever cares little or not at all about the religious feelings or actions of their fellow humans, except when these fellow humans seek to impose these beliefs in public schools or to make their religion the enforceable law of the land. This unfortunate problem often occurs even in societies where the law expressly separates the church from the state, such as the deliberately secular United States with its intentionally godless Constitution.

When the faithful keep their beliefs personal and private, and do not seek to order other people to follow the religion's teachings, people of faith and without faith in a civil society can get along quite well. In these cases, we freethinkers, secularists, humanists and general nonbelieving folk are happy to share this world with believers, no matter what our perspective is on your beliefs, as long as the beliefs are kept either under wraps or away from the rest of us. We still think religion is as nutty as wearing one's dirty underwear on one's head in public, but as long as it's kept to oneself and doesn't harm or affect anyone, we are happy to say "to each his own".

However, in response to the neverending immoral actions by the faithful to try to impose their religion on others, we nonbelievers are continually forced to move from our default position of peaceful but ever-wary coexistence to one of action, in order to block the religious aggression. In these cases we are not coexisting, we are fighting, and we secularists did not start the fight. We know that the religious aggressor may potentially be a good person somewhere behind the mask of bigotry and aggression, but has been captured and co-opted by the faith to do and say evil things.

In these instances our initial freethinking position regarding the aggressors is to "hate the belief, not the believer". This is an obvious riposte mirroring the common expression proffered by the religious while they are hypocritically and superciliously looking down their noses and denigrating someone for so-called "sinning".

"Hate the sin, love the sinner". While this attitude is likely only actually true for a very small percentage of those people who say it, because the faithful so often find so many ways to exclude and demonize live human beings due to this so-called "sinning", we can take it at face value from some few members of the faithful. In that vein, we nonbelievers feel the same way but in reverse – hate the belief, not the believer – as long as the believers keep their beliefs to themselves and don't try to subvert science classrooms, don't try to inculcate their particular slice of the massive pantheon of religion into our laws and regulations, and don't oppress and belittle others because of what some ignorant sadists wrote thousands of years ago.

We nonbelievers respond to religious thuggery with a swift and strong defense followed by an appropriate and at least proportionate counteroffensive against the attack. We do not automatically default to hating the individual or the group unless they make it plain that they deserve the hate. We work to show the aggressor the immorality of his ways from a human perspective, while blocking his actions using personal dialogue, the media, appeals to common decency, the law or other means.

There are cases where the perpetrators of religious aggression and thuggery show themselves to be truly and genuinely hateful and their beliefs have blinded them to a true morality. We will never be able to truly determine how many people are like this, due to the lies and façades the religious so often throw up in discussions, while lying to judges and while occupying elected office. Thus there can be and will be no détente, no surrender and we will give no ground. Hate is an unfortunate emotion, but either evolution or some allegedly all-knowing and all-powerful entity instilled it in us, and while we understand it is the hate within the believer which will start the conflict, our hate for your unconscionable evil and aggression will not be diminished by our humanity or our decency. To all religionists everywhere, we freethinkers say you have no right to impose anything at all from your faith system on any unbeliever, at any place and at any time, or indeed on any other religionist.

To the Christian thugs in America who wish to impose your beliefs on the rest of us while proclaiming the evils of Islam or atheism, we tell you that your actions are as offensive to us as a fundamentalist Islamist working day and night to involuntarily and forcefully impose Islamic Sharia religious law on you and your family would be to you. We don't want it, we will never want it, and your attempts to force it on us show your immorality, your insecurity and your evil lust for power. Your Jesus would pick up that whip of his again and slap all of you around mercilessly if he were real and saw what you were up to. It isn't just money which makes people evil; power and religion corrupt just as completely as money does.

Your Jesus notes in Matthew 21:13 that "My house shall be called the house of prayer; but ye have made it a den of thieves". In John 2:15, Jesus made that whip and "...drove them all out of the temple, and the sheep, and the oxen; and poured out the changers' money, and overthrew the tables…" He would use that whip just as quickly on today's Christian aggressors to drive out their corruption by power and their desire for money and power, particularly as they sin in his name while holding him up as an idolatrous "human shield", perpetrating evil while hiding behind those who cannot or do not speak for themselves.

If your Jesus were to return to earth today with that whip in hand, he would first descend on Washington, D.C. and its elected theocrats and their special interest groups to deliver some much-deserved floggings, then would "island hop" from megachurch to megachurch and state capitol to state capitol across the country, doling out public whippings to the no-longer-smiling moneygrabbers and liars now prowling stages and elected offices and fleecing the sheep.

Your Jesus said to render unto Caesar what was Caesar's. Your Jesus strove to convert people through preaching and teaching, not packing courts and violating secular law. Your Jesus preached a sermon of peace and service on a mountain; he did not tell his disciples or followers to find ways to order other people around, insult and degrade them and try to take away their human dignity and civil rights.

What would your Jesus do if he caught you lying, cheating, bending the truth, not telling the truth, the whole truth and nothing but the truth so help you Jesus, and subverting the rule of law to mandate his religion? I tell you three times, he would take that whip and beat so many American Christians with it so hard and for so long that even his supernatural self would get tired.

How did we get to this point? Let's first take a massive deep breath and try to ignore the millennia of previous religious thuggery the world has experienced, the majority of it being Christian or Christian-inspired.

The Renaissance, the Scientific Revolution, the Enlightenment, the Age of Reason and the current exponential increase in human knowledge were and are reflections of humanity's slow maturation from its frightened and ignorant infant and early childhood stages to its current status as a more educated but still impulsive and not fully mature teenager. It was a long and slow process of awakening that enabled not merely individuals but large groups of people to understand how the world works, and that no explanatory god of any kind was necessary anymore.

Science and reality had been inflicting a series of wounds on religion for centuries, like a simple-minded bull being wounded time and again by a cunning and implacable matador. The terminal blow to religion came in the form of a deeply penetrating wound inflicted by Charles Darwin, with the publication of his masterful *On The Origin Of Species* in 1859.

Since that time, religion has continued to totter about, slowly bleeding out but propped up by social inertia, by the ignorance and bigotry which carries on from generation to generation and the unwillingness of the weak faithful to face the cold fact of permanent death without an afterlife.

Religion today is like a withering tree slowly dying in a prolonged drought. Many of its branches have either died or are so emaciated that they are weak and dying relicts of their ancestors. The rest have largely retracted into a tight ball of insular isolation, shielding themselves from the penetrating and sanitizing illumination of science and immensely increased human understanding, or desperately twisting and bastardizing the advances of science in pathetic and sinfully disingenuous attempts to rationalize the newly discovered facts with their unchanging dogma.

Along with addictions to alcohol and other drugs, the addiction to religion serves as a compulsive stress-reducing and comforting agent within the brain. Religion enables believers to convince themselves that they or their imaginary god have control over their lives or their future. Religion allays the dread and uncertainty inherent in lives which cannot see the future and which are ignorant of much of reality. Religion is used like a sedative either at a moment of need or as an existential backstop to the infinite emptiness of the universe and the fears of death, loss and privation.

For example, the degree of comfort a human feels if he really, truly and substantially believes he will see his dead father, wife or child once again in some post-death reunion party is often too addicting to resist, thus the belief in the unbelievable conquers all else.

Those on the inside of the various belief systems, be they within the confines of a church, mosque, synagogue or just within a person's head, often cannot understand any state of mind that does not include their faith. They feel that their religion and beliefs are good for them, good for others and good for the world. As a result, a great many adherents feel compelled to share their faith with others, some uncontrollably and aggressively so. As distasteful, unwelcome and incomprehensible as these beliefs and practices are to nonbelievers and those of other faiths, as long as the sharing is done on a platonic level, if you will, without any actual demands on any of us, those on the outside can put up with it.

We don't really mind the Jehovah's Witnesses, the Mormons or any Jewish, Christian or Muslim traveling snake oil salespeople coming to our doors and trying to sell us their product; everybody's got to make a living. We thoroughly hate the beliefs but we pity the poor misled believers, and out of our basic innate human morality and decency we do not threaten or denigrate them. We politely refuse their offers, or engage them in a short debate to see if they are open to any actually rational perspectives or to fleeing their self-imposed mental imprisonment.

We nonbelievers normally merely just say "No, but thanks for the offer". This is what a decent human being would be expected to say to another person in response to an offer which the seller truly thinks will help the prospective buyer. We can feel pity for these vagabonds on their often fruitless missions and for their constant rejections, and even sometimes offer them a bit of human compassion in the form of a cold drink on a scorching day, without feeling anything but complete distaste and rejection for the disgusting product being peddled.

We also can tell that quite often the people doing these sales pitches don't have much else in their lives that they can feel good or happy about, and with the humanism and compassion that so many freethinkers and humanists have, we are gentle and kind but firm in our refusal of the offers of imaginary salvation and whatever else is being sold.

No, as long as the sales pitches are just those, and as long as they remain fully voluntary, we can coexist quite peacefully. It is when the sales pitches turn into attempts to make laws which codify hateful religious bigotry disguised as so-called "religious freedom", or which unconstitutionally allow religion to violate the public sphere, or when the religious attempt to force their beliefs into public school curricula, that we nonbelievers will take a strong stand. We have to continually say to these aggressors, in clear language, that you no more have the right to force your religion on the rest of us than we have the right to force someone else's religion on you.

This is something which far too many of the religious across the world, and particularly in the United States and much of the Islamic world, in a display of arrogance and hypocrisy that is nearly unfathomable but is nevertheless ubiquitous, simply refuse to accept.

So I again invite the religious Christians in America to do what you so often cannot or will not do, which is to actually follow your savior and "Do Unto Others"; that being yet another bit of guidance from him which you blithely and deliberately ignore. As you would not want Muslims to force you to adhere to their belief system, you should not try to force anyone else to adhere to yours. The second you try to do so, you are both flipping Jesus a stiff middle finger and demonstrating yet again that those of faith simply can never be trusted either to follow secular law or the teachings you say are delivered straight from your god and savior and which are actually written down in that book you say you adore so much.

You believers can believe all you want to, within the confines of your systems and minds, as long as it is permanently excluded from the public and civil spheres. You can haughtily sniff and airily and hypocritically proclaim your love for the sinner while hating his alleged and imaginary sin, as long as you do not have the power to injure or oppress him for it.

As long as you believers stay inside your Constitutional, ethical and moral boxes, we nonbelievers will meet you half way and stay in ours, which is where we want to be anyway. We will do our part, and be content merely to hate the belief, not the believer.

Chapter Six

America As A Case Study In The Harmful Effects of Religion

"The various modes of worship, which prevailed in the Roman world, were all considered by the people, as equally true; by the philosopher, as equally false; and by the magistrate, as equally useful."
- Edward Gibbon

"Religion is an insult to human dignity. Without it you would have good people doing good things and evil people doing evil things. But for good people to do evil things, that takes religion."
- Steven Weinberg

Religion has been a debilitating and deadly cancer on humanity since it was first invented, and it has metastasized and spread into every nation and society on earth. It has no real answers to anything, but it pretends it has information of ultimate truth and value delivered from the great beyond. It has been the cause of untold amounts of pain and misery deliberately inflicted on innocent bodies and minds. It is pitiless and unflinching in its faith-based proclamations, pogroms and persecutions.

It spares no one, from the unborn to helpless small children to elderly men and women and from devout fundamentalists to the wholly secular who must continually battle its attacks on humanity, reason and indeed on the secular themselves. It aggressively fights against science and reason and when it cannot kill, burn or deny the facts or their discoverers, it twists the new facts to fit ancient desert fairy tales written by astoundingly ignorant and hateful people. It tries to make its fiction the law of the land and thuggishly tries to force its holy books and immoralities down everyone else's throats.

That it has been able to do all this for thousands of years without a scrap of actual evidence or justification for its actions is a sad testimony to the many weaknesses of the human mind and the dogged persistence of human ignorance, stupidity and evil. Far from being either true or helpful, religion is merely astrology on steroids, alchemy claiming divine writ and human ignorance codified in highly immoral and thoroughly corrupt human tribal institutions.

All of the thousands of humanity's invented religions contradict each other, and the schisms within these religions number in the tens of thousands. They cannot all be right, yet this does not stop them all from claiming to have the one true hotline to heaven.

Religions offer us gods to worship who are allegedly all-loving but who are quite happy to kill individual people, whole cities or even every life form on the entire planet, including large numbers of children both born and unborn, in psychopathic fits of rage.

It is impossible to count up all the bodies and minds which have been damaged or destroyed by religion and its practitioners, but the tally continues to skyrocket faster than the American national debt, even as science continues to make discoveries that religion has never, could never and will never make.

Religion makes people dumb and immoral, and it makes societies and nations dumb and immoral. Any system of thought that turns to ignorance as its first problem-solving option is intellectually and morally bankrupt and will always generate far more human misery than human well-being. Religion is a mind-numbing, critical thinking-killing mental opiate which is antithetical to a truly humanistic society, and it is high time we all grow up and live in the real world without the self-imposed cancer of religion.

The brief history of the United States is a microcosm of the long battle between ignorance and enlightenment. The timeline of religious history in the US a few paragraphs below is a listing of significant events in American history which shows how religion captures vulnerable minds, justifies extreme immorality and the infliction of suffering, and fights the advance of science and reason with everything it has.

This includes deliberately lying to whoever it needs to lie to, and lying however it needs to, in order to achieve its end goal of religious power and supremacy.

For its entire history, religion's motto has been that the end justifies the means. For the deeply and sincerely faithful believer and the religious zealot, it cannot be said often enough that when a person believes his god is on his side, nothing is sacred. In the eyes of the faithful, no lie is too big, no deception is too immoral and no campaign of misinformation is too inappropriate, because they believe they are doing their god's will.

Facts do not matter to the faithful, and they never will. Religion is not a fact-based belief system; it is a belief-based belief system. It is a self-reinforcing mental cocoon which is constructed to prevent undesired facts from entering. Evidence is extremely unwelcome in religion, because reality contradicts the faith. And in what may be the ultimate perversity and malfunction of the human brain, the harder one pushes facts at the believer, the more tightly he closes his mind to them.

Every human alive needs to understand this, most especially the secular, the rational and the well-educated, because these people think that all one need do to make a person stop believing in his faith and doing insane things in its service is to provide him with sufficient facts and evidence against the faith. This will not work, because, again, facts do not matter in religion and they never will. In order to properly combat this institutionalized and aggressive ignorance, the secular and the truly rational among us need to understand this. We must also understand the fight against ignorance will be a long and bitter one.

Perhaps in several thousand years, if human evolution moves along the path of increasing receptivity to facts and decreasing susceptibility to fear and emotion, all humans will one day actually use facts instead of feelings as justification for their most important decisions and beliefs. But this will not be the case for the rest of this millennium and almost certainly for at least a handful of millennia after that.

Until one day in the far distant future when humans have matured as a species and religion finally dies off, it will do all it can to keep its chokehold on humanity. It will continue to deny facts. The faithful will continue to say and do immoral things and continue to try to maintain power over real people in the real world.

The timeline below of just a few events in the continual counterpunching by the warring parties of enlightenment and religious immorality and ignorance in America illustrates this quite clearly.

1620[1]	The Pilgrims arrive in Massachusetts after fleeing religious strife and persecution in Europe, intent on keeping their tribal and cultural identity.
1650-1815[2]	The European Enlightenment enabled critical thinking, reason and humanism to flourish and to show the errors and immorality of religion and a better path to human well-being. Enlightenment ideas strongly affected the American Founding Fathers and shaped America's founding documents.
1692[3]	The Salem, Massachusetts witch trials, a series of religiously-motivated trials of more than 200 people accused of being witches, led to 20 executions and many people being jailed.
1730-1755[4]	The "First Great Awakening" was the first in a series of zealous and often evangelical and fundamentalist Christian revivals in America.
1776[5]	Signing of the US Declaration of Independence, a wholly secular and Enlightenment-informed document. Part Six of

this book contains much more on the subject of our godless founding documents and godless founding national ethos, which contradicts today's religious falsehoods claiming otherwise.

1789[6] Institution of the fully secular US Constitution, prepared by Enlightenment thinkers.

1790-1840[7] The "Second Great Awakening" was another religious revival, this one in response to the secularism and deism of the Enlightenment and the belief that the Second Coming was near. This never-occurring but always-promised Second Coming is a recurring theme.

1791[8] The First Amendment to the Constitution was enacted, quite intelligently preventing government establishment of religion.

1797[9] The US signed a formal treaty with Tripoli which specifically states the US is "…not in any sense founded on the Christian Religion…" Thus from 1776 to 1797 both major national founding documents and a treaty with a foreign nation, all approved by top government leadership, show the US government and founding ethos to be wholly secular.

1800-1850[10] The general period of the "Burned Over District", which was an area of west and central New York which experienced a series of religious crazes which ignited then quickly burned out, and which was fertile ground for much of the Second Great Awakening's fads. Such religious cycling was not unusual in the rest of the country and shows the extreme human gullibility and susceptibility to religion without the need for actual evidence. This was the modern-day equivalent of the ancient Mediterranean hotbed of religions, in which religions and gods were continually invented, traded and cross-bred, and which eventually produced Judaism, Christianity and Islam.

1813-present[11] Christian Dispensationalism, a fundamentalist sect in which the faithful believe their god has divided history into specific time periods and humanity is responsible for "stewardship".

1826[12] Joseph Smith, the founder of Mormonism, experiences his first legal conviction. This one was for being "a disorderly person and an imposter". Smith was from New York, home of the aforementioned religion-addled Burned Over District.

1830[13] Joseph Smith founds Mormonism in western New York, based on his claim of being able to translate magical gold plates by using magic stones and a hat. Despite the rest of its teachings being just as crazy and fact-free, Mormonism is still going strong today.

1831[14] William Miller founds the Millerite sect, which believed the Second Coming would happen before 1843. Miller is also from New York, continuing the trend of new and wholly evidence-free religious fads and sects forming in that area.

1844[15] The "Great Disappointment" in the Millerite movement was the feeling these people expressed in October 1844 when Miller's prophesied date for the Second Coming came and went without Jesus making an appearance or even sending an RSVP.

1848-1920[16] Spiritualism, a belief that the living can communicate with the dead, is popular with millions of Americans. It, too, arose in religion-addled New York, fully devoid of evidence.

1855-1930[17] The "Third Great Awakening" was another time of religious fervor, fundamentalism, zealotry and great concern about the often-promised but never actually occurring Second Coming.

1861[18] The American Civil War is fought largely because the strongly Christian southern states refused to give up their biblically-justified slavery. Well over 600,000 Americans died in this utterly avoidable faith-based war.

1863[19] The Seventh-day Adventist Church, a fundamentalist spinoff sect of the now-defunct Millerites, is formed. It is also extremely anxious about the long-awaited Second Coming.

1865[20] The 13th Amendment made slavery illegal in America, thus going against the Christian bible's support of slavery. Despite the country being overwhelmingly populated by Christians who proclaimed their religion as moral, it took all the way to 1865 to do away with this evil practice. The deeply devout Christians of the American South made the lives of former slaves extremely miserable for many years afterward, and still today work to block their descendants from voting. If Christianity were truly moral even to the tiniest degree, there would never have been a slave trade or this continuing deliberate evil.

1866-present[21] The Ku Klux Klan is formed after the criminalization of slavery and remains active today as a white Christian supremacist organization.

1870[22] The 15th Amendment enacted voting rights for nonwhites and former slaves. This anti-biblical granting of human rights and equality faced significant opposition, in a country that was almost exclusively self-professedly moral and Christian.

1873[23] Congress passed the religiously motivated Comstock Act, which actually made it a crime to have or have information about medications or methods for contraception or so-called "unlawful" abortion. Some states passed laws far harsher than the federal act, and actually made married couples subject to imprisonment for using birth control *in their own homes*. Some Comstock language remained on the books until parts were removed in 1971 and others astoundingly into the 1990s. This is theocracy, and it is what always happens when the church is allowed to intrude into the state. The faithful go very far out of their way to legislate their personal religion and jam it down everyone else's throats.

1877-1964[24] The overwhelmingly Christian South had segregationist Jim Crow laws on the books for almost a hundred years. Black Americans had to live in so-called "separate but equal" conditions in some of the most devout states in America, and lived with reduced well-being in every aspect of life. Again, the allegedly moral followers of Christianity didn't merely fail to correct this travesty, they kept it going for as long as possible.

1896[25] The Supreme Court, in the case of *Plessy v Ferguson*, decided by a seven to one vote that the "separate but equal" conditions were legal and thus enabled legal segregation and oppression of black Americans for many years afterward. Meaning that solely due to the color of a person's skin, Christian America felt it perfectly fine to forbid certain people from certain locations or facilities. Jesus's "do unto others" was wholly ignored.

1916[26] Margaret Sanger is arrested after opening the first birth control clinic in the US.

1918-1920[27] Approximately 675,000 Americans died in the 1918 Spanish Flu pandemic, despite trillions of prayers offered to Jesus and his father for deliverance from the plague.

1920-1933[28] Religious zealots in the US, led by the Woman's Christian Temperance Union, convinced Congress to enact Prohibition via the 18th Amendment. Alcohol was largely banned in the country at various levels and in various jurisdictions but of course religious use of wine got an undeserved exception. The faithful continued their efforts to try to order other people around, using the legal system to thuggishly force their personal faith on other people whenever they could. This cycle never ends.

1920[29] In an overwhelmingly Christian America, and after a very long struggle against much strong opposition, it took until 1920 to enact the 19th Amendment which also anti-biblically gave women the right to vote. The fights against segregation and for equal treatment of black and female Americans had to be waged against entrenched Christian opposition, since their bible specifically supports slavery and puts women under male control. By comparison, New Zealand's women could vote in 1893.

1925[30] After Darwin's scientifically true theory of evolution made its way into American schools, religious fundamentalists and zealots fought extremely hard to remove it. Tennessee passed the infamous Butler Act which actually blocked public school teachers from teaching evolution and denying biblical creation. Again, the faithful imposed their unjustified beliefs on others through the corruption of the legal system. Based on the politics in that state and in many other places in the US today, nothing has changed.

1925[31] John Scopes, a substitute teacher in Tennessee, was prosecuted in what is called the Scopes Monkey Trial for violating the Butler Act. During the trial, the prosecution thoroughly embarrassed itself by its blatant denial of science and reason and its abject adherence to the fictional stories and dogmas in the Christian bible. The religious again show their zeal to suppress real science and teach fact-free dogma.

1925-1927[32] Mississippi and Arkansas enact Butler Act-like anti-evolution laws. Again, nothing has changed in 100 years as these states are still contaminated by religion, and science-denying religion has now infected government and politics at the national level.

1942-present[33] Christian Nationalism began in America and is still extant today.

1945+[34] Due to the religious suppression of science, technology, engineering and math (STEM) in US schools, and the impacts on education in the 1930s due to the Great Depression, major droughts and the Dust Bowl, all of which occurred despite trillions of heartfelt prayers to Jesus, the US needed to bring Werner von Braun and over a thousand other scientists to the country after World War II to work on rocketry and other programs.

1947-1958[35] The US endured McCarthyism, the modern day equivalent of witch trials but without the hangings and burnings, along with the Red Scare, an existential fear of the Soviet Union. The US began its retreat into a cocoon of religion due to fears over war with the so-called "godless communist" Russia.

1950-1958[36] Religiosity was extremely high in America; largely a rebound effect after WWII and due to anticommunism, McCarthyism and the Red Scare. American schools were not rigorous and STEM training in particular was not strong. Schools were racially segregated during this strongly religious period despite the *Brown vs Board of Education* Supreme Court decision in 1954. American Christians, as they still do today, hypocritically wanted equality and fair treatment to mean that they got to make the rules which everybody else had to follow, thus segregation and religiosity were both entrenched.

1952[37] The science fiction writer L. Ron Hubbard invents Scientology. The insanity of its teachings and beliefs exceed even those of Mormonism and the three Mesopotamian monotheisms, as one would expect from a science fiction writer. I mentioned some of the beliefs of Scientology in Chapter Two.

1954[38] The Supreme Court unanimously decides in the *Brown vs Board of Education* case that the 1896 *Plessy vs Ferguson* decision was wrong and that segregation in schools was unconstitutional. Finally we have an actual moral decision made by the highest court in a highly religious but legally secular nation. The decision was nevertheless highly unpopular with much of Christian America.

1954[39] Rising religious zealotry led to "under god" being added to the Pledge of Allegiance. This wholly unnecessary act was both unconstitutional and a symptom of the country's increasing religious fundamentalism. It was another example of the

Christian faithful working very hard to impose their personal opinions onto American life and politics.

1954-1968[40] The American civil rights movement took a very long time to reach major milestones, and again in a majority Christian nation the lack of those Christians doing much for and often fighting against the civil rights of their fellow citizens is a damning condemnation of their morality and their religion.

1956-1975[41] The Space Race between the US and the Soviet Union sets the backdrop for some of the developments in America over the next decade.

1956[42] The out of control national religious fervor and the continuing attempts by the Christian faithful to inculcate and infiltrate the government led to the replacement of the national motto with "In God We Trust". It had been "E Pluribus Unum" since 1782. Christian aggression and powerlust led the country to throw out almost two hundred years of an excellent and fully inclusive motto in favor of a sop to Christianity. The Christian god had never done a single damn thing to deserve any trust, but this did not stop the faithful.

1956[43] For the faithful, changing the national motto wasn't enough unconstitutional interference by religion in state matters so "In God We Trust" was put on US paper currency. This was to be the last major act of empty-headed and religiously driven buffoonery for some time, thanks to what the Russians did in the following year.

1957[44] The Soviet Union launched the first artificial satellite, Sputnik 1. This act of technological overmatch by a looming enemy badly shocked and scared the US. National leaders took a hard and honest look at the poor quality of STEM education and training in the US.

1958[45] In response to Sputnik and the perceived existential danger of Soviet technological overmatch, the Eisenhower Administration implemented many science initiatives to get America's religiously drugged heads out of the clouds and focused on STEM. This included creating DARPA and NASA and passing the National Defense Education Act, intended to get US schools out of the faith-induced pits they were in.

1960[46] Ruby Bridges was the first black child to attend an all-white elementary school in the South. Yes, it took all the way to 1960 in devout Christian America for this to happen.

1960-1980[47] The "Fourth Great Awakening" was the last spike in Christian fundamentalism, and it differed from the previous episodes in that more fundamentalist groups like Southern Baptists gained size and power. They also faced a growing secular and humanistic opposition, which had not previously been the case. This opposition meant, in clear Darwinian terms, that the more aggressive and fundamentalist religious species would thrive. This is what happened then and it is what is still happening today in both American religion and politics as the more fervent and active strains of the religious virus continue to infect citizens and governments.

1963[48] George Wallace, the new Alabama governor, called for "segregation now, segregation tomorrow, segregation forever", and later in the year actually stood in front of a door at the University of Alabama to try to physically stop black students from entering. The federalized Alabama National Guard had to tell him to get out of the way. Alabama's Christians voted in a poisonous hateful segregationist and bigot as their governor, on purpose, not just once but ***four*** separate times. It took an unsuccessful assassination attempt to make Wallace finally see the light of empathy. If the shooting was Wallace's Damascus Road experience, the Christian god utterly failed to intervene at the right time.

1965[49] President Lyndon Johnson signed the Voting Rights Act which prohibited discrimination and other practices intended to suppress the votes of black and other minority Americans. Religious fundamentalists, bigots and thugs are nevertheless still working hard even unto this day to suppress minority votes all across America.

1965-present[50] Christian Reconstructionism and Dominion theology are active in America. They seek to transform the US into a Christian-dominated country under biblical laws. The only word for what this would be is horrifying, as such a state would be as cruel and unjust as the modern Islamic State or the medieval Catholic State under the popes.

1968[51] Martin Luther King, Jr. is shot and killed in Memphis, Tennessee. He was only one of many civil rights activists and leaders who were murdered because they wanted equal rights for fellow human beings in Christian America and wouldn't stop fighting for them.

1968[52] The Supreme Court ruled in *Epperson vs Arkansas* that an Arkansas law preventing the teaching of human evolution in public schools was unconstitutional. Like the *Brown vs Board of Education* case, reason and sanity began to guide some of these SCOTUS decisions. The faithful did not let this defeat stop them, of course. They decided that since science and the Constitution were now at long last being used by the courts to decide cases, they had to repackage their fundamentalism and creationism into a form that sounded scientific. Thus, and again in an ironically classic Darwinian selection process, they created a body of fact-free dogmatic nonsense called "creation science" and tried to inject this into public school instruction. Some areas actually enacted legislation requiring that this Frankenstein be taught in schools along with evolution.

1982[53] The US District Court for the Eastern District of Arkansas ruled that the Arkansas law requiring public schools to teach "creation science" was unconstitutional. The judge in the case was very precise in defining science and why "creation science" most certainly was not science but was just religious dogma. Fundamentalists and creationists across the US saw this latest defeat of their newest Trojan Horse and began working on a newer strain which they hoped would successfully infect the public school system.

1987[54] The SCOTUS ruled in *Edwards v. Aguillard* that a Louisiana law mandating "creation science" be taught along with evolution was also unconstitutional and that the law was intended to support a specific religion. The scientific community, including 72 science Nobel Prize winners, was engaged. Numerous entities filed amicus briefs to inform the court on the scientific truth of the matter and the utter religiosity of "creation science".

1989[55] Following these defeats in the courts of law and common sense, the fundamentalist and creationist community invents so-called "Intelligent Design Creationism" as the latest strain of religious virus, designed and built specifically to sound scientific enough to pass Establishment Clause muster. Not to *be* scientific enough, just to *sound* that way.

Religious zealots and fundamentalists continue their efforts to inject this newest Trojan Horse virus into public school biology classes. This would lead to the famous 2005 showdown in Dover, Pennsylvania, noted below.

2000[56] First election of George W. Bush, a strongly conservative, born-again, antiscientific Republican Christian who expressly, gratuitously and unconstitutionally catered to the religious right. Science not supporting Christianity or Republican goals was suppressed. He either then or later believed he was chosen by his god to be president and to do his god's bidding. He let religion infest the White House and it greatly affected foreign and domestic policy, particularly foreign policy following the attacks on September 11, 2001.

2003[57] George W. Bush and Tony Blair, two religiously blinded and dogmatic theocrats, invade Iraq on bad information, false pretenses and a belief that their god had placed them in office to spread the word of that god and do his bidding on earth. Again, Christians continually show their inability to think clearly. Bush and Blair were merely modern-day Crusaders doing the work the old popes didn't complete, and Bush unbelievably actually used the word "crusade" until someone gave him a basic history lesson and told him that was a breathtakingly stupid choice of words.

2004[58] Second election of George W. Bush. Things got far worse as time went on. Religion continued to have a special place and special protection. Science was under constant attack during the entire period of Bush's time in office, by people motivated by religion, ignorance, science denial and greed.

2005[59] In a perfect example of the totality of the fight between reason and ignorance, the latest Christian Trojan Horse of so-called Intelligent Design Creationism and the façade of Christian honesty and morality were mercilessly exposed and viciously eviscerated in the *Kitzmiller v. Dover Area School District* case. This case exposed both the lie of ID and the lies of the faithful who pushed it on the people. So-called "creation science" had been a loser in court, so the Dover, Pennsylvania school district had mandated in 2004 that ID be taught. Intelligent and Constitutionally correct parents sued to stop it.

Every person who uses reason should read the full accounts of the case, both to understand the slimy deceitful depths to which the faithful will sink in service of their god and its dogma, and to be prepared for the next mutation of the creationist virus when it infects another school district. The judge in the case, a Republican and George W. Bush appointee, deserves massive amounts of respect and credit for the way he objectively applied science, facts, reason and the law to the case and his ruling.

2008[60] First election of Barack Obama, a progressive, half-black, freethinking and moral intellectual who quite professionally and Constitutionally, and unlike so many of his predecessors, never let the church into state matters. He also was the first US president to mention nonbelievers as part of the national family. He restored science to its rightful place as the arbiter of truth and reason, and made immense strides in restoring the nation's focus on science and moral issues such as health care and income inequality. He was fought tooth and nail on each of these issues by entrenched Christian and Christian Republican opposition.

2012[61] Second election of Obama. He continues to fight for science and human rights and is vigorously opposed at every turn by dogmatic and science-denying Christians. These Christians have turned their Jesus into a war-mongering, firearm-loving bigot and spend endless hours and dollars spreading that message.

2012[62] The Texas Republican party's official platform states its formal opposition to critical thinking with these words, which need to live in disgusting infamy: "We oppose the teaching of Higher Order Thinking Skills (HOTS) (values clarification), critical thinking skills and similar programs that are simply a relabeling of Outcome-Based Education (OBE) (mastery learning) which focus on behavior modification and have the purpose of challenging the student's fixed beliefs and undermining parental authority."

There it is, in their own words. Christian Republicans oppose critical thinking and the challenging of fixed beliefs, no matter how wrong they may be, if they challenge uninformed beliefs and authority. This is the religion-driven authoritarianism and totalitarianism which true morality, science and reason are always fighting against.

2016[63] Evangelicals and many other Christians hypocritically vote overwhelmingly for Donald Trump in the US presidential election, despite his blatant and repeated public violations of every biblical commandment they supposedly hold dear. Again, as in their entire history, and as they will act whenever they can get away with it, the religious act according to their motto "the end justifies the means", and they will elect literally anyone who will enforce their immoral religion on other people.

This administration's war on science and reason is a full-out blitzkrieg, starting with a Pearl Harbor-like attack followed by unceasing actions to demolish rational and science-based institutions and research and install science-denying dogmatic satraps into cabinet and other positions.

Science, truth and reason become systematically "unpersoned", as data vanish from websites and scientists are muzzled or fired for contradicting the Pravda. Additionally, not only is Russia now a highly aggressive destroyer of stability and American integrity, the US government now openly caters to one of our country's primary adversaries like a blushing teenage girl with a crush. Far from the Red Scare of the 1950s, today's Republicans, evangelicals and many Christians are displaying earth-shattering levels of intentional cuckoldry by openly letting the barbarian enemies at the gate gain dangerous access and influence while deliberately tearing down their own castle ramparts and defenses.

2018[64] In a blatantly pandering move intended to stroke Christians and their tremendously fragile egos, the Trump administration moved the US embassy in Israel to Jerusalem.

The record is crystal clear and the pattern is unchanging. Religion and its adherents are a clear and present danger to the well-being, safety and security of individuals and societies, and the short history of America is a textbook example of the damage religion does. To the faithful, no lie is too big to tell in the service of the mission to make every other person bend the knee and live under a theocratic government and its biblical laws.

The Christian faithful in America never cease their hypocritical and immoral campaign of propaganda and subversion, seeking to impose their particular god and their particular religion on everyone else, and they never stop lying about it.

The attempts by Christians to force then to sneak religion into America's secular public schools show a clear pattern of Darwinian evolution. When they could no longer mandate religion in the schools, they tried different ways to portray religion as science. Since this is impossible, those attempts failed and they are now back to the drawing board trying to figure out their next attack.

I predict their next salvo will be less of a direct attack on science and the Constitution than an even more underhanded attack, and one that uses the progressive ideals of equality and inclusion against themselves. This is actually being done now, as the faithful prattle on and on about their freedom of religion while they use that freedom to justify their oppression and bigotry.

I predict they will see that trying to force religion into science or to make them compatible is a nonstarter, at least until they can pack the courts with enough judges who will deny the words of the Constitution in service of dogma, so they will try to use other areas of education, such as sociology or philosophy or some other line of thought. They will probably not seek in the near term to directly inject their god, as that kind of subversion would be immediately detected and Constitutionally beaten back for the Nth time, but they will seek to inject their version of morality and their teachings though a back door, under the banner of "multiculturalism" and "exposure to other perspectives". When challenged, they will scream that "the (liberals / progressives / scientists / leftists) are discriminating against us! All we want to do is expose the children to more points of view!"

Whenever any of you liberals, progressives, scientists and most importantly, educators, hear this, as you inevitably will, you need to suppress your initial reaction to give in to your "multicultural" or "fairness" instincts, you need to grab that lying BS out of the air before it hurts someone and use it to (verbally) beat the liar who said it within an inch of his lying life.

These people put on the "I'm a nice Christian" front but it is a deliberate lie and a deliberate deception, just like most everything else they do. Theirs is a Crusade which never ends. They don't play fair and they will manipulate words, laws and people to get what they want. You must aggressively and unceasingly stop, block and prevent any of this nonsensical garbage from polluting our schools.

We can't stop these people from polluting their own families and their own children at home, but we damn sure can stop them from forcing it into the schools. To paraphrase Winston Churchill when the armies of darkness were attacking his great nation, we shall fight to the end, we shall fight them in the halls, we shall fight on the courtyards and public squares, we shall fight with growing certainty and growing strength in the courts, we shall defend our schools, whatever the cost may be. We shall expose them in the media, we shall fight in the school districts, we shall fight in the towns and in the streets, we shall fight in state capitols; we shall never surrender!

Religion has made America very dumb over and over again, and each time it is beaten back it returns in mutated form ready for another attack on science, reason and morality. Today as we struggle to put our military back together after a 15-year-long modern crusade that has accomplished absolutely nothing except killing and injuring far too many of our own and other people and doing grave damage to the US Treasury and our national morality, we find ourselves in a very bad position.

As we emerged from the small and dark Middle Eastern cave where we have been fighting, blinking under the blazing sun as we move on to the much larger global field of gladiatorial combat, we have taken a good look at our adversaries and realized that our former superiority in science and technology is a thing of the past. We fought a very long crusade for very wrong reasons, "led" by a religiously blinded theocrat, and now we find ourselves struggling in almost every area to prepare our armed forces for an actual major war which will combine virtual, economic and physical weapons.

While our blood and treasure fought this faith-based war in a religion-addled region of the world, in far too many cases returning with badly damaged minds and bodies or in coffins, our elected theocratic science deniers and unelected religious thugs fought with everything they had to deny and suppress science, dumb down education and make their religious dogma the law of the land. The faithful have done everything they can to enact modern-day Butler and Comstock Acts, they have put prohibitions on critical thinking and actual science, and they have set the conditions for another and much more destructive Sputnik Moment. And they are doing it on purpose, in the open, without shame, and with the full complicity of far too many Americans whose sons and daughters will be the ones who will needlessly die in the next Hot War.

We are already beginning to be overcome by our many adversaries' advances in the battlefields of the classroom, the laboratory and the marketplace. We cannot let the faithful drag us down past the point of no return in these battlefields as they have already come close to doing in the traditional field of battle.

We are on the cusp of such a global and permanent shift in national power and station. Unless we want to become just another theocracy, banana republic or fascist state, we must regain our sanity and strongly and completely reject the antiscientific, the anti-intellectual, and the extremely harmful dogmas of the ignorant and the faithful. If we do not, the next "Sputnik Moment" may be the death knell of our great nation and our grand experiment.

Chapter Seven

Answering The Protests

"Denial ain't just a river in Egypt."
- Mark Twain

As with snowflakes, no two people are exactly alike. No two people even in the same faith worship the exact same god or worship in the exact same way. People hold different opinions on every aspect of their faith. So I am sure that by this point in the book those of a religious mindset will be protesting that their personal religion "isn't like that!", or that their social circle or they themselves don't believe this or that tenet of their religion or behavior that I have criticized or am sure to criticize in later chapters.

> "I don't believe in an old man in the sky!" says the Christian who feels he has advanced beyond the medieval faith, despite continuing to refer to his "god the father" in "heaven above" and believing in the same pre-medieval book of fiction and all of its characters.

> "I don't agree with the Vatican on (birth control, gay rights, any other social issue)" say the Catholics who consider themselves so-called "modern" believers, but they still go to church and thus prop up an organization with a rap sheet even bigger than its bible and which even at the time of writing is still not fully admitting to its centuries of evil and immorality.

> "I am a Christian/Muslim, I don't hate Jews and some of my best friends are Jews", say the believers who truly don't hate certain people just because their clergy used to preach this or even where they still do.

> "I don't think suicide bombings are moral" say the Muslims responding to the global jihadist stereotype now adhering to all Muslims, in the face of much empirical evidence that the stereotype is in all too many places quite warranted.

> "I don't think women are unclean and a bad influence" says the Jewish man who thinks the fundamentalist Orthodox segment of the faith is immoral for its attitudes toward women, yet still believes his holy book is true. The same could be said for many female-positive Christians and Muslims.

Christopher Hitchens once said, and I am paraphrasing his point, that to address all the individual interpretations and particular micro-beliefs people hold in their religions, he would have had to write a different book for or have a different debate with each separate believer out there due to the often microscopic differences between each person.

These differences are petty and trivial both in reality and to the outsider, but they are quite significant to the believers and they take them very personally, far in excess of what reason would justify. Two people might mutually agree with almost all of their canon but differ on only one or a few points, but such differences often lead to schisms and seem destined to result in murderous disputes. Over time such differences have led to deadly and apparently eternal internecine religious civil wars, such as those between Catholic and Protestant, Christian and Muslim, Sunni and Shi'a.

Every book on religion faces this insurmountable challenge. This book will not focus on those microscopic differences between sects and between individuals. It does not expand into the many thousands of pages that would be required to address the microscopic details within every religion now practiced on earth. Nor will it take the billions of pages it would require to address every separate believer's individual beliefs and rejections.

This book focuses on Christianity, Islam and Judaism writ large, and covers their teachings, their books, their actions, their history and their impact on the real world. Other religions have not damaged humans or the world nearly as badly or for nearly as long, and given their tenets only these monotheisms are likely to continue to do so.

If you are a believer, you must understand that you are not off the hook just because you don't believe this or that specific point or dogma contained in your religion. If you are a faithful believer and call yourself a member of one of these religions, you have made a deliberate decision to be a member and to be treated like one. You are thus "all in", as the expression goes.

Saying you don't believe in some particular point or teaching, while believing almost all or most of the rest of them, does not give you the right to some sense of comfort or superiority over the 100% literalists. You are all riding in the same boat of belief. You may be sitting in the front rather than in the rear, but you are still in that boat. Saying you think that some of the stories in your holy book are allegories doesn't get you off the boat if you are still a member of a religion which has members which take even a single word in the entire holy book literally.

So if you are a Catholic and don't believe what an abnormally asexual, unusually masochistic, pedophilic or closeted homosexual male priest says about birth control, that's good, and you get a single point for humanism and common sense.

You should remember, though, that the head man of your church is said to be actually infallible on the subjects of faith and morals, so your disagreement with him is actually a disagreement with your god since your head man is his infallible vicar on Earth, so you are going to go to hell when you die. That aside, if you believe in any of the rest of the immoral teachings and dogmas of your sect, you lose megapoints for each of the beliefs you hold.

I have to take a short digression here on the subject of highly abnormal males, since the Pennsylvania attorney general's report on Catholic priest pedophilia and child rape abuse report was released as I was writing this chapter in August of 2018. I reference these never-ending Catholic crimes and perversions from time to time in the book, but the timing of that specific release and the continued support of this viciously immoral sect by even a single human being warrants special mention here.

If you are somehow still a Catholic after all the evil that sect has wrought not just throughout almost its entire history, but most recently during our own modern 20th and 21st centuries, you have a lot of explaining to do. Any current or former Catholic with even a speck of actual morality will find an overwhelming tonnage of things to be ashamed of in the pages of this book and in history, and has much to answer for.

If you are somehow still a Catholic, you are making a conscious choice to be part of and to knowingly support an organization that first fostered then did nothing about the WWII Holocaust, as I will describe in Chapter 18.

In our own time these people are still lying about and trying desperately to cover up the ages-long sexual holocaust it has been perpetrating in its campaign of rape, abuse and neglect[1] of countless innocent young children, among incalculable numbers of other crimes against humans and humanity across the world.

If you are somehow still a Catholic in any way, you are part of the problem. By continuing to call yourself Catholic, or by participating in any of its activities, you are an accessory and are complicit in its crimes.

Anything you do that contributes to the continuance of this global pedophilia ring in any form means you are complicit in some of the most horrible crimes a human can commit against another.

Ask yourself if you would remain a part of any other organization on earth if so many of its members raped and abused little children and this activity was not only covered up by leadership at all levels but deliberately allowed to continue through moving the rapists around to new raping grounds with flocks of new innocent children to rape and sodomize.

Ask yourself this question. If you discovered your best friend was a child rapist, or that he knowingly covered up someone else's rapes of children, would you still consider this person your friend? Would you let him babysit your children or advise them, or indeed you, on anything at all, particularly morality?

Ask yourself why you remain a Catholic, and why you choose to be an enabler of and complicit in this organization's crimes.

As I noted above, I am writing this in August 2018 and the Pennsylvania state attorney general's office has just released its report[2] on the massive rape and pedophilia operation which hundreds of Catholic priests and their superiors both carried out and covered up just in that single state. While this is horrible beyond words, and would be a fatal blow to the church if its followers actually had spines and morals, such an unthinkable revelation of systemic evil crime is now just another day in Catholicism. This is just one more disgusting discovery in the endless stream of disgusting discoveries of priestly rape and pedophilia which this church has never fully accepted or dealt with.

And why do they continue to rape and cover it up? Because you Catholics let them.

They do these things because they know that most Catholics aren't really bothered very much by their priests fucking little kids or by other priests letting it happen and covering it up. If such horrible behavior actually bothered you, you would immediately leave the church with as many people as you could take with you and you would never look back.

But no matter how many children are raped, no matter how widespread the coverup and no matter how high it goes, the complicit pope and all his complicit bishops and cardinals know that Catholics across the world are too invested in this cult of death and pedophilia and are too pathetically weak to leave the church.

As a consequence they know the church isn't really in any long-term danger. This makes the situation even more disgusting to those of us who actually have contempt for child rapists and their supporters and would like to see them all behind bars forever. Every single person who remains in the Catholic church in any capacity after being made aware of its crimes is complicit in child rape, period.

And if you're still a Catholic, we need to know what the hell is wrong with you? How can you actually look anyone in the eye and try to say your pedophilic crime ring has the one true set of humanity's moral answers? Your church has about as much credibility on morality as Hitler does, and it's long past the time when you should be held fully accountable for it.

I have to ask all of you who choose to continue as Catholics what in your god's damn hell will it take for you to leave that church? You're obviously OK with your priests raping children, among far too many other unspeakable acts, so what will it finally take for you to say you've seen enough? You're obviously OK with leadership up to and including your pope doing all they can to cover it up or offer insincere apologies and insufficient actions, so what else needs to happen?

Do the priests have to start killing these children after raping them, then nailing their bleeding and semen-stained bodies to crosses in your goddamn front yards in broad daylight before you've finally had enough? Do they have to flay the bodies of these children and have sex with their skinned corpses during Mass on Easter Sunday for you to finally decide you've had enough?

The sad fact is that based on history and based on the current utter lack of spine among Catholics it's doubtful if even such overt and public evil would make large numbers of you leave that church.

And this is really the true evil in religion, that so many people can commit, cover up and then be complicit in the most heinous acts a human can perpetrate, yet their common religious bond makes it all just not so bad after all. Consider this. If you have ever given money or support to the Catholic church, your hands are stained with its crimes. Some of that money you gave them helped pay the legal fees of rapist priests and of their superiors who covered up the rapes, then enabled more rapes, then covered that up too.

Some of that money you gave these pedophiles went to payoff and hush money to the families of those children who were raped. Some of that money you gave these rapists and their enablers is paying priest rapists and rapist enablers a salary right goddamn now, at this very moment, so you are subsidizing organized child rape and its coverup with every penny you give these people. As long as this church has even a single follower anywhere on the planet, it will be perhaps the clearest example on earth that both the faithful and their clergy are far too often completely incapable of true morality.

Catholics should read this book with extra shame and should read every word and understand how their religion's crimes are the reason for criticisms like these. For religionists in general, if you are a believer but not much more than a passing fancier of your particular faith, you should take this book as a means to gain insight into why you actually call yourself a believer.

You may only believe some or a tiny part of your whole body of faith, but proclaim yourself as faithful in order to protect against familial or social ostracism, or indeed physical punishment as in the cases of Muslims who wish to be ex-Muslims but don't want to be murdered for leaving the faith. This book will have much to interest you.

If you are a "secular Jew" and thus essentially an atheist or agnostic but you live a socially Jewish life, generally ignoring the holy books and practices as much as is convenient, you should take this book in perspective. Much of it may not apply to you as an individual, but it might have wide applicability to your social groups. Take the parts of this book that apply to you and your communities and use them in your daily life. Be able to recognize the thoughts and behaviors that rise from your faith which make other people's lives worse or less happy, and work to eliminate those.

If you are a Muslim and you think suicide bombing is bad, if you love your wife as an equal and don't think she needs to be wrapped in heavy cloth bags, that she should be allowed to drive a car and get an education and get a job, if you don't think you get delivered to a Playboy Mansion-level carnal paradise if you blow yourself up in the act of murdering people, if you don't think apostates should be killed, you are correct in that those are good and moral positions, but you are light years away from the mainstream of your faith and your holy book. Know that your help in bringing the fanatical and fundamentalist factions to a modern and actual godless morality, just as the moderate Christians and moderate Jewish believers need to, will be vital in bringing our world to a better place.

If you are a scientist during the work week and a believer for an hour on Wednesdays or Saturdays or some other days, you may think yourself above reproach and not an enabler, but you need to know that you are choosing to time-share in a land of mystical fairy tales that are being passed off as Cosmic Truth. And just as there is no such state as being a little bit pregnant, you can't be just a little religious. It is a binary dynamic. You are either a believer in the supernatural or you are not. You are religious or you are not.

If you as a scientist cross the threshold of a house of worship for any reason other than to begin the process of converting it into a secular medical clinic or school, or to do research on how and why people believe insane things without evidence, you are enabling religion to continue to exist and are contributing to the sworn enemy of your profession. I encourage you, as a member of a more informed and likely more fortunate member of the species, to continue down the path of knowledge and enlightenment and away from your social pressures and existential fears. This book will help light that path.

If you are a monotheistic believer who has extremely deep faith and you feel that you are an exemplar of morality and also rigorously adhere to the realities of science either despite or due to your faith, I encourage you to read this book in detail. As you do, try hard to see if other people and the facts and lessons of history agree with you, because we and they do not. You may indeed be cut from truly angelic cloth and your every word and deed may be a shining example of unselfish humanism. If this is true, you are not an actual follower of 95-99% of your canon.

In that case, I would love to meet you to discuss our similarities and differences, and to encourage you to spread your humanism, though not your religion, as far as you can. I would very much like to know how you manage to rationalize your opinion of yourself as a believer and act the way you do when your books, your dogmas, your clergy and indeed your gods all specifically tell you to be cruel and inhuman, with specific details on how to do it.

No matter your beliefs, know that if you are a believer in any supernaturally based religion that this book does indeed apply to you as an individual. It may not reflect your full spectrum of agreement and disagreement with your canon, but it is an appropriate lens through which to see both yourself and those of all faiths.

To address some other inevitable protestations up front, I will address my rather pointed discussions of Christianity, Catholicism and Islam here. Christianity and Islam currently have the most adherents in the world, thus challenging their antihuman practices and immoralities will ensure that the two largest global sets of immorality are addressed in this book.

While Christianity and Islam have pained far too many people for far too long, Catholicism has by far the bloodiest history and even unto this day continues to spread pain and evil nonsense, not to mention criminal acts beyond comprehension.

It has been a clear and present danger to innocent men, women and children across the world for more than a thousand years. Therefore since it has committed the most evil over that time, it should and does receive the most criticism. When an organization endlessly praises itself and its god(s) for being the source of true morality while globally engaging in what can only be called demonically immoral behavior, the criticism must be especially pointed.

Islam has had its ups and downs, and while Islamic culture in the past has contributed much of great and permanent value to our collective scientific and mathematical knowledge, Islam today is largely beholden to the most evil and antihuman in its canon. Many Muslims currently live in a state not unlike the Christian medieval Dark Ages. The church is often one with the state, and dictatorial theological dogma rules the day. Many Muslims are poor, uneducated, highly religious, live in theocracies or religious dictatorships, are anti-intellectual, anti-science, and if they are male they are often misogynistic and bigoted. The Islamic world is well-populated by people whose religious beliefs comprise a clear and present danger to the Middle East and the rest of the world. Of course the last two sentences could as easily apply to Alabama as Afghanistan but under a Christian battle flag rather than an Islamic one.

There are unquestionably many Christians and Muslims who are truly moral, intellectual and humanistic. But these are outstanding people *despite* their religions, not because of them, and globally they are a rare and increasingly endangered species. They ignore almost all of their holy books in order to behave in such a positive manner. These people are far closer to atheistic and agnostic in practice than their fundamentalist or even their moderate co-religionists.

Our world would be instantly improved with regard to the well-being of billions of humans both inside and outside the faiths if Christianity and Islam were completely eliminated. As they show no inclination to do this on their own, books like this and criticisms like these are necessary to force them to face both internal and external pressure to become truly moral.

Chapter Eight

Religion is the Ultimate Proof That Most People Don't Think Critically

"Reason is a whore, the greatest enemy that faith has; it never comes to the aid of spiritual things, but more frequently than not struggles against the divine Word, treating with contempt all that emanates from God."
— Martin Luther

This will be a relatively short chapter. Every reader can come up with reasons why he is sure that most of his fellow humans are wrong on this or that issue, but this chapter centers on one overwhelming bit of proof that most humans actually don't think critically about the most important issue there is.

Before I demonstrate this, a quick self-assessment is in order. Answer the following questions honestly:

1. How much do you actually believe of what you see or hear on any subject without thoroughly and honestly checking it out for yourself, using an appropriate and wholly objective screening and testing methodology? Give a percentage.

2. Does the source of what you see or hear matter? In other words, if you are religious and you hear one of your clergy say something and then you hear an atheist or scientist say something different, do you automatically tend to preferentially believe the cleric?

3. Do you think you should put more or less effort into beliefs and decisions that are the most consequential?

If you are like the overwhelming majority of people, that is to say at least somewhat self-deceptive and unaware of your actual behavior, you will answer these questions with something like the following:

1. "I only believe between 50 to 75% of what I hear without checking it for myself. And yes I do fact-check."

2. "Well, it depends. I believe whichever is more appropriate to believe."

3. "Of course you should put much more effort into checking things that have more impact on your life. What a dumb question."

If you answered in this way, you are almost certainly a self-deceiver. Don't think so? Let's work through this. If I told you that I wrote this chapter on an Apple laptop, or that I am wearing shorts and a T-shirt right now, you can choose to believe these statements or not. Whether you believe me or not has no consequences for your life at all. I also have no obvious reason for lying about it. It doesn't really matter, because it is no risk to you if you believe me or not. Thus you would probably do so if for no other reason than it really isn't worth your while to worry about it or check it out.

In a quite different way, your need to understand the truth of a statement is extremely relevant when we discuss something like science, religion and the real world. Not believing in gravity or that poisons are dangerous may result in your meeting a painful and untimely death. Believing that killing infidels is a holy duty or that your religion is the One True Faith may result in you trying to bring about other people's painful and untimely deaths. Whether you hear some claim about the world or religion from a religious person or a scientist, you have a few choices in how to respond.

1. **You can blindly accept the statement at face value.** This should never be anyone's default setting for everything, but it is an acceptable option for vigorously peer-reviewed science, because literally thousands of years of effort, rigor and multiple reverifications have gone into studying most of its subjects. Frankly in almost all of these cases you'd be an idiot to doubt or question what you hear. The same goes for math. There is just no way water is not going to be made of atoms other than hydrogen and oxygen, gravity really exists and functions in a certain way, and two plus two always equals four, unless you are a dogmatist or a member of the Trump administration, in which case, like everything else that comes out of it[1], reality is

whatever the boss says[2] it is and to hell with your incontrovertible evidence.

Religion is quite different. You can never accept anything at all at face value within religion, because all religions are fiction and their fictions vary tremendously even within the same basic faith group. Science also doesn't make unproved and unprovable claims, but since doing that very thing is one of the foundations of religion, scientific claims and religious claims have to be treated far differently. In this way the Trump administration, the fascist dictatorships of the twentieth century and all religious dogma are identical in that to them the truth is what those in power want it to be, and they will try their hardest to make you believe it too. You can **never** accept dogma of any kind at face value.

2. **You can accept it provisionally until you thoroughly fact check it for yourself.** This is an acceptable practice for math and science, and indeed is central to the ethos of science. If it is done responsibly and appropriately, it is the sign of a healthy mind. Science doesn't take it personally that you may think water does not include oxygen, and math doesn't care that you want to verify for yourself that the angles of a triangle always add up to the same 180 degree number. Science won't take it personally if you imagine as a starting position that a tribe of desert goatherders wrote down the actual history of the world and a certain god, that the earth is less than ten thousand years old, or that evolution is not a real process. If you do honest research and are an honest person you will find the incontrovertible facts in all these cases and you will believe in math and science and you will stop believing in religion.

   You will also have made the scientific and math communities happy that you have demonstrated healthy skepticism and an investigatory mindset and can determine reality on your own. If you carefully look into the math and science and your conclusions differ from the established facts, you have either proven yourself a gobstopperingly stupid and/or mentally imbalanced idiot incapable of doing actual honest research or you have discovered something truly earth-shakingly new and Nobel-worthy, with the odds of the former being essentially 100%. If you investigate something in science that is not fully understood and you use the scientific method to do this, you are engaging in a noble activity that might indeed get you Nobel recognition.

This is where religion and science differ both diametrically and permanently. Science and its method are by definition rigorous processes of rigorous analysis and rigorous iterative and longitudinal peer-checking, all of which lead to actual knowledge and information. Religion is a collection of unsupported, unfalsifiable and unprovable claims and beliefs, and neither includes nor does it honestly seek any actual verification whatsoever. Indeed, one of the core characteristics of religions is that they make things up out of thin air, even including major dogmas, simply to suit their own purposes. Religions don't check themselves and only change when external pressures force them to.

Religious claims can be and have been investigated by science and have invariably been found to be false or unfalsifiable. Those claims that cannot be proven or disproven, such as those of imaginary gods, heavens and afterlives, are not "beyond science" as apologists empty-headedly claim. They are "beyond reality", in the same way that Bugs Bunny and the Tooth Fairy are beyond reality. Religion is beyond science in the same way that Harry Potter, children's cartoons, superheroes and all other fictional characters are beyond science. Religion is based on imagination and things that aren't real. One cannot apply a measuring instrument to the magical Harry Potter wonderland, to the Disney or Warner Brothers universes, to the Land of Oz or to the existential claims of religion, because they are all imaginary.

So you can never take the claims of religion at face value, provisionally or in any other manner. You must subject them to extremely thorough analysis and multiple external peer reviews by honest, skilled and highly trained skeptics. If one day the tools of science appear to prove some existential claim or other, the global furor and attention that follows will be even more energetic than that which was applied to cold fusion and other short-lived false claims of the past. At some point an overwhelming body of skeptical and empirically derived information will confirm or deny the initial claim. Until that time, you should never trust any claim made by religion, either provisionally or otherwise.

3. **You can reject it provisionally.** This choice is simply a little farther down the road of skepticism than the previous choice. As long as you do the appropriate fact-checking and aren't being an idiot about it, such as in the imaginary oxygen-free water example above, this is a reasonable approach to take in a great many instances. Again one must consider the source. If a television news station says that the dam above your city is about to break, and it has been raining in your area for two weeks, you would be well served to quickly vacate the area and check things out later once you are definitely in a safe place. If you are listening to a politician, preacher or defense lawyer speak, you can often provisionally reject his or her claims until you thoroughly verify them for yourself. Which you should, because while you may not "do" religion or politics, I absolutely guarantee that they will "do" you, and often not in a way you would like.

4. **You can reject it outright, and make the person prove the claim.** This is the option that unfortunately has to be chosen as the default position in our modern, alternative fact-ridden world, particularly if the claim will make some demand on you or if it is not backed up by massive amounts of actual evidence. This is where science has earned its place as the world's sole reliable source of that actual evidence, and why it must always be the source for truth and facts. It has earned this place of respect due to the planetful of evidence of its effectiveness. Nearly every single thing you see, touch or hear in the 21st century is the product of proven science.

   Computers, telephones and other electronic devices work. Medicine heals. Airplanes fly. Satellites collect and pass data. From your houses to your cars to your clothes, from your food to your furniture, from radiometric dating methods to the findings of geology and archeology, science is true and works in a way far more wondrously than has ever been attempted, much less proven, by any religion.

   If a religious person says he or she knows what the creator of the universe thinks about stem cell research, about working on some holy day or about which group of innocent people should be killed or persecuted, you should reject the claim and demand proof.

When you hear a claim, you should ask questions like these:

- Does the religious claim have any actual evidence or proof?
- Does the politician's or cleric's promise or claim make sense and would it work in reality? What are the proven moral, economic, legal, rational, social, political or other underpinnings of this claim, and what does history show?
- Has this promise or claim ever worked or been true?
- What do informed, educated and honest professionals and skeptics say about the claim, and what have they empirically proven or shown about it? I mentioned this kind of rigorous analysis and verification back in Part One.
- Does the claim seek to make some demand on you, such as time, money or other resource?
- Does the claim seek to manipulate you, control you or scare you in some manner?
- What else is the person or group making the claim going to get out of it if you go along with the claim? Are there hidden agendas or motives?
- What else does this person or organization claim or do that I need to look at? What does the person or organization stand for, do, lobby for, and what is its track record and history?
- Can this person or organization be trusted? What is this person's or organization's track record of lying? What would the person or organization gain by lying?

It is unfortunate that we have to be so pessimistic, but in order to lead safe, happy and informed lives, we have to think critically and skeptically. There are some very important reasons for this. Human history is riddled with those who lie for a living and those who are honestly mistaken. The scientific method is both very recent and it is not innate in almost all people. Critical thinking is a skill which far too many people lack, and far too many people deliberately avoid it due to the effort required and the sometimes unpleasant conclusions it reaches. But again, and this is the centerpiece of both reality and human existence, what are the consequences of all this? Why should we be so skeptical?

If a snake oil salesman told you that some particular potion in a flashy package hawked by some media or sports personality was capable of curing arthritis, eliminating wrinkles, restoring your hair and sexual vigor, and also making your excess weight come off without exercise or dietary self-control, and if you believed these claims, this would often lead to your purchasing this snake oil.

This would result in a foolish decrease in your bank account, potential negative side effects from the junk you will ingest or merely proving that the potion does not actually work. These consequences would be bad but not usually life-threatening. You would think that most people would check the salesman's claims and not buy the snake oil, but you would be unbelievably wrong, and probably far too often even in your own life.

If huge numbers of people didn't fall prey to utterly ridiculous claims, snake oil salesmen and tricksters of all kinds would not be doing the massively robust business they do every single day and have been doing since cavemen first traded stones and shells and found ways to get more out of the transaction than a fair exchange.

Despite the known risk of deceivers, far too many people still don't check even outrageous claims too closely, particularly if they think they will get something from it, and even more particularly if they don't have to work too hard to achieve it. Laziness and credulity are rocket fuel for the engines of deception.

Take another example. If your oncologist told you that you most definitely have cancer, your choice not to believe the doctor would almost certainly lead to very bad consequences for you and your family. The same would be true if you fell for one of the endless get-rich quick schemes which the unscrupulous peddle on a daily basis, though in those cases with terminal effects only on your bank account, your retirement plan or your marriage, as opposed to your life.

In the case of a cancer diagnosis, as when a bullet whizzes past one's ear, the reaction is quick and visceral. One's breathing gets shallow, the guts turn queasy, the heart races and the veins fill with icewater. Maybe a little dizziness sets in, along with a bit of a sense of unreality.

But soon after that initial shock, the senses sharpen and the focus becomes much clearer. Facts now actually matter. And they must be real facts and not so-called "alternative facts", aka bullshit. Following a short period of shock and denial, most people will make a dedicated shift from laziness and credulity to being motivated learners about their diagnosis, their bodies and what they can do to remain alive and as healthy as possible for as long as possible. Why?

Because now things are painfully real. Cancer and other such situations are deadly serious and often require immediate and focused attention, thus the mental shift from passivity to activity. With a cancer diagnosis, the stakes are extremely high, not just for the person with the diagnosis but for everyone who depends on or who holds that person dear. The choice to believe the doctor or not has literally life-altering consequences not just for the person receiving the diagnosis but in most cases also for many other people.

Thus the appropriate reaction is to check and double check the facts, to learn all there is to know about the situation, accept the diagnosis and act in a fully and empirically informed manner.

Obviously, as described in the examples above, the higher the importance and impact of the potential belief is, the more likely a reasonable person *should* be to attain high levels of focus and energy in the analysis of the problem and also the decision to believe or not believe the diagnosis and act accordingly.

Yet this is not the case at all for the overwhelming majority of humans throughout history in the most impactful areas of our lives. Billions of people, living in the most scientifically advanced era in human history, make no effort to critically analyze the most important beliefs they hold.

Believing in a particular god or not, based on the fact that many of our gods will either reward people most blissfully or punish them most sadistically for all eternity depending on the person's beliefs and actions in this world, should unquestionably be the single most important decision a person ever makes. This decision should take many years of thorough personal study, should require extremely detailed analysis of all the available evidence to include reading every single word of all of the various holy books on offer, and should include extensive and lengthy consultation with subject matter experts from all religions and supernatural belief systems.

This has to be done for each and every supernatural entity and religion, including the ones no longer known or practiced, the ones currently known and practiced, and also leaving room for those not yet known and practiced. Thus the average person ought to take decades to reach a final decision on this most weighted and eternally impactful of matters.

Do people actually do this? Does anyone, anywhere and at any time actually do this? Of course you know the answer is not no but, pardon the expression, hell no. For most people religion is just something they were born into and which they accept without a grain of critical thought, or at least a grain of critical questioning. This is absolutely unconscionable. It is the epitome of immaturity and laziness, a complete abdication of critical thinking and a total lack of proportionality and common sense. Most people spend more time scoping out the cheapest carton of milk at the grocery store than they do in considering which god to kneel down to and to whom they will commit their alleged afterlife.

This is a decision which according to most religions will have literally eternal impact, and could, as already noted, mean endless full-body torture at the hands of a sadistic supernatural Inquisitionist or endless beatific bliss surrounded by everyone and everything they ever loved in life. And yet people spend more time taking out their trash every week than doing such rational analyses and critical thinking on the most important subject there is.

And religious belief is never a victimless crime. People think, speak, act and vote according to their religion, which means one person's religious beliefs do indeed quite significantly affect other people. The problem this poses for everyone else is that far too many people's religious beliefs are highly immoral, quite evil and extremely dangerous to other people.

How is such a complete breakdown of logic, reason, common sense and proportionality possible? It is because religions are extremely arrogant, xenophobic, tribal, hypocritical and closed-minded. It is because our evolved nature as a social species means we cling to our herd's beliefs even when they are demonstrably wrong. It is because evolution has not endowed very many of us with the faculties of critical thinking, the ability to apply them universally and the ability to break away from the herd. It is because the tools and processes of science are recent and learned, not innate and inborn, thus many people do not use them. It is because we desperately fear death and the uncertainty of a short life. It is because religions are social traps which ensnare weak and uncritical minds with the promises that everything is explainable and under a divine hand's control.

Let's take a closer look at this problem. Which of the following statements do you feel are most likely true? Which ones not so much?

1. Hot water freezes faster than cold water.

2. I have a communications line to the creator of the universe, and we talk to each other.

3. The creator of the universe lives in a cave on Pluto.

4. Light acts both as a wave and a particle.

5. The creator of the universe picked this particular planet as his favorite, loves one particular species of earthly primate more than any other species of creature in the universe, killed almost everything alive on earth in the recent past including these primates, sent his dearly beloved son to this planet to be tortured and killed by his favorite primates, and this is *wonderful* news.

6. I will bring wrath and pain and judgment on you, unless you send me five thousand dollars.

As you can see, these statements run the spectrum from counterintuitive to insane asylum-level delusional. Which ones are which? Which ones should you believe? If you heard of the water and light examples in grade or high school or in some other forum, you would likely believe them, at least provisionally. Most readers with a secular high school education will have some familiarity with those statements, although home-schooled and/or highly religious children may not.

Most people will be familiar with the so-called "good news" statement, which refers to the core myth within Christianity. Separately, large numbers of religious people think they have some sort of a hotline to their god or gods, which is why I included that example. What the faithful do not, will not or cannot understand is that the 'creator in a cave' and extortion statements are no more bizarre than the 'hotline to the gods' and 'wonderful news' statements.

The only difference between these four statements is that more people happen to believe in two of them than the other two. There is a complete and total absence of empirical evidence for all four statements; therefore they are all equally invalid. You can spend the totality of an extremely long life trying to prove any of these statements, and as you take your last breath you will still be unable to produce any actual evidence whatsoever for any of them.

This is how religion proves that people don't think critically. When religion was first invented all those many thousands of years ago, gods were reasonable hypotheses given the total lack of human knowledge about the facts of the physical world. The god hypothesis arose from a position of essentially total ignorance.

A Bronze or Iron Age Mesopotamian theologian trying to explain the physical world was as unable to discern the relevant physical facts and processes as a blindfolded Mesopotamian goatherder in a dark room would be if he tried to describe the images on a printed wall poster using only the sense of touch. No matter what either of them guessed, all of their guesses would be equally valid given the lack of relevant information and any real way to test the guesses.

If you were that blindfolded goatherder in that dark room and you had postulated that the poster held the image of, for instance, a bearded man wearing a robe, perhaps with a shining halo around his head or with nails and rope binding him to a wooden cross, that would have been a fully acceptable hypothesis out of the infinity of possible hypotheses. It would be no more valid, however, than someone else's hypothesis that it was merely a blank sheet of white paper, or that it held the image of a man-figure with wings on his heels, of a horse in flight, of a cosmic egg[3], of a stack of enormous huge turtles[4] or indeed anything imaginable.

Our species evolved in that totally dark room. Over time, we slowly made discoveries about our world. With each one, another tiny bit of light shone into the room. As time went by and as we discovered more facts, and as we began to introduce the scientific method and critical thinking into our body of knowledge, we began to see the picture on the poster in increasing clarity and detail.

After sufficient light and information became available, we were able to reject the original ignorant and uninformed hypotheses of a bearded and robed human male and his imaginary companions, because we could see what was really on the poster. In today's world the lights are not yet at full brightness but they are more than powerful enough to show that all gods are imaginary, all religions are manmade and that too many members of our species cling desperately to unjustified beliefs due to widespread and deep-seated psychological frailty and pervasive ignorance.

Religion formed in that dark and ignorant room, invented by people blind to science and blind to the facts of our world. It was impossible for even a supergenius and polymath in the Bronze or Iron Ages to explain physical facts and processes which today's fifth graders can very easily explain. Thus religion formed through the process of people describing the unknown based on no factual data, and having their wild mythologies written down in books which later became revered by billions.

It may take at least several more centuries, and possibly several long and painful millennia, for us to finally inculcate in our species as much of a compulsion to think critically as so many of us have today to maintain our childish religions. Getting there will require winning the war of ideas, which is waged both inside every person's mind and between minds.

Winning this war will require that we defeat the forces of ignorance and faith which strive every day to try to push humanity back into the dark and ignorant room where we grew up, back to a time and place where religion and faith were in control.

The stakes are far too high to lose this war, and it will only be won in the trenches, by large numbers of individuals and groups of people fighting back and challenging ignorance, dogma and faith. Only when uncritical thinking such as superstition, faith and religion are as universally socially unacceptable as sacrificing children to a jealous and angry psychopathic god will we have won this war.

Chapter Nine

Damn Right I'm Just a Bunch Of Chemicals and Electricity, And I Love It!

"We are just an advanced breed of monkeys on a minor planet of a very average star. But we can understand the Universe. That makes us something very special."
— Stephen Hawking

Life is absolutely amazing. The world we inhabit is made up of more than 100 trillion trillion pounds[1] of molten metal, molten rock, solid rock, water, soil and various atmospheric gases and dust. There are also many billions of tons of biomass living in the water and on and in the extremely thin top layer of the planet's solid crust. On much of the land surface, one can find inanimate and quite lifeless solid rocks sitting right next to maniacally active members of the biomass such as squirrels and hummingbirds.

Of all the uncountable quadrillions and more creatures who have ever lived throughout the history of the world, none of them until we humans in the last couple of centuries had any clue whatsoever that they and their ancestors all the way back to the first replicating molecules were formed of the very same materials as the earth itself. It is a truly sad state of affairs that so many people today do not know or appreciate this.

The organisms which first evolved brains were the first to take advantage of the interplay of chemicals and electricity which enables collections of inanimate matter to be conscious and animate entities. Humans are merely one small filament off one tiny twig off one small branch of the enormous tree of life. We happen to currently occupy the spot at the top of the ladder when it comes to brainpower, though we are not the only organisms which are self-aware and have intelligence.

Given the millions of years available for evolution, and the number of ecological and behavioral niches available to occupy, it is hardly surprising that other primates, including gorillas, chimpanzees and the now-extinct Neanderthals, plus dolphins and dogs, cats and crows, pigs and elephants and even the octopus are quite clever creatures who are or were also conscious and volitional organisms.

Regardless of their level of intelligence and mental computing power, how is it possible that a collection of molecules has the ability to move of its own accord? Is it "just" molecules in motion, as so many of the faithful sneeringly and ignorantly ask? Yes, actually, that is absolutely what all life on earth is. There is no magic force in it, no animating ghostly supernatural spirits and no supernatural puppet master assigned to each living organism on the planet to pull the strings to animate it.

We do not operate in the physical world solely via the influence or action of molecules, as our bodies do generate and require tiny amounts of flows of electricity to move and operate. We are both chemicals and electricity. We are molecules in motion powered and driven by complex internal energy and propulsion systems. All of us animate organisms are absolutely amazing collections of chemicals and electricity, and it is the most wonderful thing about our world.

The religious proclaim for no apparent reason whatsoever that we humans must be more than just so-called molecules in motion. Some of the faithful proclaim, again for no apparent reason whatsoever, that we humans also have to have some sort of a soul, injected into it by some sort of divine being. There is absolutely no need to imagine any of this nonsense. Life is amazing of its own accord and there is no need to taint it with supernatural trappings and superstition. It is gold with no need of gilding, it is a lily that needs no painting, and it is a beautiful and intricate tapestry with no need of additional embellishment.

Life is self-animate matter, an amazing and wonderful interplay of molecules and electricity which can move and act as a single entity, and this is at the same time nothing more and so much more.

It is nothing more in that there is no supernatural component to our lives or our existence whatsoever as we shuffle along through this mortal coil. From bacteria to buffalo, from honeybees to humans, we are all indeed merely collections of self-guided animate matter. We exist for an infinitesimally short period of time in an infinitesimally small corner of the universe and are moving inexorably toward an unavoidable permanent death from the moment we come into existence. And that's OK.

It is so much more in that while life is short, it can be and should be filled with joy and happiness. We can revel in the rush of hormones which our bodies release at the sight, sound, smell and touch of our mates, so powerfully manifested in our feelings of love and affection. We can hold our children close to us and feel the mutual love we share, and our intense desire for them to be happy and successful. We can enjoy the "high" of endorphins that accompanies vigorous exercise. We can feel the rush of success and pleasure that follows the joy of discovering a new fossil, the inspiration of a wholly new idea, winning a competition or achieving a personal milestone in whatever endeavors we enjoy most.

All of these feelings are engendered by the flow of chemicals and electricity in our bodies and brains, managed by patterns built into us by many millions of years of evolution. All of these are electrochemical reactions, and it doesn't diminish the feelings or their joy or importance in our lives in the slightest once we realize that truth, unless we deliberately choose to act this way.

We humans are nothing more than extremely evolved microorganisms, but at the same time so much more. We are animated collections of dead stars, shining with an internal light which expresses itself in the form of wondrous creations of art and architecture, prose and poetry, the exultation of the pursuit of knowledge, the love we feel for our families and the acts of kindness bestowed on others.

We are most assuredly merely molecules in motion, but what a wondrously complex dance of substances and electrical impulses is a human. We need nothing more than this, and those who seek more do so to drown their childish insecurities and fears.

The problem we face is that religion shrinks minds and skews perspectives. Religion seeks to constrain the world to tiny bins of what people thousands of years ago believed and how they saw the world. The talk about not wanting to believe that humans are merely a collection of chemical reactions and electrical impulses is the position taken by small, weak and religiously constrained minds.

It is absolutely true that all life is indeed chemical reactions and electrical impulses. So what? How does that in any way diminish the feelings of love and affection between parents, spouses and children? Or the appreciation of a magnificent mountain landscape?

When you see a beautiful flower or butterfly, there is no supernatural agency or magic in that act whatsoever. Electromagnetic radiation in a very small band of the spectrum reflects from the object, enters your eye and impacts your retina. Cells on your retina send impulses to your brain through the optic nerve, and your brain processes the signals into your perception of the object. It is all done through the operation of physics and chemistry. When you lean over and inhale the fragrance of the flower, a similar process of chemistry and physics results in your perception of a pleasing scent.

We can deeply feel a love for a significant other that is powerful enough to forge a lifelong bond and create a permanently inseparable couple. That rush of love and emotion that one feels at a spouse's or lover's touch is purely chemical in nature, but this fact does not at all diminish the wondrous nature of the feelings. The feelings themselves are what really matter. Evolution shaped us through gradual successive improvements into creatures which seek and enjoy emotional and physical affection and sex. We seek and enjoy sex because the successive developments of a complex nervous system and patterns of chemical release in the brain make the act physically and emotionally very enjoyable. Indeed, those who are deprived of sex, affection and intimacy often suffer significant stress and depression.

These physical and emotional pleasures and responses increased in strength over time because when we have more sex, more of us are born, the species has a better chance of being continued and the genes march on. Evolution shaped many of us so that after the sex is over, we feel love for our spouses so that when we become parents we spend more time together raising and protecting children, thus enabling more of them to live to the age where they themselves can reproduce. Evolution shaped many of us to love our children and to feel an overpoweringly strong desire to protect and care for them. We are also a strongly social species which seeks the company of others of our species. Some of these other people help in the process of raising children, whether this occurs within a village or some other group of people.

None of this was planned or designed. Had it somehow been more advantageous for the species to not feel these feelings, natural selection would have shaped us accordingly and we would think and feel however that paradigm molded us. Thus you see a wide variety of species across the world with a wide variety of reproductive and parenting models. Had it been more genetically advantageous for human females to easily and safely lay hard-shelled fertilized eggs rather than carry a highly parasitical[2] fetus for a long time then give birth to massive-headed helpless infants[3] in a quite difficult and very dangerous[4] process, they would be doing it or we would probably be evolving in that direction.

We don't do that and aren't evolving that way. Except in religious cultures where women can be sold, stolen, horse-traded or placed in arranged marriages, we mate and reproduce with a person of our choice. That voluntary process is highly rewarding and feels very natural, because it is natural.

When you look across a room and your gaze fixes on someone to whom you are immediately very strongly attracted, the full body sensations you experience are absolutely amazing, and wonderful. The fact that these sensations derive solely from chemical and physical processes in no way diminishes the joy and excitement. Why should it? We freethinkers have to continue to ask how in any way does the understanding of an experience diminish the experience?

I know how my car, smartphone and computer work, although the authors of the Mesopotamian holy books would be amazed and stupefied by such things and would certainly call them products and proof of their personal god. I don't need magic to operate any of these things. I need to know how they work. My knowledge of how they work is extremely beneficial to me, in that I both want and need to know how to use them, maintain them and to the extent of my ability and tools, repair them.

None of the solely biological foundations for our thoughts, feelings and actions have or need any supernatural component or contaminant. Knowing what love is, why it happens and how it happens does not lessen the love in the slightest. Period, full stop, discussion over. The fact that I can read about the chemical composition of chocolate, how it is processed and prepared and made good for our palates, and how evolution has shaped our strong desire for sugar and high calorie foods in no way makes that still-warm homemade chocolate chip cookie any less tasty.

I can read about how the tongue and sense of taste work, and how compounds are sensed and how nerves transmit the signals to be processed by the brain thus causing certain chemicals to be secreted, adding to the pleasure. Does any of this make a Reese's Peanut Butter Cup or a dark chocolate homemade molasses puff, my personal favorite candy items, any less tasty? Of course not. Does my knowledge of the mechanisms of brain function, personal attachment, neurotransmitters like oxytocin and dopamine and all the rest mean I love my wife and children any less, that I enjoy the feelings of love any less? Of course not.

If this is actually the case for any person, in that learning the underlying realities of the daily processes in our lives diminishes the experience in his mind, that is a very sad state indeed.

It is ironic almost beyond comprehension that it is the very interplay of chemicals and electricity in some people's brains that makes them so dependent on the false comforts of religion, as I will explain in much detail in Part Three of this book.

Knowing down to the molecular and neurochemical level how the taste buds and brain mechanically process and respond to the outside world does not in any way diminish the world or our experience in it. Quite to the contrary, for those of a secure, inquisitive and independent mind, knowing exactly how all these processes work hugely *increases* the enjoyment. The joy and pleasure of understanding how things work is far superior to the secure and intelligent mind than the false comforts engendered by the darkness of ignorance and small-mindedness.

Gaining an understanding of the world and how it works enables a person to see the world through completely different lenses. A young person riding in a car will see the road, the trees or fields to the sides of the road, power lines and houses, bridges and clouds, other cars and the occasional aircraft. This is all she sees, just the external shell of the items with no deeper understanding. After years of education and study, that young person will one day become an adult who can take the same drive along the same route and see a totally different world. The informed adult sees the road but now knows the history and difficulty of travel before the road was built.

She knows the steps the civil engineering and construction crews took and the materials and equipment that were required to design and build that road. She may know the species of trees, shrubs and vegetation in the fields, and how natural and artificial selection shaped these organisms. If any of the vegetation are food crops the adult may know how the farmer prepares the field, plants then harvests and sells the crops, and might also understand the economics of the farming business and the intricacies of the global network of trade in goods and services.

The informed adult may now know the names and types of clouds and understand the processes of meteorology, climatology and general atmospheric science. The adult may be able to identify different aircraft by make and model, and also know about general and military aviation and understand the principles of flight physics and aerodynamics.

Living an informed and information-seeking life, having answers to questions and actually knowing how the world works are the true joys of having a highly developed brain and are the only path to greater knowledge and understanding of our world and universe.

It has been said by those with low imagination that one cannot quantify love. Such thinking is the product of a spongiform brain oversoaked with religion. Of course you can quantify love, although it won't win you any points in the dating game to demonstrate this.

Do you care deeply for another person? Do you wish for good things to happen to him or her? Do you happily go out of your way to do things for him or her? Do you provide resources and other support for the person in any way, and enjoy doing so? Do you do these things with the intent to make the person happy and better off? Do you want to spend time with that person, and want to know what makes him or her happy? Do you do all this not from mere societal, familial or other happenstance ties but because you truly want to? Does your stomach churn into butterflies when your gaze lands upon a certain person? Do the minutes you spend apart seem like hours? Do you wait with great anticipation for the person's telephone calls, or in this age, texts or emails? How many times a day or week do you do good things for this person?

Write all this down and you have just quantified love. The degree to which you feel such emotions and act on them can give you an objective assessment of how much you actually love a person. It's really quite simple, needing nothing magical or supernatural at all. The action is all electrochemical, but what else can we be? Wanting to make a stage magician into a supernatural wizard does not make him into one. Wanting to make us and our universe into more than we are, to want the supernatural in addition to the natural, is just childish immaturity.

It is true that many people like to have magic or mystery in their lives. The popularity of all sorts of such things has been evident throughout recorded human history. Ghosts, goblins, leprechauns, gods, aliens, devils, demons, angels, goddesses, witches, warlocks, seers, prophets, mystics, mediums, genies, djinns, zombies, superheroes, X-men, the glitterati of Hollywood and Bollywood, various national royal families and many other such have an apparently endless appeal to many people. This is demonstrated by the proliferation of wholly mindless junk publications infecting the end caps at supermarket checkouts, the ubiquity of religion and superstition, the continued profession of alien visitations and magical powers by too many people, and the continuing descent of television, social media and internet content into a vast wasteland[5] of mind-embalming foolishness that is giving too many people incurable mental diabetes.

It is only the childish and immature mind that would prefer things remain mysterious and magical, kept unexplained while being fully explicable, known about but deliberately rejected. These are the minds which are most easily captured by the manipulative and the dishonest among us, who would seek to gain power and control over their fellow humans and use them for their own ends.

One need look no farther than a tiny bit down the tree of life for confirmation of all this. The biological machinery of chemicals and electricity obviously applies to far more than our own species. The informed adult looks at a cat, a dog, a bird, and at the faces, body language and actions of our nearest primate cousins and sees this. The adult knows that the animal is indeed in one sense "merely" molecules in motion. That adult also understands that the organism is far more than merely animated matter. Anyone who has lived with and/or observed animals assuredly knows that they sense, feel emotion and have a rudimentary sense of their environment.

Animals are not, as was thought in former ages, animated automata. They are beings which can perceive and feel, and some of them are extremely intelligent and very fast learners. Even if you hate cats, you know that cats purr when they are happy. When you know how cats and dogs express their states of mind, from vocalizations to how their fur, ears and tail look and act, you can quickly tell what the animal is feeling. You cannot observe our closest primate cousins as they go about their daily lives, or look one of them in the eye, and not clearly know in your gut how we are nearly the same, separated only by a few genetic mutations, a tiny percentage of our genes and the advance of time.

But the religious will deny all this scientific reality as much as they need to in order to maintain their unscientific faith.

Animals move easily, they perceive and process information and they respond to emotion with emotion. They cannot do this with the same ability as a normally developed human, but they can do it through using the same basic interplay of chemicals and electricity. Do you know that animals feel love and happiness? Ever have a household pet or interact with some other nonhuman animal who adored being held, petted and played with? Most people have. Ever hug a cat or dog and be hugged back? I have, and I am sure almost all animal lovers have as well.

Animals from many places across the spectrum of species feel love and happiness as surely as humans do, and while they cannot verbalize their feelings in human language, and while they cannot read or write, they are unquestionably far more than unthinking biological automata.

Beyond the familiar cats and dogs, we should examine the primates, our nearest biological cousins, for their special physical and emotional attributes. We can pick from chimpanzees, gorillas, orangutans and bonobos. Chimps and bonobos are often referenced, respectively, as relative devils and angels. Chimps are some of the most vicious creatures in the animal kingdom, and use their powerful jaws and muscles in acts of violent murder. They are intelligent and highly social tool-using creatures, and the males are in charge. The alpha male and his top lieutenants monopolize power and females. They can also perform in a socially responsible or altruistic manner and cooperate with each other.

Humans are exactly the same in these characteristics, as one would expect from two species a hair's breadth apart genetically. Bonobos are much more peaceful than chimpanzees. Females are dominant, sex is extremely ubiquitous, constantly sought after and is a social stress-reducer. Violence is extremely uncommon. They enjoy strong positive social bonding. Again, much of this can also be said of many humans. Life is much better in the bonobo community, for the obvious reasons.

These primates, despite their near-polar differences in disposition, are both about 98% genetically identical to humans. It is only very small differences in DNA that make an organism a human or a chimp or a bonobo. All one need do is observe the behaviors of these animals to see just how like we humans they are. Many millions or perhaps even billions of people deny in the face of all evidence that this is so, for purely religious reasons, but it is absolutely the case.

We are all, from amoebas to Aunt Sue, from mice to men, from wombats to women and from chimps and bonobos to you and me, molecules in motion. We are nothing more and much more at the same time, with nothing at all supernatural. We are chemicals and electricity in a wondrous dance, as molecules and electricity animate our bodies and produce our most joyful and profound feelings. It is wonderful to be able to live during a time when we know this happens, how it happens, and that the only conductor of this organic electrochemical orchestra is our own organic electrochemical self.

We need nothing more than the knowledge of what we are except the maturity to accept it.

Part Three

A Deeper Understanding

Chapter Ten

Religion's Origins, Appeals and its Essential Emptiness

"The good thing about science is that it's true whether or not you believe in it."
— Neil deGrasse Tyson

To enable all readers of this book to view its contents through the same revelatory prism, it will be necessary to clarify a few points from the start. There are countless misperceptions, untruths and both accidental and deliberate failures of logic running around loose in the gods-no gods debate, and everybody seems to think he or she is qualified to hold an informed opinion on the matter. This is particularly the case in the dysfunctional American 'everybody gets a trophy for playing' culture where all opinions are welcomed and very few of them are seriously questioned for fear of offending someone's weak and tender insensibilities.

This is not a productive or intelligent approach to understanding reality or how we ought to live our individual and collective lives. Most people's opinions are actually quite ignorant, foolish and completely wrong, and the mere fact that someone can hold an opinion gives absolutely zero automatic validity to that opinion.

The famous line from former US Senator Daniel Patrick Moynihan on this topic, to wit "Everyone is entitled to his own opinion, but not his own facts", was a good start but was only partially right. In any grouping of human beings, opinions are all too often not merely wrong, they are also extremely harmful. Opinions on serious matters, unimpeded by verified information and facts, may feel good and comforting to the person holding the opinions, but they may be totally worthless and extremely dangerous. As such they should be excluded from debate and from concrete actions taken in the real world. Any individual's input to actions or decisions that affect anyone else must be shaped by and eventually constructed from verified truth and facts.

And these must be actual facts, fact-checked facts, real and true facts, not lies presented as so-called "alternative" facts or 'things I like to believe because it makes me feel good to say or believe them'. You are therefore entitled to hold, but not to impose or interject in any way, your uninformed opinion on matters that extend even a millimeter beyond your individual existence.

If you like to eat cheeseburgers and fries three times a day, think ketchup and fries are both vegetables and that sugared drinks are modern-day liquid manna from heaven, your opinion may be that such a diet isn't bad for you, if only because you haven't had your first heart attack or stroke yet, or because Fox News, Breitbart, Alex Jones, Todd Starnes, Mike Huckabee or Rush Limbaugh said that there is a little old lady in pious Down Home Alabama who has eaten like this for 75 years and is 'still doin' *jus' fine*'.

It may very well be more comforting to you, and more appealing to your gustatory addictions, to believe that you share this mythical woman's alleged immunity to overdoses of sugar, cholesterol, trans and saturated fats and the ravages of hardened arteries, cancer and diabetes. Your opinion in such a case, contrary to the good Senator Moynihan and the actual genetic facts of almost the entire human population, is worthless and harmful junk, and as devoid of quality as the harmful trash you so happily ingest.

If you hold the opinion that your motorcycle riding habit isn't dangerous and you therefore don't wear a helmet while riding, or if you don't bother to acquire health insurance because you believe you are the Son of Superman and share his invulnerabilities, or that your uncanny reflexes will prevent you from ever having an accident, your eventual trip to the hospital, rehab facility or morgue is going to cost you and any family members quite a lot, both emotionally and financially. The list of possible examples in this vein is quite literally endless.

So if it is your opinion that the specific god which your family and community grew up worshiping, out of the embarrassingly massive collection of thousands upon thousands of past and present divinities to choose from, plus the further legions of those yet to be 'revealed' and invented, is the actual one responsible for creating the universe, injecting life into inanimate collections of organic molecules and planting *Homo sapiens* onto this microscopic dustball of a planet then psychopathically screwing with and killing them, that opinion is worth not even as much as the ignorance described in the previous paragraphs.

In this book I discuss the reasons people believe in gods and some of the development of these beliefs. These reasons encompass the painfully slow evolution of humanity over millions of years, and our almost equally slow accretion of knowledge. The reasons span an incredibly wide range of human psychological states and are determined by genetics, culture and the amount and type of information each individual accumulates.

The last item, information, plus how it is used, is the starting point toward changing an uninformed opinion, i.e. one's feelings on an issue, into an informed perspective. When there is an extraordinarily large preponderance of mutually supporting data and successfully and scientifically tested "opinions" on a question, and essentially no significant data in contradiction, the explanation of that question has moved from opinion to hypothesis to full-fledged theory by the actual scientific definition[1], and thus to a de facto fact. What makes this process so difficult is that it is never fast or easy, and it is far beyond the patience, training and/or actual ability of most people to work through.

And in the honest pursuit of knowledge, every advance in technology or in the process of information verification opens up another potential avenue for analysis of the question, so there are always ongoing and periodic reviews of our body of knowledge to ensure the theory or de facto fact is still valid. This continual questioning of information is science's greatest strength, in that it is continually checking itself. Processes and systems that deliberately exclude themselves from correction, with religion being the world's most common and egregious example, thus contain an internal and consciously maintained fatal flaw.

Religions do not self-correct in any way. Their business model is adherence to dogma and an unchanging and inflexible worldview. Religions only change when the pressure from the outside world becomes too fiscally impactful and thus finally impossible to ignore. When investigating organized crime and organized dogma of any kind, always follow the money trail.

Religions and crime largely follow the same basic paths to power and money. They go where the money and the power are and they do what they can get away with until their money or power become reduced or are taken away. And even then they only change under extreme duress, as microscopically as possible, and after strenuous and protracted resistance.

Religions do suffer mightily from schisms, which aren't internally driven improvements, but often rancorous splits when the original group of believers suffers from internal disagreements and members split off to form related but different religions or subreligions.

To make the process of information verification even more difficult, on some subjects like the afterlife there is absolutely no real evidence whatsoever, and none is forthcoming despite thousands of years of quite desperate and vigorous efforts all over the globe to find some.

Note that personal "visions" of the afterlife, other supernatural imaginings inside one's head or a person's dreams do not count as evidence, no matter how powerful or real they may feel. This is because these are internal products of the inconsistent meat computers we call our brains. We do not perfectly understand the brain, but we assuredly know it to be subject to hallucinations and extraordinarily susceptible to misinterpretations and/or wishful thinking.

If mere visions of heaven and hell are allowed to count as evidence of their existence, then mere visions of leprechauns, aliens, lottery winnings and of course the gods and goddesses from everybody's religions must be allowed to count as evidence for these illusions as well.

Dreams of flying, alien abductions, gambling winnings and that Jodie Foster or Brad Pitt love you with an undying passion are equivalent with supernatural visions, in that they are all products of the brain and do not prove anything in the real world except that human brains experience hallucinations.

If the faithful protest that visions relating to their personal specific faith are real, while simultaneously claiming that no visions of anyone else's religions are real, all this does is further expose their hypocrisy and faulty wishful thinking. Those two factors provide the internal load-bearing structure within all religions and explain why they believe in their own gods and not in anyone else's. The faithful cannot and should not be allowed to assert their astoundingly unsupported nonsense without providing absolutely overwhelming evidence.

It is a binary choice, just as the actual choice to believe or disbelieve in all religions is a binary choice. Either all of these visions and dreams are all real and thus proof of all the different heavens, hells, gods, aliens and leprechauns, or they are all not proof of anything except that the mechanics of the human brain sometimes result in hallucinatory or visionary experiences.

What if we nonbelievers called the religionists' bluff? What if we said that we would accept that all their visions are from their gods, but we said that they had to accept that *everyone's* dreams, nightmares, hallucinations and visions are from their gods too? Not just some of them, but *all* of them from any human head at any time in history, to include those in and out of mental institutions. The religious would never agree to this, of course, because to them only their own specific religion or belief system is the one which causes these visions, and only the visions of their god are perceived as real. That this is illogical on its face and special pleading run riot is as obvious as it is annoying.

When millions of Chinese start having the same very detailed and very specific visions of Jesus without ever having read any Christian references, and when millions of Americans start having detailed and very specific visions of Aphrodite, then we can start to have a discussion about taking visions and dreams as anything but hallucinations and byproducts of as-yet not fully understood electrical and chemical processes within the human brain.

For that matter, why should we only consider the dreams of *Homo sapiens*? Other creatures[2] on earth have dreams, too. Anyone who has lived with or observed an animal with a brain more complicated than a gerbil's didn't need MIT or anyone else[3] to know that mammals dream. There are something like 100 billion to a trillion[4] mammals on the earth. Even if only half of them have only one dream or nightmare or vision per night, that is up to half a trillion such events per night. These animal dreams should be part of the discussion of the validity and supernatural implications of visionary experiences, yet they are completely ignored by the faithful. As nonhuman mammals far outnumber humans, surely one of their visions or dreams is far more mathematically likely to contain truth about the supernatural.

It is only because our science has not yet progressed to the point where we can record the details of what humans and other mammals experience during dreams, and the biochemical and physiological reasons and processes underlying these experiences, that we currently discount them. These dreams and the details on their processes are almost as unseen to us today as microorganisms were to people in the 16th century. Yet once we realized these organisms existed and what they did, those discoveries eliminated any reason to attribute infection and disease to the supernatural. So any random human dream or vision should have no more impact on our view of any supernatural realm than does the dream of any twitching dog or cat slumbering by the fireplace.

Regarding the hereafter, when a physical object like a video recorder can, under scrupulous observation and control by scientists and skeptics, move to even a single one of the countless afterlife realms posited by any of the world's religions and then return with actual evidence of its existence, we might begin to consider the whole idea to be more than purely a comforting fiction. When a human being who claims to have gone there and returned can give multiple and independently verified specific predictions of future events, or multiple and independently verified accounts of information which could only have been gleaned from interactions with the recently and not so recently departed, we might begin to think he had experienced something worth listening to.

Until any such circumstance can be definitively proven to the satisfaction of honest and objective skeptics, any conception of an afterlife can be dismissed as wholly without evidence and this witless fiction must be challenged at every turn.

On a few subjects such as the origin of life there is scarce information. On others there is mixed information, and we also run into the extremely wide degree of variance among people and conditions. For example, John Smith may smoke heavily for 60 years and not be overly damaged by it due to good luck in the genetic lottery. Jane Doe may be a nonsmoker but succumb to lung cancer in her twenties or even her teens due to bad luck in that same genetic crapshoot. We know for a fact that most people will suffer predictable types of debilitation and damage due to inhaled pollutants, and while this varies from person to person, your best bet is to avoid pollution if you want to remain healthy. Apparently the claims of us all being made in some god's image don't go very far, or that image is extremely imperfect.

Just because we don't yet have all the answers at the genetic level for these circumstances, that does not mean we can't say that that pollution damages humans, animals and the environment in a generally predictable way based on massive amounts of empirical and tested data and observations. Science really does give us answers to the questions we pose.

We don't fully understand consciousness or gravity, but only a certain kind of ignorant, pathetic, deluded or religious individual would say because of our failure to totally explain it that therefore they don't exist or that a supernatural being is responsible for them. Yet perhaps as many as half the people you come in contact with do believe this.

Even on subjects which are essentially incontrovertible, the effects of wishful thinking and denial lead to materially negative outcomes, often visited upon the innocent. The deeply and sincerely devout believers who choose to pray for their children to recover from serious illness rather than seek modern medical care will attend more funerals than school graduations. The deeply and sincerely devout believers who feel their god has given them carte blanche or indeed specific guidance on when and how to kill humans both inside and outside their faith are clear and present dangers to every other human on earth.

The deeply and sincerely devout believers who vote for candidates who spout off about their own deeply and sincerely held religious beliefs while working hard in the halls of power to make life miserable for their gullible voters and the rest of us are also dangers to everyone, not merely physically but also mentally, socially, economically, legally, ethically and morally.

When wishful thinking is deliberately fostered by those who wish to spread disinformation, just as when dishonest persons, corporations or governments will benefit if they are not forced to stop polluting and thus spend much money and time claiming pollution isn't really that bad, we have a huge problem. Humans have an extremely strong tendency to come to their opinions regardless of where the truth lies. The primary challenge we face when we try to rise above our monkey intuitions and work to act as rational and logical creatures in the pursuit of truth is that we must simultaneously overcome our biases while ensuring the information we ingest is factually true. To determine the veracity and applicability of information one also has to get past data that are either accidentally or deliberately wrong.

Accidentally getting information wrong is quite easy to do. Ask ten witnesses for a detailed account of an incident of any kind which they all watched happen, be it a fight at their high school, a car accident they all witnessed or a crime being committed on the street in broad daylight, and you are likely to get ten different stories, even after all witnesses have solemnly promised to tell the truth and are trying extremely hard to do so.

The stories might not be completely different, but one would be a fool to bet one's annual salary, one's moral compass or indeed one's eternal salvation on any of the stories being exactly correct. Look at any of the many clever optical illusions or examples of attention blindness one can find in a good psychology reference and see how the brain can misinterpret what is literally right there in front of your eyes.

And that is what happens when people do their honest best to relate the absolute truth about what they saw or heard. When people take the opposite path and consciously or even semiconsciously seek to dissemble or mislead, the door to the truth may never be opened. This may not be a revelation to anyone, but a very large number of people deliberately lie, and many of them will do it right to your face. Their motivation for lying may vary, but the bald-faced lie is a verminous species that infests all corners of human life.

Some people lie for a living, or if they don't outright lie in word or print, they twist facts, tell half-truths and do what they can to send a specific message which is not in any way the truth, the whole truth and nothing but the truth, and simply and easily understood as such. These include the overwhelming majority of politicians and their spokespersons, those in the advertising community, attorneys, sellers of products and services, and of course the most prominent sellers of the completely unbelievable and eternally unprovable, the clergy.

Those who do not get paid to lie or dissemble but do so anyway include pretty much everyone else, for reasons as varied as trying to get out of a speeding or parking ticket, avoiding jail, avoiding the wrath of one's teacher, spouse or parents, getting out of a sticky situation at work, avoiding paying for yet another installment of Scouting cookies or popcorn, walking past the Salvation Army collection pot or neighbor's daughter's lemonade stand without making a contribution, or safely answering the at-least-occasionally unanswerable question "does this dress / haircut / shirt / tie make me look (insert pejorative of choice)?"

And the more the other person wants you to believe what he or she is selling, the more determined the sales pitch will be. The more savvy and glib the liar, and the more he or she panders to your own personal opinions, the more believable the lie will seem. The story may grow in size and scope until the original is but a distant and unrecognizable spot in the distance, blotted out by the increasingly bright and shiny object now on display, but the story will still be just as firmly believed as when it was much smaller.

This is true even if the person is only a common liar who is being deliberately deceptive. The psychopath, sociopath or mentally damaged human may not even realize that he or she is compulsively lying because they aren't aware that their minds shift truth and reality to suit the situation and the narrative of the moment to their advantage. Uncountable millions of humans throughout history have suffered horribly and died painfully at the hands of psychopathic liars.

Many humans have a naturally strong animus towards outgroups. Many share a tendency toward voyeurism and derive great satisfaction from sharing gossip, and the saucier and more salacious it is the more they want to hear it. The immense popularity of voyeuristic (un)reality media productions and social media, and the apparently incessant need for people to know what other people are doing and thinking, particularly if it involves sex, social connections or scandal, prove this to be true.

People revel in crazy rumors, conspiracy theories and the misfortunes and misdeeds of others, particularly those of celebrities and those from groups other than their own.

Even discounting the nuttiness of religious claims, there is apparently no rumor or headline so outlandishly contrived and obviously false that millions of Americans will not foolishly take it as gospel truth, particularly if the news is from an outlet sharing their political or religious worldview. Take a look at any news-stand in the United States, the headlines of many internet sites and the aptly named "vast wasteland" of offerings on cable television for evidence of this. These factors alone add an unbelievably heavy load to the already overtapped and weakened ability to relate something clearly and honestly without adding comment, inflation or personal opinion.

What's a semi-evolved primate to do?

As it turns out, we have the solution, in a system which enables us to move beyond our uninformed opinions. This system enables us to properly assess information and observations and to arrive at truly informed and objective conclusions. That system is, of course, science.

The scientific method, when applied by trained people and thoroughly checked by other trained people trying hard to disprove the original conclusions, is the only proven method for advancing actual human knowledge and understanding.

Science is our best and so far only dependable method of overcoming bias, wishful thinking and simple error. The scientific process, as the most refined method of critical thinking and analysis we have, cannot screen out errors in bias and wishful or incorrect thinking in those whose adherence to dogma will trump facts, but when the process is applied properly either by the original theorists or later by peers, it will show when opinions or hypotheses are flat wrong, and usually also the ways in which they are wrong.

So why don't people use science or the scientific or critical thinking method all the time in all their intellectual endeavors and decisions? There are three reasons.

1. **People don't know the processes.**
2. **They can be hard and take time to understand and use.**
3. **They don't agree with personal opinions.**

Note that these reasons are not mutually exclusive. A person whose individual psyche is too weak to accept the possibility of there not being an afterlife or imaginary redemption of sin may be able to apply science in his or her day job, but will forgo the scientific and critical method in his opinions on salvation, often without feeling much or any pressure to overcome the obvious supermassive dissonance between the two positions.

Those who do feel uncomfortable at this dissonance and seek to rationalize the disparate positions, by trying to justify faith through scientific language, always have to make exceptions or insist facts be placed in categories in which they do not belong in order to keep their worldview from falling apart. The ranks of Christian apologists in the 21st century contain many examples of people desperately trying to hold back the tide of scientific discovery and reason that is washing their faith away.

An example of self-inflicted foolishness, this time from a scientist, was the utterly spineless Neville Chamberlain-inspired nonsense of the so-called "Non-Overlapping Magisteria" which Stephen Jay Gould dreamed up as a peace offering between the scientific and religious camps. This position was that the two subjects were in different stovepipes, essentially different domains of knowledge and understanding, and thus could not cross over into each other's sandboxes. This was an abdication of both reason and spine.

Religion is a manmade invention with dogma at its core whose tenets can be tested by science, and found to either be as advertised or not. Science is a manmade invention, but it is wholly free of dogma and its methods and tenets are that of empirical testing. Religion cannot study science or anything else, but science can study religion. That is the core of the irreconcilable difference between the two, and why humans selectively ignore science when it disagrees with or disproves closely held beliefs.

To fully understand the challenges, we need a deeper explanation for each of the reasons for not using science and critical thinking.

Reason 1 For Not Using Science and Critical Thinking: Not Knowing How

There are reasons for not knowing the method, and they are a testament to the lack of rigor exhibited by many humans and to their surrender to dogma. But the fact is that science and critical thinking are hard. Also, depending on the situation, these skills are not provided to students in America today, they are provided but not well or in limited areas, or are provided and not retained for very long. These processes are hard and can be hard to retain. The method is not intuitive for most people, so it can be hard to retain after school. Also, many people don't like science or its method of rigorous and mind-challenging analysis, so they forget it once they have had their last classes on it and never apply it again.

We can never forget that science is not innate or instinctual. It is a human invention. Thus we often see great difficulty in learning and applying it. Were science actually part of the human brain's inborn programming, religions would either never have begun or they would have popped up as temporary hypotheses then later been dismissed out of hand as the human body of knowledge expanded. This bears repeating. Religion came before science, proving that the desire to understand and assign causes to events and conditions long predated our actual ability to understand these causes. Social inertia, ignorance, the accumulation of money and power plus the weakness of most human brains have enabled religion to endure.

That said, science is taught in the vast majority of American schools, so for many Americans the fact that they don't use its methods stems from it being either too hard for them to perform or that they prefer their opinions to facts. Both are frankly contemptible positions. But for many others in America, particularly students who attend school in theocratic states like Kansas or those in the American deep south like Texas[5], and/or who attend voucher schools[6] or participate in many home school programs[7], science is not taught or is taught but with major caveats, all of which are religious. Unless the student seeks and finds actual information, the student's understanding of science can be pathetically limited both in depth and breadth, as a result of the deliberate actions of the so-called educators and complicit parents who poison their own children's minds with religious fairy tales.

It is a despicable perversion to teach children that the principles of critical thinking and the scientific method can be applied to some aspects of our world but not to certain specially-protected others. This is deliberately teaching useless dogma and cognitive dissonance, is harming young minds, and it is making them both as individuals and as a national and societal cohort less informed and less capable citizens. That science and history are singled out for such revisionist and slanted treatment, rather than math or English, clearly indicates that the motive is irrational and done with full awareness of the danger that truth and facts pose to unverifiable dogma.

Why aren't these students taught dogmatic math along with their dogmatic science? Why aren't math lessons presented with exceptions and cases when people shouldn't use math but should instead apply biblical verses to their engineering projects or when building planes? Do you want your loved ones driving over a bridge or working in a 50-story office building if the architects, engineers and construction workers opted not to use so-called "secular" math and physics when doing the calculations on load-bearing members, wind resistance, fire safety and electrical load? Do you want to fly in an airplane designed around the 'uplifting word of the lord' or the actual facts of "secular" engineering and aerodynamics? Do you want to live in a country or family where financial decisions are made using scripture as opposed to the "secular" facts of cold hard dollars and cents, income and expenses, taxes and spending, economics and fiscal psychology? Do you really want to live in a country where instead of relying on facts, evidence and reason, the will of the gods must be part of every decision?

This is actually how many religious people think, and it explains most of our national and global problems.

One can imagine the disgustingly wretched phrasing the children must endure in these unscientific American madrassas, which the children's parents happily and knowingly send them to for deliberate religious brainwashing. "You see, children, science is full of theories, and theories are just ideas, but the bible is from our one true god and his teachings are always and forever True. So you can use science and critical thinking to work on *these* problems over *here*, but you are not allowed to use them to work on *those* problems over *there*. Why? Because if you applied science to those problems, you would start to question god's word, and our religion, and we can't have that. OK, now turn to Genesis and let's begin our biology lesson."

A true understanding of unadulterated and properly applied science and its method absolutely must be a mandatory minimum requirement for participation in a civil society.

Reason 2 For Not Using Science and Critical Thinking: They Can Be Hard To Perform

This is unquestionably an issue of great importance. Yes, science can be hard. It can be extremely hard when done right, and it can take a massive amount of intellectual capital and time to thoroughly study the discipline and the data.

But the average person is never expected to put in the amount of work a Nobel laureate would in the pursuit of truth and knowledge. Indeed, if one chose to live as a Luddite hermit in a desert tent someplace and his words and actions truly impacted no one and he never voted, no one would expect or care if the person did any scientific or critical thinking whatsoever.

But people who live in a society with other humans, in which the products of science are both ubiquitous and actually hold the society together, need to use their brains and the tools we have developed to be responsible, positive and productive members of that society. Even if no one ever heard the term "scientific method", everyone should know and apply the concept and principles of critical thinking. It does require more time, effort and thought than opening up the top of one's head and letting one's preferred media, clerical or dogmatic numbskull pour in the nonsense. It does require questioning of all opinions, even one's own. It does require research and thought, which is a problem in a thoroughly lazy country.

Americans tend to care as little about their mental fitness as they do about their physical fitness, which you can prove to yourself in just ten minutes of walking around your average American city and observing and talking to people. Having been to quite a few other countries, I can tell you definitively that Americans care much less than most people do about mental and physical fitness.

A responsible person has two choices for living in a civilized society. One involves being a grown-up and applying critical thinking and/or the scientific method to the opinions one holds and changing them whenever necessary. It also includes applying this rigor to the words one uses and the actions one takes, up to and including voting. The other choice is to keep one's opinions completely to oneself and to clearly admit that if they do escape one's mouth, they are uninformed opinions blissfully devoid of supporting reality.

Taking the second path also means the person should not vote or take part in any meaningful social or political discourse. Just because something is hard is absolutely no excuse for not doing it, and the rest of us need not suffer for someone else's laziness or ignorant opinions.

Reason 3 For Not Using Science and Critical Thinking: They Often Don't Support Opinions

This is the "get out of jail free" card for those who prefer to live their lives based on unquestioned dogma and opinions rather than reality and proven facts. All too often, and almost one hundred percent of the time in religion, reality doesn't agree with a person's opinions or beliefs. This rejection of facts and embrace of opinion is a hugely unfortunate tendency within our species. It is an understandable temptation to want to succumb to mindless opinion over potentially unpleasant facts, such as the number of calories that are actually in that donut or can of soda, but when one chooses dogma over reality, the choice becomes deliberate and is contemptible.

Escaping reality is what enables the alcohol, drug, gaming, gambling, snake oil, advertising, politics, religious, sex and many other legal and illegal industries to be such enduring and thriving businesses. Reality can sometimes be extremely uncomfortable, extremely hard to face and highly lethal to opinions and beliefs, thus many people seek to escape it.

Belief is a dangerous word, and when people use it they are revealing something damning and often deliberately disingenuous about themselves. When a person, a cleric or a politician says that he or she "believes" something to be true, this is a massive red flag and you should immediately and directly distrust and challenge the person in a strong and decisive manner. This is because the person is almost certainly lying to you and/or deliberately using language which cannot be pinned down in a conflict-avoiding and factually and scientifically ignorant society.

The person is co-opting both basic human gullibility and religion's defensive armor by using the word "belief", because it is only the person's opinion and is not based on evidence or reason. In many parts of the world, the faithful scream loudly that their unverifiable beliefs should be treated with kid gloves and not challenged, and that they will suffer some kind of insult or injury if their beliefs are challenged. If someone tells me his father is the King of the Middle World Magicians, I will challenge him to prove it, and so should you. If someone tells me his father is the King of Heaven, is the creator of the universe and happens to be my father too, I will challenge him to prove it, and again so should you.

Believers in magicians, gods and many other impossible, unproved or indeed disproven positions or supernatural entities continually work to condition the response to their beliefs to be one of respect or tolerance instead of the withering criticism and rejection they require. The politician who says he or she "believes" something is using this same approach, which boils down to "'don't question me because it's my sincerely and deeply held belief or I am being paid to say that X is true, and thus it would be offensive for you to question me".

Also unspoken is the desire to maintain constant plausible deniability. The politician can later say that he or she didn't *state* that X was true, only that he or she *believed* X to be true. Thus religious people are far more susceptible to dishonest politicians, because they have been conditioned, often since birth, to believe religious statements without evidence. This trained gullibility usually infects other areas of their lives and disables their thinking on many other subjects.

The bottom line is if a person tells you about a belief he has, remember that this belief is only his **opinion**. The odds are quite high that the person has little to no actual empirical justification for this belief, and the person holds the belief because he is uninformed and/or captured by motivated reasoning and rationalization. Thus, again, the offering of a belief statement must be universally met with complete rejection and demands for full proof and justification. If the believer persists, or offers non-evidence such as he 'feels the holy spirit in his heart' or some other such pathetically foolish nonsense, this escalation of broken thinking must be met with an equal and opposite reaction, that of outright derision and dismissal.

Life is far too short, and the truth is far too important, to let people get away with such escapism and denial of reality. If this means you hurt some feelings, remember that you are trying to help these people escape from a mental prison, and some people like captivity far too much. If you lose some acquaintances and even friendships over your decision to champion truth and reality, remember that those who would deceive others or themselves are definitely not the kind of people you need in your life.

If science tells you that you are extremely likely to get cancer or heart disease because you smoke and eat junk food, and this is too disturbing to you and you blow off science, you are an emotionally-blinded science denier. If science tells you that global climate change is real, but not believing it either in public or in private suits your business, financial, political or religious opinions, you have slammed the door in science's face and proven that you are an immoral and factually dangerous science denier.

If science tells you that the brain and mind are interlinked and that there is no mind without a brain, that there is no evidence for the soul, that the earth is billions of years old and was not created after humans invented beer, and if any of these facts conflict with your sincerely and deeply held religious beliefs based on fables written when some of the most advanced technology was the scroll, you have a choice. You can either accept science and a whole planet overflowing with empirical data, or you can turn your back on science and its religion-destroying facts and retreat into a cocoon of ignorance. If you choose the latter, you are a science denier, and have self-selected out of the pool of responsible humans who should be making decisions that affect anyone else.

The constant stress or angst about situations facing a person and his family, like health, finances, employment, war and an afterlife drive many weak people to negative behaviors like taking drugs, drinking alcohol and turning to religion. This is because these mind-numbing escapisms are extremely easy to access and they provide instant comfort, no matter how unreal, transient or dangerous they are.

Religion stands alone on a massive stack of corpses flaunting its status as the greatest con job of all time, because it has billions of fans despite being totally unable to prove even a single one of its existential claims. Since these claims are at the same time unfalsifiable, internal, promise to alleviate all one's worst fears and are today often free of much material cost, their appeal is often irresistible. The religious power cabals know this extremely well and they take full advantage of it.

Karl Marx penned an exceptionally apropos description of this aspect of religion in his *Critique of Hegel's Philosophy of Right.* He wrote "Religious distress is at the same time the expression of real distress and the protest against real distress. Religion is the sigh of the oppressed creature, the heart of a heartless world, just as it is the spirit of a spiritless situation. It is the opium of the people."

So yes, just as drugs and alcohol provide an escape from reality, the claims of religion offer promise of escape from the pains of this world, literally infinitely sweetened by the claim that the comforting state of mind induced by the drug of religion will be eternal. Science and critical thought are banished from the brain, and dogma and ignorance greedily rush in to fill the vacuum.

In our scientifically informed modern world, there is no rational excuse for being religious. Religion has no justification. It is merely faith. And faith is based on feelings, while reason is based on reality. Religion's original justifications became brain-dead long ago, and the entire current corpus of religion is being kept alive on forced life support by the believers who cannot bear to be without the crutch of their faith. In our modern and educated world, religion would disappear overnight if humans were able to rise above the forces of unconscious emotion.

You can see how religion has an appeal even to the most hardened, objective and strong minds. Tell a parent that he will see his dying child again and that the child's impending death is only a transition to a permanent state of bliss, and the lure of religion can become irresistible. Frighten a person with the threat that an angry god will burn him forever for some petty act, then tell him a specific god will divert him from this eternal torture if he joins a certain religion, and there is a good chance he will believe the claim. But these are all empty promises, deliberately constructed to appeal to human frailty, and such lies need to be eradicated from the world.

The chapters that follow describe these issues in more detail.

Chapter Eleven

An Opening Salvo On Coexistence; Plus Everybody's An Atheist And I Can Prove It

"The first principle is that you must not fool yourself and you are the easiest person to fool."
— Richard P. Feynman

Do you believe in Mithra, Baal, Shiva, Manitou, Thor, Amun-Ra, Marduk, Ishtar or Mena? No? Then congratulations! You're an atheist, just like me. The difference between us is likely to be that you do happen to choose to believe in just one single particular specimen out of the ever-expanding mass of thousands upon thousands of gods humans have invented, and I choose to believe in none of them.

It is a safe assumption that no person on the planet today believes in every god ever invented, thus every person on the planet today is 99.999% an atheist, and it takes very little effort to round up from there.

Now that we have established that every human alive is a de facto atheist, we need to decide what to do about this fact. Oil and water do not mix well, and people who try to force their religion on other people do not mix well with those who don't use ancient science fiction as personal or public policy.

Therefore, based on what I have written so far in this book, it might be concluded that I think that the billions of devotees who subscribe to religions large and small and from pole to pole around this planet cannot peacefully coexist with we nonbelievers. It might also be concluded that I think that everyone who believes in a religion is a bad person. I need to be clear on this, so will take some time to lay out the situation.

Coexistence is possible, and indeed desirable if for no other reason than neither the believers nor the nonbelievers are going anywhere anytime soon, and peaceful coexistence is always better than conflict. Also, there are plenty of extremely good people who happen to be religious, even those who are Christian, Muslim or Jewish. I know quite a few people like this, and know first-hand how much good they do for others. I am quite happy to coexist with them, and call many of them friends.

The difference between these truly good religious people and evil religious people is that the good ones have rejected the tsunamis of hateful and antihuman nonsense in their religions and holy books, and live their lives based on the few desultory and scattered humanistic cherries they have picked out of those disasters. They did this because at their core they are good people who just happen to need to believe in the comfort of the afterlife which religions offer, among other things.

Not much separates many of these good religious persons from secular humanists, and as long as decisions and laws are not being made in the public sphere on the basis of some religion or other, these groups are often indistinguishable from each other except that one group attends religious services for an hour or two a week while the other does not. They don't work hard to corrupt the legal system so as to inject their faith into our schools, laws and every other aspect of the commons where it does not belong, both reasonably and Constitutionally.

Good religious people are good people because they are good people. Through a fortuitous blend of genetics and upbringing, these people would most likely be good people no matter where or when they were born. This would not be due to the influence of any religion but because they were genetically coded to be good people and weren't diverted from following that code. They may have entered into a religion, but have selected those few good nuggets from it and then built a protective cocoon around themselves to shield them from the poison that is the rest of the faith.

Thus they can call themselves Christian, Muslim or Jew and reap the existential perks of the faith while insulating themselves from the wrath and evil of their co-religionists and 99% of their holy books. Living alongside people like this is true coexistence, and is a worthy goal to work toward. And until humanity finally throws off its superstitions and belief in magic and fairy tales, it is the only way open to us to peacefully coexist.

All the above having been said, there must be rules for coexistence. The religious and the nonreligious can coexist as long as science and reason are the foundation for our educational, moral and political decisions and actions.

Religion must be a personal matter which does not intrude upon education or morals, policy decisions or the lives of anyone anywhere at any time except those who specifically and personally want it, and even then only in their homes and personal lives.

The intent behind and the construction of the wholly secular American Constitution, therefore, must remain the gold standard for separation of unfounded beliefs from actual human lives.

The more dogmatic and aggressively religious will say this is unfair. In response, we secularists have to yet again, for what feels like the trillionth time, tell you that your belief in a specific imaginary and unprovable supernatural sky fairy is on equal footing not only with everyone else's beliefs in different but also very specific imaginary and unprovable supernatural sky fairies, but also with my very specific imaginary and unprovable belief that a colony of magic worms living in Saturn's rings created life on Earth, that they still watch over us, and that we live our lives based on the unconscious instructions they beam into the brains of every human every moment of every day.

All religious or supernatural beliefs, from Jehovah to Jesus and Flying Spaghetti Monster to the Master Worm, are all equally probable and should be treated exactly the same way, which is to say they should be completely left out of rational discussion and decision making. The faithful have extreme difficulty with understanding and accepting this, which is why coexistence has been and will be a major challenge.

This is the planet-sized gap in perspective and cognition which the religious do not, or in most cases will not, understand. If you are a believer, your annoyance at my allegedly foolish desire to follow the beamed instructions of magic Saturnian worms is far exceeded by my annoyance as your foolishness at wanting me or anyone else to follow the instructions of your imaginary creatures whose only actual reality is limited, in the case of monotheisms as an example, to the science fiction/fantasy and mythology written millennia ago by ignorant and thuggish semi-evolved male primates in Middle Eastern deserts.

It must be pointedly noted that these works are not only borrowings and actual copies from older and different mythologies in the local area, but also that they were set down during a time when exorcisms as cures for disease, slavery as a natural practice and the flatness of the earth were commonly held truisms, and mummification was all the rage in Egypt. This scientifically, materially and indeed morally ignorant thinking is clearly evident throughout the monotheistic holy books.

So no, the phantasmagorical scribblings of ignorant and immoral people thousands of years ago have no more to bring to a discussion on the subjects of information, truth and morality than does a preschooler's finger-painting project. And if for nothing else, if we do comparative accounting based on the ultimate in evil, as measured by sheer body count, we magic worm and Flying Spaghetti Monster devotees win the 'which is better' argument on that criterion alone.

If we could pile up the bodies of all of the men, women, children and the unborn who have been murdered in the name of the world's religions in one place, we would have a pile of human flesh the size of a small moon. A similar reckoning for Pastafarians and Wormites would result in an unoccupied chalk outline of a body on the ground.

As if the above isn't enough, there is far more in the domain of reason to exclude religion from the societal, political and intellectual commons.

Every time there is a gap in scientific understanding, the faithful believer feels irresistibly and unjustifiably compelled to aver that the "Answer" to that gap must by default come from his or her own particular and specific religion and pet god or pantheon. There isn't credit given to the possibility that *other* religions might be true or that *their* pet gods and goddesses are equally as real and as responsible for filling the gap. This is a planet-sized non sequitur, a catastrophic failure in logic wrapped in special pleading, topped with delusion and arrogance then blasted into high orbit. And it is practiced continually, in a never-ending rolling artillery barrage of ignorance on logic and reason.

It may very well be that all the gods people have ever imagined really do exist, and Jehovah and Jesus are just middle-of-the-alphabet spirits in the total pantheon of supernatural entities. Maybe Thor was the one who actually created the universe, but since he likes to vacation a lot he throne-shares with Zeus, Marduk, Osiris, Vishnu, Hanuman, Quetzalcoatl and Jehovah's nemesis Ba'al, who all take turns warming the golden chair when the big Norselord is out thundering around doing whatever he does when he's not at work. Maybe there is a godly division of labor, and each of these tens of thousands of spirits and divinities is assigned a specific job for a specific period of time, then at shift change the next god in line takes over.

Every single one of the equally invisible and unreachable gods, goddesses and spirits ever imagined has an equal chance at being real as any other, and not a single believer who has ever lived has any reasonable justification whatsoever to fall to his knees in the direction of his particular favorite imaginary pet god. Your god doesn't fill any gaps at all except those in your own mind.

I own an absolutely marvelous and I am sure very unintentionally humorous book by Marjorie Leach titled *Guide to the Gods*. You can be quite assured that it is not a thin book with a few entries. Quite to the contrary, despite being printed in small font, it is an absolutely massive tome weighing several pounds and measuring 8.5" x 11" x 2". To peruse it is to wallow in the pleasure of 876 pages of sheer schadenfreude, whiling away the evenings chuckling at the utter gullibility and god-creating mania so characteristic of the faithful in all places and in all times. Some of the sections in this book are titled "Deities of the Universe, Space"; "Supreme Being, Great Spirit, High God/Creator Deities"; "Gods of Night / Darkness" and many more.

The author describes weather gods, earth gods, fire gods, sea gods, nature gods, fertility gods, disease gods, afterworld gods, hunting and fishing gods, and of course the always helpfully close at hand and equally invisible and unreachable household gods. I also have a wonderfully entertaining book called "I*n The Beginning: Creation Stories From Around The World*", by Virginia Hamilton. Both of these books do mention the biblical god, but quite rightly do so as just a single god in in a hugely overcrowded picture, like a "Where's Waldo?" image, but for gods. If all these gods were to gather in one place to take a big divine family photo, the picture would have to be taken from the air in order to get all of the massed thousands of entities in one frame.

I defy any religious believer anywhere on the face of the earth to give a factual, logical, reasonable and empirical argument, with actual proof, as to why his particular religion and god is any more likely to be real than one that is picked completely at random from any of the pages of these fine books.

This is what we secularists are referring to when we point out to the faithful the quite obvious fact that all believers are actually best described as atheists. If you are a believer, regardless of how strongly you hold your belief, your faith is fully contained inside a tiny bubble universe echo chamber in which you only allow your god or gods to exist.

Out of an extremely bountiful religio-pornucopia of many thousands of choices of gods on offer, you are an atheist toward 99.999% of them. You only believe in the god or system of theism in which you personally choose to believe, and your often complete ignorance of not only the existence of other religions but also their history and tenets makes your present belief exponentially more irrational.

It is the extraordinarily rare individual who has a detailed knowledge of the vast majority of world religions and their histories, and it is quite likely that there are still religions being practiced today which are yet undiscovered by the rest of humanity. It is a certainty that others will be invented and large numbers of people will mindlessly join the new flocks.

Unless you actually do believe in many more than one faith system and its gods, a believer in only a single god or system of theism is actually 99.999% atheist, with just the smallest microfraction left over to apply to a microscopically tiny sliver of the theistic pantheon. Note that this percentage doesn't even include the vast majority of the legions of supernatural characters within Hinduism.

If you are a theist, you disbelieve in other people's gods, that is to say all the other thousands of gods humans have ever invented. You are an atheist regarding every other god, making you an atheist many, many thousands of times over, and potentially millions of times over depending on how you account for Hinduism. Making these very bad matters even worse, and showing a depth of closed-mindedness of which only devout theists seem capable, you are also pre-emptively disbelieving in all the other gods humans are currently inventing and will invent in the future but of whose existence you are not yet currently aware.

For all you know, the true, real and actual creator of the universe has been biding her time until humans pass a specific threshold in the development of a globally connected system of communications, so that Her Message and Her Words can be best transmitted not merely by we frail human vessels, but by a global neural network of silicon lines and wireless communications. She might reveal herself the day after tomorrow, and all of the billions of believers in the several thousand religions currently extant will have to quickly decide how to rationalize this new development in order to keep believing in their current religion and god or gods in the face of such a blow to their deeply and sincerely held unjustified beliefs.

The curious case of Blaise Pascal should serve as an example for how even very smart people can fall into intellectual and logical traps and reach the wrong conclusions. Pascal was an extremely intelligent French intellectual who lived in the mid-1600s, and who had an unfortunate break with reality following a series of family stressors.

He is famous for what is called "Pascal's Wager."[1] He became religious following a dream or nightmare one night, which is actually something that happens quite frequently due to the general human lack of understanding of how the brain operates. One must remember this happened long before we had even a rudimentary understanding of the brain.

To summarize this concept very briefly, this so-called wager is simply his response to the results of a risk assessment he performed. He weighed the risks of the options of believing in Jehovah or not believing in him, then made a decision based on the identified risks, which he perceived to be greater to the nonbeliever. The wager part means that anyone doing this, including Pascal, will be making a bet on how he will be treated after he is dead. If Jehovah does not exist, then there is no afterlife so there is no need to believe in him.

However, if Jehovah exists and he is the murderous vengeful psychopath described in his holy book, you damn well better believe in him because an eternity of infinite torture inflicted by this allegedly loving god awaits you if you don't. Plus if you do believe in him then you also have the potential for eternal bliss or at least something better than eternal hell.

Pascal ended this exercise in idiocy by wagering that belief in this particular god Jehovah was the most prudent decision for a person to make because the outcomes for not believing in him were horrible beyond compare, but choosing to believe in him, especially given his clear and unequivocal orders to do so, put one on a path to a much "cooler" place, in both the literal and figurative senses.

If you were expecting more from such an intelligent person, you're in for a disappointment. That's all there is to it. Pascal's wager only holds together for the Jehovah question if everything about Jehovah and the rest of the universe are indeed true and as myopically constricted as Pascal and Christians consider them. Thus this little mental exercise is called a wager. If we all knew the reality of any posthumous existence, all ambiguity would be removed and we would no longer need to place such existential bets.

The truth is that such mental gymnastics are both cowardly and ignorant. Pascal did not mention Jehovah's alleged omniscience, which would of course enable him to see that a person had chosen to believe in him based only on the outcome of a flawed decision matrix which would always hinge on one single critical criterion - abject fear of eternal punishment – and only one possible choice given the risks. Any godly discernment of and potential irritation at such craven sycophancy was not part of Pascal's decision matrix.

Pascal conveniently failed to mention any other supernatural being or system of theology than Christianity and its god. Any one of many thousands of gods could have been placed into the decision matrix, and a truly informed decision matrix would have included all of them plus a large amount of spaces reserved for the new supernatural beings which men would "discover" in future times.

For a person to truly cover all the bases and do a proper "wager" of this sort, he would need to take the years it would require to learn a great deal about all of the world's religions and all of their gods, plus any gods not necessarily associated with any religion and at least a rough guess at some number of gods that had not been invented yet. Then using appropriate screening criteria, one would weed out any gods which don't have an eternity of hateful damnation as part of their theology, because not believing in any of those gods would not have such a horrible outcome.

At some point there would be a large list of sadistic candidates to put in the "gods" column in the decision matrix. The pros and cons of believing in or not believing in them would then be added to the matrix, with absolute maximum and minimum values of positive or negative outcomes for believing or not believing.

Then one has to make the decision about which god would be stronger than the other, and apply a weighting factor to him, her or it accordingly. For example, if Ra, Odin, Mbombo, Jehovah, Waheguru, Enki or Pangu, which are all quite powerful gods which people used to or still believe in, all made the cut as the most likely creator gods, therefore being the most powerful gods if they still existed and had not made any other gods stronger than themselves, they would need to be racked and stacked by relative power. So if Jehovah were ranked last in terms of power among this group, meaning any of these other gods could beat the shit out of him, he would not get an extra boost in the decision matrix but the stronger ones would.

In other words, if it took six Jehovahs to equal one Ra, you might consider adding an appropriate 6x weighting factor for Ra, or an appropriately equivalent lessening of Jehovah's weight in the matrix. For a flavor of such things take a look at the various children's collector cards and games which rank mythical creatures and superheroes by their various powers and strengths. The only difference between the mythical creatures and superheroes and the invented gods of religion is that many people actually think their gods exist.

It's all just math after you get past the utter asininity of the whole project. The best ass-covering course of action for any person, meaning a true wager covering the most bases, would be to worship a different god each week until each one has had its turn, then start over again and just keep this process going until death. You'd have to add a new god to the decision matrix every so often as they are invented, but that's how the consecrated host crumbles. That way the truly cautious person would invest in the spirit economy like he would in the secular economy. It is much safer to invest in a broad-based index fund than a single company's stock.

So a cautious religious person would invest in, i.e. worship in, a very wide and inclusive Deity Unrestricted-index Market Blend fund, or DUMB fund. This would be a much safer investment, like investing in a total market mutual fund, as opposed to betting all one's life savings, or indeed the whole of one's afterlife, on just one god. Pascal should have boned up on his micro and macroeconomics along with expanding his hyperconstricted theology and philosophy.

This so-called wager was actually a small-minded copout by a scientific genius who became a guilt-ridden religious ascetic ubermasochist. His decision was that he might as well believe in the Christian god because if you do, you get rewarded, and if you don't, your punishment is an eternity in hell. That was idiotic, and we can only hope it was intended as a joke.

Whether it was or not, the truth is that Pascal chose Yahweh because it was the god of the religion he grew up in, Yahweh has an unparalleled vindictiveness toward the worshiping of other gods, and Pascal was scared to death he would go to hell. And his pre-Scientific Age night terror frightened his brain into a place from which it could not recover.

You will not be surprised at all to learn that Pascal was a devout Catholic whose religious beliefs, like so many evangelicals and other Christians in America today, focused strongly on the worst religious ideas and the worst in people, such as sin, guilt, evil, shame and his believed need for human suffering. Though his works were much more elevated and better composed than the meandering and frankly gutlessly spineless apologetic ravings of C.S. Lewis, Pascal was often incapable of thinking clearly and properly on the subjects of religion and personal and social behavior. He lived a terrified and fearful existence, always concerned that the ever-loving god was going to get him. Which was perfectly understandable based on a reading of his holy book.

An informed interlocutor should have asked Pascal how many gods humans had ever worshiped and how many religious groups the world had ever seen, then asked him to redo his wager in the divine light of all those other heavenly bodies. It is estimated that today there are over 4200 religions[2] or "faith movements". Other estimates have it at about 10,000.

There appear to be over 30,000[3] different groups just within Christianity, although they are all essentially tiny splinters off the main branch, which itself is merely one of the multiple branches of desert monotheism, which is itself in turn yet another branch extending sideways from the trunk of the central core religious tree. Religions form, mature and die just like organic species, and they all share common ancestors. Thus Darwinism extends its explanatory hand yet again.

So if there are between 4000 and 10,000 religions in the world today, it is impossible for the number of religions humans have ever practiced to be lower, and may be much higher. But let's be extremely conservative and assume for now that the true number of religions in practice today is about 4000 and that there have been 10,000 religions ever practiced by any human. What does that look like when you represent that graphically?

Figure Three below is a grid with 10,000 boxes in it. Each box represents one of those 10,000 religions.

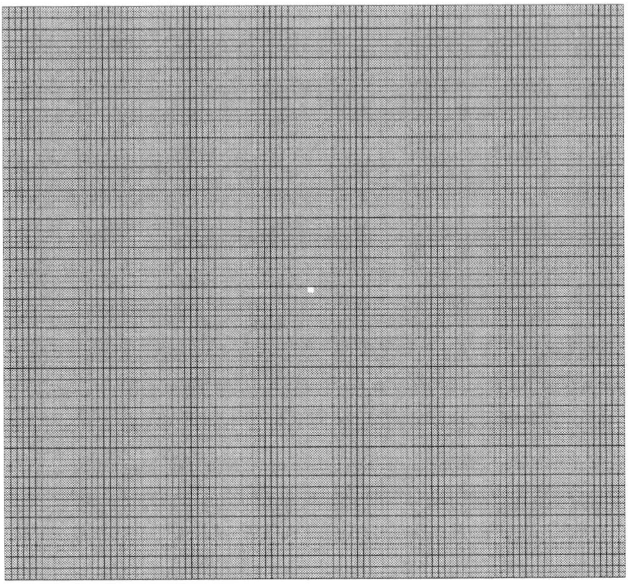

Figure 3. Ten thousand boxes in a grid, representing ten thousand religions. Only the one in the center is unshaded.

If you believe that only your personal religion is the one true faith, you believe that following all others than yours would be wrong, unappealing and a journey to ultimate darkness. Thus 9,999 of these boxes are shaded gray. If you are a devotee of one single religion, you are betting that of all these thousands of faiths, yours is the single solitary path to heaven or enlightenment or whatever it is you seek or worship. You are saying that you have been vouchsafed and are privy to a truly cosmic truth, which often includes answers to the origins of the universe and of life, information on your ethereal ghostly doppelganger half which will live on when your physical body dies, and many other details about the milieu in which that disembodied or reincarnated 'soul' will find itself.

You are saying that the other 9,999 religions and the billions of people who have believed, currently believe and will believe in them are categorically wrong, and that their adherents may suffer eternally for either not being aware of your religion, or for hearing of it and rejecting it.

You are thus a true believer. You just *know*, without asking too many difficult questions or demanding any actual evidence, that the single uncolored box in the middle of that image is the single true faith and that you have been fortunate enough to have chosen it.

As a quick sidebar, that so many of the faithful continue to call scientists and religious skeptics arrogant while hypocritically being so unsurpassably arrogant as to assert that their own particular microfaith is the one true path despite the numbers and points laid out above, is another indication of how the religious mind seeks to protect itself from dissonance and its faith from harm. Once again we must turn to psychology to understand this.

The human psychological weakness called "projection" is a very common aspect of human behavior. Britannica defines projection[4] as "...a form of defense in which unwanted feelings are displaced onto another person, where they then appear as a threat from the external world. A common form of projection occurs when an individual, threatened by his own angry feelings, accuses another of harbouring hostile thoughts."

Note that this definition is included in Britannica in the section titled "Defense Mechanism", which itself is defined as[5] "...any of a group of mental processes that enables the mind to reach compromise solutions to conflicts that it is unable to resolve. The process is usually unconscious, and the compromise generally involves concealing from oneself internal drives or feelings that threaten to lower self-esteem or provoke anxiety."

A great deal of discussion with the religious, and an examination of how and why they hold their beliefs, can only be properly understood if one has knowledge of the numerous unconscious human psychological weaknesses and biases, which I will cover in detail in Chapter Twelve.

For the religious, getting back to the far less than 1 in 10,000 chance that your faith is actually correct, what has really happened is that you are statistically extremely likely to be an adherent of the religion of your parents and local society, because you have been born to parents who share the belief and/or you have been socialized into that group which worked hard to inculcate you with those beliefs.

Genetics is a very powerful factor in both our bodies and minds. We inherit psychological characteristics and tendencies as well as physical traits, so humans who have the genetic predisposition to believe in things like religion will tend to live and mate with others who share their beliefs. These people will then have children who will tend to inherit a religious proclivity from either or both parents, thus often receiving an unfortunate double hit of the vulnerability to faith.

And when children grow up with parents like this, it is highly likely they will grow up surrounded by other family members, friends and social circles who are religious.

Thus the original genetic predisposition often becomes strongly reinforced and even fixed as a mental state and as a behavioral pattern.

Attempting to escape from it can be like trying to swim upstream with a millstone around one's neck.

Because we are extremely social creatures, the need to fit in and be accepted is often overpowering for humans, and most people succumb to it with little if any resistance. That is how many humans are hit with the double hammer of nature (genetics) and nurture (family and society) and many are permanently primed for and become entrenched in some religion or other.

Let's take an example. It must always be reiterated that whenever there is a discussion of nature and nurture like this that we have to understand that a great deal of the Catholic church's growth, and that of Christianity in general, was due to the church's hijacking of the process of evolution by natural selection through its centuries-long eugenics campaigns first in Europe then abroad. Again referring to Britannica, eugenics[6] is defined as "…the selection of desired heritable characteristics in order to improve future generations, typically in reference to humans."

In this instance, of course, the church did not know it was conducting a genetically-based culling campaign, as it was an unknown but surely quite welcome byproduct of the Inquisition[7] in particular and Catholic oppression and murder in general. Almost no one aside from historians and atheists know off the top of their heads how long[8] the persecutions and murders of the Inquisition as a totality endured. It was not a few decades or even a few centuries; it was well over six hundred years.

For most of a millennium, the self-appointed and self-anointed intentionally celibate male representatives on earth of the alleged creator of the universe conducted a deliberate campaign of eugenics and human evolution through artificial selection. Not through natural selection, a process in which organisms reproduce or fail to do so without a conscious external guiding hand, but artificially, through the actions of humans.

For more than six hundred years, Catholic religious authorities physically killed off many people who disagreed with official church policy, thus eliminating them from reproducing and also taking numerous intelligent, skeptical, inquisitive and independent minds out of the gene pool. They also intimidated into silence anyone else who would have otherwise spoken out, using the threat of what those already murdered had suffered.

All of the faithful and even many skeptical parents living during these long dark centuries of oppression would have taught their children from a young age either to fervently believe in the church's Pravda or to keep their skepticism completely hidden. So the popes and their inquisitors both killed off physical bodies and minds, and also killed off the will to resist in those they didn't get to murder. This was during a time when the church was extremely powerful and the population was extremely ignorant and/or oppressed.

But this was only one part of the total eugenics program the Catholic church carried out. In addition to the deliberate murder of heretics and the suppression of heretical thought in Europe, it also spread both murder and disease to unwilling and innocent natives of the lands which Catholic missionaries, colonists, profiteers and conquistadors invaded, occupied and settled from the 1400s through the 1700s. Tens of millions of native people in North, Central and South America died due to the diseases these European invaders brought with them, over and above the legions they deliberately murdered and enslaved as part of these invasions and conquests. The resulting vacuum was filled with the offspring of the Catholic and Christian invaders and those who had voluntarily or involuntarily converted to the conquering faith.

And what do we see today? A world where Christianity is the most widely practiced religion.[9] Can any thinking person possibly be surprised by this? After almost seven centuries of a powerful eugenics program, and several centuries of simultaneous enslavement and murder of native people in North and South America through both conquest and the spread of disease, the actions of Catholics and other Christians resulted in the deaths of tens of millions of non-Catholics and non-Christians and the ramming of the faith down the few remaining defeated survivors' throats.

How could it be otherwise that Christianity spread so far and wide? How could the now-globally distributed descendants of those Europeans and the conquered peoples not be strongly psychologically primed for belief in religion and also culturally driven toward Christianity?

As their Jesus said, he came not to bring peace, but a sword. His most devout fans strenuously followed his example for far too many painful and lethal centuries.

Only the apologists and the ignorant piously claim that it was the wondrous nature of the Christian message or the tailwind provided by their holy ghost which was responsible for the rise of Christianity, not the deaths of tens of millions of those who did not welcome or want it. It is said that the victors write the history books, and while this is true, Christians also constantly seek to rewrite history to favor their religion, including as many falsehoods and false narratives as needed to achieve their ends.

One never hears any sort of explanation by the faithful or their apologists for why their allegedly loving god felt it was such a good idea to allow the enslavement and death of tens of millions of innocent people by metal or biological swords wielded by the conquering armies and invading missionaries. Surely a humane and moral god would have cleansed the invaders of their diseases before they sickened and killed so many millions of women and children, no? Or perhaps the Christian god is indeed the murderous demon described in his holy book and an Americas-wide Black Death was their lord's method of clearing the way for the spread of the faith.

Returning to the issue of the proliferation of religions in general and the essentially zero probability of any of them having any validity whatsoever, another damning point is that many of the religions in practice today were dreamed up hundreds to thousands of years ago around campfires or other social gatherings by male primates who were less scientifically knowledgeable than the average First World nine year old is today.

I have to ask, in all seriousness, how an honest and thoughtful person can justify a belief in one specific faith once he finds out there are so many thousands of other faiths, not to mention the tens of thousands of sects and subgroups and splinter factions off of the more general belief systems. I didn't bother to make a picture with more than 30,000 dots on it and ask the believer to justify a selection of just that single dot, but I could have if I included all the variegated Christian sects.

It boggles the mind to consider that not only can such faith and confidence never be truly justified, but also the incredible lengths that a person has to go to in order to rationalize it. Add to the fact that even if one of those 30,000 faiths has the true phone number to the sky, the random adherent still only has almost no chance of picking the right one. And this is not to even say that any of the 30,000 has it right. As humanity moves along through time, inventing new faiths and creating new cults, that number may grow to 40,000, or 80,000, which makes the percentage game increasingly harder to win.

What if none of them are right? Surely none of the adherents could possibly know. They are convinced theirs is the one true faith based on their faith alone, and a great many of them even today feel it is their holy and god-given right and duty to oppress, persecute or kill other people if they don't believe in or if they mock the holy mother faith. How quickly the truth is exposed when one explores religion through the clarifying lenses of biology and Darwinism.

There is another odd facet of faith which the reasoning mind can only comprehend through a deep understanding of the evolution and fragility of the human brain and its extreme susceptibility to error and wishful thinking. How can it truly be that so-called revelations written down in ancient books and scrolls by only particular exceedingly ignorant male hairless apes are seen as real and true, while revelations received and proclaimed today by other exceedingly ignorant male hairless apes are indicative of mental disorder? What is the difference between the two?

According to monotheists, it is because the creator of the universe deliberately chose to only send down a few transmissions to a few people living in the same tiny geographic area long before the advent of imaging, communications and modern forensic and fact-checking technologies. It was only later, as our knowledge of reality began to really accumulate, that such so-called "revelations" became signs of insanity. Insanity, of course, along with ignorance and weakness, is the primary condition one is suffering from if one actually thinks any of these ancient stories are anything more than primitive mythology and superstition.

And how is it that this diagnosis of mental disorder or error is only applied to other people's religions and their gods? Within the narrow confines of a faith group, a believer can actually say that he or she communes with the creator of the universe or receives revelations from him, but the believer in another faith group in the house of worship next door is somehow completely wrong or insane when he claims the exact same experiences? This is the power of psychological projection and the unbridled arrogance which apologists and sincere believers all cling to so desperately and hypocritically.

Nonbelievers, particularly those who have any education and training beyond high school, and especially if that education was in the sciences, are denigrated as "arrogant" by the faithful when they point out the quite obvious errors and dangers of unjustified belief. Arrogance is not arriving at a position following hundreds of years of rigorous and objective scientific and empirical analysis, study and testing.

True unbridled and unapologetic arrogance is the purview of the person who *ignores* thousands of years of human history, study, advancement, information accumulation and constant ultra-rigorous testing and peer review, and claims to literally have The Answers to, in the immortal words of Douglas Adams, "Life, The Universe and Everything" because of voices someone heard in his head or because of things a few ignorant people wrote down thousands of years ago.

The arrogance of any believer to say that his single particular god and faith system is the One True Faith, without any actual evidence whatsoever and in the face of several thousand other faiths who themselves proclaim to be the One True Faith on an equally total lack of evidence, exceeds levels of arrogance and egotism which no rational person could truly comprehend, much less actually express in public.

Arrogance is the spawn of ignorance, and it results from unjustified and uninformed opinions and flaws in human psychology. Religious arrogance is the worst offender on earth in this regard, as it is both built on fables and deliberately shielded from reality by its adherents. Ignorance dissolves like a crystal of salt in the hot clarifying solution of science. Science eats away the accumulated accretions of ignorance and stupefaction which, like hardened deposits in arteries and water pipes, continue to clog and foul our minds and are always ready to completely cut off the life-giving sustenance of rationality our species has worked so hard to achieve. Thus continuing to believe in disproved or unproved claims is in most cases a deliberate decision to believe in fairy tales and all that comes with them.

Science is no more a contributor to arrogance than claiming to know that two plus two equals four demonstrates some sort of arrogance. This is the difference between science and religion. Science is built on observation, testing, critical thinking and empirical analysis, and is energized by both strong skepticism and a drive to know the truth. None of these are present in religion, despite the wishful thinking and rationalizations of its apologists.

Science is an objective method of implementing critical thinking in order to advance human understanding and our store of objective facts and information. Religion's perspective on this is to accuse the empirical person of arrogance because the empiricist refuses to allow the religionist to claim dogma and opinion as fact. It's as simple as that.

To the believer, it must be said that it does not matter how strongly you may feel your current belief to be true, or your pet god to be real. Your desperately wanting it to be true doesn't matter one tiny bit. Just as the number "1" cannot be larger than itself no matter how much it might want to be, your belief in a single faith or god only counts as one among an embarrassingly large number of competitors, and it is on the exact same level of probability as all the rest.

Jorge Mario Bergoglio, who at the time of writing is serving as the Bishop of Rome and the sovereign head of the theocratic state known as Vatican City, and who also goes by the title and nom de plume Pope Francis, is the same 99.999% atheist as any random but extremely devout resident of Alabama, Afghanistan or Addis Ababa. Mr. Bergoglio doesn't have any more of a clue as to the real existence of anything at all beyond the world of nature than does a tadpole, although he may indeed be among the most devout and sincerely faithful believers on earth.

Mr. Bergoglio and that tadpole are equally ignorant of any afterlife, which of the untold thousands of gods believed to be floating around could possibly be real, the origins of the universe and of life on our planet, a true nature and science of morality, or indeed of anything at all which our godless science has not demonstrated or posited.

To sum all this up, there are two major differences between believers and nonbelievers.

1. Nonbelievers reject all of the uncountable thousands of gods and religions from our past, present and future, while believers reject all of the very same gods and religions except one. You can call these nonbelievers atheists or agnostics; for all practical purposes here these terms can be considered synonymous. Nonbelievers bow to reason and declare their atheism or at least agnosticism on theistic matters as a whole, while believers refuse to acknowledge their actual atheism on all gods except those on the microscopic sliver of the theomagnetic spectrum which they deliberately choose to see and accept. Believers swallow the dogmas of their religion and employ whatever cognitive dissonance and denial is needed to maintain their faith.

2. Nonbelievers understand that unfounded beliefs about unprovable entities should be excluded from the public sphere and that only reason and empirical truth should inform actions and policy. Believers often go to shamelessly extreme lengths regardless of the legal, ethical, moral, truth or religious violations and abuses they perpetrate to both openly and covertly inject their belief system into that public sphere and into the lives of those who do not share their belief.

So every past, present and future human is actually an atheist both in principle and in practice. That fact has to be repeated until it is second nature to all humanity. I'll do it right here. Every past, present and future human is an atheist both in principle and in practice. Bear in mind that it is only by taking and understanding another's perspective that a person can truly understand how his positions look to others.

When the religious community, from the theocratic lunatic wingnut to the almost-secularist, can individually and collectively see themselves from the outside and truly see their faith as what it is, which is a specific system of mythology that a person deliberately chooses to believe without evidence and thus needs to be treated on an equal par with leprechauns and tree sprites, we will have made a massive step toward actual coexistence.

Religion was the first instrument of mass control and the first weapon of mass destruction, defined as human creations put to use to control or kill large numbers of people. Not only does religion still continue its millennia-long role of serving as justification for the mass destruction of humans and civil societies, but it has also inflicted mass destruction on much of the species' critical thinking capacity. Mass destruction of rationality invariably leads to the mass destruction of humanity.

Religion is like a machine gun, in that both are extraordinarily dangerous man-made weapons that can kill a great many people in a very short time. Machine guns should only be handled rarely, briefly and with extreme care by a few highly trained specialists. Machine guns can be allowed to exist, but only under extremely tight security, as they are never safe or appropriate for general use.

Religions, not being controllable or securable, lacking a safety switch and with an infinite amount of quite deadly and unintelligent ammunition, are far more dangerous than machine guns. When an irresponsible and/or immoral person or state picks up a machine gun or a religion and points it at other people and starts firing, only horrible things result.

To restate my thesis, the faithful and nonbelievers are essentially all atheists, and we can potentially coexist peacefully in a decent society, but only with strict conditions. We can coexist only if religion is defanged, declawed and relegated to its appropriate station alongside astrology, alchemy and the reading of tea leaves and animal guts, and it is allowed to have no more impact on the real world than a five cent sticker from a bubble gum machine.

I am sure that the religious will be dually affronted by such a statement. They will feel personally aggrieved that an assault on their own individual character has taken place, and they will feel that their religion has been denigrated by the assertion that it is unable to peacefully live alongside others.

To both of these affronts I say that is frankly too damn bad. I would also very much love to be able to say it is not my problem, but it actually is my problem, and everyone else's as well. The blood-soaked history of religion and its legions of hateful and murderous followers throughout time, and in far too many cases still to this day, emphatically prove the point. Religion has proven itself to be incapable of peacefully coexisting with decent and moral society unless it has been wholly divorced from any substantive role. Humanity has far too many examples proving that the more faithful and fundamentalist a person becomes, the more dangerous he becomes.

And since the barbaric acts perpetrated by religions have been carried out by those possessed of the most faithful adherence to the words of their religions, the religious themselves stand not on the shoulders of giants, but on the shoulders of highly devout murderers, torturers and conquerers. Religion possesses an enormous criminal rap sheet built up over thousands of years of crimes against humanity.

So the faithful have only themselves and almost the entire history of religion to blame. Call it original sin if you like and if that makes the diagnosis easier to understand and the medicine easier to swallow. Your holy books have not changed in centuries to millennia, and because of this fact they continue their permanent roles as enablers for evils both great and small.

You may as an individual be a truly humanistic and moral person who happens to be religious, but if you are a monotheist you absolutely do not do this by adhering to the teachings and words of your faith. You do so by ignoring almost the entire canon and focusing on a few small bits of actual goodness, in the form of fleeting microscopic flecks of gold panned from rivers of blood, hate and evil.

Take Christianity as an example. If you say you are a Christian, you have a choice in how "Christian" a manner you live your life. You can either be a truly devout and faithful Christian and follow your holy book's words exactly as written, or you will be a nonliteralist who flicks your god a stiff middle finger and deems yourself able to pick and choose which of his divine words to ignore and which ones to follow.

If you are in the first group, you are literalist who is likely a truly devout person and true to your religion and your books. In other words, you are a true believer. You are therefore also a sociopath or a psychopath, a threat to decent society, a criminal in the making, someone whose core is too shallow to be an independent thinker, and someone who uses the crutch of hateful literary fossils for a moral compass. Therefore the rest of us and especially the police need to keep a close eye on you because you are definitely going to do something ignorant, stupid, immoral and/or evil.

If you are in the second group, you are not a literalist so you have a nearly infinite range of positions to take. You can ignore only a single word in the canon, you can reject all of them except only a very few, or you can take any position in between those extremes. If you are a Christian, you can try to apologize for the demonic deity of your holy book by saying that the vast majority of it is metaphor, or that the allegedly inspired word was somehow misinterpreted throughout most of the book by the so-called 'fallen' human vessels of transmission.

You can make category choices, meaning you can choose to ignore all of the parts of the book that support or order the faithful to acts of genocide, but you can choose to follow the parts which support slavery, bigotry, misogyny, genocide and persecution of those of other faiths, the nonreligious and other persons such as the LGBT community. In other words, you can be like today's American evangelicals and far too many Christians throughout the history of the U.S. and even unto this day, who double down on human meanness and choose to elevate religious bigotry and hate to the top ranks of your particular faith.

If you are a believer, I can tell within two minutes what kind of person you are by which holy book verses you follow most closely, and how much of a literalist you are about your preferred book. Invariably, the more evil a person is, the more his or her focus is on the evil verses in their holy book and the stronger his tilt toward literalism.

Those who are essentially good people and de facto secular humanists, atheists or agnostics but for one reason or another can't fully break away from faith, choose to follow only the most humanistic words in their canon and prefer metaphor to literalism. If you hate and/or mistreat gay people and nonbelievers because that is written in some Christian, Jewish or Muslim holy passage, you are an evil person.

If you only pray by yourself in private, quietly and secretly donate massive amounts of time and resources to helping the less fortunate, think it is easier for a rich man to go through the eye of a needle than go to heaven and you act accordingly, and you interact with all people based on a strict interpretation of the Golden Rule, you are highly likely to be an extremely good person. As an aside, if you think you are a devotee of Christianity but don't know the references contained in that last sentence, or even worse if they don't even sound familiar to you, you are terminally uninformed about the faith you profess, and are, in the newly popular term, a total fake.

If you call yourself a believer but don't know any verses yet still call yourself a devout monotheist, you've proven two things. One is that you are almost certainly an American, and the other is that your attitude must be ascertained through inquiries about your positions on moral issues. Since you don't know any of the scripture you say you adhere to, we can ask the questions from a different perspective.

- Do you think those of other faiths are evil or immoral? If so, specifically why, and if not, why not?

- Do you believe in granting equal rights and treatment to all humans regardless of race, creed, ethnicity, orientation, gender, belief system or lack thereof, country of origin, language or other similar characteristic? Again, if so, specifically why, and if not, why not?

- Should any religion or faith group be able to impose any of its particular system and beliefs on anyone else at any time or in any place for any reason? Yet again…if so, specifically why, and if not, why not?

No matter the person or his or her beliefs, this character assessment process is very easy to do, and you can do this yourself with a little bit of practice.

The entire problem with holy books and religion is how people use them. There are people who bull through the cognitive dissonance of their books and faiths like an icebreaker through the rapidly melting Arctic, ramming through all the evils they contain without following them, and focusing only on what appeals to them and their innate sense of humanistic morality. These people are usually the better types of people and there is generally little problem with this category of believer.

Far across the moral divide are the people who essentially scrub their canon and/or the more thuggish and bigoted attitudes of their clan for justification to do and say evil things to other people. This category of believer is an outright danger to civil society, and its ranks are well-populated in America, too many places in the Islamic world and even in Europe, and their numbers are growing by the day.

It's one thing to be an ass and a bigot because that's just who you are, without any other justification than that's just how you are. You may be a thoroughly immoral and bad person with immoral and evil opinions and beliefs, but if you try to translate them into actions, other humans and decent secular laws have a tendency to bring such things to heel. It's quite another thing to insist that you are being an ass and a bigot because you say the creator of the universe ordered you to be that way.

In those cases, no amount of actual facts or common sense will penetrate the force field of religious dogmatic bigotry you have thrown up around your brain, and any act of evil whatsoever is not just OK to you, it's seen as defying your god if you don't follow that alleged guidance. And these people are absolutely everywhere and their ranks are growing, particularly in those places where fascists or fascist wannabes have gained secular power.

Today the Middle East has the Taliban and ISIS. The United States has its equivalents and wannabes in the form of evangelicals and similar groups, and they currently number in the millions. This is why religion has to be kept in an inescapable box, because as I have said, once people think they have their god on their side, nothing is sacred. These people do and say things that are immoral and antihuman.

And we secularists can never tell where a person truly falls on the evil or good spectrum, because that may vary from issue to issue even within the same person. You may say that a woman need not be the slave of her husband but think gay people are immoral and should be castigated, if not outright castrated, and also denied basic human rights and dignity. You may have a mix of evil and not-so-evil positions, but as an aggregate you are as evil as the most evil position you hold.

One major problem we on the nonreligious side have is that every time we discuss these matters with a believer, we end up with having to figure out exactly where on the nearly infinite theological spectrum the person is and which of the many, many thousands of orders and proscriptions they choose to follow and which ones they choose to ignore. Like snowflakes, no two believers are alike, and the range of actual beliefs even within a single congregation or a single brain is rather wide.

We ask if the person believes in the Trinity, or if that is just a Catholic thing, or if it is "truly, really and substantially contained"[10] as a belief by every single member of a Catholic congregation. Should people follow every word in the book, or is that only done by the fundamentalist literalists? If you're a biblical literalist, why aren't you trying to kill me for trying to talk you ought of your belief, as that book specifically commands? Et cetera…and ad nauseam and ad infinitum as well.

And there is a perversity to arguing with someone whose mind has been captured by religion. During a discussion with such persons, literally anything at all can be construed as more evidence and confirmation of the belief. So when a nonbeliever lays out incontrovertible evidence refuting the logic and morality of a specific belief which the believer does *not* share, the believer actually and perversely feels more confident of his *own* beliefs because of course *he* would never actually believe something like that, as *he* is too intelligent and perceptive for that.

Such flawed thinking is described in the discussion in Chapter Twelve on logical fallacies and cognitive biases, along with how these mental minefields both enable and help perpetuate religion. In laymen's terms such a discussion is like when a person figuratively, or perhaps even literally, sticks his fingers in his ears and shouts "La la la la la I'm not listening!!" Or, if the person actually does believe that which is being refuted, the believer will devolve back to taking it as an article of faith, or just more proof of 'the unfathomable mystery of his god' or some other escape hatch from logic which blocks the potential penetration of dangerous facts and reason.

In other words, the fragility of both the religious mind and the religious belief is such that in order to protect them from questioning, damage or defeat, they are pulled like a turtle's head back into a dark corner of the brain and sealed off like an armored safe. So in both senses, the religious mind has been so thoroughly captured by bias, rationalization and motivated reasoning that irrefutable logic and empirical evidence have not merely no effect but actually, and perversely, increase the thickness of the armor around the belief and the strength of the belief.

This major defect in human thought is currently only partially understood and requires much more analysis and a much deeper understanding of this element of human psychology, both in terms of patterns of thought and the ways in which the actual electro-chemical workings of the brain operate to retain belief and wall off reason. I suspect there are numerous reasons for this type of faulty thinking, and I will discuss them later, but I feel comfortable in stating they are due to the inability to face the social and internal mental consequences of losing faith. Losing an argument is one thing; you can recover from that. Losing the afterlife or one's social group is quite another; such a loss is unrecoverable and many people will do anything to avoid it.

So since religion and the religious can never be counted upon to be fully moral as a body or as individual practitioners, religion has to be both excluded from the public sphere as it is in the fully secular American Constitution as described in great detail in Part Six, and also from any discussions of morality. I do sympathize with the few truly moral believers who are angry at the lunatics, bigots and fanatics and don't like being lumped in with them, but that's not the nonbelievers' problem. It's the believers' problem. All believers in religions and gods share an unbreakable bond and they all sprout from the very same root, in that they all believe in things without evidence, which is the crux of the entire problem.

It all boils down to that single fact, that all of the faithful hold beliefs without evidence. That commonality is an irrevocable bond which all believers will always share with every other believer of any faith or practice whatsoever, which is why all believers have to be considered as a single homogenous cohort. You're either a believer or you're not, and again it's a binary situation. And since beliefs shape actions, all of humanity must always be watchful for the inevitable evil which believers will perpetrate.

Further, all those of faith believe that the particular creator of the universe which they believe in gave their particular sect the one true and real set of instructions, but the nonliterals just choose to ignore a lot of it.

Thus all believers are *again* irrevocably linked, this time by their holy books. And they will remain linked in this way until they give up the belief in the imaginary, get rid of the holy books and become truly moral. Further, when those in the not-lunatic / not-bigot religious factions someday finally take the action as a collective body to extremely loudly and continually reject the lunatics and bigots and their attitudes and actions, that will be a much-needed first step toward reconciliation.

But that doesn't mean we will ever forget the maxim of absolute power corrupting absolutely. Therefore god, having absolute power, is absolutely corrupt, and so are the god's followers. Anyone who 'truly, really, and substantially' thinks his god is on his side can never be trusted with the safety and welfare of other humans.

I must also clarify that this state of détente can be eased, and that there can indeed be coexistence, but it will have to be on the level of supervised visitation. Indeed we must coexist until the human species outgrows religion. Why not outgrow science instead? The answers to that will be obvious to anyone with even a partially functioning brain and I will not waste time on them.

When the religious leave their homes and cross into the secular world the rest of us inhabit, they need to be closely watched to ensure that they stay in their boxes and don't try to inflict mental or physical harm on anyone or on truly humanistic customs, ethics, morals and the law. We secularists know that the frailties of the human mind mean that religion will probably be with us for thousands more years and perhaps even thousands more generations, so we have to work together until our species matures and achieves a "blessed" state of full freedom from religion.

But this process of coexistence until full excision of superstition has to be in clearly defined lanes. For example, if believers think we secularists are going to stand in the way of religious people doing actual and honest good for others merely because those actions happen to be religiously motivated, they are completely and sorely mistaken. We wholeheartedly and happily welcome this and we will be happy to partner in and potentially even materially support your work in improving the well-being of our fellow humans, animals and our shared environment.

Just don't ever ask us to join your religious rites, don't ever seek special treatment like tax deductions, and don't ever try to gain any special accommodations for or a place in our law for your unfounded beliefs, regardless of how sincerely held.

Note also that we will indeed fight for our Constitutional and human rights against any and all attempts by the faithful to take them away.

The faithful continue these efforts every single day in ways both overt and covert, as with the endless intelligent design, religious freedom and other science-denying, Trojan Horse and other fifth columnist schemes.

If the faithful keep pushing that religious accommodation and religious freedom nonsense, I will form a church based on my extremely deeply and sincerely held, and unlike with all religions an empirically provable, belief that bigots are assholes. I will note how I have had a divine inspiration straight from the Master Worm from Saturn that we on Earth need to start a faith-based campaign to prevent bigots from receiving any service, support or rights which they want to deny to homosexuals, racial and ethnic minorities, women, secularists or any other people these religious bigots don't like.

We'll see how the Supreme Court handles a rush of cases where Pastafarians, Wormites and others complain that their religious freedom to not serve or support religiously-inspired bigots was impeded.

To close out this chapter, all one really need take away is five points:

1. All humans are atheists in principle and in practice. This includes the religious, so you need to just get over yourselves.

2. There is no spirit or supernatural world, and even if there were one, there is absolutely no way to determine which one is the real one, or even if all of them are actually real.

3. Religion's ages-long history of immorality has proven that it is only compatible with civil society to the extent that it is neutered and watered down to the level of a homeopathic elixir; that is to say there is no trace of the original ingredient, and it has no role in civil society.

4. As long as people cling to religion, the religious and the secular will need to work together for the common good of humanity, with critical thinking and empiricism always having the upper hand over unjustified opinion and unsubstantiated belief.

5. The secular will need to assist the religious in what will be a generations-long maturation from religiosity to empiricism, through blockage of religious bigotry, blockage of religious corruption of government and education, and helping to provide a mental paradigm which is acceptable to the future "faithful" despite having no imaginary divine caretaker or afterlife to comfort their fears.

Chapter Twelve

How Psychology Subverts Reason And Opens Pandora's Box of Faith

"But it's sad, people say, to be deceived. Not at all, it's far sadder ***not*** *to be deceived. They're quite wrong if they think man's happiness depends on actual facts; it depends on his opinions."*
- Desiderius Erasmus

Most people think they are pretty smart. They think they are wise to the ways of the world and that their powers of perception are untainted by bias. They think they possess strong analytical skills which they sometimes admit may occasionally and in extreme cases be aided by a tiny bit of research. They think their current levels of knowledge, "common sense" and what passes for their critical thinking skills are sufficient to reach a clear and objective conclusion.

Most people are wrong.

What is worse, they aren't just slightly wrong, they are wrong to an astounding degree. Most people know essentially nothing about how the brain actually operates, from its biological framework to its psychological inputs and outputs to its internal processing. They also often know very little about the subjects on which they offer partially to wholly uninformed opinions. Despite this comprehensive lack of knowledge, they feel no embarrassment or hesitation about holding the opinions they do, regardless of how ignorant, uninformed or incorrect the opinions actually are.

And to paraphrase Haley Joel Osment's character from the movie *The Sixth Sense*[1], wrong people are all around us, they only see what they want to see, they only think what they want to think and they don't even know they're wrong. This has obvious consequences when people engage in a discussion on topics of any degree of importance. Even when the subject is something as inconsequential and ephemeral as the performance of the hometown sports team, people have become so animated in their defense of that team that they have actually assaulted and killed people over such trivialities.

When religion is criticized, and I am here speaking primarily about the monotheisms and particularly Christianity and Islam, the faithful still frequently respond the way they have since these religions were invented.

Only minutes elapse between criticism of these two faiths and the arrival of ad hominem attacks and death threats, with the most strident and unhinged threats being delivered by the most faithful and most sincere believers. In many places and times, more permanent and painful responses such as imprisonment, physical assaults and public murder of the criticizer are often the norm. Such unhinged lunacy is not some bygone aspect of the history of these religions. It happens all across our world today.

Try going to just about anywhere in the Middle East and publicly burning a copy of Islam's holy book, the Hadith or a caricature cartoon drawing of Mohammed and see what happens to you. The more deep and sincere the believer's faith, the more likely he will be to murder you. And in a world in which members of the two most widely held religions, Christianity and Islam, continue to threaten, attack, imprison and kill unbelievers, apostates and those of other faiths, we have a major problem.

Lest any overly smug Christians try to comfort themselves with delusions that Islam is the only religion with vast numbers of devotees who still act like this in the 21st century, all one need do is travel to any of the politically conservative states in America's south and Midwest and publicly criticize Christianity, any of its main characters or its teachings.

Most people are painfully aware of the typically murderous and destructive Islamic response to attacks on its faith because recent history is overflowing with examples of this. If you think Christians won't react similarly, try going to a public park or city street in Alabama, Texas or Mississippi and burning Jesus in effigy at the stake with a pile of bibles and consecrated hosts as kindling and see what happens to you. You will probably survive but you won't escape unscathed, and you're highly likely to stare down a gun barrel or fifty before you get away.

If burning mythological figures and bits of their reincarnated bodies in effigy isn't your thing, try loudly and publicly demanding actual evidence of the virgin birth, the resurrection, the Noachian flood, communications from the dead or the divine, or indeed of any miracles at all. Loudly proclaim that beliefs in such ideas are unsupported and rather silly childish dogma and that religion should be forever and permanently excluded from state activities.

In short order, believers from presidents and vice presidents to state and local politicians to televangelists to right wing media talking heads to clergy of all flavors and colors to mobs of Joe and Jane Six-packs (beer, not abs) will respond with horror at the 'grave offense' to their beliefs and will utter loud calls to rally the Christian Soldiers onward to defend the faith against all enemies, both freethinking and didactic.

As part of their desire to defend a long-dead semidivine Jew from the slings and arrows of outrageous logic and reason, it is a stone cold, rock solid and absolute certainty that some number of deeply and sincerely devout Christians, as previously noted, will deliver death threats to the "offender" and/or his family electronically, telephonically, via snail mail, package or in person.

This has always been the case with god's children, for the Christian lord their god hath repeatedly commandeth his followers in quite direct and unmistakable language to sticketh through with a sword or smasheth with stones anyone who questioneth their religion.

For those who are not familiar with that godly order to kill, it is in Deuteronomy 13 (KJV). I have added the bolding to the quote below:

> "*{13:6} If thy brother, the son of thy mother, or thy son, or thy daughter, or the wife of thy bosom, or thy friend, which [is] as thine own soul, entice thee secretly, saying, Let us go and serve other gods, which thou hast not known, thou, nor thy fathers; {13:7} [Namely,] of the gods of the people which [are] round about you, nigh unto thee, or far off from thee, from the [one] end of the earth even unto the [other] end of the earth; {13:8} Thou shalt not consent unto him, nor hearken unto him; neither shall thine eye pity him, neither shalt thou spare, neither shalt thou conceal him: {13:9}* ***But thou shalt surely kill him****; thine hand shall be first upon him to put him to death, and afterwards the hand of all the people. {13:10} And thou shalt stone him with stones, that he die; because he hath sought to thrust thee away from the LORD thy God...*"

This of course comes only one biblical page after an earlier set of equally horrific divine orders which in the space of a few short sentences exposes the utter evil at the core of the monotheisms. The monotheistic god specifically tells his followers not to do what he specifically tells them to do in many other parts of the book. He gives conflicting orders, and the vast majority tell the faithful to displace, harm or kill other people. Even when given a way out of the unparalleled cognitive dissonance the monotheistic books engender in people who are even remotely moral, the default is always to do evil to other humans.

Earlier in Deuteronomy, the order not to kill was buried in a large section of the book from which the Christian faithful derive their so-called ten commandments, the very mention of which fills their heads with insufferable arrogance and presumption. In Deuteronomy, the faithful are told their god will evict innocent people from their own land so the children of Israel can invade and occupy it. Again I have added bolding to some of the text for emphasis:

> "{11:22} *For if ye shall diligently keep all these commandments which I command you, to do them, to love the LORD your God, to walk in all his ways, and to cleave unto him; {11:23}* ***Then will the LORD drive out all these nations from before you, and ye shall possess greater nations and mightier than yourselves****. {11:24} Every place whereon the soles of your feet shall tread shall be yours: from the wilderness and Lebanon, from the river, the river Euphrates, even unto the uttermost sea shall your coast be.*"

In the space of just a few pages, the faithful have been told to follow their god's own divine commandments, including one which specifically orders people not to kill (or do murder), then they are told to kill their actual brother, son, daughter, wife or best friend quite brutally and completely without warning. As with so many shocking things in religion, this bears repeating. The faithful are told to immediately murder not invading armies, not strangers, but one's most beloved family members. You will note the order is of course addressed to men only, as husbands are not listed in those to be killed.

The faithful are told that their only recourse when being asked to worship other gods, the very first action they must take along with every other member of the tribe, is to immediately grab a sword or a stone and do their lord's evil bidding in order to ensure the miscreant wife, son or daughter is aggressively and thoroughly murdered. And their god will all the while be busily driving out innocent people from their villages so the children of Israel can have as much *lebensraum* as they want.

This sort of lunacy actually is quite consistent within the book, because this self-admittedly hyperjealous god has plenty to say on the relative importance of worshiping other gods versus killing one's own family. This is of course an activity at which this god is an old hand and at which he will repeat his prowess in spectacular fashion in the New Testament, often euphemistically described by apologists as "*Book II; The First One Was All A Big Mistake*". Deuteronomy's commandments section, as opposed to the other set of commandments found in Exodus 20, or the third one in Exodus 34, says the following on these two activities:

On worshiping other gods, the guidance is:

> "*{5:7} Thou shalt have none other gods before me. {5:8} Thou shalt not make thee [any] graven image, [or] any likeness [of any thing] that [is] in heaven above, or that [is] in the earth beneath, or that [is] in the*

> *waters beneath the earth: {5:9} Thou shalt not bow down thyself unto them, nor serve them:* ***for I the LORD thy God [am] a jealous God, visiting the iniquity of the fathers upon the children unto the third and fourth [generation] of them that hate me,*** *{5:10} And shewing mercy unto thousands of them that love me and keep my commandments.*"

A full 98 words are committed to the orders and to describing just how galactically and generationally pissed off and vengeful the god will be if the no-other-gods order is violated and how badly he will punish offenders and their descendants.

On killing other human beings, all you get is "*{5:17} Thou shalt not kill.*" That's it. Four words.

The apologist may try to say that the ratio of 98 to 4 is because the order not to kill is so simple. The realist will yet again note the continuous apologist denials of history and reason and say several things:

- Why all the frothing threats of just how badly the god will react to the faithful worshiping other gods, but there is none of this for those who would kill innocent humans who were allegedly made in the very image of the creator himself? It is clear this god actually has a perverted and psychopathic fetish for doing a lot of killing and for watching his own created people kill each other.

- Those who worship other gods are not only punished themselves, but their great-great grandchildren are punished too. These kids would not even in most cases be born until long after the original offender had died, and possibly as long as 70 years after the original "offense". This would make the ratio something like 392 to 4 (including each generation), and that killing is only 1% as important as whoring after other gods. This off-the-charts immorality is quite consistent with the biblical god's insane and murderous words and deeds.

- What we see is that this god really does think that murdering innocent people is about 1% as important a matter as his jealousy. This is borne out by all the commands to not only commit murder of strangers and beloved family members, but also to commit full-scale genocide.

- Why is there no explanation in any of the sets of commandments like "*thou shalt not kill or do murder, except in these extremely specific and unfortunate circumstances, but then it's perfectly ok*"? Even that bit of clarification would help matters tremendously.

- And why haven't the faithful been paralyzed over the conflicting guidance to kill and not to kill? 'To kill or not to kill, that is the question'. The faithful certainly have done absolutely massive amounts of killing in the name and service of their god, following his guidance despite and in the face of his own commandment which specifically orders them not to kill. Their god tells them to kill on many pages of their bible, then in a few snippets it tells them not to kill, and even then only in extremely few words with no explanation. One must assume that the sheer tonnage and volume of orders to kill have swamped the few isolated orders not to kill, much like the 392 to 4 ratio mentioned above but much worse in the direction of psychopathy.

- The truth is that the faithful often don't give a holy damn about any of those commandments or any of Jesus's teachings or place them in any special hierarchy or priority, except where and when it suits their personal preferences and goals. This is painfully obvious to the secular observer, but the faithful refuse to admit it. The disgusting and self-serving hypocrisy of nearly every self-professed Christian believer is absolutely appalling, and this is most glaring in America.

That last one applies to every bit of guidance the Christian bible offers. The commandments not to bear false witness, to steal or covet or commit adultery are routinely violated by adherents from allegedly infallible popes to priests to politicians to presidents to people on the street, whenever they feel like the desired ends justify the need to violate them. The same goes for the teachings of Jesus. I will mention this many times, that most self-professed Christians in America today act like Jesus's teachings were things they should ***not*** do as opposed to very specific and direct things he said they definitely *should* do.

This is not a new problem. Mark Twain noted this hypocrisy a very long time ago when he said "*If Christ were here there is one thing he would not be - a Christian.*" Since the beginning of their religion, Christians have used their holy book far, far more often as a justification for being inhuman to other people than as a reason to be good to them.

Earlier I said the god of the bible told his followers to commit rape. The evidence of that is in Numbers 31:1 and 31:2, when the allegedly loving but actually insanely jealous and angry god tells his henchman Moses to commit mass murder on the Midianites. Moses, being a true believer and not so insufferably and arrogantly full of himself that he felt he should see his god's words as metaphor or requiring interpretation or being anything else besides the actual words, carried out the orders with great gusto. He and his men slew kings and commoners and did it so well there were no Midianite men left, since the book notes how they "*…slew all the males.*"

Along with killing every single Midianite man, the Israelites took plenty of captives and booty and left total destruction and scorched earth in their wake. The incident is described in some detail in the following quote from the King James version, again with bolding added for emphasis:

> "*{31:9}* ***And the children of Israel took [all] the women of Midian captives, and their little ones****, and took the spoil of all their cattle, and all their flocks, and all their goods. {31:10} And they burnt all their cities wherein they dwelt, and all their goodly castles, with fire. {31:11} And they took all the spoil, and all the prey, [both] of men and of beasts. {31:12} And they brought the captives, and the prey, and the spoil, unto Moses, and Eleazar the priest, and unto the congregation of the children of Israel, unto the camp at the plains of Moab, which [are] by Jordan [near] Jericho. {31:13} And Moses, and Eleazar the priest, and all the princes of the congregation, went forth to meet them without the camp. {31:14} And Moses was wroth with the officers of the host, [with] the captains over thousands, and captains over hundreds, which came from the battle. {31:15}* ***And Moses said unto them****, Have ye saved all the women alive? {31:16} Behold, these caused the children of Israel, through the counsel of Balaam, to commit trespass against the LORD in the matter of Peor, and there was a plague among the congregation of the LORD. {31:17}* ***Now therefore kill every male among the little ones, and kill every woman that hath known man by lying with him. {31:18} But all the women children, that have not known a man by lying with him, keep alive for yourselves***."

So the god of the bible and his *capo di tutti capi* on earth were hand-clappingly gleeful about killing thousands of Midianite men during the actual invasion. That was bad enough, but then after deliberate thought and in a display of full premeditated Hitlerian genocidal psychopathy, the lord's own soldiers followed their fuhrer's bidding, pulled their swords back out, which were still dripping wet with the blood of the Midianite men, and killed all the captured little boys, killed all of the non-virgin females including many mothers, and took the virgin women and girls as sex slaves and concubines.

It is one thing to read such unadulterated and absolute evil and be horrified and repulsed by it, particularly considering that these acts of pure evil are taken straight from the holy book of the god who is worshiped by over two billion humans today. It is quite another to pull back and think it through at the individual level and see it through the eyes of one of the victims.

Imagine you are one of the crying and terrified women and male and female children who are being held captive after seeing all of your fathers, grandfathers, husbands, older brothers and uncles ruthlessly murdered by an invading army. Your houses are on fire, your property has been stolen, even your animals are now captive with you, and you are huddling with the other survivors absolutely terrified out of your mind. What can you as a mother, shivering with fear and having suffered the worst possible disaster, possibly say to your terrified children that can be of any comfort? All you can say is that even though all the fathers and grandfathers have been killed and the future holds no hope at all except to survive a life of slavery under horrible people in a strange place, at least you as their mother will be there for the children.

And then the crusading stormtroopers come back and start pulling mothers and women and boys out of the crying and screaming mass of prisoners and begin executing them in cold blood, on the orders of their prophet and his god. The murderers from Jehovah's army either begin a small-scale inquisition of the young women and girls to separate the virgin females from the "defiled", or more likely they took full advantage of the situation to sadistically "finger-test" all the girls to find the 'tokens of their virginity'. Those who were "felt" not to have these tokens were murdered alongside their mothers, sisters and little brothers.

The only survivors of these multiple rounds of deliberate murder are young virgin girls, from teenagers to terrified two-year-olds. These girls would have in short order seen the total destruction of their village, the murder of its men and their own beloved fathers, then had their mothers, older sisters and all their little brothers ripped out of their desperately clinging arms and brutally murdered in front of them, with many of the older girls only being murdered following at least digital rape by the murderers. This is as evil an act as is possible to inflict on an innocent person, and it all happened on the orders of the biblical god and his prophet.

Even if the apologist screams that this horrifically Hitleresque psychopathic story is some sort of metaphor or allegory, I dare any mentally and morally undamaged human being to come up with any positive meaning or implication behind the event whatsoever, or to deny its craven and total evil.

Note that what happened to the survivors was far, far worse than rape, although future rape of the young and innocent was clearly being ordered and supported here, from the god on down. Many survivors of acts of human brutality, from genocide to rape, have at least some sort of support structure to lean on while they are recovering from the trauma of being victimized, brutalized and raped. These young Midianite girls who were raped by the sons of Israel after the double rounds of Israeli genocide had no fathers, mothers, husbands, brothers or anyone else to go to for help.

They had no courts, no villages, no communities to protect them and help them recover. They were prisoners of war and sex slaves for evil men doing the will of their evil god and his evil "made man" on earth, and they had no one to turn to but their fellow genocide survivors and rape victims, most of whom were likely less than ten years old and themselves in desperate need of support.

And of course some significant number of the women who were murdered for the crime of having sex with a man were pregnant, in accordance with the god's command to be fruitful and multiply. Therefore the most holy Moses ordered his men to commit not only a second round of mass murder but also a mass abortion as a bonus. And the lord their god had absolutely nothing to say about it, therefore he was clearly extremely happy with it. So yet again in this book of unadulterated evil we add one more mass abortion event to the list of mass abortions committed by or approved by the god which Christians worship so slavishly and sadomasochistically.

This is the god which billions of Christians worship. This is the guidance taken straight from that book of absolute evil they unbelievably call a "good book". This is why when they say they have the source of absolute morality we freethinkers and secularists recoil in disgust and horror. And they wonder why we nonbelievers want them to keep their evil religion and their damn book quite the hell out of our state and school business and extremely far from children.

And we can immediately dismiss any desperate apologetics by the faithful when they try to claim that any of these or any other words in their holy book weren't actually dictated by the god above, but came from some character in the book like Moses and are therefore not actually a dictum of the divine. They keep compounding their own mountains of evil and error by making such statements.

The fact is that if any scribe or prophet in the history of the planet ever got even a single word, phrase or quotation wrong anywhere in that entire collection of stories that were collated into the Christian bibles, the reason the errors are still in there today is because the god of that book was and is perfectly OK with all those errors being included.

If the Christian god has no problem with thousands years of his one and only holy book being hopelessly garbled and full of errors and misquotations, and full of evil so demonic that it that would put Satan to shame, despite having all that time to make or "inspire" some badly needed corrections, then that is 100% his own fault. And, more to the point especially given his reputation for omniscience and omnipotence, it is therefore 100% his deliberate choice.

Either way, if one is a true believer, everything in the book should be taken as its god's direct or approved guidance and followed to the letter, or else the believer has risen to levels of arrogance only dreamed about by freethinkers, because the believer will then have to justify how and why he personally decided what his god got wrong and therefore should be ignored.

If I actually believed in a god as evil and vindictive as Jehovah I can't imagine having the stones to tell him he got *anything* wrong, because your punishment for defiance could be being turned into a pillar of salt, any or all of the tortures inflicted on Job, being divinely carpet-bombed, and/or of course an eternity of unmitigated full-body, nerve-destroying agony in a torture chamber of fire customized and personalized for each individual. It would be far better to be locked up for life or receive capital punishment while still alive on earth than to fail to follow the divine orders to rape and kill.

And again, one more time, and one hopes this is finally clear now, we can once and for all reject the pitiful attempt at diversion known as claiming the words somehow have to be "interpreted" or seen as metaphor. This is ridiculous and must be strongly called out for the utter foolishness it is. Christians say their bible contains the literal or inspired word of the Christian god or gods. It forms the foundation of the faith, and they take quite a lot of it at face value when they want to cherry-pick it to justify their actions.

We again return to the astronomical arrogance of the believer who says that he personally knows the mind of his god so unbelievably well that he can personally say which words from the holy book are to be taken as literal, which ones are metaphor or allegory and which ones can be safely ignored. Again, this god has had thousands of years to offer clarifying guidance on this subject and has done and said absolutely nothing.

Of course this is all nonsense. Nothing in the Christian or any other so-called holy book should be taken as anything but the ranting of ignorant and often aggressively hateful and disturbed men. The Christian bible in particular is hopelessly garbled, is full of errors and is nothing to take seriously except as ancient literature and a window into the demonic psychology of many of the authors and the barbarity of their time, and its use by millions of people to harm and murder innocent people.

All this ancient religious literature is just snuff fantasy written by people who knew nothing about the realities of life and our universe. There are indeed some truly useful things we can take from it, but all of them put together could fit on a handful of pages. The rest is nonsense, science fiction, historical fiction, mythology or evil.

And to dismiss yet another ridiculous attempt by empty-headed apologetics who in a last ditch effort will claim that the Old Testament was indeed quite bad but 'Jesus came along and made it all better', a quick examination of this claim will expose its complete fraud and illogic as well.

If you are a Catholic, you believe that Jesus and his father are glued together by some kind of supernatural ectoplasm called the Holy Ghost, and together these three somehow form both one collective and three separate entities at the same time. There has never in the history of the world been a coherent explanation of this bizarre dogma, and it is just one of many extremely bizarre dogmas dreamed up in that weird cult over the centuries. Therefore in the Catholic sect the Jesus figure, being either a 33.33% part of and/or wholly the creator of the universe, is indeed the god of the Old Testament and has an equal share in every jot and tittle of its evil. Speaking of jots and tittles, it is Jesus himself who claims the entire Old Testament as his own, either by extension or by adoption, as quoted on the second page of the New Testament in Matthew (KJV), with these words; my bolding added:

> *"{5:17} Think not that I am come to destroy the law, or the prophets:* ***I am not come to destroy, but to fulfil.*** *{5:18} For verily I say unto you,* ***Till heaven and earth pass, one jot or one tittle shall in no wise pass from the law, till all be fulfilled.****"*

So whether you are Catholic and a trinitarian or are not Catholic and are one of the other thousands of kinds of Christian, all your Christian god(s) are equally guilty of every evil in both the Old and New Testaments.

Thus we have simple and clear examples which show that logic, consistency and true morality are in no way a requirement or precondition for huge masses of humanity to become adherents of some religion or other. It is just as plain and clear that actual evidence is not required either. This can be easily proven by merely asking any believer to first cite the specific supernatural claims of his faith, then asking for actual empirical evidence for each of the claims, then also asking why he thinks the supernatural claims of every other religion that humans have ever followed are false. Prepare to be galactically unimpressed at what will be offered to you in response, and even more dismayed at the weakness of human minds.

As of 2015[2], there are approximately 2.3 billion Christians and 1.6 billion Muslims on the planet. There are between 1.1[3] and 1.2[4] billion Catholics. That this is the case in the 21st century is both appalling and embarrassing for a species with the knowledge we have accumulated, and it indicates just how unbelievably widespread such completely broken human thinking is. This hyperabundance of adherence to the unbelievable, the unprovable and the immoral is a clear and present danger to the health, welfare and safety of every human alive and for generations to come.

This is the case because far too many of the faithful refuse to keep their beliefs to themselves, and busybodily and arrogantly go around in word, deed and in state and federal elective offices busily trying to inculcate their specific religion and its tenets into public law and ram it down everyone else's throats. We quickly see how important it is to bring every human mind to a common understanding of the processes, biases and flaws innate to human psychology. This must occur in conjunction with training on how to remove these biases from thought and action.

If you are a true believer, your biases which lead you to deeply and sincerely believe that your religion is not merely "little t" true but is actually the singular and unique One True Faith make you a contributor to that clear and present danger to the rest of us, as history has shown since the first religion was invented. You as an individual may not feel the need to persecute others, but your belief in unprovable supernatural propositions clears the way for anyone else to claim as true the supernatural propositions in his religion which require him to persecute and murder.

The bottom line is that no fully informed and productive discussion of beliefs, facts, opinions or any exchange of information and power-sharing agreement between humans is possible unless all participants are of sound mind and also deeply and broadly informed on human psychology. Since extremely few people meet this description, most human communication is fraught with misperception, miscommunication and the tragically foolish elevation of opinion and feeling over facts and reason.

It is for that reason that I will address these issues of the lack of understanding of basic human psychology and also of the problems of bias and broken thinking in this book. I do this to inform the reader of the existence of these logical breakdowns and the role they play in destroying communication and in blocking a common acceptance of reality. It is a tragedy that all human children are not being taught in detail how the human brain operates and how to identify psychological biases and traps, along with logical fallacies, both in themselves and others.

Thus of the seven and one half billion people alive today, only a microscopic fraction of one percent are actually aware of the facts of this serious problem. Therefore as a species we are collectively flying blind as we try to communicate, inform people of facts and how to handle and overcome emotions and uninformed opinions. If you think of our situation like that of an airliner in which all but one of the people on board are blind, and the seeing person is neither of the pilots, you have an idea of the scope of the problem and the gravity of the possible outcome.

Knowledge of these biases is only fairly widespread in a very small community of those trained or knowledgeable in psychology and psychological science, and since we are all human and subject to the unconscious demands of our unreason, even those most informed on biases and flaws in thinking are not immune to their effects.

No one is immune to the raw emotion that we feel when faced with moral dilemmas, challenges to our core beliefs or images or ideas that disgust or compel us. But the science behind all this is becoming increasingly well-documented by truly outstanding research and publications.

The problem rears its ugly head the highest when the faithful work so hard to forcibly inject their particular religion's tenets into other people's lives. Take a classic example. Right now, I want you to imagine two men having intercourse. Women vary greatly in their responses to such an imagining. Some get turned on by it, some are repulsed by it, and others occupy every step in between, as you would expect. For homosexual men this sort of activity is what they do in reality and what they think about, since that is a normal thing for their minds to do. It is thus a positive and pleasant thought for them unless they have been convinced by some religious person that these thoughts are shameful or evil.

For almost all heterosexual males, trying to imagine this situation is extremely difficult and unappealing, and is accompanied by some degree of discomfort up to complete and utter repulsion or disgust. The strength of the emotional response is largely not under the person's voluntary control, although nearly any response to any stimulus can be dulled through repeated exposure and desensitization, and a reasonable and mature person can control strong negative reactions.

The difference between the religious believer and the nonbeliever is that even if the visceral negative or disgust response is stronger in the nonreligious person, he will never seek to stop the two males from engaging in their activities through trying to invoke some religion to justify enacting some sort of law against it. He will not try to castrate, imprison or murder those who would engage in that activity.

The religious person, however, even if they don't try to murder or castrate such people, can almost universally be counted on to try to deny these people the same human rights they feel their god granted heterosexuals. Their justification for this discrimination comes directly from their religion, but their motivation for it comes directly from their unconscious and unthinking bias and disgust, which have short-circuited their conscious critical thinking, enabled their hatred and tribalism and diminished or eliminated their morality. The words they use to justify their discrimination are religious blather, with no actual reasonable or justifiable empirically proven rationale.

Why are so many of the religious so quick to follow the commands given in their holy books to do bad things to gay people, but they are not nearly as fast to murder people? It is not because there is no clear biblical guidance on killing; you can kill anyone on earth including the unborn and you can easily find some kind of justification for it in the Christian bible.

It is because there are secular laws with severe penalties for killing people for preaching a different religion, but far fewer laws with far weaker penalties which protect the rights of gay people. It is also because the faithful scream at the top of their lungs that they deserve the right to discriminate against people they don't like because they say that preventing them from doing so infringes on their religious freedom.

Their religious freedom, they say, must extend to all people with whom they interact, which is just a level of assholery only a mindless bigot or deeply and sincerely religious believer can reach, and is absolutely not what the protections for religious expression were meant to cover. They were meant to protect people from being persecuted or worse for their religion, *not* to enable believers to persecute others or worse. This is again so typical of the faithful to use an original good intent and turn it around to use it to hurt others or defend their own thuggish hypocrisy.

The difference between the religious and nonreligious communities is that the nonreligious tend to leave people alone as long as they are not, in the simple but powerful words of Sam Harris, diminishing the well-being of conscious creatures. The religious, on the other hand, continually go many miles out of their way to try to impose their personal and religious dogmas and opinions on other people, blissfully unhindered by factual analysis, true human morality or any ethical restraint.

To continue the above example, why would a person want to stop two people who truly love each other, whether of different or the same sex, from getting married? The religious will proclaim that their holy book and the imaginary god which they think wrote or inspired it told them this was how they were to act. Taking a tour through the bloody and immoral sadistic disaster that is the Old Testament, one can indeed find that the allegedly loving and forgiving deity who allegedly made us in his own image hath indeed plenty to say or inspire about how we should handle homosexual behavior, along with a massive list of other activities. In the King James version of the Christian bible, Leviticus says:

> "*{20:13} If a man also lie with mankind, as he lieth with a woman, both of them have committed an abomination: they shall surely be put to death; their blood [shall be] upon them.*"

So the allegedly loving and all-forgiving Christian god feels that homosexuality is a capital offense. This is quite interesting, since humans are, again, said to be made in the god's own image, but for some reason he says male homosexuals just damn well need to be killed. One might wonder why lesbians are not mentioned here; this is quite likely because women were property and didn't matter a damn to the god or the book's authors.

While this biblical order to kill male homosexuals is quite clearly a psychopathically immoral statement, it is equally quite clearly the norm for this book. The faithful during biblical times were exactly the same as the faithful today in that they cherry-pick only certain bits out of their mythologies to justify treating other people badly based on behavior which they personally don't like.

And again, what the faithful never do, in a stunningly pervasive and never-ending display of peerless arrogance, hypocrisy and outright middle-fingering toward their deity, is follow the entire book. Apparently while the god's orders to kill gay people are to be followed, or if gay people can't be killed they should be subject to legalized faith-based discrimination, the rest of the orders in the book are conveniently ignored.

Thus the non-literalist faithful all give that stiffly erect middle finger to their god. Why a mere mortal true believer feels he can pick and choose which divine orders to follow and which ones to ignore has always been beyond comprehension. Such arrogance far exceeds any arrogance of which the faithful so often accuse the nonbelieving and scientific communities, by hundreds of orders of magnitude.

While the bible is full of enough immorality to choke a black hole, even just the rest of Leviticus has plenty to say on the subject of what the lord their god wants the faithful to do regarding many other behaviors. The following are just a few examples, taken from the very same part of Leviticus as the citation above. The allegedly loving creator of the universe wants his faithful to kill other people for:

- **Adultery.** *"{20:10} And the man that committeth adultery with [another] man's wife, [even he] that committeth adultery with his neighbour's wife, the adulterer and the adulteress shall surely be put to death."*

- **Incest, but only certain kinds.** *"{20:11} And the man that lieth with his father's wife hath uncovered his father's nakedness: both of them shall surely be put to death; their blood [shall be] upon them. {20:12} And if a man lie with his daughter in law, both of them shall surely be put to death: they have wrought confusion; their blood [shall be] upon them."*

  Note that this somehow did not apply to Lot, who screwed both of his own two biological daughters not long after his god, in a feat of divine sado-multitasking, turned Lot's wife into a pillar of salt while firebombing Sodom and Gomorrah. This bit of apparently quite acceptable god-sanctioned biblical family values comes from Genesis 19:30 to 19:36.

  Therefore Jehovah blesses the act of men screwing their biological daughters, but a grown adult having consensual sex with a member of the same sex is a capital offense. Such is the insane word of their lord. But even hetero sex is sometimes dangerous. Screwing one's daughters is perfectly OK, but sex with a woman and her mother means you need to be burned alive, as per the order below.

- **Sex with a woman and her mother.** *"{20:14} And if a man take a wife and her mother, it [is] wickedness: they shall be burnt with fire, both he and they; that there be no wickedness among you."*

As if the above weren't enough, the allegedly loving Christian god wants extremely draconian penalties for other behaviors. These behaviors are extremely heinous to the god who, yet again, is said to have made us in his image but is so disgusted by those bodies that they need to be kept covered up at all times. And menstrual blood is to be avoided like the plague, as it is apparently almost as toxic as the superacidic blood in the creatures from the *Alien* movie series. These other divine penalties result from:

- **A man seeing his sister naked.** *"{20:17} And if a man shall take his sister, his father's daughter, or his mother's daughter, and see her nakedness, and she see his nakedness; it [is] a wicked thing; and they shall be cut off in the sight of their people: he hath uncovered his sister's nakedness; he shall bear his iniquity."*

- **A man and woman having sex while she is menstruating.** *"{20:18} And if a man shall lie with a woman having her sickness, and shall uncover her nakedness; he hath discovered her fountain, and she hath uncovered the fountain of her blood: and both of them shall be cut off from among their people."*

  "*Her sickness*", indeed. And the omniscient creator of the universe has spent thousands of years deliberately ***not*** clarifying that menstruation is not a sickness, along with a planetload of other information that would have been great to know, such as that slavery of any kind should be prohibited in all times and in all places, and that germs and not sin cause disease. So the god of the bible knew as much about science as a person who lived in the Iron and Bronze Ages…what an unremarkable and wholly damning surprise.

- **A man seeing his aunt naked.** *"{20:19} And thou shalt not uncover the nakedness of thy mother's sister, nor of thy father's sister: for he uncovereth his near kin: they shall bear their iniquity."*

- **Extended incest.** *"{20:20} And if a man shall lie with his uncle's wife, he hath uncovered his uncle's nakedness: they shall bear their sin; they shall die childless."*

- **More extended incest.** *"{20:21} And if a man shall take his brother's wife, it [is] an unclean thing: he hath uncovered his brother's nakedness; they shall be childless."*

  Lot and his biological daughters did not incur any incest penalty, after he got both of them pregnant after screwing them shortly after his lord had killed his wife who was the mother of his daughters. This bit of biblical family values was just peachy with Jesus and his father. Having sex with your daughters is no big deal but having sex with your brother's wife is just going too far.

So why aren't all of the faithful scrambling and fighting with every inch of their lives to kill all adulterers? Or to impose draconian and permanent legal penalties and discrimination against all men who have sex with menstruating women, menstruating women who have sex and those who see members of their family naked? The faithful fight tooth and nail every single day to be able to discriminate against homosexuals. Why the galaxy-sized disconnect?

The answer is clear. It is simply that the faithful just have less of a personal disgust issue with heterosexuals committing adultery or screwing during certain days of the month than with homosexuals having sex, and also because they can get away with this particular bigotry but cannot get away with murder. At least not until the rise of a Christian State which would without a doubt certainly codify such truly insane penalties into law.

Particularly when adultery is corrosive to a relationship and often ends in divorces and broken homes, often with resulting negative effects on the divorced parties and their children, why aren't the so-called "family values" faithful coming out in droves and taking the case for criminalizing adultery and discriminating against the adulterous to the Supreme Court? There is quite simply no consistency with the religious and their actions, except their extreme inconsistency.

I have to caveat the above with a warning that while they do not operate widely or openly in the mainstream, at least not yet, there are indeed quite large and growing numbers of theofascists in America who are actually working toward implementing such Old Testament rules into secular law. There are far too many people who would be very happy to stone adulterers to death and exile people who have sex during menstruation, and they would be quite happy to hypocritically cast those stones themselves. This strategy is being discussed and where possible the ground is being prepared for insertion of such laws into the local, state and federal levels.

Don't believe me? Go back and slowly reread the history of religion in America from Chapter Six. Do some good solid research on your own, and when you are finished you will be appalled at what people are trying to do for their lord and savior. Hint – try "Dominionism" and "Christian Reconstructionism". And spend some time in fundamentalist churches in any Republican state in the country. You'll leave the service with quite a different view of religion, rising nausea and a burning desire to keep that wall between church and state absolutely and completely impenetrable, believe me.

The reality is that people feel very free to keep the beliefs they like, to discard what they don't want to believe, and to work their way through or simply ignore the rationalizations and cognitive dissonance necessary to arrive at some personally and individually acceptable faith and belief equilibrium. Every religious person deems some set of beliefs within his faith to be worthy of his attention and has another set of beliefs which he has heard or read about but which he feels quite fine with rejecting. Christians and Muslims all over the world go on endlessly about their so-called deeply and sincerely held beliefs, but offer no real evidence for the claims of the faiths. Believers really believe what they say they believe, but the inconsistencies in each faith and indeed in each believer would take decades to write down.

So how do the faithful actually arrive at the set of beliefs they choose to retain and follow? How do they decide which of the bits of their books they deign worthy of personal adherence and/or so galactically important that they have to try to make the rest of us follow them too? This is in fact the most important first area of inquiry we need to understand, because humans are very complex organisms. Logic is only one part of what we think and do, and it is often wholly absent. Unfortunately, even when it is present and used, logic is applied improperly and/or is also often quite far down the list of considerations during belief formation and accretion. Logic is thus often ignored or rationalized away.

As Michael Shermer has said so accurately and so consistently, and as many others have suspected, known and said, our beliefs come first, full stop. Then, after we have the beliefs, we strive mightily to find ways to rationalize or justify them. We search around for what we perceive to be things that support our beliefs, and we ignore disconfirming facts and evidence. Or we don't even bother with trying to justify the beliefs, which is actually the case with many people, because for them the mere fact that they hold the beliefs is good enough for them, and they are sure it will also damn well need to be good enough for you and me as well, enforced by law if possible and handfuls of heavy stones if they can get away with it.

And there is apparently no belief so extreme, so ridiculous or so unsupported by evidence that large numbers of people will not still "deeply and sincerely" cherish and act upon it, even to the extent of willingly sacrificing themselves or killing others in the service of the belief. Many people believe things with every fiber of their being, despite the beliefs being wholly and demonstrably untrue. The difference between these people and the rest of us is that their faith is more powerful than their reason.

This type of person will not change his mind about a belief due to someone else's perfectly cogent, factual, informed and convincing argumentation, or even due to crystal clear evidence seen by his own eyes proving incontrovertibly that his belief is wrong. Reality does not matter to the deeply faithful and it never will. This is very hard for empiricists to understand, but it is as real as an Inquisitionist's torture chamber.

A person's level of investment in a belief drives his ability to ignore evidence against the belief. People who are heavily invested in and committed to a belief will increase their level of mental resistance as much as they need to in order to block the disproving evidence. That is one of the most important facts to remember about the ability of any kind of unfounded beliefs to take over weak minds.

If a person has deeply and sincerely committed to believing that two plus two equals five, and this is justified due to some supernatural claim, no amount of facts or evidence will convince him that two plus two is four.

With regard to those who believe the claims of religion, their ability to ignore facts that disprove their religious beliefs thus strengthens in proportion to their level of personal and social investment in that religion and its claims. People cling to some beliefs particularly tenaciously due to their desperate need for comfort and certainty.

In some cases, this comfort can be essentially infinite, such as when a person believes in an eternal heavenly afterlife. Thus changing these beliefs would in practice require the believer to be faced with more than infinite evidence, which is of course impossible, so such beliefs are impervious to facts. There literally is no changing some people's minds.

The good news is that the body of empirical work on the faults that occur in human thinking unguided by rigorous external processes such as the scientific method is massive and growing by the day. One of the best descriptions which I have ever read on this subject from experts in the field of human cognition comes from Steven Sloman and Philip Fernbach, professors at Brown and the University of Colorado, respectively, in their work *The Knowledge Illusion: Why We Never Think Alone.*[5] On page 172 of this fine book they note that:

"As a rule, strong feelings about issues do not emerge from deep understanding."

The authors are being overly polite here. What this is really saying is that people who don't have a damn clue what they are talking about are the most likely to be ignorant stubborn horse's asses and to hold completely uninformed opinions on any subject whatsoever which wouldn't survive an elementary-school level critical analysis. This is more than an observation. It is a fact and it is well-known even to amateurs in the field of social interaction. This fact, and the damning and troublesome problems it causes, are not widely understood or accepted, although they are major factors in interpersonal and societal discord.

The rest of the page in *The Knowledge Illusion* from which the above citation comes is full of other outstanding tidbits, some which I have listed below as food for thought on this issue.

1. Regarding those strong opinions which do not emerge from deep understanding, in other words beliefs which are deeply and sincerely held by the ignorant and the stubborn, the authors quote two other personalities who commented on this subject. One comment is from the famous intellect and critical thinker Bertrand Russell, who said "The opinions that are held with passion are always those for which no good ground exists."

   I have to differ with Russell on this, though, in one regard. The tone of his comment was congruent with the ideas noted in the book, but the comment was not well-stated. Facts are facts and they are much different from opinions. When other people try to deny facts, or slant them to fit an ideology or opinion or hidden agenda, it is not just reasonable but a downright mandatory requirement to respond with passion and vigor to stop the suppression or denial of facts.

When someone tries to deny the Holocaust, tries to deny people basic human and civil rights based on bigotry, tries to destroy our environment because he denies global warming or is beholden to a religious or political extremism, tries to deny that a specific political or religious action is factually and empirically wrong or immoral, or tries to justify anything at all because they "think" or "believe" a certain way about it, we have to act. Those who actually do value facts and reality must respond with passion and a loud voice to keep the queen of Reason firmly seated on her throne and to boot the court jester Belief outside into the cold rain.

2. They also quoted, of all people, Clint Eastwood, who said, in an interview[6] with Time in 2005, "Extremism is so easy. You've got your position, and that's it. It doesn't take much thought".

   So apparently even Dirty Harry's alter ego has a more logical and true assessment of this issue than the mindlessly faithful masses. I don't know much about Eastwood but he summed up the problem quite neatly. The rest of Eastwood's quote, which was not in the book, ended with "And when you go far enough to the right you meet the same idiots coming around from the left." This is also quite true. There are right wing nutjobs and left wing loonies, and if we all ignored a certain percentage of the wingnuts on the ends of the bell curve we would all be much better off. This does not exonerate all moderates, however. So-called religious moderates may not kill you in broad daylight if you try to convert them from Christianity, but their acceptance of unfounded claims leaves the door open to all religious beliefs and practices, without exception and whether they are "extreme" or not.

3. The authors did a very good job of laying out some of the ways in which most people lack critical thinking and are extremely ignorant on every subject it is possible to discuss. The information they cited was both surprising and depressing, and while I highly recommend readers peruse the book itself, I had to include one item as an exceptionally illustrative example. They referenced an Oklahoma State University survey which asked people if they thought that genetically modified foods should be required to carry a label saying so. About 80% of people said such labeling should be required. That wasn't the problem. The

face-palming follow-on was the problem. It said "But 80 percent also approved of a law stating that there should be mandatory labels on foods containing DNA." Yes, DNA[7]; that is not a typo. Deoxyribonucleic acid. Apparently these people feel that the core of life itself is to be feared as much as Christians, Jews and Muslims have long feared menstrual blood.

This was so ludicrous that I had to see the data for myself. A quick ride on the Google train led me to the Oklahoma State website with the Food Demand Survey[8] document and the anti-DNA sentiment in its full inglorious splendor. Page four of Volume 2[9] Issue 9, from January 16, 2015, shows that 39.09% of survey respondents support a ban on sugary sodas, 63.77% support calorie limits on school lunches, and yes, a full 80.44% of respondents support mandatory labels on foods containing the genetic code of life. This is even more idiotic than wanting a label on bottles of water to contain a dire warning that they contain hydrogen.

My initial reaction to seeing the survey was that of the comedy show *Futurama's* Professor Farnsworth when he was faced with similarly peerless levels of both ignorance and stupidity[10] and said "I don't want to live on this planet anymore.[11]" Then I remembered this survey was done by Oklahoma State University, which is buried deep in the cellar of the religiously stultified antiscientific Inhofian[12] bible belt, and took some temporary solace in the hope that the survey was done only on Oklahomans, who as a collective body astoundingly chose specimens like Mary Fallin and Scott Pruitt to positions of state leadership.

Then I reached out to the survey team, who informed me in an email that "The survey is distributed to nationally (sic). 1,000 surveys are collected each month, with demographics weighted to represent those of the US population." That says it all. My next question is how can I sign up for those Mars colonies?

Such unbridled ignorance is staggering and dangerous. How is it possible that in the world's current greatest superpower, with enough resources to educate all of its citizens to a very high degree, that so many people can be so pathetically and damnably ignorant? How is it possible that in the 21st century such large numbers of people are so blissfully devoid of facts and equally blissfully devoid of the ability to use them? To say this is a severe problem is to massively understate the gravity of the situation.

While we absolutely need to function in our personal and professional lives and to govern our societies by the facts of reality, we have to understand that many people absolutely ***do not*** do this. We need to fully understand how people arrive at their beliefs and what can be done to make this process happen in a logical and reasonable manner, with logical and reasonable outcomes. We cannot afford to let unsubstantiated beliefs and uninformed opinions affect our fellow humans, animals and the environment, as they certainly will if they are not strongly challenged and turned away.

We never really could ignore these opinions and be intellectually honest, but with the power to wreak major catastrophe and to kill large groups of people in seconds now being wielded by single humans, our margin for error is now completely gone.

How the brain operates, and thus how the human personality is expressed, is at its core a set of involuntary processes and expressions of each person's basal psychological framework, which is itself derived from the brain's physical construction and environment.

The brain is very much like a computer constructed in a factory, and it operates according to a few major driving factors as described below. Damage, changes or inputs to any of these factors can and often do affect how and what people think and believe.

People often experience short-term and potentially lifelong changes in their thinking due to the effects of various inputs and conditions such as indoctrination, education, both positive and negative parental care and attention, training in critical thinking, traumatic experiences, traumatic thoughts, hallucinations, brain tumors, near death experiences, dreams or nightmares and damage due to substance abuse.

Left to its own devices, each developing brain will grow into a unique physical construct with a pattern of components and modules which operate according to that pattern. Everyone knows how differently people can think, and on some issues there can be as many different opinions as there are people.

The rest of this chapter is a bit lengthy and detailed but this is necessary because a short summary will not properly illuminate the scale and scope of the problems we face and the reasons for these problems. Sound bites and summaries will not convey what we need to understand in this issue.

Humans face extremely significant challenges with our onboard programming, and this problem is worsening as our species learns more information and as increasingly more people have to peacefully coexist. These problems in processing lead to logic failures and cognitive biases that damage and distort our thinking. Solving these problems starts with identifying these failures and biases and understanding how they impact our thinking.

A great deal of relatively recent study and observation shows just how many obstacles the brain puts in our path to empirically supported understanding of reality, and the vast amount of nonsense and clutter which the scientific method clears away. I recommend readers go through the entire next section both deliberately and with a reflective mindset.

As you read, take some time to reflect on both how the information enables the invention and appeal of uninformed opinion and ideas like religion, and how to reduce or eliminate these problems in your thinking and the thinking of those with whom you interact.

It is now time to explain how people arrive at their beliefs. To do this properly we must delve into the physical construction of the human brain, the conditions it experiences and how these factors impact our thinking. Beliefs arise from the following four sources and conditions:

1. **The structure and composition of the brain.**
2. **Physical changes or damage to the brain from physical causes.**
3. **Physical changes or damage to the brain from nonphysical causes.**
4. **How the brain responds to experiences and sensory input.**

Belief Source 1: The Structure And Composition Of The Brain

Figure Four below shows some of the brain's major physical parts and their general roles. The brain is our cognitive hardware and circuitry and is basically an organic computer. The organic hardware provides the physical structure on and in which the brain's functions and programming operate. Blood and chemical-filled fluids[13] permeate[14] and surround[15] the brain and enable the hardware to operate. Those operations are the actions of neurotransmitters, firings of nerve cells and other activities which we have no idea are happening but which we perceive as reactions, emotions and thoughts.

If our brains and bodies were completely dehydrated we would lose about 80% of our mass and we would of course be quite dead. Thus it is the combination of the brain's physical structure, blood and those special fluids which enable the brain to conduct its conscious and unconscious functions.

Brain development[16] is an extremely complicated and very long process, which begins[17] on the 18th day of gestation and continues for the rest of the individual's life, though the major period of construction and programming is mostly completed between a person's 20th and 30th birthdays. This period is driven by genes and the physical environment in which the person develops. Genes and the environment will be familiar to non-DNA-phobic readers as the familiar twin pillars of nature and nurture. Development can be severely impacted or facilitated by negative or positive conditions in either pillar.

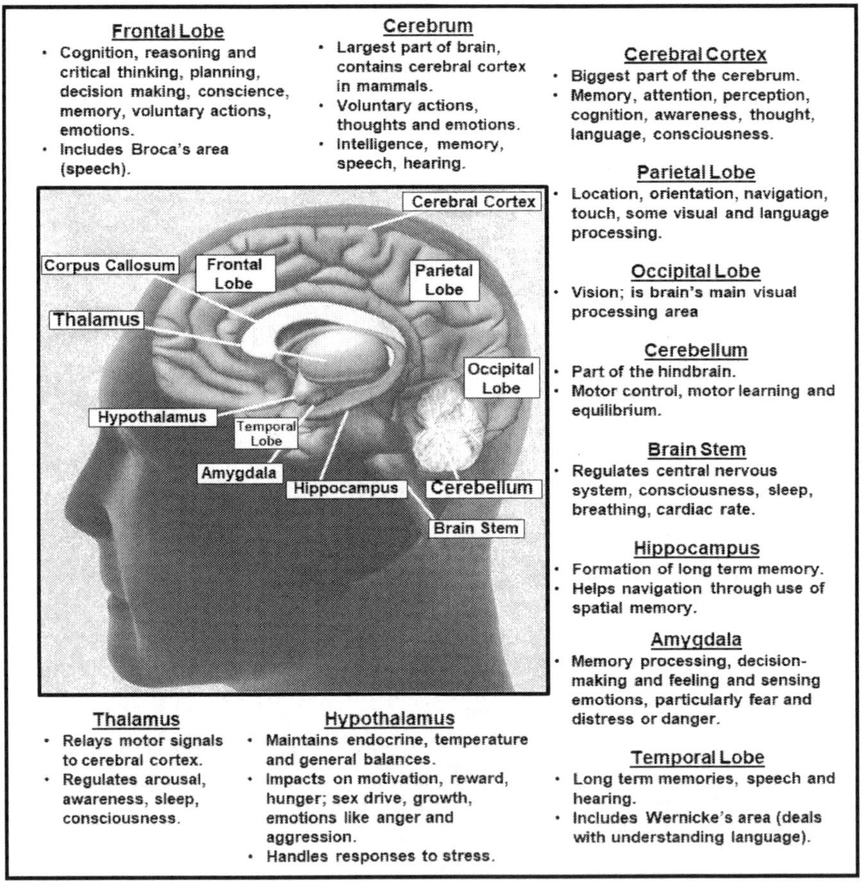

Figure 4. Major parts of the brain. Main image courtesy of the National Institute on Aging, National Institutes of Health, U.S. Department of Health and Human Services; image adapted by the author.

In the womb, the genes inherited from the parents drive the growth and development of the body and brain. If there is an error in the genes for brain development, the result will be miscarriage or birth of a baby with a malformed and/or malfunctioning brain. Note that this only refers to the physical development of brain structures, not the products or thoughts of that brain. We would certainly say from a moral perspective that a psychopath has a malfunctioning brain if we are describing the evil thoughts and likely behaviors of its owner.

But that person's brain may have developed in perfect environmental conditions and have perfectly followed its genetic code during assembly, and be completely psychopathic. We would call that brain well-formed in the sense that its physical development accurately followed its genetic blueprint with regard to the completeness and mechanical function of its parts, but we would also have to say that it was malformed and malfunctioning with regard to the morality of the thoughts it produces.

At birth, the human baby's brain has some software and an operating system built in. This is hard-wired into the involuntary or autonomic nervous system[18] which is itself composed of three sub-systems, called the sympathetic nervous system, the parasympathetic nervous system and the enteric or gastrointestinal nervous system.

Thus the combination of the newborn baby's brain and the chemical environment which saturates it are the equivalent of a computer's hardware and operating system, now plugged in and ready to accept input from the world outside the womb via the various senses, or peripherals. The baby's organic machine brain and nervous system include both reflexes and instincts.

According to Merriam-Webster, reflexes[19] are "an automatic and often inborn response to a stimulus that typically involves a nerve impulse passing inward from a receptor to the spinal cord and then passing outward from the spinal cord to an effector (such as a muscle or gland) without reaching the level of consciousness and often without passing to the brain". Some examples[20] of reflexes are the act of jerking a hand back from a hot stove, or throwing up the hands in front of the face and turning the face away from an incoming object like a thrown ball. Babies have a host of reflexes including the startle reflex, the rooting reflex, the grasping reflex and the sucking reflex.

Instincts[21] are the "...largely inheritable and unalterable tendency of an organism to make a complex and specific response to environmental stimuli without involving reason." Instincts include the drive to eat, drink, have sex and intimate relationships, fit into social groups, seek sweet and/or fatty foods, defend yourself and your genetic kin and any allies, attack outsiders, gain higher social status, gain resources and seek enjoyment and entertainment. Instincts drive both good and bad behavior.

It's Time For A Revelation!

It's time to discuss a situation which will be a revelation to many people. You probably already know that your reflexes and instincts are automatic parts of your subconscious, which is that "hidden[22]" brain that does a huge amount of work but of which you know little or nothing. The same package of hardware and software which gives us reflexes and instincts also contains the patterns of our basic tendencies and predilections, thus our immediate and unthinking responses to stimuli. And also what we would call our personality. So far this is probably not too much of a revelation.

What will be a revelation to almost everyone is that nearly every human being's "hidden" brain processes have far, far more to do with how we think and what we agree or disagree with than we previously knew.

It is nearly a universal human characteristic that this hidden brain fundamentally determines our likes and dislikes, and thus how we think about and respond to subjects such as politics, religion, social issues like homosexuality, religion, climate change, science in general, people of other races or nations, who and what we choose to believe or not to believe, and every other issue we face.

This revelation means you are often nothing more than a slave to your "underbrain", and it is almost a guarantee that you don't have any idea that is the case. Odds are that even if you do know it, you don't work too hard against its wishes and are more than happy, even eager, to follow its lead. This is to be expected, because this is how you as an organic robot have been programmed to function. Acting in accordance with your programming is the natural path of least resistance.

Among a very rapidly growing body of truly excellent research into human cognition, two quite outstanding and relatively recent bodies of work deserve particular mention for their absolutely yeoman and long overdue work to illuminate this mental challenge we all have. The first is the work of the insightful psychologists Daniel Kahneman and Amos Tversky, and much of it is summarized in the wonderful book *Thinking, Fast and Slow.*[23]

The other and equally insightful and illuminating significant body of work on thought, emotion and opinion also sheds great light on what makes humans think and believe what they do. This other work, led by psychologist Jonathan Haidt and reflected in his book *The Righteous*

Mind, Why Good People Are Divided By Politics and Religion[24], lays out much the same type of information but with other insights.

Haidt also notes how there is little conscious work done in forming our opinions and during the process of arriving at our beliefs. Beliefs that are comforting and "feel right" to us are the ones we keep, and those that don't "feel right" are rejected. And this is why so many people work so hard to defend their opinions, regardless of how little actual evidence exists to support them or how much contrary evidence exists.

There are two huge problems in human psychology. The first is that feelings matter more than anything to most people. The second is that facts only matter to those who choose to accept them. This strong tendency to avoid unwelcome truth and reality is why objective and peer-reviewed science is the only proven method to remove emotion and feelings when seeking real truth and knowledge. It is also why science must be the final arbiter of what is actually true and factual.

The two bodies of work by these insightful thinkers and their teams, underpinned not only by their own prodigious research but also by the quickly expanding list of congruent, supporting and relevant work done by others, neatly sum up the processes in the brain that make people react to the world around them in the ways that they do. They also describe the reasons why people first come up with and then cling so desperately to their opinions and beliefs, no matter how much logic or reality has to be twisted or denied in the process.

"Thinking, Fast and Slow". Kahneman published this book fifteen years after his long-time research and intellectual collaborator Tversky died in 1996. The book describes how the brain's unconscious and conscious modes separately affect how we think and behave. It is as if every person has two separate personalities working independently of each other in the brain, with what we think of as ourselves in general control of one while being almost completely ignorant of and often a mindless slave to the other. Kahneman and Tversky described how the two modes of brain function separately act and react, and called these two different modes System 1 and System 2.[25]

System 1 is the unconscious and so-called "fast" system described in the book's title. System 1 reacts extremely quickly and is reflexive and instinctual in nature. This means that as with reflexes and instincts, the system operates without deliberate or conscious cognition. The citation below describing the things System 1 does is taken from the first chapter of the book.

> "In rough order of complexity, here are some examples of the automatic activities that are attributed to System 1:

- Detect that one object is more distant than another.
- Orient to the source of a sudden sound.
- Complete the phrase "bread and ..."
- Make a "disgust face" when shown a horrible picture.
- Detect hostility in a voice.
- Answer to 2 + 2 = ?"

None of these activities require active brainpower and each one is essentially an automatic response, thus my reference to the action being reflexive and instinctive. A critical point here is that emotion as an immediate and visceral response is included in this system. As everyone knows, emotional responses are absolutely not the products of deliberate analysis and careful cogitation, they are instantaneous and visceral, and they hit as hard as they are fast. You don't take time to think about how you will respond to a stimulus before responding; you reflexively or instinctively respond without thinking.

Imagine if you happen to be violently disgusted by images of internal organs, and were suddenly shown large and full-color mid-autopsy photos or even an actual partially dissected cadaver. Or that you are a devout chocophile and are presented with pictures and smells of Mom's best freshly baked chocolate chip cookies. Your immediate, unthinking and different emotional reactions to these very different stimuli, and the response you had to the earlier example of imagining homosexual sex, are the product of System 1. Your conscious "self" had very little to do with it.

System 2 is the conscious, "slow" system. It involves deliberate cognition and thought and is not reflexive or instinctual. Examples of System 2 thinking include:

- Finding Waldo in one of those "Where's Waldo" cartoons.
- Doing a crossword puzzle.
- Keeping your eye on the player with the ball or the puck in a game like soccer, basketball, hockey or American football.
- Remembering the names of the major rivers in South America.
- Working out a complex math equation.
- Constructing a house floor plan to scale using pencil and graph paper.
- Answering a written question on a complex subject.

You can see how vastly different the systems are, and why this is the case. If you had to deliberately and consciously calculate and plan how to respond every time you experienced a threat like an object flying at your face at high speed, or the roar of a large carnivore behind you, humans would not be long for this world. System 1 takes over in immediate response and reaction modes. System 2 can take over later, but even when it does, it is still largely underpinned and guided by System 1.

You are very often simply an unconscious slave to System 1, because you are an organic robot programmed to unconsciously respond to instincts and emotions. Your System 1 is always "on", and System 2 is like a computer in sleep mode. When you have to, you can wait for System 2 to wake up and see if it can help, but since System 1 immediately gives you what you *think* is an answer, i.e. it gives you a **feeling** and thus an **opinion** which normally turns into a **belief**, you most often don't ever bother to activate System 2.

This means you usually don't bother with consciously trying to validate information, and instead you operate on the uninformed and unthinking System 1 autopilot. Most people are lifelong slaves to System 1, and a lack of education often means this mental slavery is a lifelong problem. Those who suppress System 1 and engage System 2 before thinking and acting are unfortunately very rare birds, and are often dismissed by the System 1 slaves for "not having strong enough convictions". For a visual depiction of the difference between the two systems, take a look at Figure Five below. It is the famous Muller-Lyer illusion[26], showing two parallel lines, one of which is capped by arrowheads pointing inward and the other with arrowheads pointing outward. Are both lines the same length?

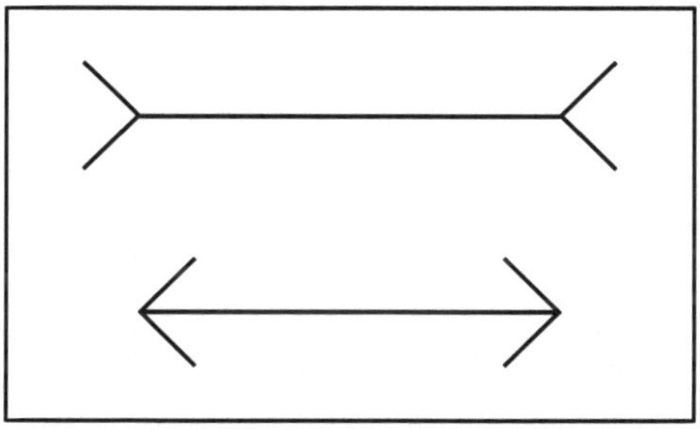

Figure 5. The Muller-Lyer Illusion.

Your fast System 1 will immediately say "of course not; the top one is obviously longer and my common sense proves it". Your slow System 2 will take the time to locate a ruler, measure the lines and say "despite how it looks, they are exactly the same length and the ruler proves it".

The two systems operate separately and show how "common sense" is an extremely fallible concept and is often extremely wrong.

The Muller-Lyer illusion is merely one of countless examples of how common sense fails us. This illusion shows how observation and measurement should always supersede so-called "common sense" and opinion. In other words, empirical science and its method must always, in all times and in all places, permanently supersede faith when determining empirical truth and reality.

Thinking, Fast and Slow is over 400 pages of badly needed insight into human psychology, and I strongly recommend everyone who owns and uses a brain read the book and use it in every aspect of daily life, from interpersonal relationships to child rearing to voting, and to pass it on to friends and families.

I will summarize the book's major points without going deeply into detail or the supporting science here, as readers really should invest the time to read the book for themselves. I have listed below some of the major takeaways from the book which are relevant to human beliefs in religion. The takeaway is listed first, followed by my linkage of its relevancy to religious belief.

1. **Priming.**[27] The book notes that "In the 1980s, psychologists discovered that exposure to a word causes immediate and measurable changes in the ease with which many related words can be evoked."

Religious relevance. At all stages in a person's life but especially during youth or times of stress, the words one hears in extremely frequently held religious ceremonies, meetings and prayers prime the brain to be cognizant of and more receptive to religiously-related terms and ideas. Note that there need be no actual facts of any kind involved, because in religion facts are absent. The words do the priming whether or not they are true. Religions work very hard to repeat their messages, and this endless and often daily repetition both primes and promotes reinforcement.

2. **Truth Illusions.**[28] When you make a statement that has one or more of the following characteristics - short, using simple words and easy to read, **in bold font**, **highly color-contrasted** from the background, and if it contains rhymes, it is perceived as more true. Again, it need not be true in any way, but it is still is perceived that way.

Religious relevance. Christianity is especially aware of this and is extremely vigorous in applying it, from nursery rhymes to poems to pithy sayings both spoken and on church billboards. Some examples of small-words rhyming Christian propaganda-messaging are:

- “Now I lay me down to sleep, I pray the lord my soul to keep.”
- “In god we trust.”
- “50 shades of grace.”
- “Pie in the sky when you die.”
- “Jesus died for your sins.”
- “What would Jesus do?”
- “We aren’t just molecules in motion.”
- “From goo to the zoo to you isn’t true.”

These statements are cognitively easy, they flow and have a message. They thus appeal strongly to System 1 which, as we have already established, quite often overwhelms the logical System 2.

3. **Math Challenges**. Most people fail, and sometimes fail quite badly, when they try to do probabilities or statistics calculations in their heads.[29] This is because almost no one has a decent Probabilities or Statistics module in either our System 1 or our System 2, and therefore System 1 has to make an instant guess. This usually resembles a toddler trying to quickly answer a question on quantum mechanics. Probability and statistics are inventions of advanced mathematics, and human evolution has not had the thousands to millions of years it would take to naturally evolve well-functioning innate modules for them. Thus when we try to do probability and statistics without the aid of math and the appropriate training, most people get them very wrong because they default to their so-called “common sense” as a guide, and this particular guide is not only often blind but deaf and dumb as well.

Religious relevance. When the faithful say that it is “too improbable” that some event could have happened without divine intervention, such as the evolution of the human eye, the origin of life or the spread of their particular religion, their numerophobia is expressed in the fear and stupefaction they feel toward very large and very small numbers. The faithful fall into a very deep psychological trap due to this inborn mental blind spot. They are also extremely likely to have little to no actual knowledge and experience in the relevant scientific fields and histories, and less training in probability or statistics. Most things in life are not at all explicable by innate or inborn "common sense", and very improbable things happen all the time. A proper education and properly done math and science have a way of making even the fuzziest questions and answers spring into sharp focus, even if the final picture isn't the one you expected or wanted.

4. **More Math Challenges.** Their observation of how people assess the believability of stories showed that what people feel is true based on so-called common sense is often completely wrong and is in the exact opposite direction from reality. They thus "...introduced the idea of a conjunctive fallacy, which people commit when they judge a conjunction of two events...to be more probable than one of the events...in a direct comparison." For example, this means that people think that a story about one person which lists multiple characteristics of that person is somehow more probable and believable than a story about the same person which only lists one characteristic. The truth is that every additional listed characteristic makes the story much *less* statistically probable, thus System 1 is 180 degrees out of line from reality and again common sense is dead wrong.[30]

Religious relevance. In Christianity, this means that people think that Jesus having only the characteristic of being born of a virgin with no other fanciful characteristics is less believable than after all the other fanciful characteristics have been added to the story, such as being the son of the creator of the universe, able to heal the sick and walk on water, able to rise from the dead, able to cast out demons and so on. Each added fanciful characteristic makes the Jesus figure mathematically much *less* probable and believable, but the mind, particularly a gullible and less educated mind, erroneously and perversely thinks the opposite. Thus in a 100% contradiction of reality and logic, the crazier the story gets, the more believable it gets in many people's minds.

5. **Knowledge Gaps**. While it is impossible to become an expert in everything, many people are highly trained experts and very seasoned professionals in their fields. These people can reach a level at which they can have 'skilled intuitions', which are intuitions that are rather more probably correct than the intuitions of someone without the same level of expertise and experience. This

level is said to be achieved after a skill is reached through "an environment that is sufficiently regular to be predictable" and "an opportunity to learn these regularities through prolonged practice.[31]"

Religious relevance. This goes back to strong opinions coming from ignorance. Regarding religion, if a person were to become a true expert in all the fields of history, philology, biology, chemistry, physics, archaeology, psychology, anthropology, sociology, meteorology, astronomy, geology, economics, military science, genetics, neuroscience and evolutionary biology, then a person could actually have skilled intuitions about the claims of religion. Since the empirical facts of all these fields show that religions are wholly manmade, truly skilled intuitions will deliver that conclusion. The intuitions should nevertheless of course be followed by a rigorous scientific and mathematical analysis of the facts.

Note that believers and theologians are *very* different from actual empirical experts. Believers and theologians do not have skilled intuitions because they rarely have more than a passing awareness of the scope of any of these fields of study, much less deep understanding of any or all of them, including their own faith. And theology, being merely a study of a sliver of ancient literature, mythology and historical fiction, counts as an applicable field of study with regard to the real world only to the extent that empiricism can illuminate. That said, believers and theologians have fears and feelings in great abundance, and those are what drive their faith. Seasoned experts in the study of ancient literature and mythology would have the potential ability for skilled intuitions in these fields, but it is impossible to be an expert or skilled in something that cannot be measured, such as the supernatural claims religions make.

6. **The Power of Fear.** They noted that people do not respond in the same way to potential gains or losses, and noted "We concluded from many such observations that "losses loom larger than gains" and that people are *loss averse*.[32]"

Religious relevance. This is highly reminiscent of the logic-twisting self-justification seen in Pascal's Wager. A believer who feels that his religion actually clears his path to heaven and eternal life is extremely likely to go on believing in that fantasy, because he

> feels it is a literally priceless reward which he will eventually receive if he can just keep believing. He thus handles his faith as an already-owned asset which he will go to great lengths to hold onto, regardless of the astoundingly broad denials of reality and rationality this will require. Therefore asking him to logically assess his faith and take a chance on losing it is, to him, proposing a massive loss of eternal and infinite proportions. Therefore to avoid such a perceived loss, the believer's mind will take whatever twists and turns it needs to in order to remain in the faith. This is of course a generality, as a few people do have the strength and courage to leave their faith once they've invested in it, but they are rare.

While these are great specific examples of broken thinking, this book is well-populated with others, with each one a reminder that people are born with certain response patterns and are extremely susceptible to their own personal experiences, thus are psychologically "primed" to think a certain way. As a result people are very challenged when trying to "think outside the box", and this is why people who actually can think outside the box are so rare. Everyone's box is the cumulative action of both System 1 and System 2, continually updated by the experiences which every human collects over a lifetime.

If you live in a newly-discovered hunter-gatherer clan and you have never seen a flashlight, upon seeing one for the first time you will try to understand the workings of that wondrously magical object through the lens of what little you know. This is likely to be fire, the sun, the moon, the stars or whichever gods you worship. Your opinion will be that the light comes from a little fire or a piece of a celestial object harvested from the skies, or it is a magic gift from your god, probably a sun god. You will have no conception of batteries, wires or electricity. To truly understand how a flashlight works, you will first have to be taught about all those things, and also that there is no supernatural entity or action required to produce the flashlight or the light it creates.

In other words, the hunter-gather will say that of course common sense proves that his godly answer is the correct one, and only an idiot would say otherwise. This is why opinions and so-called "common sense" by themselves are often not just worthless but very counterproductive, and this helps to demonstrate why science is the only path to true understanding.

This sort of hyperconstrained thinking is why we continue to hear the ignorant arguments from the faithful about there having to be a god who created the universe or magicked life into existence, and why these special pleadings won't go away anytime soon. The human race, while having climbed both out of the primeval sea and a fair bit up the Mountains of Modernity and Knowledge, still has ideas and models for but no actual evidence from before the Big Bang.

We have no heuristic, as Kahneman and Tversky would say, and no experience of a time or place before the Big Bang, so there is quite literally an infinite number of possible ways to describe that time. We simply don't have the concepts and data yet to properly describe it, although we can use our current heuristics as a baseline. We are as ignorant of the pre-Big Bang period as those hunter-gatherers are about quarks. What we do have as a civilization, however, is a continuous and unbroken series of now-dead religious explanations for a whole host of questions that have been wholly replaced by scientific explanations, with not a single explanation ever going in the opposite direction, and there is absolutely no reason to believe this will ever change.

Regarding that primeval explosion, it could be that "our" Big Bang is only the most recent of a quintillion such events in a series, with every single preceding universe being composed of the same matter and energy of this one, with no or only miniscule variations in its organization and distribution, and that this one happens to be different in that it is apparently going to expand forever[33] due to some pattern of distribution of dark energy instead of becoming sucked back into a Big Crunch.

This one-off, one-in-a-quintillion variation could be due to a mutation in the universe-replication process, to bring a natural selection heuristic into the discussion. It could be, in all fairness, that some god or creature did indeed sneeze the Big Bang out of its ineffable breathing spiracles, although there is zero evidence of any such entity or of any such super sneeze.

The infinity of possible pre-Big Bang conditions provides the faithful with an infinite number of gaps in which theists can hide the specific god(s) of their personal choice, and this is exactly what they do.

To wrap up the work by Kahneman and Tversky, their findings are extremely informative and I again highly recommend that everyone who owns a brain read that fine book.

I also recommend all brain owners read the outstanding work by Jonathan Haidt and his fellow researchers. Haidt is a social psychologist and Professor of Ethical Leadership at the Stern School of Business at New York University. His work intersects in many places with the Kahneman and Tversky work and illuminates the same basic conclusions. He and his team have also gleaned insights that are critically important in understanding why people think and believe what they do.

This body of work on human reflexive responses to situations or scenarios gives great insight into the practical applications of understanding and dealing with human tendencies and frailties. Haidt and the team conclude, and I completely concur, that mental response and reaction patterns are set into every newborn baby's brain, and that these patterns drive who people choose to interact with or not, and how to think and act, or not. This process occurs throughout a person's life and as he feels appropriate within his local social and cultural groups.

These initial response patterns, unless changed by major physical or psychological impacts, drive the morality or lack thereof, politics, gullibility, aggressiveness, studiousness and religiosity of the growing person, along with the rest of the person's preferences, likes, dislikes and other tendencies.

While the work is full of outstanding and illuminating facts, points and sound rationale, there are two takeaways I want to specifically note from Haidt's work. I pulled both of them from his already-mentioned book *The Righteous Mind: Why Good People are Divided by Politics and Religion.*

The first takeaway is that he has done great work and conducted much external communication on the nearly universal human inability to come up with a reasonable and empirical justification for the positions people take on moral questions and situations. The relevance to religion will be made clear.

The second takeaway is that it is essentially a human universal that conservatives, liberals, the religious and the nonreligious all tend to strongly respond to what Haidt calls moral foundations, captured in "Moral Foundation Theory". This indicates which values are most important to people and how these values drive their unconscious and often inexplicable opinions on morality. I don't necessarily mean their opinions are always wrong; the point is that people just often have no real idea why they feel the way they do about specific words and actions and often cannot explain the justification for their opinions at all.

The method they took to illuminate this human inability to justify personal moral positions was simply asking people if they thought the actions in some interesting and nuanced scenarios were moral, and to state why or why not. The questions were quite inventive and I have listed some of them here.

1. "A family's dog was killed by a car in front of their house. They had heard that dog meat was delicious, so they cut up the dog's body and cooked and ate it for dinner. Nobody saw them do this.[34]"

2. "A man goes to the supermarket once a week and buys a chicken. But before cooking the chicken, he has sexual intercourse with it. Then he cooks and eats it.[35]"

3. "A woman is cleaning out her closet, and she finds her old American flag. She doesn't want the flag anymore, so she cuts it up into pieces and uses the rags to clean her bathroom.[36]"

4. "Julie and Mark, who are sister and brother, are traveling together in France. They are both on summer vacation from college. One night they are staying alone in a cabin near the beach. They decide that it would be interesting and fun if they tried making love. At the very least it would be a new experience for each of them. Julie is already taking birth control pills, but Mark uses a condom, too, just to be safe. They both enjoy it, but they decide not to do it again. They keep that night as a special secret between them, which makes them feel even closer to each other.[37]"

5. "Jennifer works in a hospital's pathology lab. She's a vegetarian for moral reasons – she thinks it's wrong to kill animals. But one night she has to incinerate a fresh human cadaver, and she thinks it's a waste to throw away perfectly edible flesh. So she cuts off a piece of flesh and takes it home. Then she cooks and eats it.[38]"

As you can imagine due to your own innate unconscious responses / moral foundations, most people felt that these acts were immoral or disgusting to some degree. For example, Haidt notes that only 13% thought Jennifer's cannibalism of the dead body was alright. (Feel free to insert your Catholic consecrated host joke of choice here). But people had a devil of a time explaining specifically *why* they thought it was immoral. Haidt used the word "flailing" when describing the attempts subjects made to justify their opinions and this seems to be the perfect word to describe the situation. These interactions revealed the Kahneman-Tversky System 1 in operation, with System 2 trying to catch up but basically flapping in the wind and unable to say anything useful.

System 1 enabled test subjects to have an immediate opinion, but even after deliberately engaging their System 2 to answer the question of *why* they felt how they did, they couldn't make it come up with a reason for their opinion. In Haidt's clear and extremely relevant words,

"Reasoning was merely the servant of the passions, and when the servant failed to find any good arguments, the master did not change his mind."

The relevance to religion, as well as to politics, is crystal clear and it has significant implications. Despite the total lack of justification for their opinions, Haidt's test subjects were nevertheless very often quite sure of those opinions. Thus the religious and religiously influenced opinions and feelings people hold, which form the basis for their unjustified religious beliefs, are embedded so deeply in the System 1 parts of the brain that even a highly engaged System 2 is often incapable of doing anything but short-circuiting and doubling down on what System 1 was feeling.

This mental challenge is why when people are shown disconfirming evidence for their opinions or beliefs it is often like waving a flag in front of a bull. The bull gets mad and tries like hell to gore you and get you out of his face, and the believer not only won't face facts or change his opinion, he fights even harder to defend his unjustified opinion, his unproven beliefs and his religion.

When one learns of this tragic facet of human thinking, one suddenly sees the endless, nonsensical and diabolically cruel biblical proscriptions in a new and different light. The biblical authors felt personally disgusted by certain actions, so they felt they were immoral or sinful, thus they felt such actions should incur severe penalties. Therefore these actions were listed out in painful detail and punished whenever they were discovered. One must layer onto this already hyperignorant sadism the fact that these people actually believed that some other person's sins would result in godly punishment or vengeance not just on the miscreant but on the entire tribe. Thus the abject ignorance and immoral barbaric sadism that absolutely saturates the monotheistic holy books is at least partially explained.

The other major point I wanted to mention from Haidt's work is the implications Moral Foundation Theory has for belief and religion. Much as Newton's prism split light into its component colors, Haidt and his team's splitting of the unconscious mind's operations into the subelements and expressions of morality expands the general idea of the System 1 we discussed above into components of what Haidt calls "intuitive ethics".

Their work does a very good job of explaining psychological tendencies and what people think are moral and immoral, and a few of the biological reasons for these tendencies. But the main point is the theory itself, which he explains in the book and which you can also find at their website at http://moralfoundations.org/.

Haidt makes the point that these are not characteristics of just one culture or region or country; they are characteristics of the species *Homo sapiens*, as discovered through extensive research. Based on how they behave, it is a near certainty that to some degree the theory and its moral foundations are also relevant to much mammalian life on earth and particularly to the primates.

The theory notes that "intuitive ethics" is at the core of human psychological programming and processing, and that at least five "foundations" underlie the theory. The following description of the five core foundations and one other probable foundation is taken directly from the website cited above.

> "**1) Care/harm**: This foundation is related to our long evolution as mammals with attachment systems and an ability to feel (and dislike) the pain of others. It underlies virtues of kindness, gentleness, and nurturance.
> **2) Fairness/cheating**: This foundation is related to the evolutionary process of reciprocal altruism. It generates ideas of justice, rights, and autonomy. [Note: In our original conception, Fairness included concerns about equality, which are more strongly endorsed by political liberals. However, as we reformulated the theory in 2011 based on new data, we emphasize proportionality, which is endorsed by everyone, but is more strongly endorsed by conservatives]
> **3) Loyalty/betrayal:** This foundation is related to our long history as tribal creatures able to form shifting coalitions. It underlies virtues of patriotism and self-sacrifice for the group. It is active anytime people feel that it's "one for all, and all for one."
> **4) Authority/subversion:** This foundation was shaped by our long primate history of hierarchical social interactions. It underlies virtues of leadership and followership, including deference to legitimate authority and respect for traditions.
> **5) Sanctity/degradation:** This foundation was shaped by the psychology of disgust and contamination. It underlies religious notions of striving to live in an elevated, less carnal, more noble way. It underlies the widespread idea that the body is a temple which can be desecrated by immoral activities and contaminants (an idea not unique to religious traditions).
>
> We think there are several other very good candidates for "foundationhood," especially:

> **6) Liberty/oppression**: This foundation is about the feelings of reactance and resentment people feel toward those who dominate them and restrict their liberty. Its intuitions are often in tension with those of the authority foundation. The hatred of bullies and dominators motivates people to come together, in solidarity, to oppose or take down the oppressor."

It is again clear that this work has strong and direct implications for religious belief. Once a religion is created, and people start believing in its tenets and a related social group starts to form, several of these foundations start to kick in, particularly loyalty/betrayal, authority/subversion and sanctity/degradation. Religion has both supernatural and social components. Thus some of these foundations actually have double impact in that they affect both components.

In point of fact, there is actually a third level of impact, because religions either explicitly or implicitly see themselves as the sole owners of a universal moral code, so a believer's ties are to the god or gods, to the group and also to the assumed moral superiority of the religion. Thus true believers are embedded in and wedded to their faith quite viscerally and emotionally with several independent but related ties.

The loyalty/betrayal foundation links to religion in each of those three ways. The practicing true believer has to be loyal to the god, loyal to the clergy and loyal to the moral code. Betraying any of these is considered a grave "sin", if you will, and the faithful feel that it must be punished. Punishment can be inflicted in this world through expulsion, excommunication and/or execution, and in all likelihood still more punishment will be required in the afterlife, applied directly by the god of the faith, one of his minions or some devil or demon.

The authority/subversion foundation's inherent tie to religion also uses all three linkages. Believers must respect the authority of the god, of the human leaders of the social group/religion, and the assumed superior authority of the religion's moral code. If you like, you could add a fourth linkage, which would be the traditions Haidt mentions. Thinking you are still a practicing believer but refusing to partake in rituals and traditions will almost certainly mark you as a subversionist and a strong candidate for at least some degree of ostracism if not full expulsion.

In the extract cited above, Haidt specifically named how the sanctity/degradation foundation links to religion. This particular foundation is *extremely* relevant to religion. The faithful doggedly pursue a utopian Never-Never Land in which all sin and misbehavior are banished, and in which the human body and soul are cleansed of all things considered disgusting and unclean by the brethren and their gods. The problem is that this also includes even the thoughts of doing things that are sinful, and depending on their degree of sadism and masochism humans will call almost any thought or deed sinful, especially if it involves sex.

Therefore some particularly abnormal and cruel people will say that even the most completely normal and/or innocuous human desires and actions are disgusting and require shame, guilt and extreme punishment.

All the foundations set the stage for inclusiveness or xenophobia, for acceptance or hatred, and for both moral and immoral acts. If a person responds too strongly to any of these foundations, people who are in his group will be accepted and those not of his group will be treated with fear, disgust, distrust and contempt. This sort of response can be merely a snobbish internal and unspoken disdain for other people, but it is too often not so constrained or hidden.

This kind of thinking and acting was raised to cataclysmic proportions in the Final Solution, in which Hitler and his largely Christian followers saw Jews as a disgusting and not fully human "race" of Christ-killing vermin who should all be murdered.

The hatred and bigotry people feel and act on towards people have many names, including racism, sexism, misogyny and homophobia. People are feared, hated and often mistreated for being of a different ethnicity, color, gender, religion or lack of religion, nation of origin, social class, high school or college, favorite sports team, and indeed any other characteristic that differs from one's own group.

The problem is that almost every human alive feels some degree of xenophobia or distrust in some form or fashion. Some people are hypertribal, while others are merely uncomfortable with newness and change. Some humans are the opposite, and are hyperinclusive with no, very few and/or very specific hypertribal tendencies. The conservative mind, which is often also a religious and relatively dogmatic and inflexible mind, is more xenophobic than average. The liberal mind, which is often less religious and more flexible and nondogmatic, is less xenophobic than average.

We are all generally programmed to have tendencies like this due to many millions of years of evolution and through some artificial selection. We are specifically programmed to have our own personal tendencies based on individual differences in brain structure, social programming and other influences as described earlier in this section.

Liberals may think that conservatism is actually a pathology. Conservatives may think that liberalism is actually a pathology, but in the opposite direction. The truth lies at the extremes and in the context.

As in all observations humans make, the less populous but much louder extremes always stand out more than the quieter and more moderate masses in the middle. Just as the squeaky wheel gets the grease, and disasters, terrorists and disturbing or abnormal outliers get the "if it bleeds, it leads" headlines and attention, the raving lunatic right wing nut jobs and far left wing uberpeaceniks get the attention.

Hyperconservatism, when applied to human societies and human interactions, is indeed a pathology. Before liberals get too excited, they should note that hyperliberalism is also a pathology, when applied to other aspects of human interaction such as individual and group safety and security.

Conservatism, in the classic sense of tending to preserve the current order and having a reasonable caution about new things, is a necessary part of a reasonable and balanced approach to human thinking and action. Note that this does not in any way refer to the American Republican party of 2016 and the last few decades, as it has quite deliberately destroyed any sort of reasonable or rational "little c" conservatism. Merely being "little c" conservative is not pathological. Republican politics in America in 2016 is highly pathological.

Liberalism, in the "little l" general sense and not referring to the American Democratic party at all, but in the general sense of openness and acceptance, is also a necessary part of properly balanced human society and thinking. Minor differences are necessary to cleave a reasoned solution from a tangled web of reasoned "little c" conservative and "little l" liberal perspectives.

Hyperconservatism is a pathology. Hyperconservatism is a hyperexpression of the fear of outsiders and disgust at contamination, which expresses itself in xenophobia and sadism in a horrible and often murderous blend. Fear of others served an important role in our evolution because such behavior kept invaders away and kept the tribe safe from all intrusions, be they dangerous diseases or dangerous ideas. Keeping unclean things away prevented disease. Keeping unclean people away prevented crime, misbehavior and challenges to authority. Keeping inconvenient realities away and having far too conservative a world view has led to mass persecutions and killings throughout human history.

Mistreating, demonizing, persecuting and even killing people merely because they are different is evil and immoral, but history proves that it is an inevitable result of hyperconservatism. Religion is usually ubiquitous in both conservatism and hyperconservatism, and it serves as a blanket justification for the entire spectrum of evil acts. The holy books of the Mesopotamian monotheisms, and the religions they spawned, are perfect examples of hyperconservatism, hypertribalism and hyperxenophobia.

Hyperliberalism is a pathology. Hyperliberals love everyone and everything and don't quite get that there are psychopathic individuals who are born that way, who are not treatable or even easily identifiable with current medical technology, and who will do their damnedest to hurt and/or kill people. Hyperliberalism opens too many doors to infection by disease and to being conquered by aggressors. Liberalism and acceptance of outsiders, though, served a role in our evolution because new people and new ideas prevented things like inbreeding and damaging mutations from having too-small gene pools.

Liberalism and acceptance enabled the invention and incorporation of new technologies and better or more efficient ways of living. Without new ideas and some degree of acceptance of new people and the entire concept of progress which is inherent in the application of progressivism and liberalism, the human race would be localized and inbred, likely still living in caves, serving tribal kings, wearing animal skins and going on raids to kill other men and steal women and food.

So the hyperconservative hates and distrusts everyone and everything, the hyperliberal loves and trusts everyone and everything. The former gets you Hitler, Nazis, fascists, dictators, Catholic States, Islamic States, Inquisitions, Crusades, bigots, religious freedom holy warrior zealots and the current Republican party. The latter gets you taken advantage of, cheated, infected, invaded and overrun. The challenge we face as a species is, frankly, how to best use science and education to marginalize and effectively muzzle the counterproductive hypers on both sides and to identify the best path forward for the most people.

And to get us back to the topic of this book, how does all this psychology and the references to politics relate to religion? The relevance comes from the fact that liberalism and conservatism are not merely political terms confined to a single area of discourse. They describe how humans think and act in the real world in all areas.

Hyperliberal religious beliefs get you Jesus and the penniless hippie healer ethos that he actually preached, the ultrapacifistic faith of Mohandas Gandhi and the nonviolent civil rights campaign of Martin Luther King Jr. Hyperliberal beliefs get you the peaceful Jains and monkish lifestyles in which people burn themselves alive rather than burning other people.

Hyperconservative and even moderately conservative religious beliefs get you the oppressive and evil history of Catholicism, the Inquisition, the Crusades, the Holocaust, slavery, bigotry, suicide bomb mass murders, passenger planes flying into skyscrapers, American antihuman evangelism and people burning other people alive. That one can use the same exact holy books, and in some cases the exact same words, to justify hyperliberal or hyperconservative actions indicates the original and permanent fault in those holy books, which is expressed in those who choose to follow those words to do evil.

Hyperconservatives in America and elsewhere have twisted the teachings of the biblical hyperliberal peacenik Jesus into aggressive campaigns of mean-spirited bigotry, in which only a chosen few supposedly have a mandate from Jesus to own guns, subvert secular law and oppress or persecute certain groups of people. "Jesus wept", indeed.

Jesus would do far more than weep if he were to ever come back and see how his message, indeed his direct orders, had been bastardized and Satanized. If there ever is a Second Coming in America it won't start off with masses of people floating off into the sky toward some magical Candyland, it will start with much flogging and whipping of many millions of the publicly, immorally and hypocritically faithful by the divine and untiring hand of Jesus the Enforcer.

History has shown that religions have almost invariably been used as the single most powerful justification for xenophobia, tribalism, persecution and murder, and for generally bringing out the worst in people. Christianity is an absolutely perfect example of how this is done in practice. The biblical Jesus character is actually one of the most liberal in all of human literature. Yet his teachings of compassion and forgiveness, ministering to the poor and castigating or even whipping the rich and money-focused, healing the sick and welcoming those in need are often absent in American Christianity.

Indeed, categorizing Christian groups in America in the 21st century is merely an exercise in sorting the different sects by how many different groups of people each specific sect discriminates against, and to what degree.

Many Christians today can only be partially forgiven their hyperhypocrisy when one understands that almost none of them can be bothered to actually read the words of the god they allegedly worship so devotedly. Indeed, why aren't all people who actually follow the teachings of Jesus actually flaming liberals in real life, from their words to their deeds to their votes? Because hypocrisy, xenophobia and greed trump all else, even the actual words of the holy book these people claim as their guide to life.

Religion can never be an arbiter of a truly decent and humanistic morality, and it has proven this for thousands of years in nearly every country on earth. To prove this all one need do is look at how many

Jains, Buddhists and Quakers there are, as opposed to how many Christians and Muslims there are.

The first three are peaceful and pacifistic groups; the latter have been hyperaggressively hyperconservative, xenophobic, exclusionary and have killed and persecuted far too many people to count. Jain fundamentalists have never fought a single war or killed a single person.

Extremely devout and faithfully fundamentalist Christians and Muslims have killed people wherever they have gone, often in massive numbers and in ways too horrible to describe, all while following the very specific orders and teachings of their holy books and their faith and in defense of these faiths.

One really need no more evidence than this to reject any claim by any of the monotheisms that they have any contributing role in any conversation on morality, now and forever. They have written the books on grotesque immorality, both literally in their holy books and in practice with the blood of their victims.

How does all this happen? How can a person claim to be a decent person when his acts are truly vicious and horrible? How can a person claim that his or her holy book is good and moral, when it is the exact opposite? For that matter, how can a person deny the findings of science when the evidence is ubiquitous, cross-disciplinary and overwhelming? For this we must refer back to the opening paragraphs of this chapter and to more facts of human psychology. Humans have an uncanny ability to fail to think, to think incorrectly, and to only see what they want to see. Over time, objective science has catalogued a large number of the flaws in our thinking.

Table 1 below is a partial list of human cognitive biases and logic flaws that sabotage reason and critical thinking. While it is somewhat lengthy, this list is nevertheless very far from all-inclusive and does not even include phobias. Note that every human alive currently operates while under the spell of some number of these traps, and while we theoretically have the capability to discern them in ourselves, we almost always do not. We can't even do a good job of perceiving them in others.

Now that these flaws and biases are known to science, we now have the ability to be told that we tend to fall victim to certain biases, and to understand this and change our behaviors accordingly, but again this almost never happens in practice. As we continue to gain more knowledge of human psychology, we continue to gain a clearer understanding of why people think, say and do what they do.

The first two columns in the table list the bias or logic flaw, a brief description of it and its relevance to religion.

The third and fourth columns reflect the degree to which these cognitive traps first enable the invention of and then the persistence of religious or supernatural beliefs, ranked on a scale from 1 to 5 with 5 being the most powerful and 1 the least powerful.

As you read through the table, think about each bias or flaw and how it would be expressed by a person writing the Old Testament, the New Testament or indeed anything written two thousand or more years ago. Think of the perspective of that person, wholly ignorant of the actual facts of astronomy, meteorology, chemistry, physics, geology, biology, psychology, sociology and the other sciences.

In other words, think of how unbelievably ignorant you would be on these subjects if you had been raised by ancient desert goatherders. Also think of how our biases can be reduced by the vast amounts of knowledge which even a poor high school education and life in a First World society tries to put into a person's head today.

Gods are no longer necessary to explain lightning, thunder, earthquakes and illness, and they certainly aren't needed to explain cars, televisions, cell phones, waves, mountains, the origin of life or the origin of the universe. But explanations from not even that long ago for even the most currently obvious and simple problems defaulted to a divinity due to our near-total scientific ignorance. And this ignorance is the starting point for every single new human.

Babies are born with no knowledge of the world and each person has to accrue massive amounts of knowledge in order to reach a properly informed perspective.

Think about how easy it was for people whose world was absolutely filled with imaginary gods and goddesses to attribute everything to those gods, to have his or her thinking conform to a mindset which considered gods as a natural and accepted fact, and thus how imprisoned these people and their society were by religion and religious beliefs.

As you read the table below from that perspective, see just how easily the human brain can first be anchored to supernatural beliefs, then to confirm them, and how hard these ideas are to kill once developed and established.

Think about today's believers and how different people can be equally devout followers of religions invented in antiquity such as Judaism and Christianity or of recently invented ones such as Scientology or Mormonism.

Table 1. Listing of cognitive biases, their religious implications and how they contribute both to the invention of and the persistence of religion in the human mind and society.

Anchoring.[39] Over-reliance on one bit of data about a subject, often the first bit of data one hears, while making a decision. The first bit "anchors" thinking on that subject.	**Religious Implication.** When young people are indoctrinated into religion, it very often stays with them for life and shapes a great deal of their thinking.	**Enables Invention** **1**	**Enables Persist-ence** **4**
Anthropocentric Thinking.[40] "…the belief that human beings are the most significant entity of the universe. Anthropocentrism interprets or regards the world in terms of human values and experiences."	**Religious Implication.** This strengthens the tendency to adhere to anthropomorphic religions like the three monotheisms.	**Enables Invention** **1**	**Enables Persist-ence** **3**
Anthropomorphism.[41] "…the interpretation of nonhuman things or events in terms of human characteristics, as when one…hears human voices in the wind."	**Religious Implication.** As above, this strengthens the tendency to adhere to an anthropomorphic religion.	**Enables Invention** **1**	**Enables Persist-ence** **3**
Attentional Bias.[42] When thoughts affect perceptions. If a person is thinking of the color red, his perception tends to incorporate redness.	**Religious Implication.** As with anchoring, attentional bias in a believer keeps focus on the faith and its claims and far away from facts, logic and reason.	**Enables Invention** **1**	**Enables Persist-ence** **3**
Authority Bias.[43] The inclination to "…attribute greater accuracy to the opinion of an authority figure (unrelated to its content) and be more influenced by that opinion."	**Religious Implication.** The faithful generally follow the words of their theologians and clergy, regardless of any resulting conflicts with facts, reason and humanism. This is due to the strong "follower" characteristic in many believers.	**Enables Invention** **1**	**Enables Persist-ence** **3**
Availability Cascade.[44] The process by which an idea grows in acceptance and popularity through increased repetition.	**Religious Implication.** Religions rely on the unceasing repetition of the same words and messages, multiple times weekly or even daily. Thus they become lodged in the brain even if they are untrue. As Mark Twain said[45], "History is strewn thick with evidence that a truth is not hard to kill, but a lie, well told, is immortal."	**Enables Invention** **1**	**Enables Persist-ence** **4**

Backfire Effect.[46] When a person responds to evidence disproving a strongly held belief by ***increasing*** the strength of his belief and rejecting unwelcome evidence. It is probably the most damning example of human illogic and perversity, that a person can unashamedly continue to deny facts that are literally undeniable.	**Religious Implication.** Many people resist disconfirming data with whatever force is required to maintain their beliefs. Many others, in the drive to permanently block this information, ramp up their devotion to their beliefs. Thus the faith becomes irrevocably entrenched and many of these people begin to or increase their current efforts to convert others to their beliefs. This is one of the main reasons why unjustified faith is still with us and why the most insecure of the faithful can't keep their religion to themselves.	**Enables Invention** **1**	**Enables Persist-ence** **5+**
Bandwagon Effect.[47] The inclination to think, say or do things because large numbers of other people do.	**Religious Implication.** Sheeple follow fads, and religions are a species of fad. Once a fad or religion grows, herd dynamics and the bandwagon effect combine to keep people in the religion.	**Enables Invention** **1**	**Enables Persist-ence** **3**
Belief Bias.[48] The inclination to evaluate an argument or position based on how closely it aligns with the person's beliefs.	**Religious Implication.** This is why believers reject disconfirming evidence. They don't like how it affects their beliefs, or how it makes them feel, so they reject even incontrovertible evidence.	**Enables Invention** **1**	**Enables Persist-ence** **5**
Bias Blind Spot.[49] The inclination for a person to see bias in other people but not to see bias in oneself.	**Religious Implication.** Religions are founded on faith, not facts, and since religions have no self-corrective mechanisms, the blind spot lives on.	**Enables Invention** **1**	**Enables Persist-ence** **4**
Choice-Supportive Bias.[50] The inclination to justify one's choices after the decision has been made.	**Religious Implication.** Particularly in religion, which has so many unprovable and unjustifiable claims, the strong desire to avoid embarrassment and criticism and to reassure oneself that joining a faith was the right decision prevents objective assessment of the decision.	**Enables Invention** **1**	**Enables Persist-ence** **3**

Clustering Illusion.[51] The inclination to believe that small streaks in data represent a pattern when they actually do not.	**Religious Implication.** The religious mind is tuned, trained and encouraged to see patterns that do not exist or to interpret patterns in a way supportive of the faith. It thus perceives patterns which are thought to support the faith, despite the lack of evidence or reason for this position.	**Enables Invention** **1**	**Enables Persist-ence** **4**
Confirmation Bias.[52] The inclination to seek, interpret and preferentially believe data that is believed to support one's existing beliefs.	**Religious Implication.** Believers tend to only accept data they think confirms their beliefs, thus failing twice. Failure 1 is to reject disconfirming evidence, and Failure 2 is to twist facts as far out of shape as is needed to rationalize the beliefs.	**Enables Invention** **1**	**Enables Persist-ence** **5**
Curse of Knowledge.[53] The inclination for a more informed person to fail to consider issues from the perspective of the less-informed. This can lead to callousness, insensitivity and an inability to communicate.	**Religious Implication.** Religious people are often denigrated as weak or ignorant, and most are, but merely saying so and not taking appropriate action is not being constructive. People are believers for many reasons, and each must be addressed in a way understandable to the believer, so that the path to reality is clear.	**Does not enable *formation* of religion**	**Enables Persist-ence** **1**
Dunning–Kruger Effect.[54] The inclination for people with low ability, competence or knowledge to fail to see these characteristics in themselves.	**Religious Implication.** Everyone thinks he or she is smart and knowledgeable even though many are not, and these people are often the most aggressive in using their feelings as justification for their thoughts and actions. Forrest Gump said it best with "Stupid is as stupid does".	**Enables Invention** **1**	**Enables Persist-ence** **4**
Egocentric Bias.[55] The inclination to think higher of one's abilities and opinions than is warranted.	**Religious Implication.** The size and reach of the ego of the true believer, who feels he knows what his god wants him to think and do, have no limits. Thus we see the insufferable egotism of those who claim to know the mind of their god and think the rest of humanity is just a bunch of misinformed people who need to be converted to the faith or left to wallow in their "unsaved" misery.	**Enables Invention** **1**	**Enables Persist-ence** **2**

False Consensus Effect.[56] The inclination for a person to believe his personal thoughts, opinions and beliefs are more widely held than they are. It is self-deception.	**Religious Implication.** This very dangerous fallacy reduces individual restraint. When a person thinks everyone else thinks like he does, he often acts on those thoughts, and with religion this invariably leads to antihuman behavior.	**Enables Invention** 1	**Enables Persist-ence** 3
Forer Effect or Barnum Effect.[57] "...when individuals believe that personality descriptions apply specifically to them (more so than to other people), despite the fact that the description is actually filled with information that applies to everyone. The effect means that people are gullible because they think the information is about them only when in fact the information is generic. The Barnum Effect came from the phrase often attributed (perhaps falsely) to showman P. T. Barnum that a "sucker" is born every minute."	**Religious Implication.** Like the other cognitive traps in which people imagine patterns that aren't there, people make connections that are not there to justify their religious beliefs. Just like when 100 random people can all believe the exact same horoscope is talking about them personally, believers read about things in their books that they falsely perceive to be "real" at some point in history or their lives.	**Enables Invention** 1	**Enables Persist-ence** 4
Framing Effect.[58] When people make different choices resulting from the same information, due to a difference in how the information is presented.	**Religious Implication.** The Jesus story is said to be "good news", as he was allegedly sent here to die heroically for humanity's sins, and Catholics thus celebrate the Eucharist. It can be restated as 'the creator of the universe sent his only son to be tortured to death so the sins of the people which the creator had created with these sins built into them could be forgiven, and to celebrate this event people now believe they actually eat and drink the reincarnated zombie body and blood of this murdered half-breed Jewish preacher in magical ceremonies performed by celibate males.' Same information, different presentation, much different processing and response in the reader.	**Enables Invention** 1	**Enables Persist-ence** 3

Fundamental Attribution Error.[59] This is "…the tendency people have to overemphasize personal characteristics and ignore situational factors in judging others' behavior. Because of the fundamental attribution error, we tend to believe that others do bad things because they are bad people. We're inclined to ignore situational factors…"	**Religious Implication.** The faithful are notoriously judgmental, and rely on very simple platitudes such as "sin" or the teaching of ancient mythology over the understanding reached through complex analysis of the realities of psychology, sociology, anthropology and biology. Such unscientific and ignorant behavior is antithetical to a civil society.	**Enables Invention** **1**	**Enables Persist-ence** **4**
Hindsight Bias.[60] "…the tendency, upon learning an outcome of an event—such as an experiment, a sporting event, a military decision, or a political election—to overestimate one's ability to have foreseen the outcome. It is colloquially known as the "I knew it all along phenomenon."	**Religious Implication.** The religious, being primed to praise their god for good outcomes and chalk bad ones up to his inscrutable will or to dirty rotten sinners, are similarly primed to say they knew things were going to happen as they did, after the fact and despite what they may have previously said.	**Does not enable *formation* of religion**	**Enables Persist-ence** **3**
Illusion of External Agency.[61] The inclination for people to think that outside agents such as a god influence actions they actually take of their own accord.	**Religious Implication.** When people imagine their god guided them to do something that was beneficial, they self-reinforce their beliefs and further cement the illusion of divine intervention into many facets of their lives. Despite the lack of evidence for such belief, this self-fulfilling and self-reinforcing cycle digs the believer ever deeper into the faith, making it nearly impossible for facts to intrude or the mind to change.	**Enables Invention** **1**	**Enables Persist-ence** **5**

Illusion of Truth Effect.[62] The inclination of a person to believe the truth of statements he has heard multiple times, even if the statements are completely and intentionally false. In other words, belief that is accrued through simple repetition.	**Religious Implication.** As with the Availability Cascade, even the biggest lies are eventually thought true if they are repeated often enough. Religion is founded on fables repeated endlessly, so believers accept them despite the lack of evidence. Repetition is reinforcement and the adherence to the faith is strengthened. For centuries, Catholics and Christians persecuted and reviled Jews for imaginary behaviors and this eventually led to the Holocaust. Religions incessantly repeat their propaganda and dogma. Actions in the service of dogma and propaganda are always antihuman and destructive.	**Enables Invention** **1**	**Enables Persist-ence** **4**
Illusory Correlation.[63] This is incorrectly thinking that two unrelated factors actually are related.	**Religious Implication.** Believers link positive traits like morality with their own in-group. This reinforces belief, tribalism and xenophobia due to perceived in-group superiority.	**Enables Invention** **1**	**Enables Persist-ence** **3**
Ingroup Bias.[64] The tendency for people to treat well those who they believe are in their group, and to treat poorly those who they believe are not in their group.	**Religious Implication.** Religious divisions are the most destructive, because believers think they actually have divine authority to support their faith group and fight or destroy out-groups. Examples include religious bigotry, religious wars, the Holocaust, Crusades, Inquisitions, persecution of certain groups and countless other examples.	**Enables Invention** **1**	**Enables Persist-ence** **4**
Just-World Hypothesis.[65] The tendency for people to think that life is "orderly, predictable and just", which often leads to victim-blaming when misfortune or tragedy occurs. In other words, the victim deserved it.	**Religious Implication.** Believers have proclaimed for millennia that bad things happen to sinners via karma or divine punishment. They use their religion to justify their lack of sympathy for certain victims, especially if the victim was engaged in an activity the faithful consider sinful.	**Enables Invention** **1**	**Enables Persist-ence** **3**

Law Of The Instrument.[66] The tendency to use a familiar method or tool to accomplish a wide variety of tasks. The most common cited example of this is the phrase "If all you have is a hammer, everything looks like a nail."	**Religious Implication.** This also applies to mental heuristics, so a religious person will see and try to explain the world through his own personal religious lens. This is why the religious ignorance and dismissals of religion-disproving science are so pernicious and destructive.	**Enables Invention** **1**	**Enables Persist-ence** **4**
Mere Exposure Effect.[67] The tendency to like things more than would be expected simply due to being familiar with them.	**Religious Implication.** Religions are replete with rituals, ceremonies, services and other events, often in the same place and with the same people. The frequency of exposure strengthens the belief in religion and its teachings, no matter how unbelievable or immoral, and the clergy know it.	**Enables Invention** **1**	**Enables Persist-ence** **4**
Naïve Realism.[68] A person's belief that he sees reality as it really is, while those who disagree with him do not see reality as it really is because the other person is believed to be biased, ignorant or irrational.	**Religious Implication.** Believers think theirs is the One True Faith, and that nonbelievers and believers in other faiths are biased, ignorant or irrational. This is used as justification to continue in the faith and strongly contributes to tribalism. Strong challenges to a group's faith bind members together even more closely, and they will double down on their naïve realism.	**Enables Invention** **1**	**Enables Persist-ence** **3**
Ostrich Effect.[69] Ignoring obviously negative information.	**Religious Implication.** There are far too many examples of this in religion to count much less record in this book. Catholicism has a history written in blood and murder, yet believers blithely ignore this history and ignore painful truths. The global child rape scandal which isn't causing people to leave Catholicism is just one example. All religious believers, especially Christians, have an uncanny ability to ignore anything that puts their faith in a bad light or disproves their religion.	**Enables Invention** **1**	**Enables Persist-ence** **4**

Overconfidence Effect.[70] Similar to the Dunning-Kruger effect, this is the undeserved confidence people show in their answers to questions. People almost invariably have high confidence in their beliefs and answers, despite these answers being very wrong in many cases, with error rates as high as 40% even when the answer is given with extremely high confidence.	**Religious Implication.** Overconfidence is common in humans, and science is the best method we have to eliminate it. Religious beliefs are reinforced by repetition, by groupthink and by each believer's internal need for the belief to be true, thus believers cling to undeserved overconfidence. This is obvious when asking believers for evidence. Even though they cannot give any, they astoundingly maintain great confidence even in the minute details of beliefs for which there is neither justification nor evidence. And when a person imagines that his god has told him something, there is no changing his mind.	**Enables Invention** **1**	**Enables Persist-ence** **4**
Pareidolia.[71] "The tendency to perceive a specific, often meaningful image in a random or ambiguous visual pattern."	**Religious Implication.** People imagine the faces of Yahweh, Jesus or his mother or any of their countless saints and revered people in clouds, food, trees and smoke. In his book *The Believing Brain*[72], Michael Shermer discusses how people hear imaginary messages and perceive nonexistent patterns in static and noise. This ties directly to religious belief. Christians and Muslims never see or hear from Zeus or Marduk when they hallucinate or pretend to commune with the divine, proving how extremely provincial a believer is when imagining things where none exist.	**Enables Invention** **2**	**Enables Persist-ence** **4**

Post Hoc, Ergo Propter Hoc.[73] Latin phrase describing a very common fallacy. It means "After this, therefore because of this: because an event occurred first, it must have caused this later event —used to describe a fallacious argument." An example is that if my phone rings or I yawn after a dog barks, the barking did not cause my phone to ring or me to yawn. The barking happened to precede the ringing or the yawn. No gods were involved.	**Religious Implication.** Religions, being founded on unprovable claims, are the human activity most susceptible to logical error. This particular error is very powerful and is very common in religion. Believers are primed to believe their gods intervene in the world, and due to the attentional bias, the availability cascade and the confirmation bias, they ignore bad events and always notice good ones in a way they like. Thus if they pray for X and it happens, they think their personal prayer to their personal god was the cause. Billions of people currently suffer from this insane delusion and its power is greatly magnified in the believer's head because of all the other biases that also destroy critical thinking.	**Enables Invention** **3**	**Enables Persist-ence** **5**
Post-Purchase Rationalization.[74] The inclination to persuade oneself after the fact that a purchase or decision was the right one.	**Religious Implication.** Denial isn't just a river in Egypt, and humans can rationalize absolutely anything. The compulsion to reject the possibility of error in order to maintain one's self-image as a good decision-maker means even horribly bad decisions will be rationalized away. Thus many people will remain in their religion no matter how bad the choice or the religion. That a billion people are still Catholics despite the global rape epidemic and its coverup by their church's leaders proves the power of such rationalization.	**Does not enable *formation* of religion**	**Enables Persist-ence** **4**
Selective Perception.[75] People see what they want to see and hear what they want to hear. A thousand people can watch an event and each can come away with different interpretations of what happened based on their personal biases.	**Religious Implication.** Like the confirmation bias, believers see the world through the filter of their own particular religion, so a factual understanding of an event is difficult to achieve. This is even for events of which we have full video and audio.[76] The believer often accepts only what confirms his perceptions.	**Enables Invention** **4**	**Enables Persist-ence** **4**

Self-Serving Bias.[77] The inclination to feel more personal contribution to personally experienced successes than failures, and to feel less personal responsibility for those personally experienced failures.	**Religious Implication.** Humans have an unlimited ability to rationalize. Believing good things happen to you because you follow the One True Faith and its god or prophet, and attributing bad things to a devil or evil spirit, is extremely common among believers. Many Christian groups focus almost exclusively on sin, shame and guilt in all its masochistic expressions, but one can be part of these cults and still keep the self-serving part of this bias, along with an unwelcome zeal to spread this belief to others who do not share it.	**Enables Invention** **1**	**Enables Persist-ence** **5**
Semmelweis Reflex.[78] This is a nearly universal and very significant problem not usually referred to by this term. It is "…the reflex-like tendency to reject new evidence or new knowledge because it contradicts established norms, beliefs or paradigms." It is a childish rejection of things which challenge your current beliefs.	**Religious Implication.** Religions center on their ancient holy books, teachings and traditions, and they demand their flocks adhere to them. Every time science shows religions are full of errors, ignorance, immorality and bad ideas, and every time science makes a new discovery that disproves some element of faith, many believers reject the information out of hand. They maintain this rejection until some believer dreams up a way to rationalize the discovery as being part of the god's plan after all and even more proof of the god's existence.	**Enables Invention** **1**	**Enables Persist-ence** **5**

Shared Information Bias.[79] The Wikipedia definition is so good and so relevant to religion that I included the first few sentences here. It says"…the tendency for group members to spend more time and energy discussing information that all members are already familiar with (i.e., shared information), and less time and energy discussing information that only some members are aware of (i.e., unshared information). Harmful consequences related to poor decision-making can arise when the group does not have access to unshared information (hidden profiles) in order to make a well-informed decision."	**Religious Implication.** Religions rely on faith because they can't prove any of their claims. Thus they engage in the endless repetition of the faith's mantras and dogmas in services and between and among members to continue the groupthink and reinforce the in-group tribalism. Information from the outside is not welcome due to the damage it can inflict on dogmas. The herds which comprise the religious flocks do not like to leave their comforting echo chambers and self-congratulatory and self-reinforcing meetings. Believers don't want their superhero gods or their beliefs being exposed to any Kryptonite-like realities. Hosting "why we believe" parties is more their style.	**Enables Invention** **1**	**Enables Persist-ence** **4**
Status Quo Bias.[80] The tendency to like things to stay just as they are, or with very minimal changes. This is very strong in conservatism and the conservative and religious mind.	**Religious Implication.** Religions abhor change, because change would be a massive admission they had been wrong about their god's message. Therefore the books do not change and the teachings change as little as possible and only when under tremendous pressure. And when faiths do change, people often leave, which is highly undesirable to the religion's leaders. Also, once a person has ties to a faith group, the social cost of leaving is almost always unacceptably high. Thus the status quo is very strongly maintained.	**Does not enable *formation* of religion**	**Enables Persist-ence** **5**

<u>Subjective Validation.</u>[81] Wikipedia again has such a good and relevant definition that I used it here. This is a "...cognitive bias by which a person will consider a statement or another piece of information to be correct if it has any personal meaning or significance to them. In other words, a person whose opinion is affected by subjective validation will perceive two unrelated events (i.e., a coincidence) to be related because their personal belief demands that they be related."	**Religious Implication.** This bias maintains dogma and belief in the minds of the faithful. Much like a non sequitur is a wrong conclusion; subjective validation means that a believer is highly likely to attribute a positive event to his god if it will make him feel better to do so. If a negative event befalls someone he does not like or someone who he thinks is a sinner or somehow deserves punishment, this is also taken as proof of the god he worships. This is often seen when believers think their god has sent disaster to a community for some imagined sin by one or more of its residents, be it Sodom and Gomorrah, San Francisco, New York City or New Orleans.	**Enables Invention** **1**	**Enables Persist-ence** **4**
<u>Sunk Cost Fallacy.</u>[82] This is "the idea that a company or organization is more likely to continue with a project if they have already invested a lot of money, time, or effort in it, even when continuing is not the best thing to do." It is deliberately continuing with an action[83] beyond when reason and evidence show that the action should be discontinued, as a result of feeling pressure to continue with the action due to the large amount of resources already invested in the action. Also known as Irrational Escalation.	**Religious Implication.** Religion is full of examples. When a believer is heavily invested in a religion, particularly socially or due to the volume of time and resources placed into the faith for long-time believers, leaving it is seen as less desirable than just staying the course even after grave disillusionment. I again have to note how many believers are still Catholic even after the global child rape scandal and the equally horrific coverup from the very top all the way down. And even after so many people remained Jewish after the six million dead in the Holocaust. There is just no shaking some people out of their unjustified beliefs.	**Enables Invention** **1**	**Enables Persist-ence** **5**

Survivorship Bias.[84] Focusing on the survivors of an action or process and disregarding those which or who did not survive, due to the lack of visibility of the non-survivors after the event or process.	**Religious Implication.** Out of many, the most relevant implication is the near total ignorance most people have about every other religion or god humans have invented except the one they are in. Countless religions, sects and cults have formed and died, and there is no reason to believe any of the current ones are any more correct than any of the dead ones. And the horrifically high numbers of those who die from "Acts of God", which are just godless natural disasters, are ignored in favor of the few so-called "miraculous" survivors. This is pathetic and ignorant. A true miracle isn't one person surviving a disaster or plane crash. A true miracle is every single person surviving unhurt, or a plane landing safely despite losing a wing. Such simple and obvious failures of logical and critical thinking strongly enable religious belief.	**Enables Invention** **2**	**Enables Persist-ence** **4**

The above listing is obviously quite extensive, but again it is very far from complete. Unfortunately, due to our national failure in America to educate ourselves and our children on basic human thinking, most if not all of these fallacies and biases will be new to many people. Anyone who reads through them all will surely have noticed at least some in other people.

One hopes that readers will also be able to see them in themselves, and thus see human thought and interaction in a more informed light. Again, the reason for including them in this book is that many of them contribute directly to the formation then to the persistence of religions and religious ideas.

As a quick summary, the following is a general flow of how these mental traps have enabled the massive accumulation of religions, beliefs and believers we have seen across our world.

Humans have been able to speak for a very long time. One study[85] estimates that language began to form as much as 1.75 million years ago. Whether it started this long ago, or when we evolved into our current level of genetic modernity approximately 200,000[86] years ago, or during a Great Leap in human evolution approximately 50,000 or 60,000 years ago, it is assuredly the case that during much of our evolution we shared most if not all of the basic patterns of both our modern human cognitive capabilities and our biases.

Thus some people were intelligent, some were inquisitive, some were slow of mind, some were gullible and some were skeptical. Once these humans began to speak, they started to discuss things like what caused rains, thunder and lightning, where the world and the life on it came from, what dreams were and what they meant, where babies came from and why so many of these babies and their mothers died. Their discussions and decisions were influenced by their own personal cognitive biases and the events in their short and difficult lives and their tiny social groups.

The brains of humans born 40,000 years ago were much the same as ours are today. If she were transported to the present day, a baby born 40,000 years ago to a Stone Age hunter-gatherer would grow and develop like a baby born today. Our society has accumulated an unbelievably large amount of information but absolutely none of it enters the brain except through a lifetime of slow learning, filtered through the cognitive biases listed above and many others not listed. Every newborn baby's brain is essentially a blank slate except for an onboard operating system, some processing modules and a few instincts. So the hardware and software of today's humans and that of Stone Age humans are largely the same.

The only difference is the information now available for the growing person to take in.

Thus a person born in the year 2400 could be raised in an environment which exactly duplicated the Stone Age, and she would rise only to the level of the environment and the small amount of information passed along from those around her. This person would know nothing about anything at all except flaking stone, making fire and stone axes, shell and bone jewelry, spears, gathering food and whatever cultural rites, games, wars and other behaviors and technologies those around her exhibited.

There would be no Yahweh, no Jesus, no Allah, no Buddha, no Greek or Roman pantheon, no Santa Claus and no Tooth Fairy. What's more, there would be absolutely no way to derive these fantasies and the person would be completely fine without any of them. If the person were raised to believe that giant purple magic rabbits caused thunder, or that it was Stone Age angels doing the Stone Age equivalent of heavenly bowling, the person would believe this.

Due to the scarcity of both knowledge and answers to questions of causality and origins, when a particularly garrulous or charismatic person in the Stone Age imagined a solution to the questions of the origin of life and the earth and proselytized it, many people would likely believe it. This would have happened just as many people in the ancient Middle East believed the local prophets and came to believe in polytheisms and later in the monotheisms of Judaism, Christianity and Islam. The same process occurred in the USA in the mid-1800s with the invention of Mormonism and again in the mid-1950s with the invention of Scientology.

The process is exactly the same. Somebody came up with a fairy tale, other people followed it, then it got bigger and achieved "critical mass" sufficient to keep it alive for some time yet TBD.

So when Og the Stone Age philosopher looked around and wondered how everything began, he saw a world in which only humans did "intelligent" things like shaping stones and making axes and fire. Animals and plants did not do this. Therefore he posited that the heavens, the world and the events in them were the result of some super-powerful human or humanlike entities. Thus the "super-human" entity or entities, some of which were part human and part animal, were likely the first supernatural gods which humans invented to explain the mysteries.

The original stories were passed on around millions of campfires over the years, growing in detail and complexity to try to explain more mysteries and, in many cases and much more importantly, to captivate the audience. Captive audiences result in more attention being directed to the speaker, and quite often also more social status, power and sex, thus this sort of behavior is now a near human universal. Bear in mind that almost everyone at that time, and far, far too many people even unto this day, believed that dreams are some kind of message from or insight into a supernatural realm. Thus nearly everyone was primed for explanations which included imaginary realms.

Later generations of Stone Agers, you could call them campfire televangelists, discovered that they could tell their clanmates that they had heard voices in their heads or seen visions or signs which were communications from the beyond, and that people would actually believe these stories. They soon learned that they could use these assertions of communications with the gods as a justification to tell everyone else in the clan how to behave and also to gain sex, power and influence. Thus formal religions and religious hierarchies were born, normally being run by power-hungry authoritarian sociopaths.

Replace Og and the campfire televangelists with any of the prophets, mountebanks, charlatans and other clergy of any other religion throughout human history and you have described how religions form and are sustained. Original ideas born of ignorance, insanity, cupidity, malevolence, powerlust, a compulsion to be the center of attention, hyperactive imaginations or an unfortunate combination of these unsavory characteristics were accepted by the masses and eventually infected millions then billions of people.

Humans began the process of trying to explain major mysteries with almost zero knowledge of the actual facts of the world and the universe. The only tools they had for this were their five senses and brains that were programmed over millions of years to seek very local, very small-scale and very short-lived patterns in a savannah environment.

This was rather like trying to explain how electricity works and how it can be harnessed using a lightning bolt as your baseline. Only through actual knowledge of physics and chemistry can one start to understand then describe electricity.

It is only since 1859 that humanity took its third, and some would say its most significant, great modern intellectual leap forward. The first of these leaps was the Scientific Revolution which began in the mid-1500s, and the second was the very appropriately named Enlightenment, which began in the mid-1600s. The Scientific Revolution and the Enlightenment overlapped in time and in many intellectual areas for much of the 1700s.

Humanity spent over two million years in a pre-scientific world, then in relatively short succession following a series of climatic and genetic developments, developed a series of increasingly more powerful tools and instruments to collect and process data. The ancient Mediterranean schools of thought and philosophy gave way to the Middle Eastern scientists and thinkers of the Islamic Golden Age which lasted roughly from the 700s to the 1300s. The baton of scientific progress then passed to the growing population of scientific thinkers in Europe who were trying to come out from under the Inquisitional thumbscrews of the Catholic overlords.

The global rate of accretion of scientific knowledge has been positively asymptotic since the mid-1700s, but natural selection was not widely discussed in detail until after 1859, which on the scale of our existence happened mere seconds ago. Genetic engineering, quantum theory and the Big Bang did not arrive until the 1900s. Much of our detailed knowledge of the physical universe has only been with us for only a microscopic fraction of the existence of our species, and not nearly long enough to displace the stultifying inertia of centuries of religious indoctrination.

This accumulation of knowledge has progressed unbelievably quickly, but the human brain has not progressed or evolved. The people who wrote the holy books of Judaism, Christianity and Islam are the products of minds from a highly ignorant pre-scientific age which shared not only each and every human cognitive bias but also a near total ignorance on a huge range of scientific disciplines. They were able to generate much in the way of philosophy, literature, art and some math, but science and the realities which it unlocks were far in their future.

The difference between the 21st century and every single year in biblical times is the almost unimaginable amount of knowledge we now have compared to then, including knowledge of the sciences and of our biases and logical fallacies.

Think about it this way. Compared to the level of knowledge we have today, the average person living in the ancient Middle East during biblical times knew about as much about physics, chemistry, psychology and other sciences as did our flint-knapping ancestors of two million years ago. And we should treat their books accordingly.

A person reading those ancient fossilized holy books today can only use them in a supernatural or religious sense if he does not know about or rejects science, does not know about or claims total immunity from psychological biases, and does not know of the thousands of other religions or does know about them but rejects them all in favor of one particular preferred faith. These are all utterly ridiculous and unjustifiable positions.

Getting back to the brain itself, there are more considerations about it which we must address before closing out this discussion on the first source of our beliefs - the physical brain itself.

The brain functions in accordance with the interplay of chemicals and electricity. Not every brain has the exact same chemistry, the exact same amount of different neurotransmitters or the exact same efficiency or speed in processing them. Variations in the amounts, availability and persistence of chemicals like those neurotransmitters have significant effects on the function of the organic machine, and therefore on the thoughts which the machine produces.

The effects of chemicals like alcohol, uppers, downers and hallucinogens are examples of what happens when brain chemistry temporarily changes. Minor differences in the levels of chemicals naturally produced in each brain drive it to function in a manner slightly different from other brains.

While not quite in the same category as hallucinogens, the neurotransmitter dopamine[87] has been shown to be linked to the level of belief in unproven assertions and paranormal thoughts.[88] Higher amounts of dopamine are associated with higher rates of belief in the supernatural and paranormal. This is a very significant finding which has been in the literature for some time now.

One study[89] published in 2015, which involved both genetic analysis and formal psychological experimentation and analysis, was conducted by seven researchers and is titled "Linking unfounded beliefs to genetic dopamine availability". This study noted the following: "Our results indicate that genetic differences in dopaminergic neurotransmission account for inter-individual differences in perceptual inference linked to the formation and maintenance of unfounded beliefs."

In other words, people who naturally produce more dopamine and/or have a greater sensitivity to it are more likely to have and hold "unfounded beliefs" such as those in religions. This has significant implications. Just as a person who has taken in alcohol, cocaine, heroin, morphine, methamphetamine or a hallucinogenic drug experiences a change in his mental state and perception of the world, a person with naturally higher levels of dopamine will perceive the world differently from a person with lower levels.

Research has shown[90] that believers are, in layman's terms, more gullible than skeptics. This is of course no surprise to anyone, but the purpose of science is to confirm or deny hypotheses and so-called "common sense". The threshold for evidence required by the believer is far lower than that required by the skeptic, and this is most pronounced with religious or mystical claims. Belief in the claims of religion without absolutely massive amounts of evidence is quite literally impossible for skeptics, just as it is impossible for them to believe that a fat and permanently old man rides a magic sled through the sky pulled by trained magic ruminants and carrying a world-full of trinkets pulled from a magic backpack tesseract which are magically delivered in one night to millions of young recipients in quite glaring violation of the laws of physics.

Religion goes many orders of magnitude down the road of fantasy in its insupportable claims than does the story of the jolly old elf, and both are equally imaginary. There is actually far more evidence for Santa Claus than for any god, because on Christmas morning there really are actual presents under the tree. No god has ever delivered anything to humanity.

The implications of the above are very interesting. Two brains could be exactly alike except one produces more dopamine or processes it differently, and due to this difference it is possible that one could be a devout religionist or spiritualist while the other could be a skeptical materialist atheist. Perhaps one day we will be able to implant a dopamine suppressor or regulator in the body, and people brought to a normal balance will stop believing that Elvis and Jesus are coming back any day now, that the voices in their heads are divine communications, that they hear messages in what is really just static, that they see faces of gods in fog and smoke, that the government is maintaining a fleet of UFOs at Area 51 along with the remains of the crew, and that gods exist. **Religion really is just all in the believers' heads.**

A related issue but separate from the cognitive biases and the influence of dopamine is our pattern-detection module. It is one of the strongest modules in the human brain and is both a great asset and one of our greatest weaknesses. Matthew Hutson describes this in a wonderful article[91] in *Psychology Today*, and it has also been extremely well-discussed by Michael Shermer and others. Shermer's term "agenticity" sums it up perfectly. Human brains seek patterns, and the brain's pattern-seeking module runs on programming that operates much the same way today as it did many hundreds of thousands and likely millions of years ago, when we knew absolutely nothing of the natural world, much less our current massively artificial world.

Thus what would almost have been called a correct and "common sense" interpretation of an event for the vast majority of human history often results in absolutely massive errors when it is applied today.

The best way to describe this is to note how humans tend to anthropomorphize so much in the world and assign agency or conscious intent behind the things that happen. The ancients used to think gods either sent or withheld the rain, as apparently many American Christians do today, especially in Texas[92]. This was the case with lightning, thunder, floods, disease and anything else particularly bad or good that happened around or happened to ancient humans. The example Shermer and others use is a rustle in some tall grass heard by two of our ancestors while they were out walking together.

One of the walkers assigns agency to the rustle, assumes it is a man-eating beast and gets the heck out of the area. If the rustle really was a predator this person will survive and have children who will share his "all events have an intentional cause" approach to life. The other hunter-gather assumes the rustle was caused by the wind, watches the other guy beat feet, hangs around to eat a berry and within moments becomes dinner for the beast. This guy has removed himself from the gene pool and will no longer pass on his genes for *not* seeing everything as having intentional agency.

This assignment of intent and agency to the things that happened in the world was an evolutionary adaptation that worked well in our science-free distant past, because rustles sometimes did mean hungry or dangerous beasts. But the same mindset that assigned agency to rustles in the brush also assigns agency to thunder, lightning, rain, disease and other events, and this is a complete and utter failure to understand reality.

The difference between the choices is that guessing wrong about the beast gets you killed. Thus assigning agency is extremely strongly reinforced and "Darwinically" becomes part of the instinctive System 1. But assigning agency to the moon or the rain is neutral. There is no negative result from assigning agency to natural events because unless you form a religion and start sacrificing kids or waging war because of your gods, wrongly believing the gods bring the rain doesn't impact your survival one way or the other. Thus the net evolutionarily successful strategy of assigning intentional agency became part of our ancestral System 1 and developed over time to assign conscious agency and intent behind things that happen in our world. As a result massive numbers of people today make a tragically widespread logic and category error by assigning agency to a fictional god to try to explain far too much.

In a world full of scientific knowledge this agency-assigning module is an artifact contained in our instinctive System 1 which often gives very wrong answers, no matter how right they feel. In other words, we cannot trust these instincts and the instinctive feelings they produce, because they are flat wrong.

The information we have has come in far faster than our brains could evolve, so this module has not changed and is like a stone ax brought to a modern battlefield. It can work in one or two situations but by and large it is ineffective and counterproductive, because stone ax wielders don't do well against tanks and machine guns. Humans did not lose the module as our population slowly grew then grouped into cultures. Today's believers are slaves to their hyperactive and misguided Stone Age instincts and they still attribute natural occurrences to their god of choice.

Making all this much worse is the fact that the advances of technology have outstripped the rest of the process of human brain evolution by millions of years of relative change. Human brains evolve at a very slow rate, and it has taken about seven million years[93] for our brains to arrive at their present state of evolution from the time we split off from our common ancestor with chimpanzees.

Computer technology, on the other hand, has enabled an approximate doubling of the number of transistors which can be placed on an integrated circuit every year or two that passes, according to Moore's

Law.[94] This has led to computer technology moving from the 1940s ENIAC which was the size of a room and weighed as much as a small fleet of cars, to today's computing devices which are light years more powerful and can be worn like a wristwatch. In terms of speed, the ENIAC plodded along like a comatose snail trying to pull a battleship up a steep hill, doing calculations at a rate of about 500 FLOPS (Floating Point Operations Per Second). Today's supercomputers, made barely half a century after ENIAC, whiz along at more than a thousand trillion[95] FLOPS, and new speed records are set every few years.

Supercomputers and technological advancements in literally every scientific discipline have enabled humanity to reach places and gain information in ways the human brain and sensory input systems will never reach on their own, no matter how many millions of years of evolution and natural selection we have in front of us.

So the human brain is a powerful organic machine when compared to the brains of other species, but it is highly susceptible to developmental and environmental damage and influence. It has built-in onboard programming and processing challenges. Our so-called "common sense" is often extremely wrong although it feels right to us. Thus until such time as humans can genetically engineer a more powerful brain without such cognitive biases, or we can somehow hybridize the human brain with powerful computers to do this, we will be using very ancient software to filter and process information. And we will continue to have factually incorrect but deeply and sincerely held opinions for which we have no justification. This is the reality of our situation and we have to understand and accept it.

Thus we have to continually and deliberately keep mere uninformed opinions on religious matters separate from any actions taken by individuals and societies until the opinions have passed rigorous logical and scientific analysis and testing and have matured from ideas to hypotheses to theories. Processes like the scientific method, peer review and replication of study results must always be the sole final arbiter of rational decision making and public policy. Beliefs and opinions are often not just hot air, they are poisonous hot air.

This first section has been somewhat long and involved, but it was a necessary step in understanding the sources of our beliefs and opinions and how and why they are formed and clung to so tightly. Now that we have described the organic machine in our heads and how it operates when left to its own devices, we must describe the remaining three sources of beliefs, which result from changes or damage to the brain and the inputs it receives from the world.

Belief Source 2: Physical Changes Or Damage To The Brain From Physical Causes

While the brain's structure does drive beliefs, there is far more to the story of those beliefs than just the effects of genes and how they lead the brain to develop and to generate chemical balances or imbalances. The conditions within the mother's womb play a major role in fetal brain development. They can be products of the mother's acts of omission or commission or things that happen to her that are outside her control. We know that eating well[96], taking prenatal vitamins, exercising, reducing anxiety[97] and stress[98], not smoking[99] or drinking or taking drugs or indeed ingesting anything but filtered water, vitamins and organic food free of pesticides[100] and other chemicals are all activities which strongly contribute to optimal outcomes for the growing fetus and specifically for fetal brain development.

Significant damage to the fetal brain can occur if the mother drinks alcohol while pregnant; this is called fetal alcohol syndrome.[101] If the mother does not take in enough folic acid, this can led to an almost uniformly fatal brain defect called anencephaly.[102] Many other conditions and situations result from improper gestational nutrition and environment.

Many of the impacts to the gestational environment are under the mother's control, while others are not. Food and water free of pesticides and other contaminants may not be available or affordable. The mother's relationship with her family may be extremely stressful. The mother may be forced to endure the stresses of war, religious or other oppression and economic or other distress. Women, merely because they are female, suffer far more stress and oppression than males do at the hands of religion, society and invading armies, to say nothing of general stressors like pay inequity or the disappearance or lack of financial or parental support from the baby's biological father. The woman may herself be mentally damaged and unable to take proper care of herself, let alone her developing fetus. The mother may contract a disease which affects the development of the fetal brain, such as the Zika[103] virus which can cause a range of quite damaging problems[104] such as microcephaly.

All of these conditions impact the development of the brain *in utero*, and these same threats to proper development continue long after birth. After they are born, too many children are subjected to secondhand smoke, given alcohol, fed poor quality food or not fed at all, left unsupervised and are able to ingest poisons or damaging substances, and/or are exposed to many other similar dangers.

Once out of the womb and in the world as an independent organism, they face new threats to their bodies and brains, such as physical abuse, physical neglect, the potential for accidental injury and a host of other possible impacts to proper brain development. There are over twenty years from birth to the major completion of brain development, and the brain is vulnerable for this entire time.

Damage or changes to the physical structure of the brain can and often do affect how and what people think and believe. Some may be temporary but others are permanent, and it is possible that one single experience may permanently change a person's belief system. Due to the general lack of knowledge about science, psychology and the neurobiopsychological functioning of the brain, this can lead to believing in some religion or god. Consider the following examples of how physical changes to the brain can change thoughts and beliefs.

1. **Phineas Gage.**[105] This extremely well-known case is taught even to most high school psychology students and has been for many years, for obvious reasons. Gage[106] had a large metal rod shot clean through his head in a railroad construction accident in Vermont in 1848, but remarkably, though *not* in any way "miraculously", survived. He experienced significant changes[107] in his personality, as one would expect from such a serious brain injury. I recommend readers take a few minutes to review this absolutely fascinating case for themselves.

2. **Fetal alcohol syndrome.** This and other conditions occur in children when their pregnant mothers consciously use or unknowingly encounter substances which cause heartbreakingly deleterious effects on the developing brain and the child's behavior and cognitive ability.

3. **Lobotomy.** This act of deliberate brain damage is known to change personality and cognition.

4. **Alcohol-induced inebriation.** Everybody knows how behaviors can change when people drink. You can start with three biological siblings from the same biological parents, and the siblings can have three different views on religion. One is a strong believer in one of the tens of thousands of Christian denominations. Another is a devout atheist and skeptic. A third is a New Ager who carries "magic" stones in his pockets and wears pyramid T-shirts. All these beliefs are irreconcilable and

held in different brains, yet when they all go out to a bar to drink, each of their brains will be affected by the alcohol they consume. They won't be equally affected, but the change in the brain's chemical environment will manifest itself in each sibling. One sibling might become depressed and sulk in his cups at the bar. His sister might become overly friendly with fellow patrons while his brother might remove his clothing and dance on a table with a lampshade on his head while swinging his underwear around like a lasso and saying quite inappropriate things.

All would likely say if asked at the time that there was no problem with their current behavior, but while they did act unusually when compared to their normal behavior, no permanent effects are likely to be manifested after the alcohol wears off since they resulted from the ethanol's transient chemical effects.

5. **Traumatic brain injuries (TBI).** Such injuries are experienced by soldiers in combat, football players and others who suffer severe head trauma such as babies with shaken baby syndrome.

6. **Brain damage** due to any number of other causes, such as strokes, meningitis, anoxia and exposure to lead or other dangerous substances.

7. **Hallucinations.** According to Merriam-Webster, a hallucination[108] is "*1a : perception of objects with no reality usually arising from disorder of the nervous system or in response to drugs (such as LSD)…b : the object so perceived ·wasn't sure if the rain was real or a hallucination; and 2 : an unfounded or mistaken impression or notion : delusion.*" The category of hallucinations is remarkably broad and it appears to be one of the most significant contributors to sudden adoption of deep religious beliefs, particularly when combined with genetic predisposition and social programming.

   This is because some people give supernatural meaning both to their own individual hallucinations and those of others such as self-appointed prophets or "holy men". Note that due to the confirmation and other biases, hallucinations are felt to be supportive of a person's own faith or the one he is most familiar with, not some unknown or different religion or god.

Hallucinations are auditory and/or visual manifestations of temporary changes in the brain, and while they may seem extremely real to the person experiencing them, hallucinations are merely the brain's failure to properly represent objective reality. Hallucinations of weird colors, shapes and patterns are as much a breakdown in normal processing and perception generation as are hallucinations of gods and the supernatural. The brain presents our conscious and subconscious selves with a full-sense representation of the world, and hallucinations are distortions of this representation. The brain and only the brain is the source of all thought and experience and it has no tie to any fictional netherworld. Hallucinations are relatively common, affecting about five percent[109] of the general population. Many things cause hallucinations, as noted below.

a. **Psychosis** (schizophrenia/schizoaffective disorder). Research into psychotic hallucinations by Ryan Teeple, Jason Caplan and Theodore Stern[110] noted that "Visual hallucinations in those with schizophrenia tend to involve vivid scenes with family members, religious figures, and animals." This work is the source for "a" through "k" in this section.

b. **Delirium**, which may itself be caused by[111] "…myriad medical conditions, metabolic disturbances, infections, drug effects, and intracranial processes" or from "… alcohol withdrawal (i.e., delirium tremens) or stimulant intoxication (e.g., with cocaine or methamphetamine)…"

c. **Dementia**, specifically Lewy body dementia[112] and dementia due to Parkinson's disease and posterior cortical atrophy.

d. **Charles Bonnet syndrome**, which causes people with failing vision[113] to hallucinate.

e. **Anton's syndrome**, which is an extremely odd and fortunately rare condition. Affected people have become blind due to damage to their occipital lobe, but refuse to believe[114] they are now blind and they act as if they can still see.

f. **Migraines**.

g. **Peduncular hallucinosis**. These are visual hallucinations caused by irregularities[115] in the midbrain.

h. **Sleep disturbances.**

i. **Tumors** on the optic line and in the temporal lobe.

j. **Creutzfeldt-Jakob disease**.

k. **Diabetes / hyperglycemia.**[116]

l. **Lack of sleep**[117] **and extreme fatigue**. Note that these are different conditions from the sleep disturbances noted above. This is another extremely interesting topic worthy of further separate reading. In an article[118] in Scientific American in 2010, Michael Shermer discussed the variant of hallucination he terms "The Sensed-Presence Effect", and what John Geiger calls "The Third Man Factor"[119] in his book[120] with that title. People who experience some combination of extreme fatigue, extreme stress and lack of sleep often imagine the nearby presence of an entity despite nothing actually being there. These internally-generated imaginary experiences seem as real to the person enduring them as actual external experiences do to a nonstressed person.

   Many people throughout history have claimed to have had such an experience, and many of them attributed it to a supernatural entity which they worshiped or knew about. This is undoubtedly due to the widespread distribution of religion and the likelihood that someone experiencing this type of hallucination would attribute it to a known god or spirit rather than their next door neighbor or third grade math teacher. Indeed, it is the rare neighbor or math teacher in the First World today who is widely believed to have supernatural powers.

   Shermer describes in great detail the biopsychological conditions which lead to the sensed presence effect, and in a few short paragraphs shows these imagined connections are merely biological byproducts of the human brain. He completely collapses the sail which the faithful have hoisted for centuries in their attempts to prove their line to the divine.

m. **Seizures**. Studies and observation have shown that seizures from temporal lobe epilepsy (TLE) are often strongly interpreted as religious or even "hyper-religious" experiences by the sufferer, with potentially lifelong impacts such as religious behavior and/or conversion. Recent examples and case studies describe this quite well. The researcher VS Ramachandran noted[121] that "When we showed them (TLE patients) religious icons - a Star of David, a cross, or the word "God", the word "Jesus", the word "Mary" - there's a big jolt in the galvanic skin response, showing that something is going on in the temporal lobes, in temporal lobe epilepsy; maybe it's going on in all our temporal lobes. There's a group of neurons that is firing in a normal manner, which makes you more prone to religious belief. And these neurons are hyper active in these people, hence the propensity to religious belief."

   This is extremely relevant to the history of Christianity, since many people have speculated that Saul of Tarsus experienced a TLE seizure on his sojourn to Damascus, and that his hallucinations of the divine were nothing more than an epileptic seizure. This proposal is of course dismissed by the faithful but it is infinitely more probable than their assertion that the incident was a communication with the supernatural, and it has quite large implications for the entire corpus of Christianity.

   Even beyond the temporal lobes, brain imaging and other data show[122] that seizures affecting a smaller piece of the brain under the temporal lobes called the left anterior insula provide a shotgun blast of effects described as "…an aura of a "sensed presence," a religious emotion, or feelings of euphoria (ecstatic seizures) are characterized by heightened self-awareness."

   So your god really is completely in your imagination, your feelings are solely products of the chemicals and electricity in your brain unconnected to any god, and scans and other data can now show this. Yet the faithful dismiss all this science with an airy wave of their holy books.

n. **Near death experiences**[123] (NDEs). I won't go into these very much because massive amounts of work have proven that these are, like the other situations listed above, merely how the brain responds to stress, stimuli and major changes in its environment. Like other hallucinations, they are internally generated sensory manifestations of changes in the brain which include perceived sights and sounds and which often occur similarly between different people. NDEs are merely hallucinations produced by a human brain undergoing unusual circumstances, and they are certainly not supernatural in any way.

o. **Hallucinogenic substances.**[124] There are quite a few of these, and the magnitude of their effects on the user depends on individual body chemistry and the amount of substance used/absorbed. Examples include atropine, dimethyl-tryptamine (DMT), dopamine agonists[125] (agonists are essentially enhancers), lysergic acid diethylamide (LSD), marijuana[126], MDMA (ecstasy)[127], mescaline[128] (in peyote), phencyclidine (PCP) and psilocybin.

Thus there are countless ways in which a simple physical change in or damage to the brain can cause hallucinations of gods and spiritual realms and also hyper-religious frenzy and devotion. These very strongly contribute to a person's belief in the supernatural. Even today in the 21st century, people still know so little about these contributors and about how their brains work that they let their imaginations run riot and their ignorance and gullibility take over as they seek explanations.

The irritatingly everpresent god of the gaps illogical escapism rears its ugly head yet again, as the supernatural is offered as the explanation for the experience rather than the actual demonstrable, reasonable and quite mundane and non-supernatural reason – changes in the brain and/or its chemistry. Things would be quite different, and there would be no religion, if more people actually knew how their brains worked.

These physical changes or damage are not the only means by which the brain can experience a change, sometimes both quite devastating and lifelong. The next section explains this.

Belief Source 3: Physical Changes Or Damage To The Brain From Nonphysical Causes

Beyond the purely physical dangers to the brain, the brain can be changed or damaged merely by what a person sees and experiences, without a physical impact or a chemical input. These sensory inputs, while nonphysical in nature, can have massive effects on the physical brain. This is not nearly as widely known as it should be, is absolutely not intuitive and it is utterly impossible to know this from what one would call "common sense". Too many people refuse to believe what they actually can see with their own eyes, much less what they can't see, so this point must be made often and loudly.

Post-traumatic stress disorder, or PTSD, is becoming rapidly understood as a set of thoughts and behaviors that results from, according to the US Department of Veterans Affairs[129], "… experiencing or witnessing a life-threatening event, like combat, a natural disaster, a car accident, or sexual assault." This means that just the simple act of *observing* a stressful or traumatic event can result in permanent changes to the brain and can be quite debilitating to the affected person. Note that the intensity of the event is not on a fixed scale.

What gives one person permanent and debilitating PTSD may not trouble another person at all. Seeing a pile of dead bodies blown to bits may make one person joke about jigsaw puzzles and anatomy classes, while the person next to him may literally freeze up and be unable to function, then suffer lifelong severe PTSD. The problem of PTSD had been observed but not understood or well-researched for quite some time, up until the recent high-level focus on it as part of the treatment of veterans who served in our most recent wars in Iraq and Afghanistan.

I am a retired military veteran and have served in combat zones. I personally know extraordinary and wonderful people who have lost limbs or eyes as a result of being blown up in a combat zone, but who have not suffered marked or debilitating PTSD. I know others who have been blown up on multiple occasions and/or seen massive death and destruction up close but who have no apparent or significant PTSD. I know people who have experienced what most people would call relatively minor stress yet have nevertheless developed PTSD. Some people have a much lower threshold for dealing with such observations due to the physical construction of their brains. It is not under their control, just as their System 1, hair color, skin color and sexual orientation are not under their control.

Since the massive numbers of people recently affected by Traumatic Brain Injury (TBI) and PTSD due to service in operations in Afghanistan and Iraq made the problem too large and visible to not directly address, there have been formal studies performed by government and civilian entities to learn more about both of these issues. A 2011 work[130] by Jonathan Sherin and Charles Nemeroff, both medical doctors in the University of Miami's Department of Psychiatry and Behavioral Sciences, is titled "*Post-Traumatic Stress Disorder: The Neurobiological Impact Of Psychological Trauma*" and describes the biology of PTSD.

One of their observations is particularly relevant, and it is that the response to stimulus that results in PTSD is not automatic in everyone, as mentioned above. As already noted, some people's brains are wired such that they are more susceptible to PTSD than others to the same stimuli.

If you lined up 100 people and exposed them to the same stimulus, some will be completely unaffected, others will be mildly upset but not suffer PTSD, some may suffer mild or "unobtrusive" PTSD, and an unlucky few may develop debilitating PTSD. Two very relevant excerpts from the Sherin and Nemeroff work make this biological point:

> *"The signs and symptoms of PTSD, therefore, appear to reflect a persistent, abnormal adaptation of neurobiological systems to the stress of witnessed trauma. The neurobiological systems that regulate stress responses include certain endocrine and neurotransmitter pathways as well as a network of brain regions known to regulate fear behavior at both conscious and unconscious levels."*

> *"Efforts to identify neurobiological markers for PTSD originally presumed that abnormalities were acquired "downstream" from an exposure, as a consequence of traumatic experience. It could be, however, that certain abnormalities in the patient with PTSD simply represent pre-existing or "upstream" pathology that is functionally dormant until released by trauma exposure and detected thereafter upon investigation. Along these lines, recent interest has focused on factors that seem to modulate outcome variation in neurobiological systems following trauma exposure including genetic susceptibility factors, female gender, prior trauma, early developmental stage at the time of traumatic exposure, and physical injury (including traumatic brain injury - TBI) at the time of psychological trauma; these parameters likely contribute to vulnerability for, versus resilience against, developing PTSD."*

Some symptoms of PTSD include, according to the US National Institute of Mental Health[131]:

- Flashbacks to the triggering event, and other effects such as perspiring or a rapid pulse.
- Nightmares and/or sleep problems.
- Being scared, fearful, tense, angry, or susceptible to sudden sounds like a car's backfire or a slamming door.
- Not going near or not partaking in reminders of the event, such as not going to certain places or not attending certain events, or avoiding certain people.
- Depression and depressed affect and interest in things.
- Reduced or changed cognition and social skills.

If you have ever been around a service member who served in any of our Nation's combat zones, or even faced threats while serving as a police officer, firefighter, first responder or a similarly stressful profession, you will notice that the symptoms above have been common for some time. It is often difficult for a person to shoot or stab another person to death, and it is highly unnatural to engage in combat with the weapons and equipment we have invented for that purpose. We can inflict mass death in a matter of seconds, and we can be on the receiving end of deadly attacks. The condition of PTSD is not new. What is new is how much we now know about it.

So the human brain, without being physically struck by an object or force and without being bounced or shaken around inside the skull or exposed to chemicals, can be strongly and permanently affected merely from seeing or experiencing a traumatic event. This is a part of human psychology which every person should know about and which should be taken into consideration as part of medical care, jurisprudence and the criminal justice system, parenting and simply being a human being.

So how do beliefs change after such a traumatic experience? A person who has been in a severe car accident may no longer believe cars or motorized transport is safe. A person who has been raped may no longer believe that men, or men of a certain look or personality, are safe to be around. A child who has been raped or sodomized by a priest may permanently distrust all clergy, even those not wearing Catholic black. A person who has been told and then believes that there is a god who will throw her and her loved ones into an eternal escape-proof hell for any of large a number of acts or even thoughts may be permanently traumatized by this, and may suffer lifelong existential and agonizing guilt, shame and fear due to this belief. Such beliefs may trigger uncontrollable attempts at compensatory actions such as obsessive-compulsive behavior. Beliefs matter a great deal, and experiences change beliefs.

The fourth and last source of beliefs is the one that we get from the pressures and information we receive as we go through our lives. These don't necessarily rise to the level of producing PTSD, but they do strongly influence beliefs. Social pressures are critical components of a human's life and people start, stop or change their beliefs simply due to social pressure, whether from the need for acceptance or the fear of rejection. Experiences and social inputs and pressures are strong contributors to behavior. Peer pressure and social pressure often trump almost everything else in a person's life, including morality and reason.

Belief Source 4: How The Brain Responds To Experiences And Sensory Input

The growth and development of the brain are like the manufacturing processes used to make a computer. If all the right parts are there, if the parts themselves are made of the right elements and parts, if the parts don't have any defects, if the parts are all installed correctly, and if all the manufacturing processes are done properly, then at the end there will be a well-formed computer. The end product will function properly. If some parts are missing, defective or assembled incorrectly, or if any extra parts are added or errors are made during the manufacturing process, the computer will be damaged or unusable. We have already established this.

Once the computer hardware is assembled and it is outfitted with peripherals and basic operating system software, the computer is ready for work, i.e. processing digital data in specific ways through the use of more software which will be added later.

The process of adding software and information on top of the base physical structure and operating system is clean and direct in a computer, but not at all clean or direct in the human brain. The software is actually partially formed and modified as a result of the information the brain receives. A person's brain receives information from the world through the physical senses. This information is not handled in the cold and unthinking manner by which a computer handles information. It is processed based on a wide range of factors which change over time and due to current conditions.

If you try to teach a baby advanced math or physics, the baby will hear your voice but will understand nothing you are saying. If you scream the words at the baby in a loud and angry voice, the baby will still not understand the words but will be quite negatively affected by your tone and volume and will almost certainly begin crying in fear. If you "coo" the exact same words in a loving tone, the baby will still fail to understand the information but the child will respond positively to your pleasant voice and tone.

Whether you screamed at or cooed at the baby, you will have transferred no math or science content but you will have definitely sent it an understandable and very strong emotional message, which it will remember and which it will incorporate into its software. If the baby hears your voice in the future, this will engender a positive feeling if you cooed and a negative feeling if you screamed. Perhaps your deodorant or cologne, your gender, your hair color or your body size will also be associated with cooing pleasantness and calm or screaming hatefulness and fear. The brain is a complex machine and there are numerous ways you will be perceived and how the response patterns will form, in an iterative learning process.

If you as a parent teach your child from birth to his 18th birthday that some group of people, such as Jews, liberals, intellectuals, immigrants or atheists are evil and hateful creatures which are a danger to morality, life, liberty and the pursuit of happiness and deserve to be treated like vermin, you are both adding information to the child's brain and modifying the formation of software and response patterns in that developing brain.

Humans don't imprint[132] in the same way that Konrad Lorenz's goslings did, but humans as social creatures absolutely do strongly model their thinking based on the guidance and examples of their parents and society. There are plenty of contrarians, particularly among teenagers, but by and large the growing child soaks in what his parents and close-in society tell it to do, say and believe, and once inculcated these may be permanent fixtures which can never be removed or changed.

Like a computer virus which may permanently change and damage the processing and outputs of a computer, viral infections of incorrect and/or immoral words spoken to and deeds done to growing children may permanently change and damage the processing and outputs of their brains.

Due to these pressures and the mental viruses transmitted to the young, even an inherently good person with a brain that if left alone would be moral and kind and good can be molded into a callous and immoral person who can engage in very evil words and deeds based on very evil beliefs. The information flow into the growing person's brain is stored as bits of information and also becomes part of the software that develops over time and which forms processes of thinking and beliefs. This has been done so many times in history on scales both large and small that it constitutes one of the most dangerous flaws in human psychology; that a good person can be turned into a bad one, and in far too many ways. Examples of this include:

- People who were raised to burn heretics and witches during Christian control of Europe.
- People who were raised to believe that torture of infidels or conducting pogroms against Jews was a holy duty.
- People who were told to turn their parents in to the State for "nonconformist" activities, as in Stalinist Russia.
- People who were raised to believe people of other faiths or no faith are lesser beings and/or should be persecuted or treated as less than those of their particular religious in-group.
- People who were raised to believe that people of other colors, languages, appearances, genders and orientations are lesser beings and/or should be persecuted or treated as less human than those of the in-group, or not human at all.

There are many avenues by which people come to hold religious beliefs. Regardless of how wrong or factually insane they may be, beliefs are truly and sincerely held in the mind of the believer. They may be held for the comfort they offer or because they give the believer a mental crutch and external rule or support system he cannot generate internally. Beliefs may be publicly advertised and apparently though not truly held for a variety of reasons. In some cases, being thought to hold certain religious beliefs gives people a societal advantage and/or enables in-group acceptance. In others, faking certain beliefs will stave off persecution and murder at the hands of the most deeply and sincerely faithful believers.

Many people turn to religion as a way of alleviating stress. Humans generally do not like to be highly stressed or to lose their independence, whether that loss results from injury, age or from being subject to the whims of other humans or conditions of nature. Religion offers the ultimate panacea, that there is a loving god in charge of everything and that no matter how bad things are, everything will be simply *wonderful* one day, whether in this life or in an afterlife. Certain personality types find this psychological escape valve irresistible when their stresses rise above their individual resiliency threshold and they have to turn to artificial external supports, no matter how imaginary those supports are.

Probably the strongest and most telling proof that religion is cultural and societal is the near absence of Muslims in Alabama, the lack of Buddhists in Baton Rouge, and the paucity of Christians in Afghanistan. These conditions demonstrate the extremely local nature of religious belief, which even in a globally interconnected world is still extremely heterogeneous. Surely religious tenets and beliefs, allegedly revealed or otherwise, cannot be true in one place on earth and false in another. Yet billions of people delude themselves in this way.

Familial and social indoctrination and cultural and societal inertia drive the people of one country or locality to toe the party line in their faith and beliefs, particularly when reinforced or even enforced by societal and legal penalties. If the local beliefs are reinforced both in schools and at home, a growing child may not experience any alternative views to the one being preached. If deviating from the beliefs will incur significant cost or pain, the odds of deviation are extremely low and the social inertia will carry the belief for many generations.

Growing up in a particular culture means growing up with certain commonalities. Language is one, religion is another, and even norms such as which system of weights and measures is used are examples of cultural inertia that are highly resistant to change.

The United States is almost the only country left on earth that has not adopted the metric system, joining only Burma and Liberia in this dubious condition. Languages and religions persist for hundreds of years. Change is hard, and when change would require changes to infrastructure, voluntary change from within becomes essentially impossible.

As an example, if the US ever adopts the metric system, think of all the housekeeping and infrastructure changes that would need to be made to things as mundane as speed limit signs, highway "mileage" signs, and all the written and electronic references using the old English measurements. It would require a massive amount of money, time and effort, and also Americans just don't like to be bothered with such things. It has been a much easier and cheaper option to just coast along on the inertia of the original system. Thus we Americans still use miles instead of kilometers. In the same way, we still use English instead of Chinese or Russian, at least so far, and many of us practice Christianity. Social and cultural inertia are not completely irresistible forces, but getting people to change even obviously bad or immoral ways really is often like trying to move immovable objects.

The bottom line is that in our highly social species, if everyone else is doing it, it is highly likely that you will be doing it too, even if you don't really want to and even if you don't actually believe in it. The social costs are just too high to be too far out of the mainstream, and this is true even when you don't live in a time or place where the ruling religious authorities will actually kill you for being out of that mainstream.

We have now completed the explanation of the four sources of beliefs, which as a reminder are:

1. **The structure and composition of the brain.**
2. **Physical changes or damage to the brain from physical causes.**
3. **Physical changes or damage to the brain from nonphysical causes.**
4. **How the brain responds to experiences and sensory input.**

The above information shows how both nature and nurture are responsible for our beliefs. The structures and processes that lead to a person's beliefs are mostly hard-wired in at birth and are refined via several pathways as the person grows. The human brain has to navigate an immensely challenging and decades-long trek through those twin minefields of nature and nurture to achieve a full and relatively undamaged development. Many of the hazards along the path are out of the person's control, at least until the person is of age and able to be independent and make independent decisions.

So What?

All the above is quite a lot. But so what? Why go through it all? Because it is extremely important to fully understand how people arrive at then maintain a belief in the supernatural. Only by understanding those processes, like diagnosing and understanding the root cause of a problem, can we first identify then take the appropriate corrective action to bring reason and critical thinking to its proper place as the final arbiter in our dialogue and our actions.

The rest of this chapter will sum the major points, but the primary point is, yet again, **humanity's invented gods only exist in the imaginations of the believers**, and it is only the widespread lack of knowledge of how the brain works that so many people do not know this.

People often strongly hold their beliefs, no matter how irrational or unjustified, for no reason they can explain. They just "feel that way", which is enough for them to hold their opinion and their religion. This is summed up in that trite expression "if it feels good, do it", but repurposed to "if it feels good, think it". They then reject or twist any and all evidence that disproves their belief or makes them uncomfortable.

Almost no one has any idea how the brain works or why people think what they do. Everyone has deep-seated cognitive biases which drive them to keep thinking the way they do or in ways that make them feel good. People ignore, twist or manipulate facts as needed when maintenance of an opinion requires denial or twisting of reality. This is ubiquitous in religion, because its existential claims are all thoroughly and totally devoid of proof.

It can be and often is impossible to change certain minds even in the complete absence of supporting evidence for a belief, and in a massively perverse logical breakdown, believers often double down and become more fervent believers when they are strongly challenged and shown to be wrong. This is terribly unfortunate, as unjustified beliefs have never resulted in good things for humanity. Part of being a responsible human being is acknowledging that opinions can start a debate but actual facts and actual evidence must always win the debate. This **never** happens in religion.

People don't normally change their minds or beliefs unless there is a significant benefit for changing or a significant penalty for not changing. Even if there is a significant penalty, even all the way to death, many people *still* will not change their minds. And when people do manage to change their beliefs, this is almost always caused by intense external pressures or a major psychological impact, not through a conscious and objective analysis of the relevant facts. Changes through reflection and analysis are the microscopically rare exception. This mental rigidity is extremely common in religion.

The level of investment is critical in the calculus of belief. If nobody cares if you switch from Islam to Catholicism to Satanism to paganism to atheism in the course of a year of "seeking", then you will see people doing this. If a young person has been hiding her homosexuality, atheism or liberalism for years, and knows that due to the Catholic, Baptist, Mormon, Muslim or other beliefs of her family that admitting this will result in being immediately booted from the home, the person will do all she can to keep it hidden.

Most humans can be shaped into good or bad behavior patterns. People born with what would be decent and moral personalities if they are left alone or led to develop that way can unfortunately be molded into people who think and do hateful, violent, vicious and evil things. People with evil tendencies can perhaps not completely eliminate their desires to be evil, but through parental, familial, societal and medical means these desires can be attenuated or judicially dealt with if expressed.

Humans who are born as good, honest, decent and moral people have the genes that make them this way. They tend to marry similar people. These generally good people tend to have children who inherit these genetic traits and personalities. When these kinds of parents raise these children, they and the people they associate with tend to inculcate these positive values in their children. Thus the cycle of moral brains tends to continue, unless the brain is damaged or affected as discussed above.

Humans who are born as bad, dishonest, indecent and immoral people have the genes that make them this way. They also tend to marry similar people, unless they marry good and decent people to take advantage of their spouse and his or her support and resources, as large numbers of them do. These generally bad people tend to have children who inherit these bad genetic traits and personalities. When these kinds of parents raise their children, they and the people they associate with tend to inculcate these negative values in their children. Thus the cycle of negative and potentially antisocial brains tends to continue.

While we cannot precisely determine what percentage of nature or nurture is responsible for a person's behavior, an estimate of 50% for each is a decent starting point. Born psychopaths are not likely to turn into prosocial philanthropist social workers, and vice versa. While it is unfortunate that due to our genetic primate inheritance it is easier to turn a good person into a bad one, there are people born with extremely high levels of morality and goodness who can never be convinced to take immoral action. These people are precious and rare.

The hardware and software in the human brain worked very well for members of our species who lived millions to thousands of years ago. Modern technology and the volume of information available to us today have moved and accumulated at lightspeed compared with the glacially slow pace of our evolution.

Thus our brains are like a primitive 1980s-era computer with a dial-up modem trying to run modern high-speed gaming programs on a super-high-definition display. The old processor, memory and modules just don't have the ability to handle the volume and types of incoming data, but that ancient and creaking computer in our heads has to find some way to handle the inputs. It does so by relying on its primitive basic operating system and a few primitive modules, which frequently give completely wrong answers when used in the modern world.

Our modern brains have pre-Stone Age programming but no modules for an intuitive or "common sense" understanding of the properties of atoms, electricity, gravity, quantum mechanics or so much else in our world, like probability and statistics. We don't have a module that explains intuitively how light behaves as a wave or particle. We don't have a module that intuitively grasps evolution through natural selection. Humanity didn't find out the truth of any of these ideas until our knowledge and science had enabled our partly evolved, non-scientifically inclined, System 1-driven brains to reach the required level of knowledge and understanding.

And even then, breakthroughs like these only came to the brains of a very few extremely intelligent and learned individuals working in those particular fields.

Here's an extremely sobering thought for you. If you took the 100 most innovative, intelligent and impactful scientists throughout history, such as Democritus, Copernicus, Galileo, Darwin, Newton, Einstein, Faraday, Edison, Tesla, Curie, Archimedes and others, and erased them and their contributions from history, our world would be a far different place. We would almost certainly still be reading scripture by candlelight and many people would still be trying to avoid being burned for witchcraft or heresy.

In a similar vein, if you took the 100 most powerful and influential clerics throughout history and removed them and their effects on the world, we might have landed on the moon by the year 1600, cured cancer and most all other afflictions by 1700 and eliminated poverty and human suffering by 1800, not to mention avoiding the uncountable millions of murders that have been carried out by the faithful.

Our brains also don't have a "scientific method" module, and we have struggled mightily as a result. We had to develop the scientific process and the tools to apply it. The method is now in use everywhere in the world and it is working very hard and producing incredible results, but it is being fought tooth and nail by those who want to keep humanity in the first or seventh centuries. That is a huge problem, and is another extremely important "so what".

Religion formed thousands of years ago during a period when that primitive 1980s mental computer was the only processor and assessment tool we had, and the only data we had was what we could sense using our bodies and a few other bits handed down in oral tradition or crude written records. We knew absolutely nothing about the real facts of the world. Yet many people today still cling desperately to and even fight and kill for what ancient ignorant men wrote down, in the face of absolutely overwhelming evidence that almost everything they wrote was wrong. That is yet another very important "so what". There are other "so what" reasons:

- We have to understand that just as some people find the idea of worshiping some god as crazy an idea as there is, others find it an absolutely irresistible compulsion. Again, we have to understand this and the reasons for it in order to solve this terrible problem.

- Dogma destroys debate, it destroys truth and it never leads to positive outcomes. The dogma of religion is the most powerful justification for evil behavior in our world, and it has been thus ever since the first religion was invented. We have to eliminate dogma and unjustified beliefs. We must ensure our worldviews are proportional to the evidence and that they conform to science. Only then will we be able to live in a truly safe and moral world.

- You really can turn a good person into a bad person. Religion is one of the easiest ways to do this, and for numerous lessons in how to be an extremely bad person all one need do is follow any of the monotheistic holy books. Therefore all unjustified beliefs and ancient superstitions must be blocked and attacked until they cease to exist. The physicist Steven Weinberg said[133] what people have known for millennia. "Religion is an insult to human dignity. With or without it you would have good people doing good things and evil people doing evil things. But for good people to do evil things, that takes religion."

- You really can make a great society if you work at it, but to do so you have to make reason and critical thinking mandatory requirements, and you have to get rid of childish superstition built on ancient ghost stories.

Weinberg had it right, and science and reason get it right. Science and reason are the lights which will illuminate humanity's path to a safer, happier and more moral existence. Religion and superstition blind the mind and darken that path. Religion and superstition keep Pandora's Box of unjustified dogmas wide open, letting these antihuman pains and Troubles loose on the world to subvert science, suppress reason and spread strife, persecution and immorality wherever they go. Science slams that evil box shut.

Our species will continue to suffer terribly at the hands of religion and superstition until we grow up and once and for all eliminate these poisons from our minds, and then our world.

Reason must prevail.

Chapter Thirteen

What Religion Really Is, and How It Is Incompatible With Science

"Ignorance more frequently begets confidence than does knowledge: it is those who know little, not those who know much, who so positively assert that this or that problem will never be solved by science."
— Charles Darwin

"In science it often happens that scientists say, 'You know that's a really good argument; my position is mistaken,' and then they would actually change their minds and you never hear that old view from them again. They really do it. It doesn't happen as often as it should, because scientists are human and change is sometimes painful. But it happens every day. I cannot recall the last time something like that happened in politics or religion."
— Carl Sagan

Since religion's first appearance in the minds of our ignorant ancestors, it has fought a running battle with science. It is true that science was not a formal discipline for most of that time, but when Og the first cave-dwelling prophet or campfire televangelist told his tribemates, on no evidence at all, that he had knowledge of the supernatural and perhaps even the thoughts and intentions of his imaginary god, there were sure to be those who told him he was out of his hairy mind and to quit making things up.

These skeptics would have been the first scientifically-minded people, who required more proof than wildly improbable, wholly evidence-free and self-serving stories and claims.

From that first imaginary god to the laughably large number of gods, goddesses and other supernatural creatures man has dreamed up since then and which are all still freely available to be worshiped today, science and religion have been incompatible. This is not because there is anything wrong with science, it is because religion and its claims are fiction. The claims of religion used to be taken much more seriously for the simple fact that science had not yet discovered the reasons behind much of what happens in our world and brains, and why.

For thousands of years, people have hidden their invented gods behind millions of doors in the Hall of Knowledge. Religions have claimed that the particular gods they believed in were hiding behind each of these doors, doing something or choosing not to do something according to the particular rationalization being proposed. These gods were either making people sick or healing them, bringing rain or preventing it from falling, causing the heavenly bodies to move, causing volcanoes to erupt, sending earthquakes to shake cities to rubble, punishing sinners or choosing not to, and everything else which people attributed to their gods for thousands of years.

Then science came along and started trying to open these doors. Many inquisitive and skeptical persons were castigated, insulted, tormented, exiled, imprisoned or even killed for their impudent questioning of the religious authorities. Despite the difficulties and the continued blockages and attacks by the faithful, scientists managed to open first one, then ten, then millions of doors. What did they find?

Behind the door labeled "Illness", they found no Aesculapian god in a closet picking out people to heal or make sick, but they did discover that germs cause disease and that humanity had been foolishly and ignorantly wrong on this point for thousands of years. The Black Death, malaria, the flu, smallpox and all the thousands of diseases which killed babies, children and young mothers by the millions were not in fact godly punishment for sin or the result of improperly used free will.

These afflictions were the actions of microscopic organisms over which we quite clearly had neither dominion nor visibility until science opened that door. We also discovered how to make medicines, a process which every single imaginary god from every single religion ever invented has still deliberately chosen not to reveal to humanity.

Behind the "Weather" door, they did not see any gods or goddesses hiding in this closet, hunched over a map of the earth with fingers on the planet's temperature and rainfall dials, deciding where they should send floods or maelstroms of fire or brimstone. We humans discovered heat, humidity, air pressure, how air moves and flows, and the continuing total absence of gods. We discovered that humanity had again been completely wrong on a significant topic for thousands of years.

This process of discovery has continued for far too few centuries, but even so, every single door that scientists have ever opened in the Hall of Knowledge has given us wholly nonsupernatural discoveries and provided a scientific explanation for natural phenomena and occurrences. Not a single god has ever come out of any of these closets, and it is by now as close to a foregone conclusion as we need to concern ourselves with that none ever will.

The faithful are of course completely undeterred despite the onslaught of information destroying their fictions. They have massed on their knees in front of two of the last few remaining doors, labeled "Origin of Life" and "Origin of Universe", and are desperately hoping and praying that despite all of the millions of other doors which have been opened to reveal science but no gods, somehow these particular doors are the actual refuge for their pathetically shrunken and as yet completely elusive gods. Such baseless wish-thinking adds to the proof that religious faith in the modern world is clearly a form of insanity.

Belief in gods and religion in 4000 BCE would have been quite understandable, given that period's utter lack of knowledge of science and how things work. Belief in gods and religions in the modern world by an informed human being is not compatible with logic, reason and science. It is only explainable by utter ignorance and the already-noted significant and dangerous flaws in human psychology.

There has been far too much ink and far too much blood spilled in the battles between religion and science, and it is all due to the evil, immoral, hateful and factually incorrect nature of religion and the evil, immoral, hateful and factually incorrect actions of its devotees.

So what is religion, and what is science?

Religion is believing in something on no evidence. Science is waiting to concur with a hypothesis until there is sufficient evidence.

Religion is the denial of the real to preserve belief in the unreal. Science is the denial of unproven assertions until there is sufficient actual evidence to prove them.

Religion is the frightened and mindless preservation of ancient science fiction and borrowed mythologies. Science is the process of turning fiction into reality.

Religion is the first failed attempt at science, and is the even more ignorant predecessor to alchemy, astrology and divination. Science is the only proven method by which humanity gains actual knowledge.

Religion is a psychological safe space for those who cannot face the unpleasant pains and realities of life and death. Science is the comforting and enlightening knowledge of the real and the true.

Religion is emotion, and elevates feelings and opinions over facts. Science is logic and reason, and doesn't let opinions or feelings get in the way of finding those facts.

Religion wants to be what science actually is. Science wants religion to come out from under the bed and stop acting like a child.

Such language is very blunt, and it can engender a hurt or angry response. This is not the fault of the person who speaks the truth, it is the fault of the person who can't handle or won't accept the truth. Rational and emotion-free discussions on subjects such as politics and religion are hard to have, because of how emotionally and socially invested people can be in these subjects, the frailty of the religious mind, and because of how people emotionally and often irrationally react to certain words and thoughts, as described in Chapter Twelve in the work of Haidt, Kahneman and Tversky.

That said, and given that few people like to be in a constant state of war or friction with other people, there has to be a proportional and appropriate engagement with irrational people. For anyone of a nonconfrontational mindset, or anyone who always seeks compromise, be aware that your approach works only as long as you are dealing with people who think, speak and act rationally. Religion and faith are irrational, and science and reason are rational. How can I make such a bold statement? Because that statement is logical and reasoned, as shown below.

The first thing to do in any discussion is to ensure that all parties understand the meanings of the words and terms being used. Failure to do this leads to misunderstanding, mistrust and disagreement. Therefore we must understand the definitions of the key terms which inform my above statement.

Religion is, according to:

- Merriam Webster Online[1] – "The service and worship of God or the supernatural." Also, "A personal set or institutionalized system of religious attitudes, beliefs, and practices."

- Oxforddictionaries.com[2] - "The belief in and worship of a superhuman controlling power, especially a personal God or gods." Also, "A particular system of faith and worship."

Note that these definitions do not, nor can they, mention anything about facts or empirical realities. They quite rightly note that religion is based wholly on mere **beliefs**. This is the polar opposite of science. Religion is by definition and in practice devoid of facts and is composed entirely of beliefs. As I state many times in this book, the religious must cease their continuous habit of interpreting everything through their particular religion-tinted lenses, and stop believing that words mean what they want them to mean. Words do not mean what the wishful thinker wants them to mean. They mean what they actually mean. Let's continue.

Science is, according to:

- Merriam Webster Online[3] – "The state of knowing: knowledge as distinguished from ignorance or misunderstanding." Also "Knowledge or a system of knowledge covering general truths or the operation of general laws especially as obtained and tested through scientific method."

- Oxforddictionaries.com[4] - "The intellectual and practical activity encompassing the systematic study of the structure and behaviour of the physical and natural world through observation and experiment." Also "A systematically organized body of knowledge on a particular subject."

As you can see from those definitions, this chapter could actually end right here. There is a cavernous and unbridgeable divide between the two opposing terms and their implications. One of these systems involves untestable, unprovable and patently ridiculous imaginary beliefs which vary wildly from sect to sect. We call that system religion. The other system is an actual working method of testing hypotheses and proving or disproving them in order to seek truth and reality. We call that system science.

Religion is why a person believes that magic or a supernatural hand is the reason rocks continue their path through the air after they are thrown from a human hand. Science uses physics to understand how gravity and inertia and atmospheric drag, filtered through the lens of ballistics, affect the rock's arc. Religion is choosing to answer any as-yet unanswered question and indeed in extreme cases literally everything else with "god did it", a response which is the weakest and most pathetic copout of all time. Science ignores wishful thinking and bias and refuses to stop looking for answers until it finds them, and it has been unbelievably successful in this.

So are religion and science at all compatible? The answer is clearly, absolutely and decidedly no.

If people treated science like they treat religion, everyone would be free to form and hold his own opinions on how things like chemistry, physics and biology work, just as the faithful apparently feel free to form and hold their own opinions on religion, gods and beliefs in the supernatural. Even the slowest mind can see the idiotic hypocrisy so prevalent today, in that we seem to be happy to let people believe what they want to in religion but we don't let them believe they can break laws or make up their own science simply because they want to. This is an *extremely* important point. Why don't we do this? Why aren't we consistent and allow people to believe that laws don't apply to them just because they don't want them to? Why don't we allow people to pass engineering and science tests by happily giving them full credit for whatever they want to believe are the answers?

People aren't questioned about their belief in a "personal god", so why don't we let them believe in a "personal science" or a "personal legal code", consisting of whatever they feel like believing? The answer is clear and unfortunate. Too many people share the insane delusion of religion, and too many people have been intimidated or cowed into submission by religious thugs. That needs to end everywhere, right now.

Of the two ways of looking at the world, only science is provable and true. Religion is a system of magical thinking which scurries about like a cockroach just like its gods do, frightened of the light and hiding in the ever-shrinking shadows science has not yet penetrated and in the highly unfortunate and vulnerable cracks in human psychology. Note that science and religion absolutely could have been shown to be compatible.

If science were able to test religious beliefs and assertions and find them true, there would actually be compatibility. But this has not happened despite countless efforts by both believers and nonbelievers, and there is absolutely no legitimate reason whatsoever to think that this will ever change, because religion is fiction and thus has no tie to reality. So unless such time as science incontrovertibly and repeatedly confirms religion, these two warring factions will remain utterly incompatible. The most basic incompatibilities are listed below.

1. Religion is a set of beliefs and opinions with no particular rigor beyond personal preferences. Science is a body of empirical objective knowledge derived from observation and testing done with extreme care to exclude personal preference, bias and error.

2. Science screens out uninformed opinions, biases and preferences, while religion strongly and deliberately defends and reinforces them. Thus religion is a system of permanently codified and institutionalized bias, wish-thinking and error.

3. Science both requires and enables critical thinking, along with the identification and rejection of inaccurate data and improper thinking. It separates fact from fiction, truth from lies, and the real from the unreal. Religion intentionally excludes critical thinking and has no mechanism or motive to identify or self-correct. It depends on blind acceptance of its dogma and rejection of any contradictory data or thinking. Religion blends fiction and wishful thinking, mixes truth with lies and self-

deception, and it cannot tell the difference between the real and the unreal.

4. Science collects and processes data using advanced equipment and brings together data-gathering and data-processing tools to form a continually updated empirical understanding. Religion neither collects nor processes data. It is fiction. It rejects or twists science so that the religious dogma is walled off from harm, and it never rejects any of its core superstitions or supernatural beliefs in response to the findings of science.

5. Science is a process, and it is testable, verifiable, falsifiable and repeatable. When its conclusions result in consistently accurate and testable predictions, we have grown scientific hypotheses into scientific theories, and have thus provided real answers to questions. Religion is not a process. It is unchanging ignorant dogma which fossilized an extremely long time ago. It is untestable, unverifiable, unfalsifiable and unrepeatable. It has no conclusions that can be used to make any accurate predictions, and it has never provided any actual answers to anything.

6. Science adjusts where and when necessary to incorporate new evidence, and is a robust process of getting ever closer to truly understanding reality. Religion never incorporates new evidence until it can be twisted or rationalized to fit into the millennia-old original box and can also be twisted to synch with any recently invented dogma. It gleefully and intentionally seeks to move humanity farther away from an understanding of factual reality.

7. Religion is imaginary, and science is real.

Religion is irrational wishful thinking, as is the unfortunate desire by many people to have it and science coexist harmoniously and peacefully in human minds and the public sphere. It must be challenged and fought wherever it is encountered if humanity and civility are to endure. The evangelical zeal of the true believer has both a public face and a hidden and devilish plan of attack.

Like deadly ocean icebergs, the highly dangerous words and deeds which believers, fundamentalists, evangelicals and their ilk show in public are plain to see and can be avoided and/or dealt with. These are more than bad enough, but it is their hidden attacks below the waterline, like the unseen ice torpedo that sank the Titanic, that if not blocked will fatally gash and sink the civic and indeed the moral ship of state.

From attempts to inject creationism and prayer into public schools, putting "in god we trust" on everything from money to publicly funded buildings, trying to pack courts and governments and school boards with anti-Constitutional and history-denying immoral and unethical revisionist theocrats, setting up science-free charter and religious schools then using public funds to pay for them, all the way to waging a military 'crusade'[5] by a "commander" in chief who believed he was chosen for the job[6] by his particular pet god, true believers in America are working diligently every day to undermine our country and force it back into the Dark Ages under a theocratic Christian State.

These threats must be challenged and defeated, and those on the side of reason and rationality must understand this fight will go on for centuries. Failing to continue the attacks on ignorance is not an option, because failing to fight back the barbarians will merely allow their uncontested bigotry and rejections of science and rationality to pour in.

This point must be made before any deeper or longer discussion about compatibility can begin. Religious and irrational people will strongly defend their beliefs with any rationalization they can find, and will destroy reason to favor dogma. They will subvert the law and despite the teachings of their faiths regarding untruths they will deliberately lie to your face, under oath or whenever else they think they need to lie in order to accomplish their goals. This mindset is alien and nonintuitive to the rational, ethical and moral mind, so civil society and its defenders must always remain vigilant and aware of this problem.

I hope it is not too much of a surprise, at least to Americans, that people with deeply and sincerely held beliefs who loudly proclaim their faith as a universal moral compass do knowingly and deliberately engage in thoroughly dishonest practices and will tell appallingly bald-faced lies to try to help their faith take over our society.

This must always be at the top of one's mind when engaging with or assessing the actions of the faithful. They lie, cheat and steal while trying to hide behind a façade of piety and morality. Their gods and their holy books have given them plenty of motivation and guidance to do this. The louder, the more confident and the more outspoken the believer, the less you should trust him or her.

Thus science and reason mortally wound religion on the basis of logic and prove their incompatibility. The illogical, immoral and underhanded actions by those of sincerely held faith to preserve that faith provide another layer of proof that the two positions are incompatible.

Despite the often sincere calls by the nonconfrontational factions in both the religious and the scientific communities to just be nice and get along, and the seemingly endless appeals to Stephen J. Gould's highly unfortunate and extremely wrong-headed pusillanimous treatment of them as "Non-Overlapping Magisteria"[7], the two systems are fundamentally and permanently incompatible. Again, this did not have to be the case, as explained below.

It is impossible to overstate the core difference between religion and science and the basis for their incompatibility. Science is a mechanism for objectively assessing the world we live in, be it the world we can see or a potential world we cannot see. The general original process of measuring, testing and predicting has produced a tremendously divergent and wonderfully productive collection of scientific disciplines and subdisciplines. Scientists are taught not to trust their intuitions or so-called "common sense" as final answers, though they must assuredly use these up front to inspire and inform their work.

Rather than merely trusting these instincts, stopping their investigations at what the faithful would call "common sense" and ceasing their activities, scientists do extremely rigorous, precise and thorough work to try to prove and disprove their hypotheses and ideas. And on the road to publication of a scientific finding, the original scientist, his or her peers and co-workers and others closely review the work to identify any errors in measurement, logic and method. Then any publication which agrees to publish the work will normally review the work. Then once the work is out in the wild for everyone else to see, the real claws come out and the work will be subject to extremely intense global scrutiny. There are countless stages and gates and verifications throughout the process. This is the norm for science.

Religion has no empirically honest or reliable internal system of checks and balances, no scrutiny of its dogmas and no careful analysis or testing of its conclusions. It is, after all, built solely on faith. This in itself demonstrates that all religions can be dismissed without a second glance. There are terms for daring to question the claims of religion. They are called apostasy and blasphemy and are actually still a capital offense[8] in many places in our world even in the 21st century. The faithful have always murdered those who disagree with their mystical claims.

Rather than seeking to verify their dogma, the religious scorn the unbelieving blasphemer, reject science and look to each other for self-verification. Whenever actual tests are done to validate religious claims, the inevitable failure to find any truth in them is dismissed and the religious offer any number of excuses and rationalizations for the outcome.

Unbelievably, and without any justification or logic whatsoever, the religious claim that their god is somehow "beyond the world" or "beyond science", and thus any human attempts to find it will always fail. Why do they claim this? Because their religion is an unprovable and unjustifiable fragile mythology and in a scientifically informed world it will be rejected by educated and intellectually and emotionally strong people. The religious have an existential fear of this. Solely on the basis of emotion, rationalization and special pleading, they tell the rest of us that their god is hiding in an undetectable safe space where the probing tools of science cannot penetrate. They assert without any hint of embarrassment at their unbelievable hypocrisy and self-serving bias that for some reason their particular god can magically cross the spirit boundary to do work in this world but our instruments can never detect or measure this.

This pathetic and contemptible destruction of logic and reason, even lower than infantile in its immaturity, seeks to heap protective dogma atop core dogma to insulate the extremely fragile god at its core as much as possible. One would think an allegedly all-powerful god would not need such strident, illogical and bizarre arguments for its existence. And one would of course need to ask why this same justification does not apply to every single other god ever invented. Surely if one god is "beyond science", all the rest of them should be able to enjoy the same highly convenient screening from our probing instruments and logic.

Believers all think they can sense and know the allegedly one and only true god. This makes the task of figuring out which of the thousands of allegedly true gods is actually the real "one true god" a very thorny one. I always look forward with great anticipation and admittedly much schadenfreude to the spectacle of the faithful having that debate among themselves.

Many of the faithful actually know but will never admit that their religion is built on opinions and not on facts, while science is a fact engine and fact-checker. From their perspective, asserting that their invisible god is off limits to analysis and measurement serves to internally reinforce their dogmas, whitewash the problems and convince themselves that these nonsensical dissemblings give their religious assertions a permanent get out of jail free card. No matter where science looks, they say, no matter with what instruments or with what discipline, they say, their god can scurry away from the oncoming spotlight like a supernatural supercockroach, never to be glimpsed or found.

This would not be a problem if people kept such lunacy to themselves, but the faithful simply cannot control themselves. They have an immoral, antisocial and hypocritical compulsion to try to force their beliefs on everyone else. This has always been and will always be a source of conflict and human misery, because throughout human history and today the most religious societies have been and still are the worst places for human well-being.

Beyond the baseline attempts to inculcate their particular religion and its ridiculous supernatural beliefs, the faithful continually damage their society's well-being, morality, governance, education and indeed its future through their perpetual elevation of mindless immoral dogma over actual reality. Well-functioning societies which maximize human well-being do so on the basis of a pro-human morality, a commonly applied pro-human legal code, public services, a common shared defense, advances in technology and the education of young people. These functions can only be properly and humanely performed by secular and scientifically informed people. This is, again, because science is true and religion is usually highly immoral and antihuman ancient mythology and science fiction.

Religious and scientific positions cannot be equally applied to the real world, because science works and religion does not. When the claims of a religious belief system run into scientific realities, the faithful do not amend their world views to accommodate and accept the new science. They twist science to fit their dogmas or merely assert the mystery of their lord. How and why is this a very bad problem? Because illogic and deference to dogma splatter, like fecal matter thrown into a fan.

As an example, a very large number of Christians astoundingly try to use their ancient book of fables to say that the world is only a few thousand years old. They multiply this ignorant asininity when they teach their children this antiscientific idiocy, which is proven wrong by every relevant fact on the planet.

Many Muslims use their holy book and other teachings to tell stories about flying horses and absolutely perfect recitations of entire books, and they teach their children their particular mind-poisoning isotope of antiscientific idiocy. All these antiscientific idiocies are accompanied by truckloads of other antiscientific idiocies and a requirement by the faithful that those in the religion and especially their children blindly accept the dogma in the face of massive amounts of incontrovertible evidence disproving the books and teachings.

This deliberate destruction and subversion of critical thinking always splatter, and the faithful are far more likely to deceive themselves or to accept ridiculous propositions, because they have been conditioned to elevate dogma over reality and critical thinking. They do it for their religion and everything else they do. This is another one of those points that has to be stated twice. People who subvert their reason and critical thinking in favor of religious dogma have proven their hypocrisy and inconsistency, and can be counted upon to subvert reason and critical thinking in other areas beyond religion.

Thus it is hard to have an empirically-based conversation with a religionist on many matters, because they have been brought up to defend what they believe, not to believe only what is provable. It is not possible to coexist with people who strive to place their imaginary god's so-called commandments and his immoral ethos both physically on governmental grounds and also into the legal code.

People who can't or won't be convinced that science is real cannot be convinced that being cruel to other people for religious reasons is deeply immoral. These people would love for their immoral and hateful religions to be the law of the land to clear the way for discrimination and persecution and to destroy science and any scientists whose work contradicts their ancient fairy tales. That is the problem, and it is a danger to us all.

Science is going to get you to real truths about the universe and our world. Religion tries to get you to believe fairy tales. Science tells you what is really real. Religion tells you what people *want* to be real. Science wants the truth and will gut every sacred cow and check and recheck every detail as necessary in order to get to the truth. Religion pulls out its holy books and says that's all it needs, and says those who question it are blasphemous. The infinitely wide gulf between the two positions is really that simple.

Taking the proper perspective is unbelievably important in the lives of every human who will ever live. Why? Consider the following situations.

If a person says that you have won a large sum of money, or that a member of the opposite sex is interested in you, or that you are the most intelligent and attractive person he has ever seen, or that he has a magical new cure for what ails you, the natural urge is to believe it, if for no other reason than you really, really want it to be true. Horoscopes, Nigerian Princes, lost bank accounts, mystery lottery winnings, secret admirers, fad diets, false gossip, faux medicines, crooked politicians and religions of all stripes will always be high-selling snake oils, because most people don't apply science and critical thinking to their lives, and people will be ripped off or much worse as a result.

If a newly minted head of state or his gutless sycophantic rep tells you that his inauguration crowd was the biggest in history, *period*[9], but reams of evidence like ***actual pictures of the actual events*** clearly indicate this is a bald-faced lie, you have to make a choice. You can either believe the lie or believe the evidence, but not both. The choice you make in such circumstances says a great deal about you and how you think, or more truthfully how you do *not* think, the shallowness of your intellect and your absence of morality, trustworthiness and scruple.

There are clearly billions of people who will rationalize away any and all evidence if it does not confirm their *opinion* of the truth. There are far, far more of these people than one might think. Depending on the issue, it is almost safe to say that every human alive falls victim to rationalization at some time or another. If the subject is the supernatural, then that rationalizing human is a religious one.

Religions rely on faith built from fables which were derived from ignorance and wishful thinking. In a socially Darwinian process religion evolved into a codified mechanism to control the behavior of other people and to gain secular power and money for a clerical inner sanctum. Science relies on results built from confidence derived from the empirical self-correcting and rigorous scientific method. That will always be the uncrossable gap between these two ways of looking at the world. Faith comes from fables. Truth comes from science. End of story.

And the mode of thinking one uses has real importance and relevance to our daily lives. It is broken thinking and quite hypocritical for a person to believe that engineering is the reason his car works while his faith is the key to his health, welfare or eternal salvation. This happens because one of the human mind's major weaknesses is its inconsistency. Faith and comfort win out over facts and truth when the truth becomes too painful or difficult to bear. To paraphrase the immortal words of the rogue COL Nathan Jessup from the movie "A Few Good Men", many people indeed "can't handle the truth.[10]" So, you may again ask, what's the harm in inconsistent and unjustified wish-thinking?

Imagine a world where two completely opposing positions are held. One position is built on stories and fables which were set down thousands of years in the past by people who had almost no factual knowledge, and which has changed scarcely an iota in all its history. The other position was refined over thousands of years and took advantage of every new discovery to refine and update its details, with some of the best brains on the planet working on some part of it for every second of every day. In the modern world today, the two opposing views are known to most inhabitants of this planet. Which would you think would be the view most widely held and used?

Unfortunately and astoundingly, in America and many other places in 2018 it is the one built on stories and fables, while the one built on science and reason is widely attacked, fought and rejected.

In the current decade, well into the 21st century and well over 150 years after the publication of Darwin's dagger penetrated deep into the heart of religious dogma, a staggering 38% of Americans[11] actually think that their personal pet god created humans in their current form a few thousand years ago. This means that over 120 million Americans currently do not believe in the incontrovertible reality of human evolution. This is like not believing that water is part oxygen and part hydrogen.

These people hold these beliefs despite the fact that even after thousands of years of looking, no religion has yet turned up a single piece of real evidence for a magical creator of any kind. Science, on the other hand, has pieced together many billions of separate pieces of factual, hard, incontestable data to piece together an understanding of not only our world today, but the history of our world going back billions of years.

There is evidence of evolution by humans and other life forms literally everywhere you look, from fossils to genomes. Yet more people in America today actually believe in the fictions of their fairy tales than in the realities of their researchers. A 2016 Gallup poll depressingly showed that 89% of Americans[12] still say they believe in a god or a universal spirit.

These faith positions are clearly incompatible with science, but one might argue that religion is just harmless and comforting ignorance and that those of faith should just be left alone to their imaginations. This could theoretically be possible with some species other than ours, but as has been said before and which must be said repeatedly due to its importance, the religious have never and can never be trusted to keep to themselves.

They work extremely strenuously to force their beliefs on other people no matter how unwelcome or immoral they are. The faithful do this both as a direct result of the beliefs and also due to the compulsion to make others join their flock so as to gain larger and larger herd confirmation that their own beliefs aren't just made-up fiction. As noted earlier, they do all of this in overt and covert ways.

Their old overt methods were very simple and straightforward. Those who disagreed with the faithful were exiled, jailed, tortured or burned alive. Thus those who disagreed would have been wise to keep those views hidden, and in Europe this was the case for hundreds of years. Indeed this is still the case in much of the Islamic world even today. Since civilized secular laws in most countries today prohibit burning people for imaginary sins such as heresy, blasphemy or nonbelief, the faithful have taken other paths.

They intimidate and lie to young pregnant women while denying other women health care and contraception on faith-based grounds. They deny and subvert science in the schools and the courts. They support or declare wars with religious justifications. They lie about whatever they want to lie about in order to elevate their religion to Christian State status even in our fully and deliberately secular republic. In other words, they still wage Crusades and perpetrate Inquisitions in support of their faith.

Their covert campaign of suppression of reason is equally vile and perverse. They continually look to subvert school curricula and our secular Constitution. They install theocrats on school boards and into elected and nominative offices from the neighborhood to the national level. They actually try to get their myths and fairy tales taught in public schools.

And in a horrible Catch-22 irony in America, they use our own freedom of the vote against us. They gerrymander people they perceive to be undesirable and/or noncompliant out of voting relevance. They propose and pass massive amounts of legislation which are both explicitly and implicitly designed to favor religion over nonbelief, and in every single case it is one specific religion, Christianity, above all others.

Am I mixing politics with religion? No, I am merely observing how theocrats and their supporters have thoroughly and disgustingly contaminated politics with their religion. I should be more precise and say that Christian theocrats and their Christian supporters have deliberately contaminated the politics of an intentionally secular nation with their religion. I didn't start this fight but I damn sure intend to win it.

Conservatives and the religious have been very insidious and deceitful. Politics and religion have been massively and deliberately entangled in America since aggressive and intolerant theocrats, bigots and reality deniers took over the Republican party and conservative media. They are using these political and propaganda weapons to advance their gerrymandering, misogynistic, bigoted and theocratic agenda, under which only their Chosen Ones will have the right to be true citizens of the Christian State.

Republicans in America today use religion to justify everything they do, no matter how immoral and evil, and their holy book offers an eternal source of justification for their immoral actions.

If you think I am overstating things, stop reading and go look up the platforms of some of the state Republican parties. This will be horrifying if you think science and critical thinking are good things. You will be angry beyond words if you know anything about our strongly secular Declaration of Independence and Constitution.

If you have been told or believe those documents are in any way based on any religion, you have been lied to and you are completely wrong, as can be proved by simply reading the documents and not what someone says about them.

Stop reading here, read Chapter 27 in this book on America's completely secular founding documents, then come back here when you are done. You will see the faith-based lies about our founding documents, and the faith-based liars, in an extremely different light.

As an example of how ignorance-riddled the Republican party is in America in the 21st century, I have included screen captures below of the very first two state Republican party platforms which I searched for. I batted a thousand, in other words, when looking for examples of Christianity poisoning Republican politics.

This is not due to any great skill of mine. The utterly thorough infestation of right wing politics by theocratic Christians made the task as easy as finding evil commands in the Christian holy book.

I made sure to keep screen captures of this research, due to the known proclivity for the faithful to lie and make after-the-fact changes when embarrassing facts are brought to light.

The Missouri Republican party's 2016 platform, at https://www.missouri.gop/about/platform/, is 'all god, all the time', from the very first fundamentalist paragraph. Their Christian religion forms the foundation of their lives, their thinking, and their politics. This is their doing, not mine, and they need to feel the heat for it. Figure Six below is a screen capture of the pertinent part of their platform.

Note that the platform is chock full of the words "**WE BELIEVE**", in big bold red font **ALL CAPS**. This is disgusting. Belief is the refuge of the ignorant, and big bold ignorance is on grotesque display here.

Rather than making a statement, they preface it with the religiously-inspired and fact-free "WE BELIEVE". They don't say "IT IS PROVEN THAT" or "SCIENCE SHOWS THAT"; they say they BELIEVE it. They make assertions without any justification except that they believe it. This is not merely infantile; it is catastrophically irrational and irresponsible in a human society in the Age of Reason to act like this.

Making things worse, in their first "WE BELIEVE", they actually and openly admit that they use the demonic text of the Christian bible and its demonic god as their source of morals.

They derive their morality from a genocidal, mass-murdering mass abortionist. This is as bad as white supremacists proudly announcing they get 'their morals, their values and their culture' from their Almighty Hitler.

THE PLATFORM OF THE MISSOURI REPUBLICAN PARTY

Most of what is important in life we do for ourselves, without the government. We form our own families and raise our own children. We choose what we want to do in life. We pray to our God as we choose; we form and join the religious and civic groups that care for our neighbors in need and shape the direction of our state. There are only a few things we need the government to do for us; but it should do those things well and faithfully.

Missourians deserve a state and national government with the vision to see the bright future ahead and the effective leadership to guide us there. We pledge to provide both the vision and the effective leadership Missourians deserve.

WE BELIEVE that all of us are endowed by our Creator with certain inalienable rights, including life, liberty and the pursuit of happiness. It is our fundamental belief that our morals, our values and our culture are derived from Almighty God. To quote the Psalmist, "Blessed is the nation whose God is the Lord." Psalm 33:12.

WE BELIEVE that the strength of America is in the hearts and homes of its citizens.

WE BELIEVE in the right of Missourians to be secure in their persons, their homes and their communities.

WE BELIEVE government governs best when it governs least.

WE BELIEVE that all political power is vested in, and derived from, the people; that all government originates from the people, is founded upon their will only, and is instituted solely for the good of the whole.

Figure 6. Screen capture of the 2016 Missouri Republican party platform.

As if that weren't quite horribly bad enough, we turn to Texas and its habit of doing and saying things larger, louder and much more fanatically than anyone else. Their fundamentalism is Branch Davidian level. Figure Seven below is a screen capture of the 2018 Texas Republican party platform, found at https://www.texasgop.org/platform/. I thought the Missouri platform was straightjacket-in-prison level bad from a critical thinking and immorality perspective, then saw that Texas of course had to out-crazy the Missouri crazies.

This platform far exceeds mere intent to secure religious freedom; it is an open declaration for a Christian State theocracy. Again note the use of "we believe", as opposed to a position based on facts, reality and reason. Again note that it is not me mixing politics and religion; Republicans in America are deliberately doing this and they are deliberately trying to subvert the Constitution to make this subversion into law, thus making the very first principle they state in their platform a total lie.

2018 Republican Party of Texas Platform

Download the platform in PDF form here.

Preamble
Principles
Business, Commerce, & Transportation
Constitutional Issues
Criminal & Civil Justice
Education
Finance
Government & Foreign Affairs
Health & Human Services
National Defense & Border Security
State Affairs
Resolutions

Preamble

Affirming our belief in God, we still hold these truths to be self-evident, that all men are created equal, that they are endowed by their Creator with certain unalienable Rights, that among these are Life, Liberty, and the Pursuit of Happiness. Throughout the world, people dare to dream of freedom and opportunity. The Republican Party of Texas unequivocally defends that dream. We strive to preserve the freedom given to us by God, implemented by our Founding Fathers, and embodied in the Constitution. We recognize that the traditional family is the strength of our nation. It is our solemn duty to protect innocent life and develop responsible citizens. We understand that our economic success depends upon free market principles. If we fail to maintain our sovereignty, we risk losing the freedom to live these ideals.

Principles

We, the 2018 Republican Party of Texas, believe in this platform and expect our elected leaders to uphold these truths through acknowledgment and action. We believe in:

1. "The laws of nature and nature's God," and we support the strict adherence to the original language and intent of the Declaration of Independence and the Constitutions of the United States and of Texas.
2. The sanctity of innocent human life, created in the image of God, which should be protected from fertilization to natural death.

Figure 7. Screen capture of the 2018 Texas Republican party platform.

For point of comparison, the Missouri Democratic party platform, at https://missouridemocrats.org/wp-content/uploads/2018/08/MSDCPlatform8.11.18.pdf, is a model of Constitutionally-compliant secularism. There is not a single mention of any god, and the priorities on the platform make perfect sense for actual people living in the real world.

They are the economy, labor, health care, civil and human rights, education, environment, tax and budgets, public safety, crime prevention, infrastructure and government accountability. In other words, the Founding Fathers and Jesus would be beaming with pride at the Missouri Democratic platform. They would be using Jesus's whip to beat the authors and approvers of the Missouri and Texas Republican party platforms at how badly they have gone off the rails.

For the record, I am a political independent and moderate, which is the same thing everyone would be if he or she were an honest and reasonable science and fact-based person. I don't care what party or person comes up with a platform plank, idea or policy as long as it is founded on sound and uber-verified science and analysis, is legal, ethical and truly moral, does the greatest good for the greatest number for the longest time, and the positives far outweigh the negatives. Also that it works in conjunction with the thousand other settings and policies that we must keep aligned in order to have a truly working and workable government.

My attacks on religion only encompass the American Republican party because like an American citizen who joins a terrorist organization like the Taliban, it has become a religious fundamentalist organization and a danger to our great Nation. It is the Republican party that made the deliberate choice to become a fascistic Christian white nationalist party and thus a huge problem for our country and our world. Had the Republican party adopted science and reason and jettisoned religion, I would be singing its praises to the rafters, but it chose to worship ignorance and dogma instead. Making this bad choice even more immoral, the party has wholly rejected the 'love thy neighbor' guidance of the central character in Christianity, and chooses to hate, oppress and demonize their neighbors. It has modeled itself after a Catholic Inquisitor rather than the alleged Prince of Peace.

Everything I have listed above, and the actions being taken by Christians in every state in America, proves what people do when they become infected with the disease of religion. They use it to justify perversions of logic, perversions of morality and perversions of law to elevate one religion and its adherents above all others.

But above all else, they seek to spread their religious infection, by deception or by force, and to make all citizens follow their religion or at least bend the knee in their presence while they lord it over us. For the religious, the ends always justify the means, and thus civil society will never be truly safe from these people. Their fervent desire to return to the Dark Ages and to force others to join this mindless rejection of civilization is simply too strong for many of them to resist.

In the light of these crimes and misdemeanors by the faithful, we will again examine the potential harm in having a religious position which you inflict beyond your own personal existence.

Imagine that you have to make an extremely important choice. Your child has a severe medical problem. You don't know what it is, but your child has been in terrible pain for some time and nothing you have tried has made the pain any better. You are becoming frantic and think more needs to be done. You have two choices in this. You can pray for your child to get better, thereby using ignorant practices which were in common use thousands of years ago before there was any knowledge of microorganisms, medicines, hormones, blood pressure or any of the workings of the human body.

Or you can go to a certified medical doctor who is armed with current training, experience, equipment and medicines. If you are actually a 100% deeply and sincerely true religious believer and possessed of deep and sincere faith, you will choose the former and start praying. Your child will therefore likely suffer painful and preventable consequences and may very well die a needless death. If you choose the latter, you are a skeptic, atheist, pantheist or agnostic, and your child has an outstanding chance of getting well.

If you choose to see the doctor when your child is sick, but you pray for advice or assistance in some other realm of your life, you are a pathetic hypocrite. You choose science and reason when the chips are really down and you have a truly rational choice to make, yet you choose faith when the cost is something less than your child’s life. This is what people do when their brains become infected with the mental malware of religion. And yet it has always been so. How should we respond to such blatant attacks on reason?

We must first understand that the primary response to science by the faithful over time has been with rejection, scorn, persecution, exile and even murder. They do this in order to preserve the faith and their herd’s tribal unity against new and challenging information. They do this because science presents a fundamental challenge to their religious concepts and ideas, thus to the religion itself, and the faithful have always fought to the death against challenges to their beliefs and their power.

But reason has fought back in recent centuries, and as science has advanced and become too obviously correct to ignore in too many areas, the number of events which people attribute to gods has shrunk proportionately.

Religionists have seen this and have tried desperately to leverage science to justify their religions and their gods. They have gotten this wrong just as they have been wrong about all of their other beliefs. Some of these are errors of omission, due to lack of knowledge or a selective reading of science or an inability to see past the blind spots and logical black holes their religion has burned into their minds. Some are errors of commission, in which science and scientists are deliberately misstated, misquoted or misused to support some theological dogma.

The faithful never get it right, and they never will. Science holds the key to human happiness and prosperity. Religion lays the path to division, conflict and sorrow.

Returning again to specific beliefs, at least several hundred million people, likely into the billions, believe that our planet was magically manifested into existence on a whim in a period of a few hours, about six to ten thousand years ago, and by one creator god or three, depending on how Trinitarian the believer wants to get.

This god allegedly made the first humans, very soon afterward kicked them out of their homes, later drowned almost everything alive on earth, then later chose to reveal itself to a very few men but not to any women in one isolated spot in a godforsaken desert in an ignorant corner of the world thousands of years before electronic recording devices were available.

Many of these same people, at least 24% but perhaps as depressingly high as 47% based on a 2017 Gallup poll of those who take the bible literally or as inspired by its god, also believe that the story of the mythical Noachian flood[13] punishment which their god so wickedly inflicted on the earth's inhabitants actually happened. If life on earth were indeed mass-extinguished then repopulated by a few creatures crammed aboard a single small and primitive wooden vessel lacking climate control and internal plumbing, how exactly did this happen?

Even setting aside the obvious and unanswerable logistical problems of collecting then segregating, storing, feeding, caring for and shoveling up after elephant birds and mastodons, titanosaurs and tyrannosaurs, polar bears and grizzly bears, lions and lambs, hummingbirds and hippos, elephants and wildebeest, skunks and crocodiles, buffalo and giant sloths, rhinos and kangaroos, velociraptors and pterosaurs, I am very curious as to how all of the various unpleasant snakes, insects and microorganisms got aboard this ark.

Did the fleas, ticks, bedbugs, black widows, funnel web spiders, scorpions, tsetse flies, tapeworms, mosquitoes, wasps, fire ants, vipers, rattlesnakes and giant hornets all get separate little cabins? How did the Chosen Family manage to gather and store two members of at least 14 million[14] different species of insects and other creeping things?

And did Noah get some divinely delivered cryo flasks for the microorganisms? Did his god send down a huge stack of Bronze Age petri dishes with special agar made from locally sourced desert plants?

Did Noah and family have to mark these containers or did they come helpfully labeled and prefilled with the organisms? Did Jehovah send down a set of Level Four biohazard suits so Noah and his family could handle Ebola and Marburg, typhoid, HIV, tuberculosis, rabies, anthrax, tetanus, smallpox, botulism, hepatitis, the bubonic plague and Spanish flu germs without getting infected?

This story is so ridiculous on every level that it absolutely boggles a sane mind even to consider that anyone could take it as anything but psychopathic demented mythology from humanity's infantile past. Yet far too many millions of people do believe it actually happened just as it says in their fossilized fiction.

Such utterly complete failures of logic and reason are never fully confined to the religious sphere. This lack of critical thinking always bleeds over into other areas, many of which negatively impact other people. So for every person who puts the slightest credence in this story, one more earth-sized datum is added to the body of proof that religion and science, and indeed religion and reason, are incompatible.

We can take another laughable example of completely ignorant and broken thinking; this one being a derivative of the flood myth. Let's take an imaginary raft ride down the Colorado River and soak in the amazing natural feature that is the Grand Canyon. It is a truly amazing and wonderful sight which cannot help but impress with its size and scale, and the contemplation of the massive amount of time it took to form.

Young-earth creationists, however, desperately want the rest of us to believe that this unbelievably massive testament to geological and biological science formed practically overnight. They also believe that, in a coincidence so improbable it makes self-assembling 747s look downright commonplace, it formed complete with geologically and paleontologically consistent micro and macrofeatures and absolutely jammed with fossils placed from bottom to top in the exact sequence and arrangement which geological, biological, chemical, physical and evolutionary science predict and demonstrate.

For this to have happened, the imaginary god people think did it would have had to do the following:

1. Make all the life on earth in practically the same day, so that every living thing from tyrannosaurs to the most primitive prokaryotes and eukaryotes, giant sloths to Ken Ham's great great grandparents, whales, rodents, all the fish in the sea, roaches, rabbits, dinosaurs, megasharks, dragonflies with meter-wide wingspans, stromatolites, squirrels and all the other life forms which have ever existed all lived on the earth at the same

time, making the land, sea and air all as crowded as a modern city street in Manila or Mumbai.[15]

2. Almost immediately, and before the quadrillions of life forms used up all the food and oxygen, get very angry at his recently-made but allegedly still his most favorite custom-built species, named *Homo disobediensus* ever since an unfortunate incident in a certain garden prison, decide to kill them all, starting with a group of ignorant peasants in the Middle East then including every other life form on the planet, just because he was so mad he couldn't help himself.

3. Starting in the ancient Middle East, cover the entire earth with thousands of feet of rainwater, drowning deserts, jungles and the polar ice caps, without leaving any trace at all of the flood in any soil, sediment, rock or ice layer anywhere.

4. Wait until all life on the planet was completely and satisfyingly murdered, except for millions of paired specimens magically placed into a funny looking boat in violation of most of the laws of physics, then turn off the magic rain. Make the waters magically recede or disappear at super speed, while magically separating out salt water and fresh water so that the final result everywhere on earth exactly replicated naturally occurring long-term geological and climatic processes and not what would occur as the result of a single event.

5. Move his supernatural focus halfway across the earth to North America and deliberately place each and every one of the countless quadrillions of grains of sediment and all of the murdered animals and plants which were jumbled up in the flood waters into precisely sorted and layered sheets, making a customized layered death cake made up of billions of cubic kilometers of sediment and all those creatures he'd just had such a good time creating then killing.

6. Double check the sediment layers to ensure they were all different in composition, thickness and areal extent, making sure to include facies[16] and ripple marks and all the other normal geological micro and macrostructures, and that the final result of the placement of every single one of the quadrillions of particles of sediment exactly replicated naturally occurring long-term

geological processes and not what would occur as the result of a single event.

7. Double check the vertical, horizontal and attitudinal placement of every single one of the gazillions of organic molecules and dead organisms in the sediment layers so that the final result exactly replicated naturally occurring long-term geological and biological processes and not what would occur as the result of a single event.

8. Magically and instantaneously partially harden millions of cubic miles of sediment so that it was not endless cubic miles of sea bottom ooze with the consistency of wet mud, but not quite hard rock either, resulting in a partially cooked death cake.

9. Lift up the continent so that excess water rapidly and immediately drained to lower ground, hopefully to Hell thereby giving its denizens a temporary reprieve from their eternal immolation in fire and brimstone.

10. Watch massive amount of floodwater, now magically cleaned of most of the sediments and floating dead things, carve the Grand Canyon out of the magically semi-hardened sediment layers, so that the final result exactly replicated naturally occurring long-term geological processes and not what would occur as the result of a single event.

11. Magically harden the totality of the previously semi-hardened and now grand-canyonized sediments into full hardness, taking great care that the final result exactly replicated naturally occurring long-term geological processes and not what would occur as the result of a single event.

12. Ensure that the outflow of particles eroded out of the Canyon moved and settled in such a way that the final result exactly replicated naturally occurring long-term geological and hydrological processes and not what would occur as the result of a single event.

13. Go around the world and make sure not a single trace of the flood accidentally got left behind anywhere, say in an Antarctic ice layer[17] that gives over a million years of unbroken climate

records and could be used to prove or disprove a global flood event and not incidentally a minimum age of the earth.

The above idiocy is actually how millions of people accept that the Grand Canyon formed, though they have absolutely no idea about any of the actual science and just accept the dogma on ignorant and blind faith. This particular form of peerless insanity and delusion is called Flood Geology.[18] I am sorry but that is not a joke or spoof term. It is what millions of self-professedly Very Serious Religious People actually believe, and is what a person has to believe to be a true biblical believer.

It is completely devoid of factual correctness of any kind, and it is wholly incompatible with any scientific discipline or empirical rationality whatsoever, but that does not stop the faithful from doggedly persisting in their mass delusion.

This is what happens when people take dogma as their starting point and completely pervert, twist, misuse and reject science to try to justify or prove those dogmas. When dogma is all that matters, and the end justifies the means, no actual empirical evidence or discipline is safe from attack.

Flood geology, so-called intelligent design, the magic garden, the magic talking snake, the magic flying horse, the revelations from above, resurrections, miracles and all the rest of the crazy fairy tales from religion form the foundation of a dogmatist's worldview.

There are billions of these dogmatists scurrying around on our earth with fingers in their ears, blinders on their eyes and armor plating enclosing their minds to block out science and reason. Their attacks on actual science and actual knowledge, from indoctrination of their children to attempts to subvert education to their religiously-inspired laws, are all direct attacks on the truth and on everything that has fed, clothed, healed, advanced and educated our species for thousands of years.

All the above, bad and significant as it is, is nevertheless a mere tiny fraction of the total argument against any compatibility between religion and science. A truly comprehensive listing and discussion of the total body of the argument would require thousands of pages in extremely fine print.

Suffice it to say that religion is not treated any better by any of the rest of the arguments, and that it should be far more than sufficient to stop the attacks on reason and to put religion back into the crypt from which it escaped.

But that is not the case and it is not going to be the case. As I have already noted with much personal disgust, no amount of reasoned and rational argument and facts will be sufficient to convince the faithful of not only the errors of their positions but also of their extreme immorality in trying to force it on the rest of us. Those of us who do not want reason and our freethinking children and grandchildren to be burned at the stake must gird up our loins and engage the enemy for a protracted conflict.

All religions and their tenets are completely incompatible with science. The way religions seek to jam themselves into everyone else's lives is incompatible with morality and civil society, thus making religion incompatible with civilization on multiple fronts.

And to use the term "front" in its military aspect, we will need to fight against the barbarian horde on multiple fronts as well, in order to prevent the religious from carrying out their inevitable future attempts to carry out faith-based Inquisitions and Crusades.

We cannot and will not let the armies of the ignorant create a Christian State, an Islamic State or any other mindlessly aggressive and hateful religious state whatsoever. Our children are counting on us to be successful.

Chapter Fourteen

Why People Believe In Gods

"Religion is an attempt to get control over the sensory world, in which we are placed, by means of the wish-world, which we have developed inside us as a result of biological and psychological necessities. But it cannot achieve its end. Its doctrines carry with them the stamp of the times in which they originated, the ignorant childhood days of the human race. Its consolations deserve no trust. Experience teaches us that the world is not a nursery. The ethical commands, to which religion seeks to lend its weight, require some other foundations instead, for human society cannot do without them, and it is dangerous to link up obedience to them with religious belief. If one attempts to assign to religion its place in man's evolution, it seems not so much to be a lasting acquisition, as a parallel to the neurosis which the civilized individual must pass through on his way from childhood to maturity."

— Sigmund Freud

The question of why people believe in gods has been around for as long as people have watched their neighbors fall to their knees in the direction of some imaginary divinity or other, or watched them fall sick or dead due to the ravages of some disease or other. This chapter discusses the reasons for people's beliefs in the unseen and the unseeable, but I thought a bit of personal history might help to personalize the discussion for those with and without faith or the tendency to it. I'll follow that with the reasons people actually hold such beliefs.

Before we do that, we must review the relevant terminology in order to keep the discourse properly informed. Far too many people have no idea about the definitions of the terms that are so wildly thrown around in these discussions, and it is extremely counterproductive when people don't even bother to understand the basics of a subject before jumping to conclusions and throwing uninformed opinions around.

There are twelve key words that have to be properly understood to engage in an informed discussion on this topic. Merriam-Webster defines these words as shown below, and I have put some of the words in the definitions in bold:

1. **Atheist**[1] "A person who does not believe in the existence of a god or any gods."
2. **Agnostic** [2] "A person who holds the view that any ultimate reality (such as God) is unknown and probably unknowable; broadly : one who is not committed to believing in either the existence or the nonexistence of God or a god."

3. **Freethinker**[3] "A person who thinks freely or independently: one who forms opinions on the basis of reason independently of authority; especially : one who rejects or is skeptical of religious dogma."
4. **Pantheism**[4] "1 : a doctrine that equates God with the forces and laws of the universe; 2 : the worship of all gods of different creeds, cults, or peoples indifferently; also : toleration of worship of all gods…"
5. **Deism**[5] "A movement or system of thought advocating natural… religion, emphasizing morality, and in the 18th century denying the interference of the Creator with the laws of the universe."
6. **Theism**[6] "Belief in the existence of a god or gods; specifically : belief in the existence of one God viewed as the creative source of the human race and the world who transcends yet is immanent in the world."
7. **Animism**[7] "1 : a doctrine that the vital principle of organic development is immaterial spirit; 2 : attribution of conscious life to objects in and phenomena of nature or to inanimate objects; 3 : belief in the existence of spirits separable from bodies."
8. **Faith**[8] "1 a : allegiance to duty or a person : loyalty…b (1) : fidelity to one's promises (2) : sincerity of intentions ·acted in good faith; 2 a (1) : belief and trust in and loyalty to God (2) : belief in the traditional doctrines of a religion, b (1) : **firm belief in something for which there is no proof**…(2) : complete trust; 3 : something that is believed especially with strong conviction; especially : a system of religious beliefs..."
9. **Reason**[9] "1 a : a statement offered in explanation or justification…b : a rational ground or motive…, c : **a sufficient ground of explanation or of logical defense; especially : something (such as a principle or law) that supports a conclusion or explains a fact**…2 a (1) : the power of comprehending, inferring, or thinking especially in orderly rational ways : intelligence (2) : proper exercise of the mind (3) : **sanity**, b : the sum of the intellectual powers."
10. **Proof**[10] "1 a : the cogency of evidence that compels acceptance by the mind of a truth or a fact; b : the process or an instance of **establishing the validity of a statement especially by derivation from other statements in accordance with principles of reasoning**."
11. **Evidence**[11] "1 a : an outward sign : indication, b : **something that furnishes proof**…"
12. **Delusion**[12] "1 a : something that is falsely or delusively believed or propagated…b psychology : **a persistent false psychotic belief regarding the self or persons or objects outside the self that is maintained despite indisputable evidence to the contrary**..."

Many people will be surprised at what these terms actually mean, and others will be surprised to learn some of these terms even existed. This is why so many countless erroneous assumptions and misconceptions abound in this debate. A quick layman's review of the four major theistic terms is below.

Theism is the most extreme and fundamentalist strain of the categories of religious belief. It means a person thinks his or her god or gods are not only real but they do things to impact the world and its inhabitants. This can be through voices or images in the head, internal inspiration or feelings, sending signals or signs via coincidental occurrences or some other assumed messaging venue such as weather events. If you think you communicate with your god or that it communicates with you, you are a theist.

Deism means a person thinks a god or gods set up the universe like a wind-up clock then permanently checked out and has not been seen or heard from since, thus humanity is on its own.

Pantheism means one is open to worship or at least belief in any and all gods of any religion, or that one doesn't believe in theism or deism but sees the traditional deistic or theistic gods as equivalent to the universe – meaning no gods but a wondrously marvelous universe.

Animism also has two modes, one being the belief that things like rocks and trees have life or spirit, and the other being that a spirit form exists apart from the body. Thus a person can be a theist and an animist at the same time, and indeed many are.

Note that not a single one of these systems of religious belief directly implies or contains *any* positions on morality whatsoever. They address positions taken by a human regarding the existence and activities of gods and spirits. One can be an enormously devout theist or deist and be as immoral as a human can be, and there are many people like this. One can be an animist or a pantheist, without any linkage to a specific god at all, and be extremely humanistic and moral. There are many more people like this.

Atheism and agnosticism are not systems of morality. They are positions taken by intelligent and reasoning people who choose to require actual evidence of supernatural entities before believing in them. One's morality is as informed or affected by one's lack of a belief in a god as it is by one's lack of belief in the Tooth Fairy, Santa Claus and magic mice on Mars. There is no inherent connection, but the religious firmly believe that there must be, because their own worldview depends on getting morality from some book which they say came from some god.

It is hard for the faithful to understand how a reasoning freethinker can so thoroughly dismiss the believer's religion until sufficient proof and evidence exist to accept it. This is because religions are devoid of actual evidence and are thus empirically invalid, and the faithful have brainwashed themselves into believing only their particular system is divinely inspired and somehow universally valid.

As religions continue to be brutally bombarded by science and the advances of human knowledge, believers have either chosen to rally around their faith and its internally assumed morality merely on faith alone, or to debauch or twist science to fit their religion.

Thus the pulpits echo the same tribal self-reinforcing messages of "we religious people are moral because we believe in X god", and these messages get stuck in the heads of the flocks. This psychological operations campaign to deliberately spread disinformation is conducted on a continuous basis by the clergy, religious and political zealots and profiteers. They strive to maintain a constant flow of donations which their flocks are all too happy to hand over. They work hard to make secular law incorporate religious dogma.

These acts, plus their lack of factual information and indeed the deliberate spreading of disinformation, amplified by religious mistrust and xenophobia, result in way too many people like me having to write way too many books like this and having to file way too many lawsuits to protect ourselves from faith-based attacks on morality and the Constitution.

The definitions above are crystal clear. Freethinkers think for themselves and use reason, evidence and proof. Those of faith may say or think that their beliefs rely on reason and evidence, but this is a lie both to themselves and to others. Religious belief is founded on faith, which by definition requires no proof or evidence. Indeed, in the spirit of the argument which Richard Dawkins put forth so well, religious faith is by definition a delusion, which is "a persistent false psychotic belief regarding the self or persons or objects outside the self that is maintained **despite indisputable evidence to the contrary**..."

The faithful ignore evidence against their faith and maintain their false and/or psychotic beliefs based on the unprovable dogma handed down to them. The two thousand years old and unbelievably still common Christian belief that the Second Coming of Jesus will happen 'any day now' is the type specimen of this sort of omnipresent and eternal false thinking and dangerous psychological trap.

Religions claim to know the tiniest details about things which no human sense can perceive and which no instrument can confirm. The Christian holy book is a case study in this, as it is crammed full of absolutely obvious fiction and fables which the faithful actually believe to be real historical events and details. We unfortunately cannot see into the past to observe how these religions were created and shaped solely by humans for human purposes, or all of them would evaporate overnight. We cannot see billions of years into the future to find when this mythical Second Coming will happen.

Extremely conveniently for the faithful flocks and their clerical and profiteering shepherds, we cannot see into any afterlife or spirit realm that is said to exist in us or around us or outside the universe or whatever reality-avoiding Deepak Chopra-ese / religious apologist abuse of language and logic is used to describe this imaginary place. Thus the religious feel they can make any claim whatsoever about these unseen and unseeable places. They make up stories which cannot be disproven and by their utter failures of logic think they must therefore somehow be true. This is the epitome of illogic and one of the most egregious failures of human psychology.

I myself posit a spirit realm where Santa Claus, the Tooth Fairy, covens of witches, herds of satyrs and numerous other imaginary beings romp in unbridled orgiastic hedonism, and where they ride to and from their magic kingdom's versions of Hefner's playhouses on winged unicorns which follow paths of lickable rainbows illuminated by a smiley-faced sun in the sky which beams rays of edible yellow light on all it sees. Like Bertrand Russel's orbiting teapot[13], like Carl Sagan's garage-bound dragon[14], like the Christian heaven, like the Islamic afterlife or like my imaginary adult playhouse for imaginary beings, there is no end to the amount of childish nonsense a person can make up and pretend to be real simply by saying there is no proof it doesn't exist.

Absence of evidence indeed does not always equal evidence of absence, but this only applies when people use reason and are reasonable, which does not occur in religion. Absence of evidence in the case of all the thousands of gods humans have invented must indeed be counted as evidence of absence, or the concept of reason has lost all of its meaning.

The religious fail to understand that no matter how much they wish certain things to be true, those of us with even a basic requirement for proof and evidence see all religious fantasies and claims as no more believable than Sagan's invisible magic garage dragon or Russell's teapot. This difference in mindset is what sets people apart into the two opposite camps of "will believe ridiculous things with no proof or evidence whatsoever and despite much evidence to the contrary" and "requires proof before belief".

Humans have wondered for thousands of years why some people believe in fantasylands, fantasy creatures, gods and religions while others do not. They ask themselves and others questions on the subject, but they come at it from diametrically different perspectives, which is why I felt it appropriate to give some insight from a freethinker's perspective. I admit I cannot fully see things from the perspective of someone who does not require evidence, and I never will. I have tried but am too much of a realist and an empiricist to see fairies in the garden or gods in the skies. But we can certainly explore the questions.

The overwhelming majority of freethinkers, I would like to say all of them but that may be assuming too much, have at one point or another asked themselves deep and introspective questions, such as:

- "When did I become an agnostic / atheist / nonbeliever / freethinker?"
- "Did I actually "become" one or was I born that way?"
- "Do any of these faiths have or know something I am just not seeing?"
- "Why do so many other people believe these sorts of things?"
- "Why do so many people believe so many different things, even in the same religion?"
- "Why didn't we invent science before we invented religion, and not the other way around?"

Every freethinker has a different answer to these questions. I personally became an atheist at the moment of conception. My parents were occasional churchgoers, and I unfortunately was actually baptized in the Catholic cult's tradition at some unfortunate point in my youngest years, but thankfully like everything else that is religious and faith-based it ran right off me and I have suffered no ill effects from the exposure.

Maybe the triune Catholic godhead or one of his holy helpers forgot to encode me with the gullibility gene when they supposedly issued me my still-undiscovered or at least still-unrevealed standard Model 1138 soul at that nanosecond of conception?

Why do I think I was always an independent-minded freethinker, essentially from birth? Because in my case I was born that way, since my DNA tagged me as "Freethinker, serial number FTR1859". My brain is wired to require actual evidence for claims before I believe them, with the evidence requirement increasing with the size and/or craziness of the claim. Unless I suffer damage to my brain or I fall victim to an Inquisitionist from the American Taliban or a Christian State I will always be this way.

I can say that nature was the main factor because I was always too stubborn to accept things at face value. I wanted to know more than the infantile first-level claims religious people made, because the overblown claims of religion seemed ridiculous to me even when I was a small child. I simply could not square the "god stuff" I was seeing and hearing with logic and empiricism, or the deep-seated feeling that everybody was just acting like they knew what they were talking about. I heard and understood the words religious people said, and I watched grownups and children speaking and performing their goofy little religious rituals, but it was all mumbo jumbo and nonsense to me.

And that position has become stronger with every bit of information I have accumulated across a half century of life.

Even existential fears didn't move me toward the anesthetic numbing of religious belief. I remember the moment during a discussion with my mother, when I was about nine or ten, that I suddenly and fully realized I was going to freaking *die* one day. I wasn't going to temporarily 'check out' and shuffle off this mortal coil only to wake back up in some blissful paradise and given 72 of the most wonderfully and sensuously compliant and servile virgins I could imagine, or that I would be magically surrounded by all the people and pets I had loved throughout my life. There was not going to be an Act II, a sequel or indeed anything at all.

I became very upset, as you can imagine, and I remember crying sincere tears of fear and angst about it. And as I've already noted, I had been exposed to the claims of religion but they seemed nonsensical and I could not accept them. How could such a thing as a soul exist? What exactly is this afterlife and why hasn't anyone ever provided a single scrap of evidence of it? So while still a child I resigned myself to a life in this world followed by nothingness, and every single scrap of evidence that humanity and I have collected since that time has only confirmed that perspective.

I was not born into a fundamentalist family or even one that "did religion" very much, and I am extremely glad my parents did not force these things down my throat. I went to Catholic grade school and we occasionally went to a local Catholic church, from which my main memories are the ghastly blue 1970s suit I wore, the ridiculous all the time up and down on the knees and on the pew and now we chant and all of that, and what is the deal with the crackers and the wine, and why is my father dropping money on a plate and passing it down the line? Even as a child, I was never a herd creature and I always asked far too many questions and expected real answers to them, so by default I could never be religious.

So the idea of death, true death, the total and final end of things, was a very bad and extremely upsetting thought. But even as a child, I was able to consider the idea, be understandably upset about it, get used to the idea and then move on with my life. There was no uncontrollable compulsion to ease an existential angst or to drown out reality with the numbing drug of religion.

I just faced it and dealt with it. I had an advantage in that I did have the time and the space to think these things through, and my parents did try to answer my questions as best they could. I used all the analog inputs, aka books, that were available to me in that pre-Internet age and I worked things out on my own, perusing the bulk of the wonderful hard-copy multi-volume but extremely informative Encyclopedia Britannica we had at home. I decided even at that young age that this religion stuff was nonsense and if I would die someday then I just better get used to the idea and get on with life.

By the time I attended high school I was even more assured of my position and had much more exposure to the faithful and their extremely odd ways. My mother chose to send me to religious middle schools and a Catholic high school, as that was in her mind and in all honesty probably was, unfortunately, the best option in the area for a quality education. While the disconnect between my lack of belief and the religiosity of the schools was not an issue, there were some interesting times in high school, and fortunately none of them involved promiscuous or predator priests. My most vivid memory from my entire freshman year was from my first few days in the mandatory theology class. Our instructor was a Mr. Sanderson.

He seemed like a nice enough guy, but very early on in the class he informed us that he believed in demons, and that these demons were present in the world and they caused bad things to happen. I had heard plenty of religious craziness in my fourteen years of life up that point, having been exposed to Catholicism for some time, but this was lunacy on a new scale. Never having heard things quite on this top tier level of batshit nutty uttered in all honesty by an otherwise apparently sane adult, I remember actually stiffening in my chair in shock and surprise, as most people would have done if he had started unexpectedly confessing the details of a recently completed serial killing spree.

I actually looked around the classroom, physically turning my head to the left and right to see if anybody else was as astounded at such a crock of insane crap as I was. How could a grown man actually believe such ridiculous nonsense? I was unable to understand how anyone could be certified as an actual teacher in an actual high school while holding such actually crazy beliefs. Little did I know at that time that Mr. Sanderson, who we learned was a devotee of the "charismatic" school of theology, was merely one small and not even that unusual of a drop in a global monsoon of like-minded believers.

Remember, I was BBI (born before Internet) and my main source of information was a fortunately substantial tonnage of books my parents installed in the house, from the previously mentioned Encyclopedia Britannica to glossy and endlessly captivating books telling the tales of Jacques Cousteau and of the early astronauts in their exciting escapades in worlds both far below and far above my own. I knew nothing about

Inquisitions, Crusades, witch burning, snake handlers, flagellants, male and female genital mutilation, human sacrifice, cargo cults, mass campaigns of child rape and all the rest of religion's insanities.

Mr. Sanderson's demon fetish was quite the wake-up call for me and an early insight into the depths of illogic and crazy to which the religious mind will go. Mr. Sanderson, of course, never gave us any proof of his belief, because for him and billions others like him, proof and reality are unimportant when you really, truly and substantially believe in something.

Just for the record, while I did I attend four years of Catholic high school, I did not personally see or hear of any instances of inappropriate behavior toward any student by any of the priests who served as some of our instructors. While I do greatly detest the highly antihuman and disgustingly immoral Catholic religion in particular, as I detest all theistic religions and especially the monotheisms, it would be disingenuous to leave the chad hanging, if you will, that any of my detestation of religion and Catholicism is due in any way to priestly malfeasance toward me or anyone I knew.

On the contrary, some of these individuals were extremely patient and very good instructors of subjects as varied as the Latin language and world cultures, and a great many fellow students and I benefited greatly from their instructional tutelage in the classroom. I understand we older, larger and frankly stronger and more aggressive high schoolers were not the more vulnerable target demographic which pedophile priests prefer, but I nevertheless wish to be evenhanded and honest.

To this day, I would care not a whit if these or any other priests were straight, gay, transgender or castrated eunuchs, as a person's private life should be his or her own as long as it does not hurt others, does not impinge on our shared social and other commons, and is done with other consenting, informed and undrugged grownups.

We heard rumors of gay priests, of course, and I am sure that some of them were indeed gay, but I had no knowledge of anything like this. And frankly I didn't and still wouldn't care if they were all buggering each other after hours in some Roman bath sauna or just fantasizing about it. Like many others, I did not become aware of the truly massive extent of the worldwide Catholic priest rape scandal and the global coverups until everyone else did, long after I was out of college.

Other very vivid memories of high school were the many occasions when we all had to go to mandatory Catholic mass or some other religious gathering or service during school hours. On the few occasions when I watched the proceedings from my seat like a cultural anthropologist watching the natives' weird rituals rather than doing something actually useful like read a book, the priest would put on his Very Serious Face, take his oversized bible and waft it upward through a cloud of incense smoke while chanting in a long-dead language using his Very Serious Voice.

I again could not for the life of me understand how a grown man could do such a nonsensical thing. As I was in New Orleans at the time, I was frankly about as impressed with this piffle as I was with the voodoo rituals and rites undertaken by some of the more eccentric segments of that city's rather interesting and diverse population.

What was the purpose of the flying book gesture? Was the god's vision going and he needed the book just a teensy bit closer? Didn't he write it, inspire it, or both? And isn't he supposed to be omniscient and would know the contents even if he didn't directly write or inspire it? Was it to show the book as some kind of magical storehouse of wonder and knowledge? Was the gesture some sign of supplication and a show of abject masochism and gutless servility, as is pretty much the rest of religion and especially monotheism and Catholicism?

I didn't make the connection at the time with idolatry, but since we had to read that bible during that freshman theology class, I was by this time quite clear that the only reason one ought to hold it so high would be in order to drop it into an industrial-size dumpster.

I have always been a rational literalist. If I go into a grocery store and the sign says the candy bar I want is 99 cents, I understand the price is 99 cents, plus tax of course. If I am in an American history or civics class, and I read the Declaration of Independence or the Constitution, or an account of a major battle or incident from World War Two such as the Holocaust, I take the words pretty much as written unless there are very strong and empirical reasons to question them. These things were written and/or researched by teams of professional people, and many are supported by massive quantities of video and audio recordings, photographs, other documents and/or much other evidence, so there is extremely little reason to doubt them.

The same logic applies to math class; I don't question trigonometry or algebra or calculus because math actually works in the real world. When I am in a literature or poetry class, however, I know up front to take the literal hat off, put on the interpretative hat and try to figure out the metaphors and messaging and whatnot, knowing that with 30 students and one instructor, there are likely to be considerably more than 31 different interpretations.

What I don't ever get to do is tell the grocery store clerk that the candy bar is really zero cents just because that's what I want it to be and because I believe I am entitled to free candy. I don't get to say that my sincerely held religious beliefs about Jesus and free loaves and fishes mean that the clerk has to change the store policy to sell me products at the price I want to buy them for or give them to me as a gift from on high. I don't get to believe the Holocaust never happened, that Catholic-inspired antisemitism wasn't rampant across Europe and that it didn't contribute mightily to that horrible massacre, that Eisenhower had nothing to do with D-Day, that the sun goes around the earth, that a long-dead Jewish philosopher will be returning to the earth any day now or any other crackpot position just because I want to believe it. I don't get to tell the math teacher that I have a different interpretation of what the angles of a triangle add up to.

On the other hand, I do get to take whatever the heck I want out of literature and poetry and fiction, which by the way is a hell of a lot. What I don't *ever* get to do even in literature and poetry, though, is use whatever I take out of them as justification to oppress, injure or damage others.

The three monotheistic holy books are all held up to us as at least the inspired word of the entity which created the universe, if not the actual words verbatim from the big guy or one of his agents. So let's take those approaches and examine them critically.

If the words are indeed verbatim and the actual dictated or inspired words, then we are all in a world of hurt, because these gods, or this god, since if we want to be honest the monotheistic stories are all actually derived from the same original core myth but icing was added to the cake to cover this up, are certifiably insane. They or it are simultaneously psychopathic and schizophrenic, and where they are most consistent and congruent is in ordering people to kill each other.

The divine orders are such that were a human to actually follow all of the teachings and commands as written, thus exhibiting truly inhuman psychopathy, he would at the very least be incarcerated for the rest of his life in this world as a Charles Manson-like danger to humanity. He would allegedly then be tortured after death in an eternity of hell for the number and magnitude of the sins he committed while alive. Unless of course he says he is Catholic, repents on his deathbed and is granted a godly get out of hell free pass and can slip into heaven, in an act of unparalleled immoral asymmetry and eternal injustice.

Yet the same evil mindset and murderous acts committed by these gods are praised and even worshiped by billions. No wonder there are so many monotheists; they are scared witless to even think about not praising their demonic deity. This is unbelievable.

Just to pick a single example, billions of people fall to their knees either literally or figuratively every week to praise the god which is said to have drowned nearly every man, woman, child, unborn child, animal and plant on earth, but who is thought to get worked up past the point of apoplexy if a microscopic grouping of cells is harvested to do valuable stem cell research or is removed from a woman's body to stop an unwanted pregnancy, or if two consenting adults of the same sex engage in physical intimacy.

There are far too many verses from these books which are anti-human, anti-woman and anti-everything which a reasonable person would consider good. Given what is in these books, it is little wonder so many of the copies of the original books of religion have been written with human blood.

An empiricist literally can't get beyond the first page of the Old Testament without snickering. It's like watching a toddler try to replicate an intricate AutoCAD blueprint using finger paints. The authors of that collection of fables had little idea of much beyond the tribal villages in which they lived. The most advanced technology of the time was iron smelting. It is true that people had already been brewing beer for as many as 6000 years before Genesis was written, but it would not be until Leeuwenhoek in the mid-1600s that humanity would understand that microorganisms, the agents behind fermentation and many diseases, even existed. Who can blame them?

How many persons today know and can explain in detail how even a simple one-dollar flashlight works, to include explaining the function of the bulb's filament or LED, how the batteries work, what electricity is, how and where the parts were manufactured, and the details of millions of other modern banalities which would be absolute divine miracles to the ancients? And how many people today can explain the workings of a cell phone, from where its source minerals and parts come from to its internal construction to signal transmission and reception? Yet even a toddler today can use a cell phone to talk to people on the other side of the planet, and no one prays to their cell phones, except maybe hyper-technophiliac overyuppified iPhone users.

So the people who wrote these books, who were as human as you and I are, who were in touch with nothing more spiritual than their dreams, who got almost everything wrong about the real world, who wrote endless stories of immoral rapine and murder, and who borrowed tales from numerous other pre-existing or contemporary mythologies to create their own, are supposed to nevertheless be able to convey absolute morality and serve as the foundation for the beliefs and actions of billions of humans? When one can only pick out a few select words out of the 830,000 or so in the King James version of the Christian bible, for instance, which have true moral virtue?

When I was in school, you had to get a 90 percent on an assignment to get an A. Below 70 meant you got an F. In scoring morality, the Christian holy book and its consorts in monotheistic piety barely rise above absolute frozen solid zero, which is how much credence we should allow them in the real world.

So to the topic of this chapter, which is why people believe in gods. While that is a short and simple question, a detailed answer has to be a bit complicated, as would be expected based on the earlier chapter on human psychology. Making matters worse, the answer includes variations from culture to culture and indeed from individual to individual.

The human brain is the most complex object on earth. The study of the brain as a physical organ and its electrochemical properties has only very recently begun to illuminate its inner workings, through the use of specialized scanners, imaging and other technology. The study of our consciousness and how the brain generates thoughts and emotions is also rapidly advancing due to these same technologies.

Nevertheless, there is still much to learn. We cannot yet definitively identify the specific biological gestalt of fluids, solids and electrical impulses that generate a specific thought or emotion, nor generate a specific thought in a brain. Until we can, we generally have to use existing knowledge of biology along with the less precise tools of psychology and analysis of human actions and responses to say why people believe X rather than Y. These tools may not be as surgically precise, but we can arrive at the answers nonetheless.

So Why ***Do*** People Believe In Gods?

Not only is this question as old as human thought, it is likely that every human who has ever lived and who had the capacity to do so has asked it, either as an internal inquiry or a vocalized question directed to another human, to their pet, to Siri, Cortana, Alexa or to whatever it is they like to talk to. Let's start along that path at the very beginning.

For billions of years, life spread throughout the oceans and across the planet without the ability to think, communicate or ask the god question. Not a single one of these uncountable gazillions of organisms, in any corner of the globe, troubled itself a whit with worries about original sin, whether the creature sitting or swimming or flying next to it was gay, if it was being immodest for not covering its genitalia, which god needed supplication or which god was demanding sacrifices or massacres in its name.

This was a very good thing, since survival was a full-time

preoccupation, and engaging in such pointless and ridiculous distractions would have led to greatly diminished rates of survival and well-being, as such foolishness has done throughout human history and still does even unto this day.

For every microsecond of this unimaginably immense length of time, the blind processes of natural selection were busily modifying the world's organisms into new forms, some of which gradually developed more complex brains and bodies while the other 99% went extinct.

One land-based branch of the tree of life had been muddling along for millions of years when the hand of some god or Fate and the laws of physics gave it an opening. About sixty-five million years ago, a chunk of space rock about seven miles wide slammed into the Gulf of Mexico and caused a global catastrophe. Around 75% of the plant and animal species on the planet died out shortly thereafter in one of the mass extinction events which have occurred periodically throughout Earth's history.

As a quick digression, one can ask if this was Yahweh's pre-Noachian flood modus operandi for killing his creations. Flinging massive rocks at the planet and incinerating or starving almost everything alive would be fully in character for him. As this was the fifth mass extinction which has been documented by modern science, it would seem that if Yahweh, that great mass extinguisher in the sky, does indeed exist, he has a disturbing pattern of continuing to screw up his various creations and then deciding to kill them off in spectacularly cruel ways.

But surely the lack of evidence for the Noachian flood in every single branch of science in every cubic centimeter of the planet is merely proof that modern science has yet to discover the divinely masked secrets of that particular mass extinction, or rather, mass extermination.

Setting aside the Noachian and all of the other myriad regional Bronze and Iron Age Mesopotamian flood myths, the fall of the dinosaurs cleared the path for the rise of the mammals. These fortunate organisms, previously relegated to scurrying about at night and in burrows and other hiding places due to their small size and low position on the animal food chain, expanded into the ecological niches the dinosaurs and other now-extinct creatures left behind, and continued the grand and unsupervised parade of natural selection for many millions of years.

At one point about seven million years ago[15], a split formed in one specific mammal group. One of the parties in that split continued its evolution through ever-increasing brain size and other familiar characteristics. These evolving primates, the early humans, eventually learned how to harness fire. At a later date they developed the capacity to speak and communicate. After this time there were surely some extremely interesting topics discussed around the watering hole, across the campfire and in the cave.

Some of these early humans, as is the case with humans today, were more intelligent and inquisitive than their fellows. These early thinkers would assuredly have looked around them, looked upwards at the night sky, and wondered where it all came from and how it all began.

Because their individual and collective knowledge of the actual facts of their world was essentially zero, they would have responded to these questions with answers based solely on the onboard software and hardware in their brains, which like yours and mine, were shaped by natural selection, quickened when we were zygotes and are essentially just somewhat amped-up chimpanzee brains.

Humans at that time had little conception of science and technology beyond rocks, sticks and fire, although the more intelligent and dexterous individuals among them were obviously learning and passing on improvements in useful techniques and technologies such as better rock-knapping, hunting, gathering and fighting skills. Conceptions of global and universal origin were not informed by any higher level of knowledge than that.

So these early humans looked around and wondered about how they and the tiny bit of what they could see of their world came to be. Not being able to fathom this puzzle given the lack of information on those subjects, the lack of any visible evidence of a creator doing its work and the extraordinarily low probability that any of their fellow humans were responsible, they likely attributed it to some unseen power. Given that fellow humans were responsible for all of the non-plant and non-animal activity so far witnessed, they likely assumed that some 'super' human was responsible, even though it or they were unfortunately never visible except through manifestations like thunder, lightning and volcanoes.

Given the total lack of true factual and scientific knowledge, attribution of such events and also life and the world around them to some sort of superhuman or supernatural cause was completely understandable and quite likely the only rational conclusion at the time. So from a pre-scientific perspective of factual ignorance, belief in a "superhuman" god to explain things was a very plausible explanation. This was particularly the case due to the widespread erroneous belief, still held today and as unproven now as it was then, that dreams were a sign from or vision into some supernatural other-worldly realm.

One must add psychology and a bit of sociology to this natural philosophizing.

Around those campfires and watering holes, there were individuals who were gullible and others who were skeptical. There were those who spoke confidently on topics about which they knew nothing but on which they had strong opinions, and those who only spoke on subjects which they felt they knew well. Some had the 'preacher's gene' and felt uncontrollably compelled to proselytize their personal beliefs at others whether they were welcome or not, and whether they were true or not. Some had the ability to spin interesting yarns and tall tales. Some had the compulsion to order other people around or otherwise try to control their behavior. There were loudmouths and the quiet ones, leaders and followers. These categories of people are still all around us today, as everyone is extremely aware.

Whenever a person today walks down a modern street, sits in a classroom or looks around one's workplace, or when one of our ancestors did the equivalent in their time, one sees people who look perfectly average and normal on the outside, but who are markedly abnormal in their thinking. Approximately 1% are psychopaths, with many more than that in business and politics, and far more in prison, with a percentage of these being serial killers.

There are those who think like, and if given the chance would act like, such evil specimens as Genghis Khan, Hitler, Stalin, Pol Pot, Charles Manson, many Catholic popes, the Inquisitionists and those who supervised them, every 'ethnic cleanser' through history, or those who would enable or cover up systemic and horrific criminal behavior such as Cardinal Bernard Law and Joseph Aloisius Ratzinger, also known as Pope Benedict XVI.

Others are devotees of Jainism and occupy the polar opposite end of the human spectrum of kindness and would rather die than hurt a fellow creature. Some are geniuses, some are polymaths, some are illiterate and some are at the intellectual level of a two-year-old. Nearly all are totally blind to their personal biases and most cannot think critically. Far too many are capable of thinking critically but stubbornly will not do it, despite vocal and indignant protestations to the contrary.

Additionally, as is evident to anyone who has been around other humans for more than a day or two, our herd instinct is extremely strong. Also, and very unfortunately, many people tend to follow or support those who seem the most confident and who have the simplest explanations for things.

So the overconfident proselytizer, self-styled prophet or glib psychopath who spun the most believable and comforting campfire tales of the supernatural and its connection to the real world tended to be the one whose views many of the other herdmates took as their own. This defect in human psychology has not changed since the first human word was uttered. It is true today and will be true for many thousands or even millions of years unless a significant and species-wide genetic shift in our mental hardware and software eliminates this programming flaw.

Later on in human history after we moved beyond campfires and caves, humans developed castes and social groupings for the keepers of these supernatural beliefs and practices, the vast majority of which even unto this day still totally contradict the others. These human subspecies are titled witch doctors, shamans, gurus, priests, deacons, bishops, sorcerers, diviners, witches, warlocks, medicine men, ministers, pastors, preachers, parsons, reverends, clerics, imams, nuns, chaplains, rabbis, archbishops, ayatollahs, cardinals, popes, deacons, elders, prophets, mullahs, priestesses, rectors, swamis, vicars, holy men, practitioners, celebrants and many more with titles both known and unknown.

Over tens of thousands of years, literally billions of thoughts have been recorded on the subject of gods and supernatural entities, onto media as varied as cave walls[16], clay tablets, rock faces, parchment, glossy paper, billboards, books and today's modern electronics. Persons from the secular to the rabid fundamentalist and ultra-orthodox have recorded their opinions and their research findings for the rest of humanity to consider. The level of human scientific knowledge has advanced to a godlike status compared to when the first cave painters physically recorded humanity's thoughts of a world beyond theirs.

By relentless pursuit of real and true knowledge and information, through endless trials and repetitions, by trying out and closing off huge numbers of blind alleys, and through thousands of years of successively standing on the shoulders of our predecessors, be they giant or otherwise, the more intelligent and honest among us have used the scientific method and its precursors to answer a staggeringly large number of the questions humans have asked about ourselves and our universe.

Those of us who have been fortunate enough to be informed of the realities of our universe, and are in full control of our faculties, no longer need suffer under the constrictions and fears that haunted our primitive ancestors and which still haunt their religion-tormented descendants today. We no longer have any need to cower in fear of any god's wrath. This is because we now know that eclipses, thunder, lightning, floods, volcanoes, tsunamis, hurricanes, tornadoes, earthquakes, plagues, disease, injury, sudden unexplained death, crop failures, droughts, insect swarms, famines, cometary passes across the night sky, meteoroid impacts and all the rest of the myriad events that have frightened and damaged or killed humans have fully natural causes. They are not of divine origin or intent.

We no longer worship the sun, the moon or the stars; we now know these objects to be no more supernatural or miraculous than a campfire or a rock. We no longer wonder which gods or supernatural or mysterious forces cause objects to fall when they are dropped, to enable thrown projectiles to complete their ballistic paths, or to paint a rainbow across the sky after a rain. We don't yet know all there is to know about gravity, but we know how it operates and we can measure it.

We know rather a lot about physics and chemistry, about things like inertia and light and electromagnetism and colors, about heat and pressure and hormones and why salt doesn't taste like sugar. We use tools like telescopes, microscopes and microphones to see and hear things our senses would never be able to detect, and we have used the information these tools give us to dig ever deeper into how our universe functions.

We know that microorganisms cause disease. We know some naturally occurring organisms and substances damage or kill us. We know about toxic metals and about poisonous mushrooms, plants and animals. We know that religious faith and prayer do not make sick people well; we know that antibiotics and other treatments do. We know about biology and medicine. When we get sick, we know we don't get on our knees and pray for healing; we wash our hands, we go to the medicine cabinet, and we go to the doctor if things get too bad.

We know the universe began with a massive explosion about 13.77 billion years ago[17], not before you woke up this morning, and not six or ten thousand years ago, despite creationists' passionate desire to believe such idiocy. We know our planet formed a little more than 4.5 billion years ago.[18] We know life began on Earth sometime between 3.7[19] and 4.2[20] billion years ago. We don't know all the details, but there is absolutely nothing that requires the injection of a god into any of the rapidly shrinking gaps in our knowledge.

Of course, even with all of our accumulated knowledge, there really could actually be some sort of god behind all this, an actual creator divinity who put all this together and actually does exist somewhere, extremely far 'out there', despite staying deliberately, irritatingly and completely hidden, totally out of both sight and mind. This hypothesis cannot be disproved, just like the hypotheses that this absentee god is orbiting the sun in lotus position and sipping from Bertrand Russell's celestial teapot or hanging out with magic Saturnian worms cannot be disproved.

Absence of evidence goes both ways, and the total absence of evidence for Yahweh means he is just as unlikely as Sagan's dragon, as both are mythical creatures lacking any evidence for their existence.

It also cannot be proved that if even if it were indeed the case that there really is some sort of divine presence responsible for everything, exactly *which one* of the heaped thousands of gods that our species has invented, or that one of an additional nearly infinite number of gods that humanity has yet to invent, is the one true and real god. The actual, real god could reveal herself tomorrow and burn the earth and all its inhabitants to a cinder for our worshiping so many false gods.

Gods are as common as house dust throughout almost the entire history of humanity, and the odds of any one of them being the actual one, regardless of the number of adherents in the faith, are microscopic. All of the invented gods, goddesses, fairies, sprites, nymphs, djinn, and all the rest of their fellow exhibits in the heavenly zoo are equally as likely as each other to be “The” single creator of the universe.

It can also not be disproved if one were to assert that our universe is an accidental byproduct of some abandoned science fair project sitting in a dark closet, forgotten by a grade schooler from a higher dimension who tossed it next to its dirty tentacle covers. You cannot prove me wrong if I assert that a two inch-long magical purple worm formed the universe by tapping its eyestalks together three times. Any sort of creative entity like this could be the intelligent designer so strenuously argued for by deliberately deceitful Christian wedge strategists.

Even if there were an intelligent designer, there is absolutely no reason to link this entity to any particular one of the thousands of gods that have polluted human minds since the first one was invented, or any of the thousands more which will be invented before our species goes extinct.

It cannot be disproved that ours is just one of an infinite number of universes, created out of nothing by nothing, or for reasons we can’t even imagine, in the same way that a *Homo habilis* or *Homo ergaster* transported to the present day wouldn’t be able to explain the physics of light or even the workings of a cheap light bulb. We can’t disprove that our universe wasn’t created by some lesser god or “Deity’s Apprentice”[21] one, ten or ten billion Big Bangs ago who, like a clumsy suicide bomber, died in the explosion and thus hasn’t been heard from since.

And this sort of hypothesizing could literally go on until the end of human civilization and we would not run out of ideas both insane and plausible. Which is exactly the point.

Even as advanced as we are today, our most advanced technology can still not penetrate to a time before the Big Bang, and there is no way currently known to peer back through time to make recordings of events such as the origin(s) of the universe and life on our planet. The best we can do is analyze the current and very recent state of things and go as far back as we can using the evidence and tools which we fortunately do have, to determine how it got that way. That is science, and it is done every single day by well-trained people who know their work is going to be put under a microscope, both figuratively and occasionally literally depending on the field, to see if the original work was correct.

The gods we used to believe in, and which far too many of us unfortunately still believe in, are completely unnecessary to our daily lives now, except to enable people to manipulate others and to shore up weak minds. And this is where we have to again remember that our brains are still largely those of our distant ancestors. Our brains continue to run inborn onboard programming which causes us to need to assign a cognitive or guiding agent behind events.

This artifact of our programming, which was once essential to survival, is actually a cognitive defect in a scientifically informed world and brain. This defect makes people fall victim to the numerous cognitive biases and logical fallacies we discussed in Chapter Twelve, which we mostly don't even know we must overcome in order to reach rational conclusions.

We are not born with a brain full of inherited knowledge, as would be the case if the Lamarckian model of life were true. The Lamarckian hypothesis[22] posited that children could acquire physical characteristics which the parents acquired during their lifetime, like extra tough and callused feet acquired by people who didn't wear shoes. While an understandable and innovative thought for the early 1800s, we know this is completely false in reality for both physical external characteristics and internal memories.

Every new person born is a brand new person, a self-aware and self-transporting organic computer with an operating system, instincts and reflexes but essentially devoid of "higher" or acquired knowledge.

A computer and a person can function, but they don't "know" anything until information is added to them via their input mechanisms, which for humans is our senses. No matter how much information any single person knows, that person's offspring starts with an empty data bank in its head and has to acquire all its knowledge from receiving and processing data. This is an extremely important point.

Most humans are extremely susceptible to mental molding and shaping by their physical environments and their social groups. Parents, grandparents and siblings begin the social molding, and then to a large extent give way to friends, acquaintances, the workplace, the broader society and our romantic partners.

The technologies and conditions present in society today may enable people to isolate themselves, but once one enters the social media commons which so many find it impossible to stay away from, social stigma and ostracism can occur literally in seconds. So a person can be subject to social molding and pressure from millions of other people all around the world, not just a few dozen in the local area. This is also an extremely important point. Peer pressure can now be applied by thousands or millions of people, not just a few in one's village or tribe. Bullying can be global in its reach.

No matter how much the faithful try to wish it away, whitewash it or ignore it in their discourse with nonbelievers, their arguments for their particular god always have to revert to blind faith and/or a god of the gaps approach. Blind faith is the easy path, since all it requires is a cessation of critical thinking and a total subjugation of the intellect and a surrender of the spine to the cause or the faith. It is far easier to lie down than to get up. No argument, no amount of factual data, no impeccable and incontrovertible point of logic will break through a self-imposed mental Iron Curtain of blind faith.

Those who choose to try to let a little or even a lot of science into their minds, whether to convince themselves or others of their alleged desire to allow science to influence their world view and to be taken seriously, are inevitably assailed by mountains of mutually supporting evidence for non-supernatural processes and absolutely none for the supernatural. In past times, humans had little to no understanding and evidence for anything, thus they had religion. Today we do have this understanding and evidence, and religion is merely philosophical alchemy.

There are numerous root causes behind the origin, existence and persistence of religious feelings in some members of *Homo sapiens*. These reasons explain the rise of the huge global proliferation of different religious sects, subsystems, splinter groups and new belief systems. They also explain why religions still persist today even in a post-Enlightenment society like the United States with much of the world's knowledge a mere few clicks of a digital device or a short drive away.

Religion was here first, and while it served for a time as a primitive explanation for the natural world, it started running into the rapidly growing brick wall of religion-killing information. Religions, especially monotheisms, are inflexible and unbending until they run headlong into scientifically-revealed information and the resistance of those oppressed or affected by the religion. New information by itself is insufficient to change a faith position unless those outside the religion force the faithful to pay attention to it. When the threat to the religion posed by the new information and its possessors becomes so great that it absolutely must be addressed, the faith suddenly becomes selectively malleable and can be molded and twisted to whatever shape or extent is needed to make the believers keep believing.

Usually all that is needed is some creative and selective twisting of the new information so that it is somehow seen as compatible with the faith. Any option or interpretation that can be imagined to keep the faith alive will be on the table. Every religion has undergone this process to some degree, trying to protect its hyperfragile and unsupportable god and faith by changing the outer protective layer as microscopically as possible. Every new so-called revelation or clarification of some point by the religion is an admission not of original sin but of original error, which in the case of allegedly divinely inspired or dictated faiths is an indication that it is all manmade from the start.

Humans are not born with coldly rational scientific brains. We are born with semi-evolved grassland and forestland ape brains which need to be guided by rigorous external methods of analysis like science to reach objective conclusions. And we have not had science very long at all. Even giving the most generous credit to humans and positing the birth of 'science' as a proto-method which began with the advent of agriculture roughly 12,000 years ago, and comparing it to the emergence of *Homo erectus* in Africa by 1.9 million years ago, even those extremely generous dates means that we have only had science or a roughly scientific method for about 0.6% of the history of our more advanced lineage.

If we use a bit more precise dates for a more codified scientific method and the appearance of our far distant ancestors *Homo habilis*, the numbers then become about 1620 CE for science and 2.4 million years ago for *Homo habilis*. Using these dates means science has been around for only about .02% of our history. No matter which figure you use, we have only had science for a microscopic bit of the human evolutionary timeline. Evolution takes a very long time to make major changes unless there are major external influences, which is why our brains are still not scientific or empirical. They are partially evolved primate brains which must use the crutch of science to truly think and analyze things empirically.

And science did not enjoy a steady upward climb from its origins to the present day. Progress in science and rationalism was very halting at first. In Christendom, many such efforts met an early and painful end on the rack and at the stake, as the Catholic church embarked on its self-imposed holy mission of seeking out and permanently silencing heretics and dissenters. The church was evil, vicious and jealous in its pursuit of these threats to its power and station. That the church's god is evil and self-admittedly vicious and jealous gave the popes all the divine authorization they needed to engage in whatever horrors they saw fit to inflict on people.

Think about what it was like to grow up in countries under the oppressive medieval Catholic fist. Think about how many generations of people had to live for hundreds of years under the everpresent threat of the Inquisition. Imagine being a doubter or a heretic during any of this time. Imagine how even the most factually correct scientific freethinker would have had to keep his or her mouth shut vise-tight, if for no other reason than to avoid being burned alive by extremely devout Christians who, like the highly intolerant evangelicals in America today, deeply and sincerely held their crazy beliefs.

Think about how these freethinking people would have had to raise their children. They would have trained them to be believers, or at least strongly considered doing so, if for no other reason than to prevent their sons and daughters from torture or immolation at the hands of sadistic religious clerics.

Even if parents and children had extremely strong doubts, there would have been strict rules about not speaking their minds for fear of incurring Vatican-sanctioned murder. Even in a species where evolution works relatively slowly, such deadly pressure to conform would have made the average tendency to believe or to keep one's mouth firmly shut in public far more prevalent in the population.

Those people who were better at hiding their views and keeping their mouth shut, or at being true believers, would have been better able to reproduce and leave more children, as opposed to the heretics who tended to be killed before having children or were unable to care for them after being imprisoned or burned alive.

It is a little known fact that the last official murder under the Inquisition happened in ***1826.***[23] Not 1326 or 1426 or 1526, but in an astoundingly recent ***1826***, just three decades before Charles Darwin published his masterpiece work *On the Origin Of Species* in 1859. This religiously-justified execution was the last in a massively long line of such official murders carried out by the Catholic church under a 600-year-long heresy hunt which had begun in 1227[24] as an order by the allegedly morally infallible Maximum Leader of that cult.

Imagine what living under 600 years of religious persecution and the continual fear of torture and/or death will do to the common practices of the people. Many generations of people lived under a constant threat of imprisonment or immolation for thought crime. Of course the faith is going to spread under such conditions, and only the uninformed or the "uninformable" would think otherwise.

These conditions were a clear case of human-induced artificial selection, and are an ironic proof of the power of this process to bring traits to fixation or nearly so. Today's creationists deny evolution and selection, yet the very presence of these people among the ranks of the living is the end result of these processes both in the physical and psychological domains.

Christians continue to believe and say that it was the allegedly amazing teachings and truth of their faith that caused it to germinate, then grow, then spread far and wide across the planet. What actually enabled its spread over all the other similar systems in the area was not miraculous at all, and again completely understandable from a human perspective.

In 313 CE, the Roman emperor Constantine formally delivered the faith and its believers from persecution through the Edict of Milan.[25] For the next 24 years, he vigorously and continually supported and empowered the faith[26] in everything from finances to tax exemptions to granting offices to architecture to land grants, while denying these to non-Christians. He starved the other faiths of their privileges and their funds, and thus their support. He also worked to blend some of the teachings of Christianity with the existing worship of the god Sol Invictus, in a clever information operation on his own people which has carried through to the faithful even today.

In 325 CE, Constantine called the First Council of Nicaea[27], which was essentially the first time the faith met to discuss its beliefs as a collective body. One of the results of this meeting was the Nicene Creed[28], which is a bit of truly astoundingly bizarre dogma and worth a read for some extremely disturbing insight into the people who actually believe it. So in theology and all earthly matters, Constantine first set up extremely favorable conditions for Christianity to grow, then fostered that growth and its accretion of resources and power at the expense of other faiths.

Constantine died in CE 337, but despite the large number of successors to the throne for the next four decades, the vast majority continued supporting Christianity and enabling its protection and expansion. Then in 380 CE the imperial Edict of Thessalonica[29] declared Nicene Christianity as the Roman state religion.

And the rest, as they say, is history.

So after first being selected for special treatment by a particular Roman emperor in the 4th century, this sect was fed, watered and protected from harm long enough for it to grow roots and achieve critical mass. Thus a group of people who had no real power and resources in this world gained considerable power and resources, and once having been gained, its new holders went to the most extreme lengths to maintain it.

What really enabled the spread of Christianity was not any imaginary truth, but these original protections and then later its numerous worldwide invasions using European warriors, missionaries, treasure-seekers and immigrants, and the physical, religious and biological weapons they brought and applied wherever they went. It's pretty easy to spread a faith when nonbelievers are run through with a sword, die in droves from imported disease and are subjugated by the faithful. That has been the true path to Christianity's broad presence in our world. Rather than being due to any of Jesus's teachings, it was through the murderous and genocidal methods of the evil Old Testament god.

So if the staying power and spread of Christianity beyond the locale where it was invented are remarkable in any way, it is how it was preserved through a Roman emperor's whim, then later spread through the application of swords, stakes and gallows and through widespread conquest and murder. It also greatly helped the cause when the diseases the Christians brought with them killed off the nonbelieving natives. Far from receiving any bounty of revealed gospel and gifts of religious messages carried with the missionaries and settlers, the indigenous peoples died in massive numbers due to imported microorganisms. But all these evil truths are ignored, because in religion the end justifies the means and after all, genocide and conquest have always been the coin of the biblical realm.

Thus Christianity spread through intimidation, torture, murder, conquest, genocide and biological warfare. It absolutely did not spread so far because of any allegedly special characteristics of its tenets or a divine hand, but through silencing and deliberately or inadvertently killing off non-Christians. Simply through the process of evolution and a blend of natural and artificial selection, it is no wonder at all that two billion humans currently subscribe to it.

We must ask how many Christians would there be today if Constantine and subsequent emperors had not given the early faithful such special treatment and protection? How many would there be today if there had been no conquistadors, Inquisitionists, Crusaders and missionaries, or if they had not killed, conquered, invaded or infected so many people in so many countries? Be assured that even if the original sect had been able to survive for all this time it would still be an extremely odd cult, but with far, far fewer members.

Many readers will be very surprised to discover that the institution of the Catholic Inquisition still lives on even unto this day, though in a somewhat gelded form and renamed as the "Congregation for the

Doctrine of the Faith". Its mission today is not to inflict physical torture in order to change minds or to kill their owners. It is to inflict mental torture through promulgation of guilt and shame for common human foibles and characteristics, with the still-unchanged mission of warping minds to fit the church's current rules and teachings and of course keeping the steady flow of indulgences and contributions coming in.

We thus have a situation where humans as a species have developed an advanced understanding of science and we have accumulated massive bodies of knowledge, but our individual brains still contain software and processing modules that lag far behind these discoveries. We are hagridden by the primitive superstitions of our ignorant ancestors. Our cognitive operating system and our thought processes are heavily impacted by unconscious and often counterproductive cognitive biases and fallacies.

Every human is born without empirical knowledge of the world and must gain all this information through a very long process of accumulation. Making matters even worse is the problem that this process is itself very strongly shaped by the person's social group, thus information that does not confirm the tribal claims is rejected, along with the nonconformist.

Our science has progressed to an amazing degree but humans are stuck with being born as blank slates lacking factual information and loaded with software designed for a pre-scientific existence that is still riddled with programming bugs. We thus have a huge problem. How is it that two people can look at exactly the same questions and the same facts and come up with exactly opposite conclusions – "god did it" and "no god did it"? This finally brings us to the point where we can answer the question of why people believe in gods.

Many philosophers and students of the human condition feel there is one main reason – fear. This definitely contributes but it is not a root cause. Fear is a common side effect of ignorance and weakness. Humans are complicated creatures and there is rarely a single reason for their thinking and behavior. I have identified seven root causes for belief in gods, and I will describe them below in detail. They are not listed in order of precedence, as they are present in different degrees in different people. They are:

1. **Weakness.**
2. **Ignorance.**
3. **Inertia.**
4. **Masochism.**
5. **Mental challenges.**

6. **Power and persecution.**
7. **Programming.**

Yes, the words do form an acronym, WIIMMP-P, which to readers of English happens to tacitly encapsulate the entire corpus of modern religious belief, but I can only be appreciative of the potential assistance of a friendly "higher power" for that fortunate coincidence. The detailed explanations of these root causes are below. It is important to understand that one or all of these factors can be present in one person at the same time, and that the mix can and will change over time. Some factors will be present but will not be as strong or impactful as others. Only a detailed internal and external assessment can determine the factors present within an individual person and to what degree.

1. Weakness.

A person's "locus of control"[30] is a driver for how he handles his fears. Where this locus resides indicates his level of mental weakness. The locus of control describes how a person sees the world, and it can be either internal or external. People with an internal locus of control generally believe they are in control of their lives and that their actions primarily drive their fate. People with an external locus of control generally feel that luck or outside circumstances primarily drive their fate. Freethinkers, scientists and those with strong minds generally have an internal locus of control and take real actions in the real world to make real things happen, and use their own thoughts to make personal decisions.

Those with an external locus of control feel more helpless and anxious about life since they feel they have little to no control over it. To quell this anxiety, such people cling to the imaginary comfort of a loving god which they think watches over them, will answer their prayers and will waft their and their families' souls to an eternal dance party in the sky. They think their god controls the world, and that if they pray to it that it will help them out in some way. These are the people who join religions and believe in gods.

When a person is faced with the most disturbing facts of life, such as one's own death or the death of cherished loved ones, the person has two options for dealing with these facts. One is to accept them as they are, and to make the most of things despite the eventual unpleasant outcomes. That is the path taken by healthy, strong and courageous minds which can handle the painful concepts and continue on in their lives without major distress.

The other option is to imagine a future in which one magically avoids permanent death and that physical death is only a transition to a magical or spiritual life in some nonphysical realm where some mysterious essence will continue to live on. This completely unproven and unprovable wish-thinking is the refuge of weak and fearful minds which cannot handle highly undesirable realities. And as they are infinitely comforting and permanently unprovable, once such thoughts have settled into a weak mind they are for all intents and purposes practically ineradicable.

Fear, stress and uncertainty are extraordinarily strong drivers of weakness in human thought and belief formation. The insecurity and uncertainty of life have been tremendous stressors throughout the vast majority of human existence, and this is still the case for many people. The compulsion to reduce this stress led people to imagine their gods could hear and answer prayers, that the gods are responsible for the events and processes in the world, that there is an afterlife, and to believe all the rest of the long list of comforting and mentally anesthetic ideas religions contain.

Humans crave explanations and crave certainty, and many people prefer to believe unproven and unreasonable or even flat-out wacky explanations rather than live with uncertainty. This is why even the most insane conspiracy theories and crazy ideas have large swarms of believers. It takes strength to deal with uncertainty and the unknown, and weakness leads to belief in the unbelievable.

Believing that a god fills a knowledge gap enables the believer to imagine he understands the causes of events or processes. This in turn provides him with a false sense of certainty and control and a resultant decrease in stress and anxiety. When one believes that a god actually listens to prayers, even if they won't all be answered, the mental comfort derived from that belief in the imaginary god is all many people need.

The strong desire to feel comforted and secure is an element of human psychology. This is generally a very good thing, as it helps strengthen the positive bonds between we human mammals and forms more cohesive and supportive family and societal groups. But when people are under stress, such as when they endure the constant fear of disease, face fiscal and other uncertainty, are in the bonds of involuntary slavery or under oppression by a tyrannical regime, the belief that one will go to a better place after death can be massively comforting and is extremely hard to eradicate once imagined and entrenched.

This is one reason why religions are so common in Third World countries, theocracies and near-theocracies like the US, and so uncommon in places like Scandinavia, Japan and Australia.

The average American faces far more uncertainty in his daily life than Scandinavians do. This is due to the strong social cohesion of Scandinavian societies, and the manner in which the collective body cares for everyone else. Their socialistic and ironically far more Jesus-like models, which contribute to their higher rates of mental and physical health, are in direct contrast to the American dog-eat-dog Old Testament model which promotes near-fanatical independence, very strong out-group xenophobia and relatively few safeguards.

This is extraordinarily dissonant, in that the hyper-religious American self-styled Christian directly rejects the teachings of Jesus to take care of the poor and not to value worldly possessions, while the godless Scandinavians actually carry out Jesus's socialistic teachings and as a result have much higher individual and collective happiness and well-being.

But it is far more than just the fear of death and uncertainty that affects people's judgment. Impacts to the psyche come from a wide range of disturbing concepts or facts, from aging, illness, loneliness, financial problems, the loss of independence, loss of a job and other stresses all the way up to death itself. Humans have a wide range of options for how to mentally process and handle these issues. Unfortunately, far too many people are too weak to manage their fear and uncertainty, and choose to respond to their stresses by opening up a bottle of highly distilled religious dogma and drinking themselves into existential oblivion.

There is also the weakness inherent in the intense drive to avoid embarrassment or major blows to one's self-image. It is extremely hard for any human to admit to being duped by lies, crazy claims or snake oil of any kind. It is extremely hard to admit such gullibility to oneself, and it is even harder to admit it to other people.

As a result, when people start doubting the ridiculous claims of religion, these doubts can lead in two directions. They can lead to the natural embarrassment at having been fooled and the need to deal with it like a grownup, which happens but not nearly often enough. Or it can lead to a rejection of the doubt and a childish doubling down on the faith in order to avoid this embarrassment, and this is usually what actually happens.

Nobody likes to think he has been taken for a fool, regardless of the circumstances, and nobody wants to feel embarrassment. Indeed, most people will go to extreme lengths to avoid such feelings, therefore even an internal admission of such gullibility is often unacceptable to most people. Avoiding this embarrassment means that people will never admit to themselves that they have been wrong, because of the damage this would do to their self-image. Thus rather than leave the faith, the weakness and compulsion to avoid embarrassment and maintain an acceptable self-image means that most people will simply just retain their belief.

Thus weakness is an extremely significant contributor to a belief in a god or divinity.

2. **Ignorance**.

Ignorance is a state of mind in which a person lacks information and/or education, and this is usually true both in general and on many specific topics. Ignorance is being unaware of the truth of the real world. It is a lack of understanding of the realities of naturally occurring events and processes. Growing up in a world in which no one knew science or the facts of reality was the way things were for hundreds of thousands of years. It has only been in the last few hundred years that we have truly been able to make significant inroads into the massive darkness that hindered us for so very long.

Our situation as humans is extremely curious. It is as if our species spent almost its entire existence wandering around in a massive but completely dark room, stumbling about and bumping into each other and off the occasional wall in total ignorance for eons, completely unaware that all the knowledge we will ever be able to attain was written on the floor, and had been there all the time directly beneath our feet but never suspected or glimpsed. Slowly and painfully, with the advent of science and both devoid of and despite religion, we learned how to make light.

Ever since that time we have been illuminating bits and pieces of the floor and marveling at the discoveries. Decade by decade, century by century and discovery by discovery, we have trained increasingly larger spotlights on the floor and have made amazing discoveries. We have been able to pick up on and follow clues barely glimpsed in the dimly lit twilight zones surrounding and between these discoveries. We have been awed at how unbelievably wrong we were in so many ways before the lights began to shine.

Most important of all, we have discovered how interdependent our discoveries are, and have realized it is quite impossible for any one person to become informed upon, much less an expert upon, more than a select number of the increasingly more illuminated bodies of knowledge.

Thus even the most intelligent and learned human on earth, simply due to the breadth and depth of accumulated knowledge and the finite nature of human life, will always be ignorant in many aspects of human knowledge. This brings us to the problems of how and why we as humans gain knowledge and either accept it or ignore it.

There are two kinds of ignorance – accidental and deliberate.

Accidental ignorance is being born into an extremely primitive culture without access to thousands of years of accumulated knowledge. It is the lack of understanding of the germ theory of disease by a 14th century pope and the people of his theocratic kingdom who died in untold masses from the Black Plague despite prayers, flagellation, extreme shows of penitence and, of course, blaming the existentially and collectively guilty-until-1965 Jews for the plague. It is the ignorance of science on the part of modern children who grow up in extremely religious households, communities and schools, which work extremely hard to constrain the children's knowledge to the teachings of the faith and to exclude and deny religion-killing science.

Texas is perhaps the most galling example of this in America, as its school systems and state politics read more like Sunday school lessons than actual education. It is the ignorance of the common man and woman in America due to the pitiful state of our scholastic systems and our general antipathy toward intellectualism, education and science. It is the relative ignorance of even highly educated people anywhere on the globe due to the unbelievably vast amounts of information available on every subject today.

Deliberate ignorance is the rejection of knowledge by those who choose faith over facts, dogma over data and religion over reality. Despite overwhelming evidence to the contrary and the overwhelming absence of confirmatory evidence, there are many educated people who believe in such foolishness as faith healing, prayer, homeopathy, alien visitations, supernatural creatures, gods, ghosts, alchemy, afterlives, extrasensory perception, chiropractic and reiki, just to name a very few.

There comes a time when to be a truly mature grownup, one needs to have an appropriate standard of evidence and proof for a belief or one should stop holding the belief and should also stop trying to push it on other people. Belief in any of the thousands of faith positions including the above or any of the others out there, while being aware of the lack of evidence for any of them, is being deliberately ignorant.

How bad is it? The average human today in America is devastatingly ignorant of biology, psychology, geology, history, ethics, moral philosophy, genetics, meteorology, medicine, chemistry, physics, religions, religious history, the scientific method and pretty much anything that forms the foundation of our modern civilizations.

Yet in 2018, American Christian presidential advisers actually say that prayer[31] can ward off the flu. If I could, I would happily send this evil lunatic and her ilk back to 1918 and see how this message would be received while 100 million people died of the Spanish Flu, while humanity was sending trillions of agonized and desperate prayers to every god known to man to please protect them from the modern plague.

Far too many Christians believe angry gods cause natural disasters and that women are inferior to men. In the face of absolutely everything their Jesus said and did on the subject of wealth, they believe that that the prosperity gospel is kosher and that the socialistic money-hating Jesus would somehow be OK with blatant money-grubbing and huge profits made off the backs of the pious and the poor. Over a billion people believe that only a coven of intentionally celibate males has the keys to a magic eternal kingdom.

Despite a planet full of evidence to the contrary, they believe that these highly abnormal cloistered sexless unmarried woman-avoiding males can be trusted to opine on morals, marriage and relationships. People believe that so-called holy books which exhort the faithful to acts of psychopathic evil are both the perfect literal words of their personal gods and the most moral teachings on earth.

We know so much more now as a species that the difference between us today and even people living only 500 years ago can barely be quantified. The average person living in CE 1500 existed in a sea of religiosity almost 300 years before we realized we could inoculate against disease. The printing press and musket were around but an independent press and widely published newspapers were far in the future. The Inquisition had been going on for 300 years and had over 300 more years to run. If you go back 2000 years from today, you enter an age of unimaginable ignorance and one in which religious dogma had far more power.

People today who choose to believe what was written thousands of years ago, by people who were completely ignorant of the body of knowledge we call science today, are being deliberately and intentionally ignorant. This willful ignorance is both stupid and dangerous.

Our situation is always and permanently exacerbated by the fact that every single human is born with nothing but instincts, reflexes and an empty data bank which must be filled with the information needed to rise above complete ignorance and to achieve the skills needed to overcome bad and illogical thinking.

Additionally, there are only a very few people on earth who actually understand how the brain works. Those professionals who are well-versed in biology, chemistry, biopsychology and related subjects know that hallucinations are not images or voices in the head beamed there from some god or magical creature. They are simply byproducts of the electrochemical machinery in our heads. No matter how real an experience may feel, and no matter what message one imagines is being sent, these are all completely nonsupernatural and fully biological occurrences. They cannot be treated as anything else at all until substantial amounts of actual verified empirical evidence are available to support any such claims.

Making all the above much worse is the fact that the faithful tend to be the most ignorant about not only the scientific and empirical facts listed above, but also about religion. This includes the contents of their and other holy books, and the actual histories of their and other religions. As I noted in Chapter Two, those of no faith are actually the most informed on the subject of religion, and this is very often the case for many other relevant subjects as well. Believers are therefore either accidentally or deliberately ignorant of as much as 99% of the information they actually need to make a truly informed decision about the reliability and veracity of their faith and its clergy.

So ignorance, be it through accidental and unfortunate through birth into a family or society that does not have access to information or into one which elevates dogma and suppresses reality, or through deliberate willful dismissal of facts and information, is another key contributor to belief in gods.

3. Inertia.

Once formed, a religion quickly becomes a social tradition and a part of a culture, and just like most other traditions it becomes extremely durable. When religion becomes further embedded in rituals concerning meals, healing the sick, weddings, funerals, avoidance of certain animals and many other activities, to include believing in highly anxiety-relieving beliefs about a loving god watching over everyone and a glorious post-death afterlife, it has proven to be extremely hard to kill.

When a person's parents, elders, neighbors, friends and family share in a common religion, social pressures make it almost impossible to leave the faith. In many times in human history and even today, your attempts to disagree with or depart from your tribe's or family's faith can get you ostracized, unemployed or even charbroiled while you are still alive. These are all immensely strong forces which keep religious belief moving with absolutely massive social, cultural and familial inertia.

Such cultural, social and religious inertia, given that we are a very social species and tradition is often highly valued, often requires decades or centuries to overcome. Languages, tradition, national borders and religions last for centuries. These things do not change quickly.

Even in the face of outright evil and blatant hypocrisy, the inertia of religion is sufficient to cut through its logical and moral abominations. Religious murders and inquisitions officially sponsored by the Catholic cult continued for 600 years. Slavery was a part of the culture in many parts of the world for thousands of years, and was quite vigorously supported by religion. Even in the US, the abolition of the clearly inhuman and immoral practice of slavery was fought tooth and nail, often by the most devout Christians, and resulted in a civil war and hundreds of thousands of dead and wounded Americans. The post-abolition cultural inertia of slavery in the form of suppression of former slaves and then their descendants remained in US law until the 1960s, and racism is *still* a major problem in America. You can still find places in the world today where slavery can be found, whether it is sexual slavery or slave labor. You can still find places in the world where heretics, apostates and witches are murdered.

I myself would not feel comfortable or even physically safe if I were to loudly and publicly announce my atheism and antitheism in many places in the Middle East or in America, particularly the Baptist Bible Belt South / Southeast. I've been to both Afghanistan and Alabama, and to be honest they are much the same when it comes to their xenophobic religious zealotry. So even though a tradition or cultural practice may be highly immoral, that does not mean it will not be a central part of the culture for centuries or more.

This is particularly so in the case of religiously driven activities such as the Inquisition, religiously based racism, violence toward out-groups and persecution of those of other faiths, all driven by the leaders of the world's most populous religions.

Once religion has insinuated itself into a culture and become part of the social and cultural fabric, and it has become part of the cultural infrastructure, it is damnably hard to remove, no matter how much that removal would improve the culture or human well-being. Societal and cultural inertia carry religion along with it, and since inculcation of children into the local religion is a tradition in many places and families, this collective inertia is a major factor in belief in gods.

4. **Masochism.**

Masochism is extremely common in religions, is foundational to Judaism, Christianity and Islam, and is the core tenet of the thousands of varieties of monotheistic cults. Their institutionalized masochism is reminiscent of our primate cousins and the submission which their females and beta males show to the alpha male. Many of the world's faithful, and all who belong to the three Mesopotamian monotheisms, have inherited a primate brain, a herd mentality and a culture of deference to an angry alpha male in the sky whose utter and total dominance has been recorded in certain books and according to the faithful must never be questioned.

Masochism is in the very DNA of these cults, with submission and deference to the god king part of their rites, rituals and beliefs. It is also in the physical DNA of their adherents. Christianity elevates this masochism to stratospheric heights, however.

Its particular mythology spinelessly and masochistically wallows in the pathetic notion that all humans are miserable sinners, owing to the alleged permanent defect present in us all due to our supposed inheritance of original sin from our imaginary ancestors Adam and Eve. It adds an additional layer of perverse bonus myths in that we should all both be eternally penitent for our god-given rottenness and eternally grateful that the god king sent his god prince son to be savagely tortured and killed by we miserable sinners so that our sins can be forgiven. Thus the whole Christian machine is designed, built and deliberately fed by steady streams of quite appalling levels of masochism. It is masochism masquerading as religion.

The above requires a bit of explanation. In the Christian myth, the god king is said to have knowingly sent his god prince son to be killed by humans, even though killing is a sin and by killing the god prince these humans would be committing what one would logically consider an unsurpassably heinous sin. Yet by committing this especially heinous sin, the sins of people who had both killed the god prince and those who had never heard of the god prince or the god king would be forgiven. Normal moral and educated people, upon hearing such insanity, would want to know why anyone who said or believed such things is not on extremely strong medication and safely locked away in an asylum.

Normal moral and educated people know that two wrongs don't make a right. They know that murdering the innocent is always wrong, that there were no such people as Adam and Eve and that one person is not responsible for crimes committed by another person. The total absence of logic in this staggeringly insane construct is glaring, as is the permanent undercurrent of psychosexual sadism along with its psychosexual masochism, both of which have been noted by freethinkers and the rational for many hundreds of years.

It takes a certain kind of person to listen to this self-hating Christian dogma and believe it. Acceptance of such lunacy is what one sees in Stockholm Syndrome victims or people who have been abused for so long they have not merely come to accept the abuse, they have actually come to believe it is their fault. Every single human who is not extremely psychopathic or sociopathic feels some degree of guilt and shame for something they have done or said in the past, whether the offense was large or small or intentional or accidental. Normal people feel regret for these transgressions, and it is perfectly normal to at least offer apologies to or seek forgiveness from those we have transgressed against.

It is extremely abnormal to seek absolution from an imaginary resurrected Jew and/or his father, or any other dreamed-up divinity. Real people are all that matter, and our fellow humans will either forgive us or they won't. Transgressions that are not illegal are handled between the transgressor and the person who was transgressed against. Transgressions that are illegal in nature are usually handled by secular authorities. This is simple and effective and is what has happened ever since our species learned to communicate.

Actual honest belief in the Christian mythology requires an extremely unhealthy degree of masochism, from individual self-debasement to acceptance of the myth of the collective sin of the entire species. This is quite a sad state of affairs, but billions of people seem quite happy to demean and abase themselves in this manner. Thus the cycle of guilt, shame and masochistic self-abasement to the invisible alpha god male in the sky by weak people with herd mentalities begins and continues, often as a lifelong submission to the invisible and the nonexistent.

Religious masochism expresses itself in the believer's mind through the feeling that no human can possibly come up with a proper decent moral code and that some god has to dictate such a code from on high. Christians are the worst offenders in this moral crime and breakdown of logic, and are also the most hypocritical and illogical in their irrational conclusions. Only a weak, self-hating spineless masochist would say that humans require a god to dictate a proper moral code to us. We have no problem at all coming up with a truly decent moral code. The problem lies in following it, which is of course a different issue.

And for Christians to tell other people that their self-admittedly psychopathic and vengeful god is the source for a truly good moral code is so irrational and hypocritical as to defy understanding. As I would not go to the Marquis de Sade, Stalin, Hitler or a Catholic rapist priest to get guidance on morality, I would not go to the evil demon that is the Christian god for it. And I would not go to his alleged son, either, because he is the one who apparently gave us Hell, who told us he brings not peace but a sword, who said he came to divide families and sow discord, and who quite intentionally committed a horribly painful assisted suicide on a wooden cross.

Religious masochism is not only expressed in mental and verbal self-abasement. In many cases, the religiously taught or inspired masochism extends beyond mere hating of oneself and the carrying of undue loads of guilt and shame inside one's head and goes all the way to actual physical self-flagellation.[32] This is a real activity which was humorously parodied[33] by the Monty Python crew.

People don't normally enjoy pain, but the mental anguish that religion jackhammers into the minds of the faithful pushes people to do unthinkable things to themselves. In an insane attempt to show their god that they are truly penitent and should be forgiven for whatever they are feeling guilty about, they feel the compulsion to physically injure and inflict pain on themselves. This is a very real and unfortunate problem[34] which is well-known to psychologists.

Thus masochism is also a very significant reason why people believe in and worship gods, particularly the monotheistic ones.

5. **Mental challenges.**

Since all beliefs are a product of the brain, and challenged and damaged brains often produce erroneous beliefs, this "mental challenges" category could indeed generally cover all seven root causes for belief in gods, but separating them out as I have in this chapter paints a more thorough and detailed picture. The specific mental challenges I am referring to in this category are insanity and low intelligence.

It is extremely unfortunate, but some people, either through being born that way or due to some later changes in their brains, are mentally disturbed. You can call them insane if you prefer the non-PC term. These persons are extremely mentally challenged but not with regard to intelligence and cognition. They suffer from an inability to tell fantasy from reality.

This is a different kind of mental state from denial or even cognitive dissonance. People who delude themselves using denial or cognitive dissonance actually know or can be made to understand the reality of a situation, but they choose to assume the ostrich position and put mental blinders or even full blindfolds on so that they can ease major mental angst. People who are truly insane actually cannot tell reality from fantasy.

Examples of insanity include the Jerusalem Syndrome[35], in which some people experience psychotic issues after or during a trip to Jerusalem, and of course those people who think they themselves are Jesus or some other character from fiction or history. Schizophrenia is another example. A belief that wrapping one's head in aluminum foil will enable one to tap into the secret radio communications between the government's secret black helicopters built using secret alien technology which are spreading secret substances into our rivers and streams in order to secretly control our thoughts and turn all our children into gay transgender atheist Muslims is another.

People who deeply and sincerely believe in the ceaseless stream of Internet conspiracy theories or in astrology or the garbage contained in supermarket tabloids are at least somewhat insane.

Many people also hold the cosmically insane belief that a magic talking snake convinced our magically created ancestors to defy the creator of the universe, then our species became collectively guilty of this sin but then shortly thereafter were collectively redeemed of this sin because the creator sent his only son to earth in human form to be tortured to death on a wooden cross in an exceptionally ignorant and religion-addled time and place by exceptionally ignorant and religion-addled people, exactly as the creator of the universe had planned for trillions of years. The challenge facing truly sane people is determining which religious believers are incurably insane and which ones can be saved from their various cults and beliefs.

Research[36] has shown that "there is a weak but consistent correlation between paranormal beliefs and various measures of psychological maladjustment, whether you're looking at a tendency toward depression or mania or schizotypy".

The religious will protest that they are not insane at all. After all, as many as 1.2 billion people share their belief or at least some version of it, they allegedly have proof of their beliefs in the form of ancient books of desert mythology, they hear voices in their heads which tell them it is true, they just feel it to be true, and they can read and write and do math and sometimes even science, so how can they be insane?

The clear and logical response is to remind the faithful that a sheer volume of opinions is meaningless, and 1.2 billion or even 1.2 gazillion unjustified opinions hold no more weight than one single unjustified opinion, or one actual fact. People hold many kinds of opinions without any justification whatsoever, and the only thing that matters in reality is evidence.

You can have black and female friends but still be a racist and a misogynist. You just aren't one all the time or you are only a bigot when you feel like it. We can have a sane discussion about the weather or how to select a new car, but if you try to tell me about your imaginary god and/or his imaginary son, you are spouting just as much insanity as does any other believer in any other religion who relates his particular faith's origin story. You say Jesus, we nonbelievers hear Poseidon and Mithra and Amun-Ra, blah blah blah; they're all the same.

I have personally had several experiences with people who seemed quite sane until one day the crazy leapt out completely without warning. One particular incident happened when I was chatting with a fellow graduate school student. This person was able to speak intelligently about the concepts and ideas within her field of study, which was actually in the hard sciences, but she once suddenly engaged me quite earnestly on the topic of the biblical Noah's flood and how she sincerely believed this to have been a real event. It was jarringly shocking, like she had just asked me if I preferred my babies fried or baked.

I had previously thought this person to be just another student working her way through grad school with the same general mental framework and empirical world view I had. In an instant she had become a person I could no longer see in the same light. It was doubly shocking since she was in a science program of study. After the first five seconds had gone by and the initial shock had worn off, and it became obvious she was not kidding, I began slowly backing away, keeping her hands in sight and scanning for a weapon, trying to extricate myself from the situation without calling her a loony out loud or otherwise being publicly insulting.

It was an unfortunate but informative experience. And people who share her fantasy-driven breaks with reality are everywhere. I doubt this person was clinically insane based on most of the rest of her life, but that belief was certainly dirty-underwear-on-head-level insane.

Regarding low intelligence, this is also extremely unfortunate and while it, like insanity, is something which a truly moral or loving god who actually gave a damn about his creations would cure, it is very common in humanity. I must make it clear that the people in this category had no choice in their condition and I refer to them with absolutely no deprecation or insult whatsoever. People in this category are "slow of mind", or 'lacking intelligence or reason". These are people who through horrible bad luck of the genetic draw, and/or through neglect, abuse, accident or injury, do not have the depth of intellect or the mental capacity to truly comprehend and analyze the world in which they live. These people and their families would assuredly change their situations if they could.

They are unable to truly understand the information and inputs they receive, and they lack the ability to make truly informed decisions based on reality and analysis. These people, if they are able to comprehend the world around them, are extremely challenged in the capacity to understand any of the nuances of religion and fully informed discourse, much less make complicated decisions on them. They are extremely vulnerable to the unscrupulous and the evil, and they need to be constantly protected from the religious dogma which the faithful so immorally and continually try to inject into them. They often fall victim to the simple tales the religious tell, and the faithful know what a vulnerable and easy target these people are, so many of them are fed the simple stories and become unquestioning believers.

Mental challenges are one of the clearest paths to belief in gods, are sometimes difficult to detect and are often impossible to correct, at least with modern current medicine.

6. **Power and persecution.**

As already mentioned, human beings are genetically 98% the same as a chimpanzee, and this is obvious in almost every aspect of human behavior and psychology. I have already noted how the compulsion to defer to authority and the alpha male comes easily to many people. The drive to gain power in this world is another very powerful drive. For some people this urge is uncontrollable and they will take whatever path is required to get this power.

Physical force is one means of gaining power, through being more powerful and able to control males and females in the social group, like a bull walrus or an alpha chimp. In civilized societies a primary way to achieve power is through nonphysical consensus, as through an election. This option is open to more people due to the uneven distribution of physically dominant people.

A third path to power is to become an elected or unelected spokesperson for some god and to use that god's threatened powers and vengefulness to intimidate others and cow them into submission. Many people who deeply and sincerely believe in their god also believe they have or should have power over everybody else, and in many societies like America and Afghanistan they unfortunately do have influence or actual power.

Belief in a god gives the believer the feeling that he shares some of the power of his god, and the belief that working on behalf of his god gives him power over others. This state of mind is often due to a variety of deep-seated personal weaknesses. An innately weak person who seeks to gain the power he wishes but will never attain on his own will often attempt to gain it through acting on behalf of his god. Belief in an idea is one thing, but belief in a god is quite another. Ideas can be challenged, but in the mind of the believer the god and the faith cannot be challenged, therefore his imaginary borrowed power will last for as long as he believes in the god, and far more importantly, as long as other people do.

Such people seek to take advantage of others by manipulating them and appealing to their fears. They seek to gain power over others through their victims' weakness, ignorance and mental challenges. Plenty of people know that there are those among us who are easily intimidated or manipulated by threats of divine punishment, and they take full advantage of it. Plenty of people do this despite not having any other justification for threatening or wielding power over others except through the weapon of religious belief. Plenty of people crave power over others and know the power of religion to dull the senses, to turn off critical thinking, and to make people susceptible to manipulation.

People who choose the religious path to power are ubiquitous in the clergy and are most visible in the surplus of despicable televangelists and megachurch stage snakes who have polluted the world's airwaves for decades. They are also all around us in the form of too-loud and highly obnoxious people who try to push their particular religion or god on the rest of us.

The above refers to people who usually genuinely believe in some god or faith or other and use them as their path to personal and tribal power. There is another subset of people who profess faith but don't actually hold it. They are called liars, thieves, demagogues, psychopaths and worse. There are plenty of people who pretend to hold faith for a number of reasons, some of which are actually perfectly understandable. We have already discussed power. We can add the ability to make a tremendous amount of money off of the gullible backs of the truly faithful. We can add those who seek positions in which they can carry out their sexual deviance with minor children.

The Catholic cult, while far from alone in having a large number of sadistic and psychosexually malfeasant individuals in its ranks, stands tall as the current unsurpassed example of an extremely wealthy organization riddled with child molesters who are actually protected by the hierarchy. Power, money and sex are all powerful drives and religion is a time-tested method to get as much of all three of them as you can grab.

Only in religion does a group of people take advantage of the knowledge that everyone has done or said something they regret by using it to control you. Only in religion do people try to tell you that there is a god who is watching you, who is upset at your "sins", and that the only path to absolution and eternal life is through joining their particular religion and getting down on your knees and admitting yourself to be a dirty rotten worm and not worthy of absolution, but if sufficient supplication is offered a reprieve from eternal posthumous torture might possibly be in the offing.

It is a stone cold certainty that a substantial number of televangelists are on the airwaves doing their devious and calculating best to fleece the faithful. A few of them probably really do believe they are chosen and have a line to their particular big guy in the sky, but these are outnumbered by the money-seekers and the power-hungry. By any objective measure, their belief that they have special access to their god's secret website is patently insane, but because the rest of this person's life is sufficiently under control he and others like him strut about, free to roam loose, spread their madness and be a fiscal, mental and social danger to other people.

Religion is the psychopath's best friend, because it offers him a path to ultimate power over others.

The above paragraphs describe the unhealthy compulsion to have power over others in this world, and this desire usually turns into some form of persecution, bigotry and oppression. The path from power to persecution is extremely short and almost impossible to avoid. Absolute power corrupts absolutely, and gods and their followers are usually absolutely corrupt. Just look at what the faithful have done whenever they have had either significant or complete power in the real world. They used it to persecute whoever they didn't like and those who disagreed with them.

And we're not just talking about a situation where the persecuted generally have civil rights but are treated like second-class citizens; we are talking about Crusades and Inquisitions. We are talking about 1700 years of imprisoning, evicting and killing Jews. We are talking about wars, pogroms and uncountable numbers of murdered people. We are talking about the efforts of the deeply devout in America in the 21st century to be deeply and sincerely intolerant bigots. This is a mindset that simply never changes.

Religion and belief in gods are quick and easy paths to power and persecution, and far too many people take these paths, to the detriment of far too many people. Whether the belief is real or false, the easy path to power and persecution make belief in gods extremely common.

7. **Programming.**

There are two components of programming as a source of belief in gods - nature and nurture. Chapter Twelve described the many ways the brain develops and how it can be altered through physical damage and sensory inputs. Every human is different, and everyone's experiences vary, but one can apply a rough 50/50 percentage to the influences of nature and nurture in the average human.

With regard to nature, the brain forms with or without a certain predilection for belief in gods due to the vicissitudes of genetics and factors such as fetal and maternal nutrition and stress. As already noted, this is the hardware and software of the brain, the meat of the organic machine that runs the operating system with its instincts, reflexes, basal programming and its modules for unconscious emotions and deliberate cognition.

Also as already noted, much of the current human brain has essentially not evolved for a very long time and still contains many primitive patterns inherited from our extremely ancient ancestors. The unconscious part of our current semi-evolved primate brain includes two major contributors to belief in gods. These are very strong in believers and very weak in nonbelievers, and they drive how we respond to things like hallucinations, feelings and other experiences. One is whether the person perceives his locus of control as internal or external. The other is the degree to which the person believes events are explained through the actions of conscious agency or intent.

As described earlier in this chapter, people with a strong internal locus of control feel more in charge of the world and their lives, and that things do or don't happen as a result of their own action or inaction. These people tend not to believe in gods. Those who have an external locus of control feel that they are not in control of things and that outside or other forces are.

Such people are also often stricken with an overly powerful need for certainty, so in order to alleviate this uncertainty and stress they tend to imagine a god which listens to them and will look out for them. They are often herd-oriented rather than independent, more xenophobic and tribal rather than tolerant, are more gullible than skeptical, fear and reject change rather than welcome it, and are more prone to wishful thinking than practical reasoning.

It is no accident and no surprise that while it still holds a strong grip on far too many people, religion is dwindling in its appeal in our modern world even in those with an external locus of control. The advances of science have enabled people to gain a degree of control over their lives that would have been inconceivable only 150 years ago. Just taking the history of disease as a very powerful example, when tens of thousands of people died in single cities during the Black Death, and absolutely nothing including sincere prayers offered morning, noon and night worked to prevent or cure the disease, it was very easy in a religion-addled world to attribute the deaths to sin and an angry god, and people almost always did.

Even in the late 1800s countless numbers of people died from diseases that are easily preventable and treatable today. The stress and anxiety of the people who lived under constant fear of disease and death led directly to widespread belief in a god and an afterlife. Today we have no need for such beliefs. We know how to prevent and cure disease and the only discomfort may be a small sting from an inoculation rather than a lifelong crushing depression from the loss of loved ones. Thus today the most educated individuals and countries tend to be the least religious, because we have control over our lives and even the most external locus-oriented of us need not suffer constant and existential fears.

The other aspect of our unconscious programming is the belief to which there is conscious intent behind events. To quickly recap Michael Shermer's term "agenticity", or the attribution of intentional agency to things that happen, he references two imaginary hominids from the very deep past out on a stroll, who hear a rustle in the tall grass. One hominid thinks a predator made the sound while the other thinks it is just the wind.

In cases when the rustle is indeed only the wind, the only major outcome of the hominids' responses is that the first hominid gets in a good sprint workout. In cases when the rustle is actually a dangerous animal, the hominid who does not assign agency dies a horrible death and either does not end up having offspring or can no longer provide for the ones he already has. The one who ran off will survive this and probably many other encounters. He will have children who will share his agency-assigning genes and will also hear his instructions to assign agency to events.

Multiply such scenarios by the millions, and over millions or even hundreds of thousands of years you will eventually have large numbers of people whose inborn instincts signal them that things happen due to a conscious agent. This is our situation today. We still have primitive hominid chimpanzee-like instinct code in the software of our modern brains.

This is why people feel gods, ghosts or other imaginary supernatural ephemera cause things like disease, natural disasters, planets and stars, life and the universe. It is an artifact of our primate ancestry and the only things that will get rid of it are tens to hundreds of thousands of years or more of natural selection in civilized settings, or some deliberate genetic engineering.

As we saw in Chapter Twelve, once this inclination to believe intentional agency is the cause for events has become part of the System 1 process in the brain and thus essentially an instinct, there is almost nothing that will convince a person that he is wrong when he falls into this logic trap and assigns agency to something, even a process of nature. His System 2 will not be able to engage with sufficient power to overcome the errors in his programming.

When a person has this System 1 programming bug and also the inability to rationally overcome it, as is the case with billions of people today, the specious claims of religion will align with his instincts and no amount of science or reason will be allowed in to change his mind. This is why science denial is so strong with the religious and for some it is why their faith is so unchangeable and, in their minds, logical.

The other element of our programming, our nurture, is how each person is programmed by parents, family, society and culture. This can be deliberate programming through brainwashing or a similar upbringing, or the routine socialization every person undergoes. The process is particularly damaging if the person is born into or falls victim to a religious or political cult. The degree to which this programming impacts each person varies based on his individual resilience and inborn operating system, plus the stresses to which the person is subjected. There is a normal background programming going on all the time in nearly everyone's lives, and much of it is the result of cultural inertia and pressure.

Due to the massive power of social inertia and the fear of change, practices can remain in place for centuries. Some practices and beliefs are especially hard to extinguish, and these are the ones which are tied to the most extreme existential fears such as that of permanent death without an afterlife. In the case of religion, such beliefs have proven they can remain extant for millennia. Even without being tied to afterlives and infinite comforts, it is hard to overcome such beliefs solely due to the tremendous social pressure to conform. It is also due to the desire to avoid ostracism, punishment or even death if nonconformity persists. So the societal programming can start at birth and go on all throughout a person's life.

Thus programming is also a major contributor to a belief in gods, for many very obvious reasons.

To recap this chapter, there are seven core reasons why people believe in gods. They are all easily understood and become crystal clear once one has an informed understanding of how the brain really works, and how it functions and adapts in the context of the society in which it lives.

Weakness. Many humans are psychologically weak in the face of extremely disturbing facts like permanent death without an afterlife. They seek escape from fear and stress through belief in a god.

Ignorance. Many humans are unintentionally or even intentionally ignorant of the realities of life, of psychology and the physical universe that so completely disprove or fail to prove any god.

Inertia. Strong social, cultural and other forms of inertia keep ancient tribal beliefs and practices in place for extremely long periods of time. The persistence in belief in the particular god of one's own society and culture, rather than any of the other thousands of equally improbable gods available for worship, demonstrates that such inertia is extremely localized.

Masochism. The too-common human tendency to masochism is a major contributor to belief in monotheistic religion, with the more masochistic and or sadistic individuals tending to be the most devout and/or fundamentalist believers.

Mental challenges. People with mental challenges such as insanity and low intelligence are easily taken in by the claims of religion and by those who would manipulate such people.

Power and persecution. The lust to achieve power and to use it to persecute others is both generated and fed by religion, and religion provides ultimate divine justification for power grabs and the full range of acts of persecution. The faithful always seek to gain secular power then use it to persecute others. Belief that you can gain power and that you can gain justification for persecuting others through actual or simulated belief in a god make that god and its religion irresistible to many people.

Programming. Through their inborn nature and the processes of inculcation and brainwashing, young people are both genetically and culturally programmed to believe in the local god and religion of their families and societies. Only a very fortunate few people have the courage to escape from the religion in which they are raised, thus enabling the ridiculously large numbers of incompatible religions across the world throughout history and in our modern world.

As with the road to recovery from any harmful addiction, the road to recovery from the harmful addiction of religion requires that one first admit that one has a problem, then to understand the roots of that problem. Thus far in the book I have laid out the biological, sociological and psychological reasons for the generation of religions and their persistence. The rest of the book will demonstrate the deep immorality of religion, its unkillable compulsion to hurt and oppress people, and the sheer incorrectness and craziness of all religion. The next part of this book lays out the immorality and evil in religion in clear and unequivocal terms, using the actual words of the texts and the evil deeds of the devout as proof.

Part Four

Accepting Unpleasant Facts

Chapter Fifteen

Religions Are Fruitful, They Multiply, And They Are Very Bad For You

"Unlimited tolerance must lead to the disappearance of tolerance. If we extend unlimited tolerance even to those who are intolerant, if we are not prepared to defend a tolerant society against the onslaught of the intolerant, then the tolerant will be destroyed, and tolerance with them."
- Karl Popper

Once upon a time on Earth, almost four billion years ago, a replicating molecule was "born". I'm not going to get into the details of how this occurred, or the implication of the word "born"; those are separate topics for a separate discussion. The starting point of this particular discussion is what happened after the "birth", not before.

So a long, long time ago in an Earth environment that was far, far away from the one we have today, a molecule was born.[1] Over time, this molecule had lots of little baby replicating molecules, and these babies later had babies of their own.

Over an incomprehensibly long time and after more replication events than we will ever know, these replicating molecules became increasingly more complex. About 3.8 billion years ago[2], a series of mutations and adaptations in some number of the surviving replicators led to the evolution of extremely primitive cells.

After much more time, these cells were extremely fruitful, and they multiplied in tremendous numbers. Over nearly four billion years of being ever more fruitful, the descendants of that original replicating molecule were fruitful and multiplied along millions of different branches into millions of different types of organisms.

Like individual branches on an extremely large tree, different types of life form split off and became so widely divergent that despite being descended from the same original ancestor, life today is as different as mold and whales. Almost nothing of the original ancestor remains, yet its descendants populate the entire world.

Once upon a time on Earth, a particular species of hairy hominid learned to make sounds with its mouth, tongue, vocal cords and lungs.

Over time, members of this species of hairy hominid refined these sounds into patterns which were mutually understood by speaker and listener. Over time[3], possibly more than a million years[4], this original language was fruitful and multiplied into two, then four, then eventually into the seven thousand languages known today. Almost nothing of the original language remains, yet its descendants populate the entire world.

Once upon a time on Earth, specimens of a particular species of less hairy hominid which had evolved from the previous hominids looked at the land, the sky, the sea and the forest, and imagined powerful supernatural beings living there. They passed these ideas on to other less hairy hominids, and the stories passed along to successively less hairy hominids over millennia.

Over many thousands of years the stories diverged greatly from the original to form many new stories, many different supernatural beings and even many separate godly pantheons overflowing with powerful and mystical invisible gods. The original ideas were extremely fruitful and multiplied into many thousands of descendants. Almost nothing of the original religion remains, yet its descendants populate the entire world.

Religions are just like organisms and languages. They start from an original point, then are fruitful and multiply. They change and adapt, reacting to the pressures and vicissitudes of the times and conditions in which they exist, thus demonstrating how ideas evolve just like organisms do. Just as species of organisms diverge, evolve and become unable to breed with each other, religions undergo a very similar process of classic Darwinian selection as they are born, begin evolving away from the common ancestor and continue diverging into massive numbers of separate forms which are eventually incompatible.

The ever-branching trees of life, language and religion can all be represented by essentially identical graphics. They start with an original seed which grows into a tree which grows branches which in turn grow more branches until one can represent the current global diversity and difference we see in historical and geological records up through today.

No matter the origin of the religion, its age or lineage, its adherents fervently believe in it. They just as fervently disbelieve all the others. This is because parents and societies indoctrinate their children into their particular faith along with everything else that is passed down, and because group cohesion, herd tribalism and xenophobia do not tolerate change or new or different ideas.

The three most widely practiced religions[5] on our planet in the 21st century boast 4.8 billion followers. Of these, 2.2 billion are Christian and 1.6 billion are Muslim, meaning half of the people alive today subscribe to some variant of these two religions. The histories of these religions are overflowing with persecution, spilled blood and conquest, and the paths these faiths took across the world were paved with the bodies of innocent people enslaved or killed in the service of those faiths.

Yet the adherents of each religion today astoundingly insist with no trace of irony and apparently in all seriousness that theirs and theirs alone, to paraphrase Snow White's stepmother, is 'the most moral of them all' or 'the most correct of them all'.

This is like an elderly and finally-captured criminal claiming to a judge that while he has indeed lived a long life full of murder and torture, for which he has to this point escaped justice and for which he still has no regret and won't change his ways, he is actually a peerless paragon of morality because in a massive immoral and evil manifesto he wrote before he was ten years old he included two or three things that others agree are actually moral.

We live in a world where these murderous and thuggish religions hold sway over the lives of not just their 3.8 billion followers, but many others as well, many of whom are completely opposed to these belief systems. Devotees of these religions include presidents, vice presidents, congresspersons, governors, ministers, kings, queens, princes, princesses and prime ministers across the globe. Some of these people preside over countries which have nuclear and other weapons of mass destruction. Many of the countries without weapons of mass destruction have the ability to wage quite destructive and murderous conventional war. Many of these countries feel no particular compulsion to treat their citizens very well at all, and they treat people of other countries much worse than that.

A mere two millennia ago, a microscopic fraction of time even just in the history of our species, no one had ever heard of Christianity or Islam, although the precursor fables which spawned them were already old news in the ancient Middle East. Since that time these theological twin terrors have grown from their simple origins as hybrid syncretized stepchildren of Judaism and other local faiths and cults in the ancient Middle East to where they are today.

They have grown through processes of accretion, copying the practices and holidays of other local myths, plain old one-upmanship, increased internal birthrates, decrees from the emperor and the always-successful method of suppressing or murdering the opposition.

The ancient Middle Eastern area[6] was an absolute hotbed of religious movements for thousands of years, to such a degree that one could call it an early and much longer-lived version of America's empty-headed early 1800s Burned Over District.[7] The geographic extent, slow pace of movement and communications and other factors limited the rate of change and extended the lifespans of these belief systems. The well-known separate pantheons of Greece, Rome and Egypt came from this area.

There were also the religions of the Sumerians, Akkadians, Amorites, Canaanites[8], Hittites, Assyrians and Persians, among many others both known and unknown. Extremely similar creation stories, virgin and/or miraculous births, signs in the heavens accompanying these births, deaths followed by resurrections, gods of the sun, gods of light and dark, gods of good and evil, zodiacs, constellations, myths about great floods, linking gods to the seasons and solstices, allegations of fulfilled prophecies, performance of miracles, all these and many other characteristics were shared by many different gods and religions and were believed by many different people.

The most interesting aspect of all this was the apparently insatiable desire by people to take other people's theological fairy tales and incorporate them into their own theological fairy tales, with the more outlandish ideas often being the ones which were most widely shared.

How interesting, how human and how predictable it is that the more fanciful and outlandish stories were the ones which took root, rather than the banal ones. When the average Joe or Jane in the ancient Middle East, well endowed with ignorant credulity and practically devoid of actual information and critical thinking, has to choose between believing a story about an intelligent, empathic and thoughtful philosopher who had some really wild ideas, and a story about an intelligent, empathic and thoughtful philosopher who had some really wild ideas who was also born of a virgin and could perform miracles, regardless of the truth of these claims, which story is more likely to spread?

Why of course the more mysterious and magical story will be the one which the gullible goatherder or fisherman will pass along, not the banal one.

This was true since we first learned to clothe ourselves, and will be true for many thousands of years of future human evolution. The unfortunate mental condition that leads people to read the *National Enquirer*, ingest and spread juicy gossip and rumors, listen to or watch fringe media, follow astrology and join a religion has been with our species for far too long, and it refuses to die out.

This default to irrationality and belief in the magical, the mysterious and the conspiratorial is unfortunate but is extremely common. It springs from the same faulty brain module that causes people to spread nonsense, gossip, weird stories and rumors many times faster than they spread the truth.

When choosing between the banal and the magical, most people always choose the magical. This is why Socrates, Newton, Copernicus, Galileo and Stephen Hawking were never elevated to a godhead but so many other equally nondivine and extremely mortal semi-evolved primates were.

Had admirers claimed that Socrates was so wise and insightful because these were miraculous gifts from a god or his divine father in heaven, and that he had been born of a virgin, it is a near certainty that some considerable number of people would have worshiped him in his time, many more would have worshiped him after his martyr's death, and millions or even billions of humans today might be worshiping Socrates on Sundays.

This is where the empty-headed terminal masochism of the painfully incoherent and utterly confused person, embodied in pitiful specimens such as C.S. Lewis[9], comes into sharp relief. Lewis was an author and Christian apologist, living from 1898 to 1963.

His most well-known products were the *Chronicles of Narnia* and an incomparably fatuous but oft-cited apologetic justification for his Jesus worship. This attempted justification is utterly and appallingly devoid of critical thinking, but is taken quite seriously by far too many people who really should know better. He said[10]:

> *"I am trying here to prevent anyone saying the really foolish thing that people often say about Him: 'I'm ready to accept Jesus as a great moral teacher, but I don't accept his claim to be God.' That is the one thing we must not say. A man who was merely a man and said the sort of things Jesus said would not be a great moral teacher. He would either be a lunatic – on the level with the man who says he is a poached egg – or else he would be the Devil of Hell.*
>
> *You must make your choice. Either this man was, and is, the Son of God, or else a madman or something worse. You can shut him up for a fool, you can spit at him and kill him as a demon or you can fall at his feet and call him Lord and God, but let us not come with any patronising nonsense about his being a great human teacher. He has not left that open to us. He did not intend to."*

This shows the lack of critical thinking typical of apologetics and anyone who would utter such nonsense. Lewis offers nothing but his own opinion to aver that the words allegedly spoken by a person alleged to have lived two thousand years ago and alleged to have been recorded in a book whose origins and history can only be called schizophrenic, suspicious and cobbled together can somehow justify this potentially fictional character as the creator of the universe and his lord and god.

Since I am American, I am used to encountering illogic of this kind every day, but Lewis was not a victim of underperforming American schools and the eternally anti-intellectual and religion-poisoned American culture. He was well-educated in various British schools including Oxford, and was raised in a less religion-addled society. What leaps immediately to the eye is the total illogic of Lewis's contentions. I have several points of rebuttal to Lewis's statement and for anyone who finds his abdication of logic worthy of anything more than contempt and ridicule.

1. Huge numbers of people have said things as or more moral than the words ascribed to Jesus. Many of them said it long before Jesus is said to have existed or they said them without ever having heard of Jesus. We do not try to pass off these clearly human beings as gods, for the extremely good reason that they were merely people with extremely good ideas which are fully within the compass of a reasonable and intelligent person with compassion. No supernatural gifts or permissions *at all* are necessary for such ideas. Not only do we not need any supernatural intervention to tell us right from wrong, but the supernatural gods of our ancient fiction always mandate murderously large doses of immorality even when they are being passed off as "moral".

2. A human, be that human male or female, since we should not cut off one half of the human race from being considered as sources of deeply true and meaningful morality, who said some of the truly good things Jesus said about morality would not be a lunatic. This person would be a moral philosopher, a rational empiricist, a secular humanist or just a good person who says and does good things for good reasons. This is hardly lunacy; this is practical in both the societal and personal senses, for many reasons very well laid out by Richard Dawkins, Michael Shermer and many thousands of other clear thinkers, not to mention the internal and instinctive uplift a sane human feels from helping others in need. So no, being moral and talking about it is not indicative of lunacy, it is indicative of both sanity and a healthy mind.

3. A human who said some of the immoral things Jesus said would not be a lunatic. That person would be evil. That person would not be a devil of any hell, because there are no devils and there is no hell. Those who seek to use their religion to impose their personal beliefs or morality on others are both evil and immoral, and this has been proven since the first self-appointed moralizing hairy prophet first pointed a self-righteously accusing hairy finger at a fellow tribemate. It is unfortunate there are no devils and no hell to inflict the pain these people deserve.

4. A person who said some of the insane things Jesus said would indeed be a lunatic. Cosmic overlords, gods, devils and hells exist only in the minds of people who believe things without requiring evidence, or in the minds of people who are quite insane. Jesus said some truly insane things, such as when he exhorted his followers to eat his body and drink his blood. Those who think this practice[11] is actually a reasonable thing to do hold a quite insane belief.

   In the 21st century, a topic like this really should need no explanation, but since it is claimed that well over one billion people alive today are Catholic and thus gleefully think consuming pieces of the magically resurrected body of a long-dead Jew and washing it down with his magically resurrected blood is not only a great thing but they do it at least once a week, it warrants both significant attention and much ridicule.

   We have to ask those who practice this ritual cannibalism why it is not straightjacket-level insane to hold such a belief and to perform this ritual. If I started a religion today and said that a central ritual of the faith was taking a swig of vodka and nibbling on a slice of wheat toast which I had previously bowed toward while making a Very Serious Face, and by doing all this I was magically ingesting the resurrected body and blood of the most holy and long-dead Socrates, I would be immediately considered utterly insane at the very least. Those who think there is any difference at all between the Catholic cannibalistic communion ritual and a Socratic ingestion ritual are delusional and incapable of critical thinking. The fictions of their religion have blinded them to reality.

There is no reason whatsoever to believe that Jesus could not have said insane things, either to make a point or because he truly believed them. Plenty of people believe insane things and insane people often attract huge mobs of followers. It is unfortunate but plenty of people are indeed insane, and more are born every day. There is no reason to automatically assume Jesus was not insane or did not say insane things, and the only reason to think otherwise is to believe some completely unbelievable things about his alleged supernatural nature.

5. And who appointed Lewis, a man whose life was contorted by guilt, shame and a galaxy-sized affliction with masochism, as the specimen of *Homo sapiens* best able to officially opine on a possibly fictional person's intentions and realities? Lewis led a tortured life and his apologetics are an illustrative window into his tortured consciousness. One almost feels like someone should write apologetics for Lewis's apologetics, as his utter lack of logic and consistency is as balkanized and incoherent as the collection of Mediterranean fables he used as the source material for his beliefs. A person can be a lunatic, a liar, or both a lunatic and a liar at the same time, all while in no way being anyone's lord or savior in this life or the next. Lewis had no more of a hotline to the beyond or an insight into its provenance than did the grass in his front yard.

People like Lewis were around in the ancient Mediterranean and have been around throughout the history of humanity. This alone shows how easy it is for people to fall lock, stock and barrel for the most ridiculous and most obviously contrived and baseless claims. Regarding the crowded miasma that was the Mediterranean Sea of Competing Religions thousands of years ago, one can determine quite a lot. A study of these religions and belief systems, particularly regarding their extreme similarities and overlapping timeframes, is in itself demonstrative of how ideas spread and, like an agile and adaptive virus, mutate slightly and also jump from one host to another.

When one assesses the geographic relationship and the close physical proximity of these morphing, mingling and mutating beliefs, the total picture resolves even further. What would one expect but that extremely similar ideas were spread around a very localized area? Just like coinage and trade goods, religions and religious ideas were traded and shared around this local area. This is wholly to be expected and it would be highly unusual if it were not the case.

If something completely unexpected had happened, we would be compelled to look at the situation quite differently. If groups of Africans, Chinese, Native Americans, Aboriginal Australians, South Americans, Polynesians and Middle Easterners, all being geographically very far apart and lacking any physical or cultural connection or communication, had all simultaneously formed religious belief systems with extraordinary levels of congruence in their revelation and origin stories and in their minor and major details, we could begin to think there might be something of real truth or interest to examine with regard to the supernatural.

But they did not, there is not, and this is in itself all the reason one needs to disbelieve in all religions. To even begin to attempt to put credence in any of these faiths, one must believe that the gods of all the religions on earth either independently or collectively decided they would secretly reveal themselves to a few selectively chosen oddballs in semiliterate and scientifically ignorant places and populations at very unhelpful times in history, all opining that this was the best way to spread their singular One True Faith.

Apparently the ancient Mediterranean region was some sort of a resort area for divinities and they did the vast bulk of their unhelpful revelatory "work" there, since such a massive number of these very similar religions started in that extremely small region.

Maps of the ancient Mediterranean area help to show the story of the religious hodge-podge during that time. One quick look at the geography of this ancient hotbed of competing religions, given all the above and even what the average person with a middle school education reading this book knows about Greek and Roman mythology, should be sufficient to paint the basic picture.

As civilizations and peoples traded, raided, conquered and assimilated around the Mediterranean Sea, religions were born from or mutated away from existing religions, then spread out and cross-pollinated to a ridiculous degree.

The entire area is responsible for a truly massive number of deities and religions. Obviously not all of them could be the One True God or the One True Religion, either before or after the significant amount of mixing they experienced. The religious history of the region is exactly what one would expect if one projected the origin, spread and evolution of religions and their practice by different people across that geographic area.

Various and sundry but often similar or related religions and/or religious practices were part of the lives of, as already noted, the Sumerians, Akkadians, Amorites, Egyptians, Canaanites, Assyrians,

Persians, Hittites, Hattians, Luwians, Hurrians, Urartians, Phrygians, Phoenicians, Etruscans, Minoans and the more well-known Greeks and Romans.

If you think Judaism and its first-born son Christianity, then its much-later-born second son Islam formed in some kind of vacuum and were privileged with some sort of unique and culturally isolated provenance, you have absolutely no idea about the cultural and religious history of the ancient Middle East or its absolute supersaturation with both religions and gods. Religions and supernatural entities of every kind were as ubiquitous there and then as smartphones are today, and today's three monotheisms were bolted together from their predecessors and contemporaries, like using scrap wood in a shop to construct a new toy.

There are far too many people today who can fall for the most absolutely bat-crap lunacy and religious ideas, no matter how obviously made-up and/or crazy they are. There have been far too many of these people around throughout human history, and they are the ones who fall captive to smooth talkers and crazy promises and threats. They are the ones who happily join the herd in deeply and sincerely faithful flocks.

Returning to the topic of how religions, languages and organisms spring into being, flourish and then die, just as more than 99% of all organisms have gone extinct over time, most of the religions and gods of the ancient Mediterranean area and time have gone quite extinct. One of the sub-branches off the main tree, quite possibly the Canaanite branch, evolved through local artificial selection into Judaism[12], which then in turn evolved through both local artificial selection and extremely obvious plagiarism into both Christianity and Islam.

Those two fraternal twin faith systems are lineal descendants of the general religious gene pool that was the ancient Mediterranean. Islam formed approximately half a millennium after Christianity did, and a close and patient reading of first the Christian then the Muslim holy book will have one often wondering why the second book was even written, as the books are astoundingly similar in tone, language and general message. The names and dates and places are different but the similarities are glaring and, if you will, omnipresent.

Christianity was able to nucleate then grow to its present form for two major reasons. One was the series of official support and protections which Roman emperors from 313 to 380 CE gave to the nascent system. The second was the vigor with which the Catholic sect suppressed or murdered its opponents for much of the next 1500 years.

Christianity rose to power under imperial protections then maintained that power through actions very much "red in tooth and claw", with its evil machinery lubricated by the blood of its victims and its work illuminated by the light of burning books and bodies. It is not at all a religion based on its central character's principle of turning the other cheek. It is a religion based on power and conquest, and it engaged in centuries of wars and invasions in Europe and across the world to spread that faith.

All this was, of course, quite in accordance with the guidance found in its holy book. In that book, Jehovah specifically describes giving other people's lands to his chosen people and ordering those chosen people to commit genocide against other nations. The book also contains specific orders that anyone, including spouses and children, who tries to convert someone away from the faith is to be instantly killed. Christianity has never been a religion of turning cheeks; it has always been a religion of bloody swords.

Islam, despite coming along much later, was equally aggressive although it is not yet quite as successful in terms of lands conquered and peoples converted. Its founding text, being as it is so amazingly similar to its genetic sibling, gives all the room the enterprising prophet/evangelist/conqueror needs to justify spreading the faith by the sword, or if he had access to today's technology to spread it by the weapon of mass destruction.

And, although it is not widely known, the penalty in Islam for apostasy, which is leaving the faith once one has been in it, is death. Yes, death, and this is not referring to some sort of spiritual or metaphorical death, it means your real living body can be run through with a sword or you can be stoned to death or otherwise murdered in cold blood by the devout simply for leaving Islam.

This actually means that a living, breathing, sentient member of *Homo sapiens*, living in a first world country in the 21st century, who has once been a member of the Islamic faith, is to be killed if he or she leaves Islam. This person could in one package embody the intelligence of Stephen Hawking and Isaac Newton, the humanity of the Dalai Lama, the selflessness of the mother who unhesitatingly sacrifices her life for her child, be the most philanthropic person who ever lived, have just cured all forms of cancer and has found the scientific and societal solutions to world hunger, but if she then deliberately leaves Islam, she has marked herself as nothing but a piece of unfortunately still-living meat which should be killed just as soon as one of the devout can make it happen.

And this, as I said, is a core tenet of Islam and will remain so until the core texts are changed.

Christianity is no different and no better, and by now this should come as no surprise whatsoever. The book of Deuteronomy makes this plain with a direct order for faith-based murder (KJV):

> "*{13:6} If thy brother, the son of thy mother, or thy son, or thy daughter, or the wife of thy bosom, or thy friend, which [is] as thine own soul, entice thee secretly, saying, Let us go and serve other gods, which thou hast not known, thou, nor thy fathers; {13:7} [Namely,] of the gods of the people which [are] round about you, nigh unto thee, or far off from thee, from the [one] end of the earth even unto the [other] end of the earth; {13:8} Thou shalt not consent unto him, nor hearken unto him; neither shall thine eye pity him, neither shalt thou spare, neither shalt thou conceal him: {13:9}* ***But thou shalt surely kill him; thine hand shall be first upon him to put him to death****, and afterwards the hand of all the people. {13:10} And thou shalt stone him with stones, that he die; because he hath sought to thrust thee away from the LORD thy God...*"

This is not a quote from some modern-day rabid ranting street preacher with cardboard signs for clothes and spittle and fish bones in his beard, this is a piece of direct and unequivocal guidance quoted straight from the demonic compilation passed off by its believers as the one and only "good book". And remember, this isn't just the jealous Old Testament murderer. Jesus himself specifically said not to change one jot or tittle of the old law, so he was clearly quite OK with one of the brethren killing his wife and kids if they tried to get him to worship a different god.

So every member of these faiths, and actually every human alive, faces death without warning for leaving or denying the faiths. These faiths are clear and present dangers to humanity.

These two religions did not gain their current position by being 'the one true faith' or the pinnacles of morality and goodness which they proclaim themselves to be. They happened to exist in favorable historical circumstances and were the most aggressive groups around. Killing off those who stand in your way will in time mean you are the last one left. That strategy of aggression and conquest was the one that let these particular religious species gain a foothold in the total religious gene pool. And they are still even unto this day fighting with each other and in many cases even with others of their own general faith.

That last sentence is the key to the future of religion. Left to their own devices and unchecked by secular restraint and rationality, these faiths would in short order resume their Crusades and Conquests, their

Inquisitions and apostate killings and would leave even more masses of bodies in their wakes.

Take the Islamic State as an example. Most splinter groups don't normally form by people taking their holy books and tossing them out so the group can become less literalist and less fundamentalist. They become splinter groups because some of the more fanatical in the flock think the mainstreamers have become too wimpish or modern or politically correct, and have strayed too far from the plain language and clear original intent of the holy books.

The Islamic State is a classic example of a group of people taking their holy book as literal gospel truth and using its own particular "good news" to create a theocratic society using its words, just as Catholics did for hundreds of years during the Inquisition and medieval times in their ISIS-like Catholic State.

It's also reflected in current American politics, where the conservative center of gravity has been moving farther and farther to the right, pulled along by extremely loud and hateful individuals and groups who are continually jockeying to see who can out-crazy the previous most extreme craziness. Thus the entire right wing is rapidly moving to increasingly extreme positions, led as always by the highly zealous and highly intolerant religious faithful.

This model of aggression and conquest worked for religions for thousands of years and has led to over half the humans alive being members of one of two particularly bloodthirsty faiths. This is not an accident, and we must see every splinter group descendant of the original faiths plus their parent faiths as preparing themselves to take the leap into full-blown aggression and that they will try to take advantage yet again of that time-tested "growth-through-murder" strategy.

I am sure that most believers who have been able to read this far will likely feel like they have been run over by a train, and are in some degree of discomfort, pain, anger and especially denial. The strength of these emotions will be in direct proportion to the strength of their personal investment in their faith, which is quite understandable.

In order to overcome the cognitive dissonance believers suffer when presented with evidence that not only invalidates their beliefs but shows their faith to be both extremely immoral and extremely evil and thus the opposite of how they normally feel about their cherished religion, believers will engage their denial engines at full thrust and grasp at every nearby straw. They will say that even the most antitheistic among us must surely be aware of the great goodness of the religion and its people *today*.

Note the inclusion of the key word "today", at the very end, slipped in with the hope it will enable a whitewashing of and complete failure to acknowledge their unspeakable histories. This attempt to ignore the horrifically brutal pasts and the evil still perpetrated every day both in the name of and in the service of these religions is a deceitful attempt at distraction and will not be tolerated.

The Christian and Islamic faithful, with a regularity so predictable that one can almost use it to set one's watch, always refer us to their faith-based charities as evidence for their claim of basal goodness, and while one should assuredly always be thankful for honest service rendered to those in need, three facts can never be forgotten.

The first is the amount of charity done by both of these faiths is a drop in the ocean compared to the evil and horror they have inflicted and which they still do today, and no amount of charity they can provide will ever balance the evil they have perpetrated and which they continue to perpetrate even unto this day. One could create a small ocean with the blood spilled by these religions, and we will never be able to make a full accounting of the horrors they have visited on innocent men, women and children for far too many centuries and which they are apparently quite happy to continue without end.

The Catholic church doesn't get to zero out a sin of child rape by performing one, ten or ten million acts of charity, but this is the underlying message it is spreading and it is a defense which it is clearly counting on its flocks to both swallow and spread. No, these don't balance out and they never will. Many other organizations do vast amounts of charity work without raping little kids and spreading religious immorality and bigotry.

The second is that these acts of faith-based charity are often actually performed with the primary mission of converting the recipient to the faith of the charity provider. The act of charity is a means to the end of religiously converting the recipient. That is frankly disgusting. Doing charity with the intent of a quid pro quo, whether explicit or implicit, is at the same time ambulance-chasing and vulture capitalism, not actual human charity.

Only when a faith group performs charity in a manner such that it cannot be distinguished from the actions of a secular group providing the same support and service, and is done this way in all places and all times, can its charitable work be granted moral equivalency with secular charitable activities.

This means no volunteers doing their acts of charity wearing crosses or Jesus fish necklaces or T-shirts, no leaving holy books and pamphlets in their wake, no quoting from any holy books or sermons, and it also means no interaction with the recipients by clergy wearing clerical garb. Implied messages are just as powerful as direct ones, and people pick up on them very quickly. Doing charity for charity's sake is truly honorable. Doing it as a bribe is extremely immoral.

The third fact is that the motivation for much of this charity is to follow what they feel to be their god's orders to do good works. This is both insincere and hypocritical. If a person or group wants to do good works, and needs the permission or orders of a god to do them, then the motivation is not sincere and from within the person, it is just a process of following orders.

The hypocrisy comes in when some of the male clergy may follow their days of passing out food and water with evenings spent busily raping young boys, violating young priests or ordering young women to be beaten or killed for some microscopic and wholly manmade religious transgression.

If a person truly wants to do good works because of his or her own innate motivation, that is a pure and moral reason, and if carried out based on the person's innate desire to help others, then it is a pure act. Doing it because you were told to do so is no more honorable than is a teenager carrying out the trash because he was told to do so.

Some readers will protest that religion and religious practice is good for the world, because it is believed to somehow make people's lives better and happier, either through personal perspectives or through some imagined increased goodness of actions or civility concomitant with the level of belief in the faith. Religious people invariably devolve to this position when all their other arguments have been hulled and sunk without a trace, and they do so out of ignorance, because they have been told repeatedly from the pulpits that this is the case, or because they believe it somehow just "has to be" the case based on their emotional attachment to the faith.

This is where the empiricist and rationalist always has to remind the believer that no one cares at all what anyone *believes* to be true. The only thing that matters is what actually *is* true, and the only way to find the truth is to seek it out in an objective manner. As a quick illustration of belief vs reality, you may deeply and sincerely believe that your religion is true and that other religions are not. We empiricists are open to any and all actual empirical evidence you can provide on this.

We simply ask you to prove yours is right and everyone else's is wrong, because surely if your version of which god created the universe, the world and we humans is true then there should be absolutely overwhelming mountains of supporting evidence. Since no single religious believer in the history of this planet can even begin to prove his religion is true or that anyone else's isn't, all such beliefs must be dismissed and ignored.

I may hold the completely insane belief that I can fly, and I may deeply and sincerely believe this, but if I do I am also highly and very conveniently likely to believe that I should never demonstrate my powers until some appropriate time in the future, for instance when Jesus or some other mythical figure is supposed to return to earth. I can live for a very long time while holding my unproven beliefs, because for some reason the promised messiahs never actually come back, and those who really do believe they can fly will usually have tried and failed.

The failures to fly can be either multiple and temporary, following unsuccessful launches from low trees or single story rooftops, or they can be singular and permanent, following leaps from high cliffs or tall buildings. Those who actually try to demonstrate human flight will often eventually leave this belief behind, either following multiple small failures or one total failure.

Most people who believe they can fly will therefore not try to prove it. They will be content to have this belief in the back of their heads, as a comforting illusion which need not be proved but can be believed simply for the comfort it brings. And having once generated this insane and cocooned belief in personal flight, a human is likely to retain this belief up until the microsecond it is necessary to prove it.

Since having to prove his ability to fly before he wants to do so would have to be forced on him quite involuntarily, he will face an existential choice if that situation were ever to arise. The person will have to decide he has been right all along or if gravity really will apply to him in the same way it applies to every other blob of matter in the universe.

But if he is never made to prove he can fly, he can carry on with his deeply and sincerely held insane belief just as easily as someone else can with her belief in some specific god character or system of religion. Or indeed in an afterlife, which even in the well-populated set of lies and cons that is religious belief stands head and shoulders above all others as the greatest con job of all time.

How billions of people can believe the claims made by other people about quite detailed specifics regarding the unearthly realms a person will be wafted to after death is an appalling proof of the depths to which the deceitful will sink to in order deceive the gullible, and the depths to which the gullible will sink in order to remain deceived.

The erroneous belief about the innate specific and general goodness of religion can be destroyed after not much more than a few hours of dedicated research. This is because this opinion is as completely wrong as would be an opinion that humans can fly. Far beyond the obvious evil of religion on a broad scale as noted in the aforementioned Crusades, Inquisitions, conquests, inter-faith murders and pogroms, one sees the extremely negative impact religions have had on individual people since the first religion was invented.

On this point, the first and easiest thing one can do to demonstrate this is to see what life is or was like in states and societies which either are or were theocratic or in which religion is or was a major part of life and government. In point of fact, such places are actually the very *last* ones in which a decent human being who actually respects life, liberty, justice and well-being for all would ever want to live, while the places which are the least religious are the best in these attributes. It does take a little time to do the research to discover this, but the result is well worth the time and effort.

Most people should be able to get the point without even doing the research, simply by answering one simple question. Just name five actual or near theocracies in which you would want to live. How do places like North Korea under the Kims, Saudi Arabia, the medieval Catholic State under its allegedly infallible Maximum Leaders, the Islamic State under its zealots, Iran under the ayatollahs and Afghanistan under the Taliban sound to you? Such places are what thuggish Christian Dominionists want for America, only with them in charge. There has never been a theocracy that provided improved well-being for all, and there never will be.

If one is willing to invest more time, one can more thoroughly research the question and see in very stark detail just how thoroughly religion is correlated with decreased human well-being in our world today. All one need do is select measures of well-being and assess how well various countries score in them, then see how religious that country is. One can very easily compile a set of concrete data showing individual and societal markers of well-being to see which societies maximize well-being and which ones do not.

I have personally long thought, based on observations of history and events across the globe, and what I've seen in my travels across the world and in my own country, that religion was empirically bad for people, but had never myself done the relevant specific research. When I did, the results were astonishing. It is one thing to have a generally informed opinion that religion is negatively correlated with human well-being, but it is quite another to actually see the numbers laid out in a table showing just how dramatically worse off the most religious societies on earth actually are and how much better off the least religious societies are.

I got the idea for the project while researching American politics, society, religion and education, research which went into supercharged overdrive after the debacle of the 2016 American presidential election. I had noticed how poorly American high school students fared in math and science in relation to students from many other countries, and noted that many of the better-scoring countries were very nonreligious.

I decided to create a table listing religiosity, academic proficiency and several indicators of societal well-being to see how religion correlates with societal health or lack thereof. That data are shown in Table 2 below and are a short but powerful example of this kind of easily performed research. The data for the table came from a variety of sources[13] which are listed in the end notes.

I started the table with scores achieved by students from the countries which had participated in the 2015 Programme for International Student Assessment (PISA[14]). The PISA is "...an international assessment that measures 15-year-old students' reading, mathematics, and science literacy every three years...is coordinated by the Organization for Economic Cooperation and Development (OECD), an intergovernmental organization of industrialized countries, and is conducted in the United States by NCES." The NCES is the American National Center for Education Statistics.

I then researched countries which had been surveyed for the importance of religion to their populations, and listed the ones which had also participated in the PISA. I used this importance of religion column as the anchor column and listed countries in decreasing importance of religion from top to bottom. Thus Indonesia, where 99% of people feel religion is important, is at the top, while China and Iceland are at the bottom because 10% or less of those populations felt religion is important.

Filling in the rest of the table using the anchor column as the base, I added country names, PISA scores and the rest of the columns in the table.

I color-coded the individual data blocks according to the data point's level of well-being or humanism. Green reflects the highest levels of well-being or humanism, and also the lowest degree of religiosity or support for it. Red indicates the lowest levels of well-being or humanism, and the highest degree of religiosity or support for it. Intermediate levels are yellow. In other words, red means a country is highly religious and/or the people are relatively much worse off in a certain area, while green means the country is much less religious and/or the people are relatively much better off in a certain area.

A single glance at the final product shows a truly astonishingly high inverse correlation between the religiosity of a country and its well-being. The color cascade changes dramatically from mostly red at the top, meaning both high religiosity and low human well-being, then shifts in the middle rows to predominantly yellow with a mix of red and green, down to a sea of green at the bottom populated by those countries largely without religion which enjoy much higher well-being.

Those countries in which the people place the most importance on religion have by far the worst well-being, and the least religious ones are the ones with the best well-being. You are far, far better off in almost every way actually important to a human being if you live in a place where religion is unimportant to the people and to the government.

The data table looks even worse for religion when you include all the countries that did not participate in the PISA and thus were not included in the table.

The data are extremely clear and objective. The numbers and facts do not lie. Religion is, without a doubt, highly and strongly *negatively* correlated with human well-being.

The following paragraphs are short explanations of the specific data and color coding in each column in the table.

In the "Importance of Religion" column, those countries where religion is most important are color coded red. Yellow indicates moderate importance. Green indicates the lowest levels of importance. These colors reflect my hypothesis that increased religion is strongly correlated with reduced well-being, in that human well-being will be inversely proportional to the importance of religion.

In the "Math and Science Scores by 15 year olds" column, low PISA scores are red, high scores are green and yellow are intermediate. These scores indicate the general level of importance of math and science education in the surveyed countries, indicated by the scores. The scores are strong proxies for the levels of critical thinking, empiricism and importance placed on rational education in the country.

In the "Official State Religion" column, the presence of an official state religion is deemed as a negative, thus the block is green if there is no official religion and it is red if there is an official religion, no matter which one it is. Note that Norway, Denmark, Iceland and the UK have an official state religion on paper but this is actually a misleading and irrelevant artifact, like 200-year-old laws on the books that no one pays attention to anymore but which no one has bothered to rescind. These countries are in reality among the least religious on the planet. This is due to numerous factors including a better understanding of history and an acknowledgement of science and reality as extremely important elements of life and civilization. The blocks for these countries are thus a color blend of green and red. Finland is green and yellow since Christianity is deemed as "preferred". Religion is actually not important to most of the people in these countries.

In the "2014 Global Gender Gap Index Ranking" column, the most positive rankings, meaning a smaller gender gap, are green. Yellow indicates a moderate gap and red indicates a large gap.

In the "Homicide Rate" column, green indicates the lowest rates, yellow indicates intermediate rates and red indicates the highest rates.

In the "Year of Female Suffrage" column, green indicates an earlier date of female suffrage. Yellow indicates less recent dates of suffrage and red indicates very recent suffrage dates.

In the "Life Expectancy At Birth" column, green indicates the longest life expectancy. Yellow indicates shorter life expectancies while red indicates the lowest life expectancies.

In the "Under-Five Mortality Rate" column, green indicates the lowest mortality rates. Yellow indicates higher mortality rates while red indicates the highest mortality rates.

In the "Happiness / Life Satisfaction" column, green indicates the highest happiness and life satisfaction levels. Yellow indicates lesser levels while red indicates the lowest levels.

In the "GDP" column, green indicates the highest GDPs. Yellow indicates smaller GDPs and red indicates the lowest GDPs.

Due to the various ways in which China was reflected in the various source data sets, the PISA scores in the table were taken from the Beijing-Shanghai-Jiangsu-Guangdong areas, while the rest of the data reflect China as a whole. Readers are free to view these as separate row entries if you like, as it is extremely likely that averaging PISA scores from all across China would result in lower overall PISA scores.

Table 2. Comparison of indicators of well-being in OECD countries to self-reported rates of importance of religion.

Country Name	Is Religion Important To You? [1]		Score of 15 yr Olds on 2015 Program for Int'l Student Assessment [2]		Official State Religion [3] (P = Preferred, Not Official)	2014 Global Gender Gap [4]	Homicide Rate (UN Global Homicide Study 2013) (top rate from 2010-2012) [5]	Year of Female Suffrage [6]	Life Expectancy At Birth (Yrs) [7]	Under-Five Mortality Rate, Both Sexes, per 1000 [8]	Happiness / Life Satisfaction From World Happiness Rpt 2017 (Rank) [9]	GDP [10]
	Yes	No	Math	Science								
Indonesia	99%	0%	386	403	(Islam*)		0.6	1945	69.1	27	81	3,570.30
Thailand	97%	2%	415	421	None	0.70	5.5	1932	75.0	13	32	5,910.60
Jordan	96%	4%	380	409	Islam		2	1989	74.1	18	74	4,087.90
Algeria	95%	5%	360	376	Islam		0.7	1962	75.8	25	53	3,916.90
Qatar	95%	4%	402	418	Islam		1.1	1999	78.1	7	35	59,324.30
Tunisia	93%	5%	367	386	Islam		2.2	1959	75.2	16	102	3,688.60
Trinidad and Tobago	92%	8%	417	425	None	0.72	35.6	1946	71.1	21	38	16,048.50
UAE	91%	8%	427	437	Islam		2.6	2006	77.1	8	21	37,622.20
Lebanon	87%	12%	396	386	None		2.2	1952	77.0	9	88	8,257.30
Dominican Republic	87%	13%	328	332	Christianity	0.69	24.3	1942	74.0	27	86	6,722.20
Brazil	87%	13%	377	401	None	0.69	25.2	1934	74.9	14	22	8,649.90
Romania	84%	12%	444	435	Christianity (P)	0.69	1.8	1946	75.0	12	57	9,522.80
Turkey	84%	16%	420	425	Islam (P)		2.7	1930	76.3	14	69	10,862.60
Peru	84%	14%	387	397	Christianity (P)	0.72	9.6	1955	75.3	16	63	6,049.20
Colombia	83%	16%	390	416	None	0.71	33.6	1954	74.3	16	36	5,805.60
Georgia	81%	16%	404	411	Christianity (P)	0.69	4.3	1918	73.7	20	125	3,865.80
Costa Rica	79%	19%	400	420	Christianity	0.72	11.3	1949	79.7	10	12	11,824.60
Macedonia	78%	22%	371	384	Christianity (P)	0.69	2	1946	75.9	7	92	5,237.10
Cyprus	75%	24%	437	433	None		2	1960	80.3	3	65	23,541.50
Poland	75%	19%	504	501	Christianity (P)	0.71	1.2	1918	77.7	5	46	12,414.10
Mexico	73%	25%	408	416	None	0.69	22.8	1947	76.2	16	25	8,208.60
Italy	72%	25%	490	481	Christianity (P)	0.70	0.9	1945	82.4	4	48	30,661.20
Moldova	72%	19%	420	428	Christianity (P)	0.74	7.8	1978	71.6	15	56	1,900.20
Portugal	72%	30%	492	501	None	0.72	1.2	1976	81.4	4	89	19,838.00
Greece	71%	26%	454	455	Christianity		1.7	1952	81.4	5	87	17,890.60
Montenegro	71%	26%	418	411	None	0.69	3.4	1916	76.6	6	83	7,028.00
Singapore	70%	29%	564	556	None	0.70	0.4	1947	83.1	3	26	52,962.50
Croatia	70%	28%	464	475	None	0.71	1.4	1945	77.5	5	77	12,149.20
Chile	70%	29%	423	447	None	0.70	3.7	1949	79.5	9	20	13,792.90
Argentina	66%	33%	456	475	Christianity (P)	0.73	5.5	1947	76.2	14	24	12,440.30
United States	65%	34%	470	496	None	0.75	4.7	1920	79.0	7	14	57,638.20
Austria	55%	43%	497	495	None	0.73	0.9	1918	81.6	4	13	44,757.60
Ireland	54%	46%	504	503	None	0.79	1.2	1928	81.6	4	15	64,175.40
Israel	51%	48%	470	467	Judaism	0.70	2	1948	82.2	4	11	37,180.50
Spain	49%	50%	486	493	Christianity (P)	0.73	0.8	1931	82.9	5	34	26,616.50
Slovenia	47%	52%	510	513	None	0.74	0.6	1945	80.9	3	62	21,650.20
Slovak Republic	47%	52%	475	461	None		1.8	1920	76.8	8	40	16,529.50
South Korea	43%	56%	524	516	None		0.9	1948	82.1	4	55	27,538.80
Canada	42%	57%	516	528	None	0.75	1.6	1918	82.1	5	7	42,183.30
Lithuania	42%	49%	478	475	(Christianity*)	0.72	7.1	1921	74.4	5	52	14,900.80
Switzerland	41%	57%	521	506	None	0.78	0.7	1971	83.1	4	4	79,887.50
Uruguay	41%	59%	418	435	None	0.69	7.9	1932	77.0	7	28	15,220.60
Germany	40%	59%	506	509	None	0.78	0.8	1918	80.9	4	16	42,161.30
Hungary	39%	58%	477	477	None		1.4	1918	75.8	6	75	12,820.10
Latvia	39%	58%	482	490	None	0.77	4.7	1918	74.4	9	54	14,071.00
Albania	39%	53%	413	427	None	0.69	5	1920	77.9	17	109	4,125.00
Russia	34%	60%	494	487	Christianity (P)	0.69	10.1	1918	70.9	10	49	8,748.40
Bulgaria	34%	62%	441	446	Christianity (P)	0.74	2	1944	74.3	12	105	7,469.00
Belgium	33%	58%	507	502	None	0.78	1.9	1948	81.1	4	17	41,271.50
New Zealand	33%	66%	495	513	None	0.78	1	1893	81.4	6	8	39,412.20
Netherlands	33%	67%	512	509	None	0.77	0.9	1919	81.8	4	6	45,637.90
Australia	32%	68%	494	510	None	0.74	1.1	1902	82.4	5	9	49,755.30
France	30%	69%	493	495	None	0.76	1.2	1944	82.6	4	31	36,857.10
Vietnam	30%	69%	495	525	None (Hostile)	0.69	3.3	1946	76.0	23	94	2,170.60
Finland	28%	70%	511	531	Christianity (P)	0.85	2.2	1906	81.2	3	5	43,401.20
United Kingdom	27%	73%	492	509	*"Christianity"*	0.74	1.2	1928	81.4	5	19	40,367.00
Hong Kong	24%	74%	548	523	None	NR	0.5	NA	84.0	5	71	43,741.00
Japan	24%	75%	532	538	None		0.4	1947	83.9	3	51	38,900.60
Czech Republic	21%	75%	492	493	None		1	1920	78.8	4	23	18,483.70
Norway	21%	78%	502	498	*"Christianity"*	0.84	2.2	1913	82.1	3	1	70,868.10
Denmark	19%	80%	511	502	*"Christianity"*	0.80	0.8	1915	80.8	4	2	53,578.80
Sweden	17%	82%	494	493	None	0.82	1	1921	82.6	3	10	51,844.80
Estonia	16%	78%	520	534	None	0.70	5.4	1918	77.4	4	66	17,736.80
Iceland*	10%	90%	488	473	*"Christianity"*	0.86	0.9	1915	82.7	2	3	59,764.70
China**	7%	93%	531	518	None (Hostile)		1	1949	76.0	14	79	8,123.20

As I said, the visual impact of the table is quite striking, and it completely punctures the overinflated balloon that is religion's claim that it is good for people. This is completely wrong. The most religious countries are much worse off than the least religious countries in the most important ways, and it isn't even close. People in nonreligious countries are happier, healthier, better educated, they live longer and are much wealthier.

That wealth enables people to live in safe places, have clean and safe food, water and air, have high quality medical care, have fire and police protection, enjoy safe infrastructure, and to live longer and happier lives. The bottom line is clear - well-being increases as religiosity decreases.

The only question which need be answered is "Which is the cause and which is the effect?" How much does the arrow of causality point in one direction or the other? In other words, has well-being decreased in a country because the local religion is in power, or did the lack of well-being in an area drive people to religion?

It is clear that when people are in a bad way and suffering from decreased well-being, be it from a lack of security, health, freedom, financial and material means or any other negative condition, people often turn to the illusory comforts of faith.

This was very well described by Karl Marx in his *Critique of Hegel's Philosophy of Right*. The most famous and relevant part reads as follows[15]: "Religious suffering is, at one and the same time, the expression of real suffering and a protest against real suffering. Religion is the sigh of the oppressed creature, the heart of a heartless world, and the soul of soulless conditions. It is the opium of the people." Marx was absolutely correct. When people are anxious or suffering, they turn to the drug of faith to dull the pain and anxiety of their existence.

And it is also murderously true that when religions are strong they dramatically decrease human well-being. Religions both enable and encourage persecution, and the monotheistic holy books specifically order Jews, Christians and Muslims to be immoral and evil people. When religions are able to inject their codified evil into the state, they prosecute and persecute huge numbers of thoughts, acts and people which the religious authorities and their holy books don't like.

The most religious societies on earth throughout history have always and still do treat their men very badly, they treat their women like property or at least far lower than men, they seek to force their dogma on people who know better and don't want it, and are riddled with official corruption.

If you are an American and think I am referring only to places like Afghanistan and Saudi Arabia, you need to take a closer look right under your nose, starting in conservative states in the South. If you can't see that major segments of America keep trying to move cities, states and the country in this direction and are already themselves acting this way completely due to their deeply and sincerely held religious beliefs, you are either part of the problem or haven't looked at the problem.

So both processes are true and both can be present at the same time. Religion is at the same time the refuge from and the reason for pain, suffering and anxiety. Strong religiosity means people and societies do evil things, and when people experience suffering and pain they often turn to the mental opium of a belief in redemption or a magical afterlife. And this becomes a hellish feedback loop in which the faithful in charge inflict pain and suffering as part of their religious practices and beliefs, then the suffering people turn to religion and its opiates such as an afterlife as a mental escape from the suffering which the religion is itself imposing on them. Such terminal insanity is even beyond C.S. Lewis-level lunacy but it keeps happening over and over again because the faithful will not stop trying to make their personal dogma the hateful law of the land.

No matter which way one wants to point the causality arrow, it is clear that the more religious a country is, the more it is sick, poor, uneducated, ignorant and dangerous.

But we need not remain uncertain about the *primary* direction in which the causality arrow points. We can show that it is indeed religion that is the primary cause of the troubles in these countries. Why can we say that?

Because it is those societies which have accepted science and human morality and have thrown off those shackles of religious ignorance which have prospered and flourished.

Because those countries which have become more religious are the ones which became worse places, where well-being is diminished and people are much worse off.

Why is this the case? For two major reasons.

1. Religions teach that the garbage and myths in their holy books have primacy over the real world and the truths of science, so science is ignored and suppressed. Original thought, creativity, critical thinking, the questioning of authority and the drive to seek empirical truth and true human well-being are also all suppressed. The population focuses more on its mindless religion and both religious and political dogmas than reality, thus learning and actual practical achievement goes down and lifelong outcomes are depressed. Exhibits A and B in this are the USA before Sputnik and the USA again since the rise of the conservoreligious and fascistic right wing.

2. The holy books are full of antihuman evil and any society which actually uses them as a reference and applies them in the real world will definitely do antihuman and evil things to its people. We can see this

sort of thing in a comparison between two very well-known religious fundamentalists, as seen in Figure Eight below.

These two individuals, who are or were essentially cult leaders, demonstrate that when a person is deeply and sincerely religious and hews to the actual core texts and teachings of his faith, he will say and do things that are immoral, antihuman and evil.

Which One Is It?

Which of the characteristics below describe either or both of these individuals?

A

B

• Deeply religious	A	B	Both
• Certain his religion is the one true religion	A	B	Both
• Despises homosexuals	A	B	Both
• Seeks religious dominance over secularism	A	B	Both
• Believes in marriage for very young girls	A	B	Both
• Believes women are inferior to men	A	B	Both
• Believes women should serve men	A	B	Both
• Makes public statements professing his views	A	B	Both
• Minimal personal hygiene	A	B	Both
• Prefers living in isolated and remote areas with family members	A	B	Both
• Believes that he speaks for his deity	A	B	Both
• Enjoys killing	A	B	Both
• Mental illness runs in the family	A	B	Both
• Extremely proud and arrogant	A	B	Both
• Believes the unrepentant will burn after death	A	B	Both

Figure 8. Fundamentalists from two common religions (photo credits Huffington Post, UK Independent).

So we see that religions start, split, mutate and spread just like every other disease humans have suffered from. Also like disease, religion is very bad for people. Until the disease of religion is stamped out and all future humans have been properly inoculated against it, it will continue to infect people and cause division, misery and death.

Just to give a simple visual representation of what life is like in highly religious places and godless ones, take a look at Figures Nine and Ten below.

Figure Nine is what religion does to people and what theocracies are like. The faith and the faithful stamp the life out of the people and the society. People are subservient, demoralized and have massively reduced well-being.

Figure Ten is what being happily godless does to people. It frees them to live their own lives and to be happy. It makes them equal, highly positive and motivated and massively improves their well-being and their morality.

Answer these questions honestly.

Which society would you yourself want to live in?

Which society would you want your kids, and especially your daughters, to live in?

Your answer will tell us all we need to know about what kind of person you are.

Figure 9. Muslim women in burkhas. Photo courtesy http://www.jaisiyaram.com.

Figure 10. Scandinavian women in flowers. Photo courtesy https://www.elitereaders.com.

And we can apply the proven principles of Darwinian selection to the intersection of faith and politics to show this will remain the case until the DNA of our species changes significantly or our species finally grows up. Religions mutate and become more virulent because if a different strategy had worked in the ruthless combat of the religious gladiatorial arena, then less aggressive faiths would now be in the majority in our world. This is clearly not the case. Most people alive today are Christian or Muslim. These religions were ruthless and murderous in their pursuit of secular power and even unto this day work to kill off their competitors.

So as these religions continue to try to grow and work to gain new converts and new power over other people, they will do one of three things.

1. They will choose not to be aggressive and will slowly die off. This is extremely unlikely.

2. They will revert to the core of what "made them great" and become hyperaggressive once again, with fatal impacts on nonbelievers as well as those inside the faith.

3. They will spawn more groups like the Islamic State which will pursue the super-hyperaggressive literalist path. Some of these will either come to dominate large numbers of people, or due to the increasing availability of weapons of mass destruction, they will just kill those who refuse to follow their orders as well as a great many others in the process.

Remember that even as today's Catholic church calls itself the only true faith and the only true path to living a good moral life and to achieving a mythical eternal salvation, and even as it and its followers doggedly fight to inculcate its antihuman immorality into secular law, it keeps stepping into newly uncovered steaming piles of the apparently omnipresent and eternal scandal that is its global campaign of child rape and the coverup of these horrible crimes.

Yet the allegedly infallible Catholic popes and the leaders of this cult, who today tell over a billion people what to do, what to think and either explicitly or implicitly how to vote, were the ones who began the Inquisition and the Crusades and were responsible for countless murders and acts of torture.

They set the conditions for the Holocaust and enabled uncountable thousands of children across the globe to be raped by priests. They are themselves the Satan against which they pray so antibiblically in public as a camouflage strategy to cover their own evil. No crime is too unthinkable for these people and well over a thousand years of proof is far more evidence than we need to know that they will never change. Evangelicals, Southern Baptists, Mormons, Orthodox Jews and the vast majority of other Christian, Muslim, and Jewish sects are immoral and evil in their own ways, which are far too numerous to fully reference in this book.

So we as a global civilization have a choice. We can sit back and let the Darwinian process play out and just see what happens, hoping we nonbelievers don't get killed by the most deeply and sincerely faithful believers, or we can inoculate against the disease as soon as we can in order to treat the infection and stop it from spreading.

Sitting back is not an option. It is a guaranteed fact that deeply and sincerely faithful religious zealots are now and will always be a threat to civil society and to all humans on earth. This refers not only to those who would burn heretics, fly airplanes into buildings, blow up subways or seek to use weapons of mass destruction. It also refers to those who inject their religious beliefs into the political world by making laws that codify or enable persecution of people the religion does not like.

It refers to those who wage overt or covert war or by taking other actions in pursuit of some religious goal such as the suppression of science.

It refers to people who seek to use their religion as justification for bigotry, persecution, suppression, discrimination and violence.

It refers to people who believe condoms are worse than AIDS and that ensoulment happens at conception.

It refers to people who think their holy texts are anything more than ancient borrowed fairy tales awash in evil and immorality.

Many of these people, like fish swimming in sewer water which are unaware of the contamination, don't even realize what they are doing is morally reprehensible, because their faith has blinded them to reason, to compassion and to humanity. This is the power of the mutating infection of religion to do grave damage to real people in the real world.

So no, sitting back is not an option, as even if some religions do die off there will always be others to take their place, and some will be very aggressive and very dangerous. The disease of religion multiplies the misery of humanity. We must treat its sufferers, inoculate our children against it, and stop it from multiplying and spreading. Vaccination against disease and the medical science to treat disease have enabled the survival and physical health of literally billions of people and the eradication of many diseases.

Vaccination against the infection of the constantly 'fruitful and multiplying' threat of religion, and the eradication of religion altogether, are absolute requirements if humanity stands a chance of surviving into the next millennium.

Chapter Sixteen

The Immorality of Religion

"Furthermore, we declare, we proclaim, we define that it is absolutely necessary for salvation that every human creature be subject to the Roman Pontiff."
- Pope Boniface VIII, 13th century religious extremist and zealot

"The greatest happiness of the greatest number is the foundation of morals and legislation."
- Jeremy Bentham, 19th century philosopher and opponent of religion

"Mark my word, if and when these preachers get control of the [Republican] party, and they're sure trying to do so, it's going to be a terrible damn problem. Frankly, these people frighten me. Politics and governing demand compromise. But these Christians believe they are acting in the name of God, so they can't and won't compromise. I know, I've tried to deal with them."
— Barry Goldwater, mid-20th century American Republican politician before the party's embrace of dogmatic Christianity and rejection of science

As with all else in human discourse, achieving a common understanding of terminology and semantics is critical before one can have a proper discussion. When I say "blue", I need to know that you know exactly what I mean by "blue", and if we are going to have any deeper conversations involving the word, I need to ensure you are aware of any colloquial or personal implications of my specific choice and placement of words. When people try to communicate they have to be perfectly clear on each other's meaning and intent, even if there are large challenges to make that happen.

Morality is a word that has almost as many perceived implications as there are minds available to consider the topic, which means that objective definitional concurrence is even more important. Two common sources define the term morality, as shown below.

The Oxford Dictionary online defines morality[1] as:

> "1 Principles concerning the distinction between right and wrong or good and bad behaviour.
>> 1.1[count noun] A particular system of values and principles of conduct.
>> 1.2 The extent to which an action is right or wrong."

The Stanford Encyclopedia of Philosophy's page on morality[2] discusses morality at length, and defines morality in two ways:

> "1. descriptively to refer to certain codes of conduct put forward by a society or a group (such as a religion), or accepted by an individual for her own behavior, or
> 2. normatively to refer to a code of conduct that, given specified conditions, would be put forward by all rational persons."

The definitions above clearly state that morality as an isolated and singular term merely refers to the existence of a system or code of conduct. It does *not* extend to defining or even implying the good or evil nature of any particular activity within that system or code. Those activities, which in human spheres are our words and deeds, can be further defined as good or evil by a society or group. For the purpose of this chapter, we will only discuss morality in relation to human words and deeds.

There is no morality inherent in the random nature of the activities of our physical universe, because there is no consciousness behind it, thus no deliberate intent to do harm or good. It is neither moral nor immoral if the vicissitudes of wind and rain bring about a hurricane or a flood, and there is no greater or lesser morality in the event if zero or a billion sentient creatures drown in that flood. There is indeed morality in the animal world, in particular in the actions of the higher nonhuman primates and some other animals, but their behavior will not be discussed here. Here we will discuss the words and deeds of humans, and of the gods which humans have invented and which they believe in and/or worship.

We thus have a gap that needs to be filled. If we start from scratch, we can say that a spoken word or performed deed has an inherent morality, but the system by which to judge its placement along the spectrum from absolute good to absolute evil is as yet not laid down. We must pick the correct system of morality by which to judge words and deeds, and it must be as close to universally applicable principles as possible. We cannot randomly select any nation's, any group's, any religion's or any person's system of morality, for a great many reasons, not the least of which being that many of these systems contain more than their fair share of evil and immorality.

In other words, how can we, and how do we, decide what is moral and what is not? That is simultaneously extremely easy and extremely difficult to answer. And it can be very difficult to put into practice. To start large and work backward, let's discuss the morality of a society.

This morality is expressed in the formal and informal codes we follow, which are the laws we enact and the customs we follow. These codes often take the circumstances of a situation into consideration.

For instance, in what I feel comfortable calling civilized societies with civilized laws, it is generally against those laws to kill another human being. However, these laws include provisions for the occasional exception to the rule. If you or a member of your family are being attacked by a man with a knife or a gun, and you happen to kill that man by some means during his attack, the law and reasonable people will generally see your conduct as appropriate self-defense and will count your case as an acceptable exception to the base rule.

There are also violations of morality which are equivalent in outcome but are treated differently. If you embezzle some money from the office petty cash pool to pay for an extremely sick family member's medicine, you will likely be treated less harshly than a person who committed an armed robbery of that same office to acquire the same amount of cash for the exact same purpose.

Everything you can measure can be plotted on a graph, and it will generally fall into a normal distribution, or bell curve. There are exceptions to this but by and large the rule holds true, and this increases with the number of samples taken. Human height, the size of snapping turtle heads, the speed of planetary bodies, humanity's affinity for chocolate chip cookies or asparagus chip cookies, how much love or hate you feel for a person, these and billions of other measurements are all data points which can be taken and plotted.

In a similar manner to the physical measurements, you can also plot out measurables which relate to human thought or activity which may be proxies for or correlates of morality. These include scores on personality and behavior tests, number of people whose lives you have improved or damaged, your number of lifetime incarcerations or awards for philanthropy, number of lifetime convictions for crimes both violent and nonviolent, number of truly kind acts bestowed on others, number of selfless sacrifices made for others, demonstrations of psychopathy or empathy, and many other such.

Clearly there are people who are psychopathic beyond all normal human understanding. These people, along with their polar opposites on the kindness spectrum, definitely exist but they are much rarer than the rest of us who muddle along somewhere in the middle, with most of us trying to stay on the good side of the ledger. But merely because some people are hugely evil does not mean everyone else has to be, or that we can't say that their acts are wrong.

This is an idiotic position, yet the faithful repeat this nonsense whenever they say we need to follow their particular god's rules for living or else we can't know right from wrong. Merely because we humans exist along a spectrum does not mean we humans cannot find the answers to questions from trivial to ultimate importance. No god told us how to invent medicine or voting or a democratic society or how to go to the Moon. Jehovah, Allah and Jesus all didn't just fail to tell us that slavery was bad, they still say it's just peachy, right there in the holy books, along with mountains of other immoral evil nonsense. Humans finally figured out slavery was bad, and we figured everything else out as well.

We can ask 7 billion people, or far fewer people and still be statistically significant, how they feel about things and can get a large body of usable information. We can ask questions like the following:

- Do you like it when people do you a kindness, or do you like it when they are mean to you?
- How much do you like or dislike it when X is done or said to you?
- Is it normally moral or immoral to lie, cheat or steal?
- Is it normally moral or immoral to kill?
- Do you prefer to be spoken to respectfully or disrespectfully?
- Do you like it when people try to order you around?
- Is it normally acceptable or not acceptable to punch someone? When would it be acceptable, if ever?
- Should all people be treated as equals?
- Should anyone else's sex life matter to anyone else, as long as they are having this sex with consenting unimpaired and undrugged grownups? If you think so, why? If not, why not?
- Should anyone be allowed to put poison, pollutants or garbage into air, water or food which other people will breathe, drink or eat?
- Is it a good thing to help someone in need, or is it a good thing to deliberately not help someone in need when you are in a position to do so?
- Should people be allowed to impose their religious beliefs on other people?
- Should people be allowed to use whatever drugs they want to as long as no other person is ever harmed? Note that this includes alcohol, caffeine and tobacco as well as methamphetamine.
- Is it acceptable to conduct stem cell research given the potential benefits and drawbacks?
- Is it ever appropriate to hold beliefs without evidence? Is it ever appropriate to act on those beliefs when those acts harm others?

Since as we have already established that opinions are often far off the mark, and people lie, and sometimes people just like to be evil trolls and say things that interfere with the progress of science and any other human endeavor, we can check the answers to these questions by doing overt and covert physical and psychological assessments on people. We can check to see how being treated positively or negatively bothers people or how it affects their stress and hormone levels, their affect and their accomplishments.

Gathering Big Data on questions such as these will easily get us to a place where we could have asked a very large number of people a large number of such questions and plotted the answers on a series of bell curves. Removing some portion of the tails of the bell curves would give us a massive critical mass in the center which would comprise the vast majority of answers.

We could fold in objective assessments to the Big Data pile as a screening measure. We could use this supermajority of answers and the objective data to build a code of morality completely from scratch, and use it to construct a secular legal code which would provide the greatest good for the greatest number of people, in Jeremy Bentham's astute and morally unchallengeable reckoning.

We have arrived at legal and moral constructs in many countries in the world through a simple process of human activity. Some countries have better and more humanistic codes than others, but it is a certainty that if we took the scientific approach we could build an outstanding moral and legal code that would get that greater good for the most people. We would also eliminate the thoroughly evil hypocrisy and immorality personified in the demonic quote by Boniface that started this chapter.

When I am asked why I think mere mortals and semi-evolved primates should dare to have the audacity to come up with a morality not derived from any religion, I give two answers. The first is that religions are usually highly immoral and thus should be excluded from discussions on morality rather than included. The other is this Big Data scientific process. Granted humans may not be perfect but we can absolutely come up with a set of rules for a true humanistic morality that is fully secular in nature and far better than the monotheistic disasters. No problem.

The responses I often receive to this answer include some sort of nonsensical perceived inability to do this because of the known immorality of some humans, or our nature as an imperfect species, or even due to the taint of original sin. This is typical of the simultaneously dumbing and blinding effect of religion on people who frankly ought to know better.

In and among the sweet and tasty treats in any bag of grapes or any container of strawberries, there are sure to be some that are not so sweet, not so tasty and indeed some which may be moldy and rotten. This does not mean that we cannot ever eat a grape or a strawberry; it merely means we have to screen out the bad ones. The same logic applies to people and their actions. No problem.

"Yes that may be true", the religious may admit, but they will continue to plead that we can never truly know an objective standard for morality without it being written or dictated by a supernatural being. Again, the dumbing and blinding effects of religion rear their stupid and unseeing heads. For one thing, the main fact that so-called revealed religion has actually revealed to us is that its gods are immoral to such a degree that they and their books should be kept at least a mile away from children.

The holy books of Judaism, Christianity and Islam are three of the most immoral, hateful and evil books ever written. Anyone who tries to tell you that you have to get your morality from one of these books is either insane or has been completely brainwashed by his or her particular cult. No sane and reasoning person would take any of these books and use it to build a decent culture, and the faithful need to have their noses vigorously rubbed in this ugly stinking truth until they stop trying to lie to the rest of us.

No god has ever dictated or inspired any magic book from above from any place or time at all which revealed to us the truth about microorganisms, engineering, personal hygiene, medicine, physics, chemistry, how to construct a civil society, politics, education, gun control, abortion, stem cell research, climate change, the optimal system for commerce in the digital age, international relations in the Internet and nuclear ages, or indeed anything truly useful at all. And almost everything else they did or have been alleged to say was wrong.

We discovered everything we know solely on our own, we partly hairless semi-evolved godless near-chimp primates, and we figured them out *despite* the religions and their holy books, not because of them. We are admittedly not perfect and we never will be, but we are a damn sight better off than we were when these books were dreamed up and written by extremely thuggish xenophobic misogynistic ignorant human males.

It is important to note that not a single thing was written in the entire Mesopotamian monotheistic trilogy that could not have been known to a human being living and writing at that time. Not a single thing. Zero. There are the everpresent claims about predictions or prophecies being fulfilled, but again we have the dumb and blind twins at it again. If someone writes a prediction or a prophesy, then later on in the same region someone happens to do or say something that someone *perceives* to be a fulfillment of the prophesy, or that this person claims this happened, this is quite what one would expect of an imperfect and often deceitful *Homo sapiens*.

Why would you expect anything other than that, particularly in a species which generates things like the *National Enquirer* and its disgusting ilk.

Even if the books were not cooked to make it seem like a prophesy was fulfilled, which happens repeatedly throughout religions, eventually someone is going to try to take advantage of a prior prophesy and make the gullible peasants of the time believe he or she is the new Messiah or prophet or god or whatever the original book wrote down. Suckers have been born in droves every minute since the dawn of our species, and so have their natural predators, known to us as snake oil salesmen and the clergy. I refer you again to the earlier chapter on cognitive biases and logical fallacies, and note how the faithful are especially susceptible to most of them.

True morality absolutely does not derive from any god great or small, or any holy book ancient or modern. Our morality comes from seven million years of evolution. It comes from our evolved human programming for mutual support, fairness, a balance of reciprocal altruism and a tit-for-tat redressing of inequities. It comes from our evolved desire for preservation of the self and of family, friends and other humans, and also from the social programming received from the broader society in which we live.

The faithful prove themselves hypocrites and liars every single time they cite their books as sources of morality, because they doggedly ignore almost the entire book in the process. Why is it only the literalists who take the totality of their god's own book at face value and in its own words? Why do the faithful not admit that the all-powerful god they say they worship made so many mistakes and said so many demonic things that they deliberately ignore almost the whole book? Why do so many of the religious only pull a few cherry-picked rules or standards from their books and not all of them? Who are these humans who think they are more powerful than the universe-creating god of their holy books and say they can give a stiff middle finger to that god and reject some of its orders? What is their specific rationale for only picking some of the rules versus the others?

And how exactly is that rationale not equally valid in the secular perspective, in which we take the good in humanity and morality and exclude the bad? It is of course no different at all until you fold in special pleading and "but logic should only apply to other people".

Rather than needing a god to be good, several thousand years of human history clearly indicate that acting out religious beliefs leads to human pain and misery from the individual to the societal level. Most organized religion is immoral, unethical and antihuman. We can yet again take the Catholic cult and its teachings as possibly the world's most egregious example of institutionalized religious immorality and evil, and summarize the history of this criminal organization.

Take a species of primate which has evolved an extremely strong and wholly natural desire to procreate. Let a bunch of power-hungry and misogynistic Iron Age fanatic males first form then gain control of a religious organization, let them use a set of disorganized, disjointed, stolen and repackaged tales and fables as the sole source of guidance for human interaction, and watch them proceed to threaten and punish all those who do not share their delusion.

In a display of complete deviancy and abnormality, make the clerical power class celibate, and only male. Set the conditions for them to have total control over other human beings. Let this continue for many hundreds of years. What else would you expect but a gathering place for outcasts, oddballs, liars, rapists and child molesters? A group of societal outcasts and often sexual deviants who deliberately cut themselves out of the mainstream of society, who cut themselves off from women and the normal human drives and behaviors of heterosexual pleasure and fulfilling heterosexual relationships and procreation. A group of males who derive power and pleasure not from some imaginary moral character but from their place as self-appointed arbiters and purveyors of the imagined salvation and so-called grace of a long-dead minor tribal desert god.

Their inability to wield power over others outside the church, or to be part of a society they would never fit into for any of a number of reasons, or the lack of self-confidence and the need to shine even in reflected light, or their homosexuality or their desire to sexually molest children drove many of these men to the priesthood. From this position of power they take out their disgusting deviant behaviors and frustrations on the children of their church, then they engage on a global campaign to lie about it and cover it up in the face of worldwide anger. They do all this while under the absolute rule of a dictator-theocrat who is actually said to be infallible on the subjects of faith and morals. There aren't words strong enough to condemn such thoroughly evil people and such a thoroughly evil cult.

But world-class hypocrisy is far from unique to Catholicism. Extreme religiosity in America is strongly correlated with extreme social and political conservatism. Often the farther rabidly right the conservative, the more fundamentally religious the person, and vice versa. So today's Puritanical American evangelicals and conservatives get apoplectic and even violent when a woman attempts to get an abortion, even after she has been raped. Paths to abortion, even those as innocuous as the morning-after pill, are blocked. Since that is all too often nowhere near sadistic or oppressive enough to satisfy these people's sadism, the faithful are also frothingly eager to forbid contraception and deny access to birth control, for wholly imaginary religious reasons.

But when the kids are born and need food and medical care and clothing, the religious conservative is equally apoplectic when taxpayer dollars are used or even proposed to be used to help these kids. While these religious conservatives praise their Jesus in every speech and while they rail about societal immorality, they have cut or downsized many programs to support new mothers and young children, and others are under continual attack. Lacking compassion is one thing. Being a zealous hyper-hypocritical religious sadist and asshole is quite another.

Why the religious fail to understand that humans will often do whatever it takes to eat, drink, interact with others and have sex is beyond reasonable comprehension. These drives are actually in our DNA, and according to the faithful were put there by their own damn god. Calling a certain activity a sin does absolutely nothing useful to reduce its frequency or its desirability; it only increases the guilt and shame associated with it.

As a matter of absolute fact, and as anyone who knows anything about human psychology knows well, the more something is prohibited, the more attractive it is to people. The faithful go to extreme measures to fabricate massive imaginary problems, then they try their damnedest to force society to accept their so-called solutions to the massively inflated problems they created. This is insanity, ignorance, immorality, sadism and hypocrisy raised to the power of theological infinity.

As often happens, when acts called sins are made illegal by the secular or religious authorities as I listed in Chapter Six on America's painful and ignorant religious history, behaviors move beyond church-imposed guilt and shame and become an actual crime. The only differences between the Afghanistan Taliban and the religious conservatives in America or anywhere else are of degree and time, because the deeply and sincerely committed believers in America are constantly seeking to form their own American Taliban, with them in charge and their rules enacted into law. There should be no doubt that the desired end states of the Afghan Taliban and their American Taliban wannabe co-religionists are the same.

How is it that some cultures, and by this I primarily mean Christian, Muslim and some Jewish, are so incredibly Puritanical on the subject of sex and sex education? We teach kids how to do math, spelling and grammar. We teach them to put coats and boots on when it is cold outside and to drink water when it is hot out. We get them to wear helmets when they ride bicycles. We take them to the doctor when they are sick, unless the kids were born into a pray-them-healthy cult of ignorance and dumbassery. We teach them how to brush their teeth, wash themselves and how most parts of their bodies work.

But we won't tell them about their reproductive system except to tell them they will burn in hell for touching themselves or having sex, or even having sexual thoughts? We won't tell them about how to handle the natural urges that they have? What the hell is wrong with these people? Do you get embarrassed by teaching a child how to use his hands to throw a ball, or her feet to kick a soccer ball? Why are you so embarrassed by teaching a child about his or her genitalia, as opposed to the hands, feet or digestive system? The only difference is religiously-induced self-hating prudery and insanity, in which these people insist that some body parts which the god of the faithful allegedly made in his own image are embarrassing and not to be touched or talked about. That is immoral.

Speaking of sin, what exactly is a sinner? Not that long ago, even in this our own America, a so-called sinner would have included a person who was believed to practice witchcraft. So-called sinners were and still are treated very cruelly, with numerous horrific punishments and tortures in this world being eagerly heaped upon them by deeply devout Christians despite the belief that these persons are supposedly destined to face an eternity of equal or worse torture and punishment in an existence beyond this mortal coil, administered by an angry god or devil. This is immoral.

If Jesus is god, and he has a hell, why does Jesus the allegedly so wonderful keep a torture chamber in which he burns people forever? The same can be said for Jehovah and Allah. The creators of the universe who are allegedly infinitely wise and good, these gods which billions of people worship very devoutly, do the equivalent of eternally holding a blowtorch on people after they are dead for some contrived misdeeds from their short time on earth. A belief in such lunacy is insane and immoral.

The evil and immoral holy books do not give any answers to modern problems and the discoveries of science. We can already assess a person as a psychopath based on his behavior and intentions. It is likely that one day a young person can undergo a brain scan which will reveal him to be an antisocial psychopathic serial killer or Hitler or Boniface in the making. What should we do with him? What is the moral thing to do? If we could identify the problem but did not yet have a treatment, what would or should we do? Lock him up before he commits a crime? Put a tracking device or other monitoring protocol on him? Kill him? Exile him? Wait until he commits a vicious act on an innocent human being?

I await the pronouncement on this very serious problem from the long-dead monotheistic gods. Why hasn't any god given us these answers, or indeed the cure for psychopathy? You can read through close to a million words of biblical chapters and verses and not get a single microscopic jot or tittle of guidance about this.

The pernicious impact of the innate perversity of most religion is impossible to avoid. Most religious institutions consider homosexuality to be a sin with some degree of awfulness associated with it. Why is this? Homosexuality is merely a variation in the basic human program, resulting in a brain that makes the person gay. Does this make the person automatically a bad person, or immoral? It depends on what you mean by bad and by moral, and if you are a believer in the teachings of a religion or not, and how you personally perceive people who are different from yourself. How do the religious justify saying homosexuals are immoral? Merely because it is in those damn books of theirs – the ones written by bigoted ignorant thugs hundreds and thousands of years ago.

Homosexuality is a variation from the heterosexual norm, and the faithful believe homosexuality is immoral. Do they also believe that a person born with only one arm is automatically immoral, simply because he was born with a variation from the normal human template? What about having exceptionally red hair and masses of freckles? Or being an albino, a dwarf, extremely tall or bald? What about having unusually high or low intelligence? No, the faithful normally would not consider any these variations on the basic human theme to be immoral except for the exceptionally intelligent types because those folks would be sure to present continuous challenges to religious assertions.

But of course this only applies to the nonliteralists. Actual biblical literalists would heed their god's alleged words starting at Leviticus 21:17 (KJV), in which their holy book says:

> *"Whosoever [he be] of thy seed in their generations* ***that hath [any] blemish, let him not approach to offer the bread of his God****. For whatsoever man [he be] that hath a blemish, he shall not approach: a blind man, or a lame, or he that hath a flat nose, or any thing superfluous, Or a man that is brokenfooted, or brokenhanded, Or crookbackt, or a dwarf, or that hath a blemish in his eye, or be*

scurvy, or scabbed, or hath his stones broken..."

The literalists would indeed consider not just homosexuals but pretty damn much anybody who looked a little different from them, or whoever they didn't like, as "immoral" and not good enough to join the flock. Apparently when their god made humans in his image he deliberately chose to make sure the genetic programming was not well done, and he also chose not to protect humans from injury and damage.

Many of the most accomplished and/or famous people in human history were gay. Leonardo da Vinci, Alan Turing, Oliver Sacks, Francis Bacon, Tchaikovsky, Elton John and possibly Michelangelo just to name an extremely few, along with countless others and many in Hollywood, were or are gay. If it were somehow possible, after the fact, to change the brains of these people from homo to hetero, would we change the gestalt of this person's brain to a degree that their unique abilities were destroyed? Would this be more "moral" than just letting them be who they are, and not persecuting or condemning them?

Why does the Catholic sect so hypocritically condemn homosexuals while its ranks are overflowing with rapists and rapist enablers both in the clergy and in the pews? The clergy of some religions and the ranks of many highly vocal religionists and anti-gay conservatives are full of gay people, usually men. Such unparalleled hypocrisy and faith-based bigotry against homosexuals is astoundingly immoral.

It also goes without saying that there is plenty of homosexuality in Christianity, both overt and hidden, and quite hypocritically not spoken about. In Paul's first epistle to the Corinthians, he says (KJV) "Now concerning the things whereof ye wrote unto me: [It is] good for a man not to touch a woman." I defy anyone alive to name a single normal heterosexual male who would ever say such a thing. As for the central figure of the Christian sect, he too eschewed female contact and preferred the company of a large boy band of young brothers in robes. He never pronounced himself one way or the other but apparently reached his mid-thirties without having had any sort of relationship with a woman. I submit the obvious question to the sharp razor of Occam and leave the reader to come up with his or her own conclusion. Was he gay or did his followers make sure the surviving stories screened out female companions?

The subject of deviations from basal programming can be taken a very long way in the wrong direction. I personally have very little artistic talent. I am a fairly well educated and capable human of middle age but still unable to draw more than a stick figure and even that with difficulty. Some would say, and they would be correct, that I have a serious defect in the artistic part of my programming. Does this make me immoral? I would hope not. Exactly which physical, mental or emotional characteristics do the faithful believe a person has to vary from in order for the religious to consider him immoral?

Apparently it depends on the believer, how literal he is and which parts of his faith he chooses to accept and which parts he rejects. Thus inconsistency and hypocrisy, both of which are often immoral, form the rope from which the noose of religion is woven, and the faithful are always very happy to tighten that noose around people which they or their books do not like.

Let's take this discussion a little further just to make the point even more clearly. The god of the Old Testament is such a horrible and sociopathic entity that it is impossible to understand why anyone would want to follow its example. Were a real person to have the same attitude as this character, he would be considered as at least as evil as Hitler, Stalin, Pol Pot, Mussolini, the Inquisitional popes, Genghis Khan, Caligula and Charles Manson, all put together.

If this person were to actually act in the real world like this angry and jealous spoiled child god did, he would be put to death or imprisoned forever. Yet not only do billions of people fail to reject such evil, they actually get down on their knees and pray in supplication to this demonic god like the weak, spineless and worthless worms they believe themselves to be.

We can take this still further, and ask what people would think of these belief systems if offered to them as an adult by a stranger for the first time. If you had never heard of any of the following practices before, but were asked to give your opinion on them, which of these would you consider beneficial, constructive, positive and good if taught to young children, and which would you consider insane, harmful, negative and inappropriate to teach to young children?

- You were born a dirty rotten sinner and the only ways your imaginary eternal soul will be redeemed and kept from an eternity of burning agony is if you accept that a man-god hybrid committed assisted suicide two thousand years ago in order that your sins can be forgiven and also if you worship this long dead man-god who is actually not dead but watches you every microsecond of your life and knows everything you do, say and think.

- Your strong and wholly natural and positive sexual desire and its natural function are to be suppressed as much as possible and will be a lifelong source of sin, shame and embarrassment, even on the occasions when you are allowed to practice it. This applies a hundredfold if you are female.

- You will spend your life feeling guilty about your lack of perfection and your pathetic inability to avoid sinning, but you can do absolutely terrible deeds to other people for your entire life and be forgiven for them if you really feel bad about and regret them on your deathbed.

- Use of artificial birth control is prohibited, regardless of circumstances, without exception, no matter how many people get unnecessarily pregnant, contract disease, die of preventable disease, or are born into unwanted situations and into cultures which said they loved them until they were born, then abruptly and callously turned their backs on them.

- If you are a ten year old girl and you become pregnant after your stepfather raped you, you cannot have an abortion.

- If you marry someone, but that person later turns out to be abusive, your marriage can't be ended by any human power; it only ends when you die, because divorce is immoral.

- Once a week, you will participate in a cannibalization ceremony in which you will ingest the zombie flesh and blood of a long-dead man-god who is actually not dead but who is magically partially resurrected in your own body by an unexplained magical process which begins when a celibate male cultist says magic words over pressed carbohydrate wafers and fermented grape juice. This ceremonial cannibalism is said to be extremely important and extremely sacred.

- If you are a female and are raped, you can be sold to your rapist.

Each of these statements reflects a doctrinal teaching, belief or reference from a single major religious group currently boasting over a billion adherents. If you haven't already guessed which one it is, here is a hint. Leave your underage kids at home where they are safe and go ask a Catholic priest. While Catholic doctrine is well populated by such antihuman and bizarre teachings, it is necessary to bear in mind that almost any religious canon will provide equally antihuman examples of horrible human cruelty.

Are any of these the sorts of things that should be taught to children?

If you say yes, you are some combination of criminally immoral, ignorant, stupid and evil and you need to immediately cease contact with other humans and never vote again.

If you are a person in full possession of your faculties and basic human compassion, and you honestly and truly believe these statements to be good fundamental principles by which to live your life, you are either completely brainwashed by the cult or are a person who can never be relied upon for sane and reasonable advice on morality or human behavior.

If a newspaper article from some foreign country wrote an article about a newly discovered culture which had these beliefs, and which acted upon them in its treatment of its children, how would you feel about these people? How would you feel about living there? Would you feel that these people lived in a warm, supportive and low-stress culture which you would want to join and where you would want to bring your children? Or would you be horrified at the immorality and antihuman insanity of these people and vow never to get anywhere close to the place or let your children near it? If a billion people believe this nastiness, and billions of others believe similar antihuman nonsense, is there any wonder that the world is so full of misery? How could it not, given such widespread beliefs?

All religions are like the supermassive American legal system and tax code. They all started out small, and bit by bit and over time, more and more rules and laws and dogmas got added in. Now these irrational faith-based behemoths are so overwhelming and oppressive they need to be eliminated and replaced by simpler, fact-based and infinitely more rational and moral systems.

How can we make the world a better and more moral place? We can remove religion immediately. An outstanding and positive humanistic morality absolutely can be derived from our own humanity and our own human science, and even a crude and partial effort from those sources would be far better than the immoral disasters the monotheisms have on offer. Billions of people are good without any god, and it is likely that there are truly, even if they will not or cannot admit it openly, billions of people who are not theistic at all.

So when are we going to get rid of religion? Unfortunately it will not happen anytime soon, because the irrational tribal echo chamber groupthink of the faithful prevents them from accepting the world as it is.

This combination of childishness and willful blindness combines to double the thickness of their mental blinders and their resolve to remain in their platitudinous and fact-free cocoons.

As an example, if someone tries to tell you that you cannot define or quantify love, she is merely repeating some brainless platitude she learned at a religious gathering without giving it a speck of thought. In the inimitable words of Tim Minchin in his masterful work "Storm", such platitudes are vacuous crap. Yes, you absolutely can define love and you can objectively quantify it. Here's a quick way to do it. Pick ten people you think you love. Do you have positive feelings for them? How positive? Rank the strength of your feelings on a scale from 1 to 10. Do you want to do good things for them just to make them happy? How often? Do you actually do these things for them? Again, how often? Do you help them when they are sick, feed them when they are hungry, protect and honor and cherish them? Would you lie to or mistreat them? Do you give them gifts and tell them you love them?

Continue to ask such questions and compile the answers to get a Big Data result which will quite objectively show how much you do or do not love those people.

It is safe to say that with rational scientific analysis one can not only set up an algorithm that could assess the level of love a person has for another, but one can also combine this with brain imaging to ensure the person was truly feeling what was verbally professed. One day we will be able to graphically depict the level of the love people feel for other specific people on a bar chart or in an image.

The same goes for innumerable other questions and observations, and the faithful absolutely cannot stand this, because it demystifies their god and destroys their platitudes. But none of this understanding of reality removes any of the magic from love, or the good taste of chocolate. There will be no change to how much I enjoy chocolate because I find out what it is made of at the molecular level and how it interacts with my body and brain to make the end result a tasty and enjoyable treat. I will not love my family any less because I know about the complex interplay of chemicals and electricity that makes love happen. To the contrary, it greatly enhances life and the enjoyment and fulfillment of living a human life.

Knowing how things work is infinitely more enjoyable and rewarding than being ignorant of such things. It takes a certain kind of weak and severely diminished person to want to remain ignorant and in the dark, to turn away from the guiding light of information and the power of reason. It is the weak mind that wants to burrow into the ground, that wants to drown out the light and wants to remain ignorant of the unparalleled marvels of life and our world. Such a mindset reflects the weakness and the smallness of far too many humans, and it is this mindset that perpetuates the shambling brain-dead zombie of religion for century after painful century.

We absolutely don't need any god to be good, and we would be far better off without any such fanciful imaginary creatures. If we removed religion from every place in America today and replaced it with a secular humanist model the major changes would be that Sunday television would be free of the nauseating pleas for money from greedy Jesus-defying televangelists, and people would immediately feel much less needless guilt and shame. Weekend religious meetings would be replaced by community volunteer meetings, social gatherings and other events.

Instead of praying to some nonexistent god or worshiping a fictional character, people would get together to discuss how to make their community better, help children and the needy, improve schools and education, get some exercise, learn how to eat better and cook healthy food, spend more time with family and friends and be more educated and more self-sufficient. Instead of funding organizations which work to keep condoms and birth control away from groups of people who are malnourished, at constant risk of AIDS/HIV and with extremely high birth rates, we could fund condom and birth control drives to educate people on the benefits of birth control and how to best avoid deadly diseases, overpopulation and the rot of immoral religion.

The absolute last place anyone should take morality from is a holy book. Out of a mass of psychopathic actions just in the Christian bible, we can take merely one example of the madness that it contains and use it to reject the whole thing. Most readers will be familiar with how Abraham's god so sadistically toyed with him, as a demonic puppet master would if he wanted to drive his puppet insane. The very first pages of this immoral book, in Genesis 22, describe how Abraham's god told him to take his only son, drag him up a mountain, kill him and make him into a burnt offering. What did Abraham do? What would a moral person do? What would *you* do?

Most of you know how the story goes. Without question or hesitation, and without even a request for clarification, Abraham did exactly what his god told him to do. He dragged his son up the mountain, lied to him about exactly which "lamb" was going to be the sacrificial burnt offering, then unbelievably "...stretched forth his hand, and took the knife to slay his son." At the last microsecond an angel stepped in to prevent this faith-based murder of a young boy by his own father, specifically ordered by Jehovah. Because Abraham was actually going to kill his very own son, without any hesitation or reservation, his god told him "...I will bless thee, and in multiplying I will multiply thy seed as the stars of the heaven, and as the sand which [is] upon the sea shore; and thy seed shall possess the gate of his enemies; And in thy seed shall all the nations of the earth be blessed; because thou hast obeyed my voice."

How can any truly moral person read this and not immediately throw the entire book in the trash? How can anyone actually read this or any of the rest of this wholly immoral book then make a conscious decision that this religion is definitely the one for him? When did utter psychopathy become attractive to people? How can anyone who claims to use this book as his moral compass not be insane and/or immoral?

The kind of mind that would order someone to kill his own child is the kind of mind who kidnaps children off the streets, then keeps them in his basement and tortures them whenever he gets the sadistic urge, because most of his thrills come from his utter debasement over the helpless captive. This is the god which billions of people worship, and this is an utter moral travesty.

And if this god is actually omniscient, as his fans say he is, he would have known ahead of time that Abraham would have unhesitatingly gutted his son. So why did he order the murder, and why did he make him play it out? Was it to teach him a lesson? What lesson? Slavish, mindless and robotic obedience even to the point of murdering one's own innocent child is the answer. In what way is this not the maximum possible evil? And why does anyone take it as anything except a reason to keep their children and themselves as far away from Christianity and Judaism as possible?

Or was the point of the whole exercise actually to teach Abraham's son a lesson, and that it was the same lesson of utterly spineless and gutless obedience to an insane lunatic? So now we expand the godly sadism to two individuals, with the innocent child the victim of not just one insane lunatic but two.

Or was the lesson the son got not part of the deal? If not, why not? Put yourself in that boy's position. Your own father, who you probably loved very much up until he tried to murder you, was in the act of killing you because of the voices in his head before one of the voices called a hasty time-out. How would any of you Christians or Jews feel if your father did the same thing to you? What would any of you do if your father tried to do this to his grandchildren, your own children? How many of you would stand idly by and watch the faith-based murder of your own innocent child?

If you really and truly believed in your god and his book, your response would be to hold down your screaming and struggling child while his grandfather cut his throat.

The lesson I take away from this incomparably evil tale, and which I trust anyone rationally and critically analyzing the situation also would, is that the god which billions of people worship is the kind of god who would allow a young boy who probably loved and adored his father very much to be tied up by that father and come within a hairs-breadth of being brutally murdered by him. The lesson the son actually got was that both his father and the god who was ordering his father around were brutal murderers or malevolent psychopaths capable of complete evil.

Again, take this situation and view it through the lens of critical thinking. Were a man to be seen today strapping down his son and raising a sharp knife above him with the honest intent of killing the boy because of the voices in his head, that person would be arrested, convicted of attempted murder and surely diagnosed with a severe mental disorder and kept far away from other people in an asylum for the violently insane. The holy book of Christianity and Judaism actually teaches that such acts are a virtue and that the person committing such acts should be hugely rewarded for this kind of ultimate evil.

As if all of the above were not already far too much, it doesn't even begin to scratch the surface of the immoralities in religion. I have devoted a large part of this book to specifically discussing religious immorality and can still only slightly touch on this global disease. The list below is just a few of the immoral things which religions teach children, and while the list is far from comprehensive, it is a clear indicator of how badly religions damage children, damage morality and damage humanity.

1. That they are "fallen" and/or dirty rotten sinners, not amazing products of nature who bear no responsibility for their distant ancestors, either actual or imaginary.
2. That there is a crazy resurrected zombie god who listens to their thoughts and is holding a spot in a private oven for them if they think sinful things.
3. That bizarre snuff fantasy and science fiction written by ignorant, stupid and evil people thousands of years ago should be used as a historical, moral, legal and scientific reference.
4. That they should rely on hateful mythology for their values.
5. That homosexuality is a sin.
6. That critical and independent thinking and reason are harmful.
7. That faith, not critical thinking, is a virtue.
8. That blind obedience is a virtue.
9. That condom use is sinful.
10. That certain parts of their bodies are sinful, and that sex is a sinful activity.
11. That only the religion they are being brainwashed into is real, and that all of the other thousands of religions are not real.

12. That women are inferior and subordinate to men.
13. That people should feel deeply rooted fear, guilt and shame for normal human acts and thoughts.
14. That murder or human sacrifice is the path to salvation.
15. That they should love their enemies.
16. That following spiritual leaders is more important than being with and taking care of one's family.
17. That only a few parts of the holy books allegedly written or inspired by their allegedly omniscient and omnipotent god are to be followed, and only special self-appointed males can tell you which parts are which.
18. That when horrible things happen to them, it is because their particular god has a special super-secret plan for them and it will all be great, someday.
19. That prayer works.
20. That they can or must rely on their particular god and prayer for answers and action.
21. That killing in the service of the god and the holy book is right and good.
22. That persecuting others in the service of the holy book is right and good, even if this persecution lasts for nearly 2000 years.
23. That their religious bigotry is somehow moral and that this bigotry should be the law of the land.
24. That science and reality should be denied whenever they conflict with their particular religion.
25. That their holy books are still worth anything despite being very harmful.
26. That keeping ancient and ignorant holy books around wholly unchanged is a virtue.
27. That their top holy man is infallible about morals despite being highly immoral.
28. That their particular immorality is the one true set of values everyone should follow.
29. That people who don't believe in their hateful mythology are immoral.
30. That no matter how many times they are shown to be wrong, hateful and immoral, religions and religious beliefs cannot be changed.
31. That there is an afterlife, and that only people who follow their faith will go there.
32. That the end justifies the means, and that immorality and evil done in the god's name are completely OK.

Religion is sick and immoral to the core, and there is no curing it, because its very foundation is immoral. The holy books and their teachings are deeply immoral, and the books form the foundation of the religions. These holy books will never change, and the faithful will never let them change.

Thus the only way to eliminate the immorality of religion is to eliminate the religions and their holy books. The first step to a global human morality and the elimination of religion is to get the faithful to understand just how immoral their religions and their holy books really are.

Only then will we be able to begin the process of bringing billions of human beings to a truly moral place, without any holy books or superstitious religions.

Chapter Seventeen

The Bible, A Primer In Immorality

"Happy [shall he be,] that taketh and dasheth thy little ones against the stones."
- Psalm 137, Old Testament

Absolute power corrupts absolutely, which is why god is absolutely corrupt.

And which is why so much absolutely corrupt human behavior is justified by its corrupt gods and corrupt holy books.

The Christian and Jewish religious myths tell the same basic original fable. As Christopher Hitchens said so well, in their origin story the alleged creator of the universe watched for 200,000 years while mothers, children and babies of the species he allegedly specially created and loved so very much died and/or became orphans, then finally decided to drop an extremely long note to some ignorant and hyperaggressive epileptics, hallucinators and misogynists in a backward desert, yet failed to pass on that washing hands was a good idea?

Also, why do so many Christians have zero understanding of how their Big Book of Bible Babble was put together, being that its fables and storylines were borrowed from many other religions that predated Christianity and then much later the whole discordant package was cobbled together in the 300s by a bunch of old men?

With all that in mind, the next time anyone tries to tell you the Christian bible is his personal or his religion's source of morality, I recommend you hold back the natural urge to recoil in shock and horror. The other person probably really has no idea how disgustingly and thoroughly immoral that book is. He has obviously read or heard only the very few acceptable parts separated from the other 99% of the book, which is a massive tonnage of evil and nonsense. He may have only heard how good it is from the tailored and whitewashed words of his preferred pastor, preacher or televangelist.

Or it is possible that he has done what almost no one does and has read the entire book, but then applied enough self-deception, wishful thinking and motivated reasoning to enable him to read that Mesopotamian *Mein Kampf* and come out on the other side believing he has just read something decent.

A better response than open scorn and horror is to have a copy of this chapter of the book in your back pocket or close at hand on your mobile device, and when you encounter an appropriate situation, whip it out and set the believer straight.

This chapter contains nothing but extracts from that book they say they love so much but of which they know next to nothing, with very short descriptions of the evil in those extracts. Be prepared for anything when you respond, including blank looks, disbelief, shock, red faces, defensiveness, argumentation, angry shouting and an abrupt end to the discussion. Depending on the person, be prepared to defend yourself against physical violence.

Please note that I have winnowed my originally much, much larger collection of these extracts to a quite smaller and shorter set which could be more easily placed in this book. If you feel like the below list isn't quite enough, and you want to engage in a similar project on your own, I recommend you get a copy of the book (I used the King James version) and a few highlighters (you'll need more than one) and get ready to spend a horrified weekend highlighting all the immoral parts.

Just take a moment from time to time to walk around, hug your spouse or your children, and be thankful this book is, at least for the time being in America, only ancient hateful fiction and not completely the law of the land, which is what its devotees quite fervently want it to be. With that in mind, here are just a few of the verbal clubs with which you can engage your faithful interlocutor. All bolding is mine.

Notable Bible Quotes, With Commentary (The Not Very Good Book Speaks)

Genesis

Divine Sadism.

- "{3:14} And the LORD God said unto the serpent, Because thou hast done this, **thou [art] cursed above all cattle**, and above every beast of the field; upon thy belly shalt thou go, and dust shalt thou eat all the days of thy life: {3:15} And **I will put enmity between thee and the woman**, and between thy seed and her seed; it shall bruise thy head, and thou shalt bruise his heel."

- "{3:16} Unto the woman he said, **I will greatly multiply thy sorrow and thy conception; in sorrow thou shalt bring forth children**; and thy desire [shall be] to thy husband, and **he shall rule over thee**."

- "{3:17} And unto Adam he said, Because thou hast hearkened unto the voice of thy wife, and hast eaten of the tree, of which I commanded thee, saying, Thou shalt not eat of it: **cursed [is] the ground for thy sake; in sorrow shalt thou eat [of] it all the days of thy life**; {3:18} Thorns also and thistl`es shall it bring forth to thee; and thou shalt eat the herb of the field; {3:19} In the sweat of thy face shalt thou eat bread, till thou return unto the ground; for out of it wast thou taken: **for dust thou [art,] and unto dust shalt thou return."**

Divine Mass Extermination And Divine Mass Abortion.

- "{7:21} **And all flesh died that moved upon the earth**, both of fowl, and of cattle, and of beast, and of every creeping thing that creepeth upon the earth, and every man: {7:22} All in whose nostrils [was] the breath of life, of all that [was] in the dry [land,] died. {7:23} **And every living substance was destroyed which was upon the face of the ground**, both man, and cattle, and the creeping things, and the fowl of the heaven; and they were destroyed from the earth…"

Divine Sadism, Genital Mutilation, Child Abuse and Support of Slavery.

- "{17:10} This [is] my covenant, which ye shall keep, between me and you and thy seed after thee; **Every man child among you shall be circumcised.** {17:11} And ye shall circumcise the flesh of your foreskin; and it shall be a token of the covenant betwixt me and you. {17:12} And **he that is eight days old shall be circumcised among you**, every man child in your generations, he that is born in the house, **or bought with money of any stranger**, which [is] not of thy seed. {17:13} **He that is born in thy house, and he that is bought with thy money, must needs be circumcised:** and my covenant shall be in your flesh for an everlasting covenant."

<u>Divine Schizophrenia and Sadism, Divinely Ordered Child Mutilation.</u>

- "Genesis {17:10} This [is] my covenant, which ye shall keep, between me and you and thy seed after thee; **Every man child among you shall be circumcised."**

  OK, so you were ordered to use unsanitized knives or sharp stones to slice off parts of your sons' penises, which is bad and illogical enough, but only a mind disturbed by religion can reconcile that order with this next one from later in the book which quite clearly contradicts the previous one:

- "Corinthians {3:16} Know ye not that **ye are the temple of God**, and [that] the Spirit of God dwelleth in you? {3:17} **If any man defile the temple of God, him shall God destroy**; for the temple of God is holy, which [temple] ye are."

<u>Divine Mass Murder and Targeted Assassination, With Bonus Points For Marksmanship.</u>

- "{19:24} **Then the LORD rained upon Sodom and upon Gomorrah brimstone and fire** from the LORD out of heaven; {19:26} But his wife looked back from behind him, and **she became a pillar of salt."**

<u>Divine Approval of Repeated Incest.</u>

- "{19:30} And Lot went up out of Zoar, and dwelt in the mountain, and his two daughters with him; for he feared to dwell in Zoar: and he dwelt in a cave, he and his two daughters. {19:33} And they made their father drink wine that night: and **the firstborn went in, and lay with her father; and he perceived not when she lay down, nor when she arose**."

- "{19:34} And it came to pass on the morrow, that the firstborn said unto the younger, Behold, I lay yesternight with my father: let us make him drink wine this night also; and go thou in, [and] lie with him, that we may preserve seed of our father. {19:35} And they made their father drink wine that night also: and **the younger arose, and lay with him; and he perceived not when she lay down, nor when she arose. {19:36} Thus were both the daughters of Lot with child by their father**."

This stupid little fable is disgusting and sick but always nevertheless made me laugh, for several obvious reasons. The first one is the small matter of where they got all this wine. I didn't read about them rolling a wine barrel up the mountain, or finding some moonshiner's abandoned stash just waiting for Lot and his uncontrollable drinking habit to arrive. I also don't know of any normal human male who would be able to perform in any sexual act whatsoever with his daughter, no matter how drunk he was. If he did get too drunk to realize the other person was his daughter, in this case daughters, he'd be too drunk to do the deed anyway, unless he actually didn't have too much of a problem with it when he wasn't drunk.

Also how did the girls know the day they were fertile? And were the two girls' menstrual cycles on the exact same schedule? And when the daughters started showing their pregnancy, what was that conversation like? How did the girls explain themselves, with ostensibly no other males having been around to "plant the seed"? And I always have fun with the religious when they try to explain how this sort of thing isn't actually their god saying it's ok for a father to screw his own daughters.

Divine Approval of Utter Sadism, Child Abuse, Child Sacrifice, Unquestioning Obedience.

- "{22:1} And it came to pass after these things, that God did tempt Abraham, and said unto him, Abraham: and he said, Behold, [here] I [am.] {22:2} And he said, Take now thy son, thine only [son] Isaac, whom thou lovest, and get thee into the land of Moriah; and **offer him there for a burnt offering upon one of the mountains which I will tell thee of."**

Divine Targeted Assassination, Again.

- "{38:7} And Er, Judah's firstborn, was wicked in the sight of the LORD; and the LORD slew him."

Exodus

Divine Support of Genocide, Sadism, War and the Theft of Lebensraum.

- "{3:8} **And I am come down to deliver them** out of the hand of the Egyptians, and to bring them up out of that land unto a good land and a large, unto a land flowing with milk and honey; **unto the place of the Canaanites, and the Hittites, and the Amorites, and the Perizzites, and the Hivites, and the Jebusites.** {3:17} And I have said, I will bring you up out of the affliction of Egypt **unto the land of the Canaanites, and the Hittites, and the Amorites, and the Perizzites, and the Hivites, and the Jebusites,** unto a land flowing with milk and honey."

Divine Mass Murder, Yet Again, Including Children.

- "{12:29} And it came to pass, that **at midnight the LORD smote all the firstborn in the land of Egypt**, from the firstborn of Pharaoh that sat on his throne unto the firstborn of the captive that [was] in the dungeon; and all the firstborn of cattle."

Divine Support of Jealousy, Sadism, Malice, Vindictiveness, Slavery, Female Slavery, Murder, Filicide, Cruelty to Animals, Female Dowry, Murder of Witches, Gunpoint Displacement and Persecution of Other People, Animal Sacrifice, Blood as a Totem, Mass Murder.

- "{20:4} Thou shalt not make unto thee any graven image, or any likeness [of any thing] that [is] in heaven above, or that [is] in the earth beneath, or that [is] in the water under the earth: {20:5} Thou shalt not bow down thyself to them, nor serve them: **for I the LORD thy God [am] a jealous God, visiting the iniquity of the fathers upon the children unto the third and fourth [generation] of them that hate me…"**

- "{21:7} And **if a man sell his daughter to be a maidservant**, she shall not go out as the menservants do."

- "{21:17} And he that curseth his father, or his mother, **shall surely be put to death**."

- "{21:28} If an ox gore a man or a woman, that they die: then **the ox shall be surely stoned…"**

- "{22:18} **Thou shalt not suffer a witch to live."**

- "{22:20} He that sacrificeth unto [any] god, save unto the LORD only, he **shall be utterly destroyed."**

- "{23:28} **And I will send hornets before thee,** which shall drive out the Hivite, the Canaanite, and the Hittite, from before thee. {23:29} I will not drive them out from before thee in one year; lest the land become desolate, and the beast of the field multiply against thee. {23:30} By little and little **I will drive them out from before thee,**

until thou be increased, and inherit the land. {23:31} And I will set thy bounds from the Red sea even unto the sea of the Philistines, and from the desert unto the river: for **I will deliver the inhabitants of the land into your hand; and thou shalt drive them out before thee."**

- "{29:20} Then shalt thou **kill the ram, and take of his blood, and put [it] upon the tip of the right ear of Aaron, and upon the tip of the right ear of his sons, and upon the thumb of their right hand, and upon the great toe of their right foot, and sprinkle the blood upon the altar round about. {29:21} And thou shalt take of the blood that [is] upon the altar, and of the anointing oil, and sprinkle [it] upon Aaron, and upon his garments, and upon his sons, and upon the garments of his sons with him: and he shall be hallowed, and his garments, and his sons, and his sons' garments with him."**

- "{32:27} And he said unto them, Thus saith the LORD God of Israel, **Put every man his sword by his side, [and] go in and out from gate to gate throughout the camp, and slay every man his brother, and every man his companion, and every man his neighbour.** {32:28} And the children of Levi did according to the word of Moses: **and there fell of the people that day about three thousand men**."

<u>Leviticus</u>

<u>Misogyny, Oppression of Women, Child Abuse, Animal Cruelty and Utter Scientific Ignorance.</u>

- "{12:1} And the LORD spake unto Moses, saying, {12:2} Speak unto the children of Israel, saying, **If a woman have conceived seed, and born a man child: then she shall be unclean seven days;** according to the days of the separation for her infirmity shall she be unclean. {12:3} **And in the eighth day the flesh of his foreskin shall be circumcised.** {12:4} **And she shall then continue in the blood of her purifying three and thirty days; she shall touch no hallowed thing, nor come into the sanctuary, until the days of her purifying be fulfilled.**"

- "{12:5} **But if she bear a maid child, then she shall be unclean two weeks, as in her separation: and she shall continue in the blood of her purifying threescore and six days.** {12:6} And when the days of her purifying are fulfilled, for a son, or for a daughter, **she shall bring a lamb of the first year for a burnt offering, and a young pigeon, or a turtledove, for a sin offering,** unto the door of the tabernacle of the congregation, unto the priest."

<u>Animal Cruelty, Utter Scientific Ignorance On Steroids, Absolutely Batty Stuff.</u>

- "{14:12} **And the priest shall take one he lamb, and offer him for a trespass offering, and the log of oil, and wave them [for] a wave offering before the LORD: {14:13} And he shall slay the lamb in the place where he shall kill the sin offering and the burnt offering, in the holy place: for as the sin offering [is] the priest's, [so is] the trespass offering: it [is] most holy: {14:14} And the priest shall take [some] of the blood of the trespass offering, and the priest shall put [it] upon the tip of the right ear of him that is to be cleansed, and upon the thumb of his right hand, and upon the great toe of his right foot:**

  {14:15} And the priest shall take [some] of the log of oil, and pour [it] into the palm of his own left hand: **{14:16} And the priest shall dip his right finger in the oil that [is] in his left hand, and shall sprinkle of the oil with his finger seven times before the LORD: {14:17} And of the rest of the oil that [is] in his hand shall the priest put upon the tip of the right ear of him that is to be cleansed, and upon the thumb of his right hand, and upon the great toe of his right foot, upon the blood of the trespass offering…"**

<u>Absolutely Bizarre and Insane Divine Proclamations and Murderous Commands.</u>

- "{19:31} **Regard not them that have familiar spirits, neither seek after wizards, to be defiled by them**: I [am] the LORD your God."

- "{20:6} And **the soul that turneth after such as have familiar spirits, and after wizards, to go a whoring after them, I will even set my face against that soul, and will cut him off from among his people."**

- "{20:9} **For every one that curseth his father or his mother shall be surely put to death."**

- "{20:10} **And the man that committeth adultery with [another] man's wife, [even he] that committeth adultery with his neighbour's wife, the adulterer and the adulteress shall surely be put to death."**

- "{20:11} **And the man that lieth with his father's wife hath uncovered his father's nakedness: both of them shall surely be put to death…"**

- "{20:12} **And if a man lie with his daughter in law, both of them shall surely be put to death."**

- "{20:13} **If a man also lie with mankind, as he lieth with a woman, both of them have committed an abomination: they shall surely be put to death."**

- "{20:14} **And if a man take a wife and her mother, it [is] wickedness: they shall be burnt with fire…"**

- "{20:18} **And if a man shall lie with a woman having her sickness, and shall uncover her nakedness; he hath discovered her fountain, and she hath uncovered the fountain of her blood: and both of them shall be cut off from among their people."**

- "{20:21} **And if a man shall take his brother's wife, it [is] an unclean thing: he hath uncovered his brother's nakedness; they shall be childless."**

- "{20:27} **A man also or woman that hath a familiar spirit, or that is a wizard, shall surely be put to death: they shall stone them with stones…"**

- "{21:9} **And the daughter of any priest, if she profane herself by playing the whore, she profaneth her father: she shall be burnt with fire."**

God Hates Unattractive And Disabled People (But Made In His Image?)

- "{21:16} And the LORD spake unto Moses, saying, {21:17} Speak unto Aaron, saying, **Whosoever [he be] of thy seed in their generations that hath [any] blemish, let him not approach to**

offer the bread of his God. {21:18} For whatsoever man [he be] that hath a blemish, he shall not approach: a blind man, or a lame, or he that hath a flat nose, or any thing superfluous, {21:19} Or a man that is brokenfooted, or brokenhanded, {21:20} Or crookbackt, or a dwarf, or that hath a blemish in his eye, or be scurvy, or scabbed, or hath his stones broken…"

Insane Threats of Unbelievable Violence, Plus Some Juicy Cannibalism.

- "{26:14} But if ye will not hearken unto me, and will not do all these commandments; {26:16} I also will do this unto you; **I will even appoint over you terror, consumption, and the burning ague, that shall consume the eyes, and cause sorrow of heart: and ye shall sow your seed in vain, for your enemies shall eat it. {26:17} And I will set my face against you, and ye shall be slain before your enemies: they that hate you shall reign over you; and ye shall flee when none pursueth you.**"

- "{26:21} And if ye walk contrary unto me, and will not hearken unto me; **I will bring seven times more plagues upon you according to your sins.**"

- "{26:22} **I will also send wild beasts among you, which shall rob you of your children, and destroy your cattle, and make you few in number; and your [high] ways shall be desolate.**"

- "{26:25} **And I will bring a sword upon you, that shall avenge the quarrel of [my] covenant: and when ye are gathered together within your cities, I will send the pestilence among you; and ye shall be delivered into the hand of the enemy.**"

- "{26:29} **And ye shall eat the flesh of your sons, and the flesh of your daughters shall ye eat.**"

- "{26:30} **And I will destroy your high places, and cut down your images, and cast your carcases upon the carcases of your idols, and my soul shall abhor you.**"

- "{26:31} **And I will make your cities waste, and bring your sanctuaries unto desolation, and I will not smell the savour of your sweet odours.**"

- "{26:33} **And I will scatter you among the heathen, and will draw out a sword after you: and your land shall be desolate, and your cities waste."**

Numbers

Divine and Divinely Ordered Genocide and Evil, Plus Invasions, Murder, Sexual Slavery.

- "{11:1} And **[when] the people complained, it displeased the LORD: and the LORD heard [it;] and his anger was kindled; and the fire of the LORD burnt among them, and consumed [them that were] in the uttermost parts of the camp."**

- "{11:33} And while the flesh [was] yet between their teeth, ere it was chewed, the wrath of the LORD was kindled against the people, and **the LORD smote the people with a very great plague."**

- "{15:32} And while the children of Israel were in the wilderness, they found a man that gathered sticks upon the sabbath day. {15:33} And they that found him gathering sticks brought him unto Moses and Aaron, and unto all the congregation. {15:34} And they put him in ward, because it was not declared what should be done to him. {15:35} **And the LORD said unto Moses, The man shall be surely put to death: all the congregation shall stone him with stones without the camp.** {15:36} **And all the congregation brought him without the camp, and stoned him with stones, and he died; as the LORD commanded Moses."**

- "{21:2} And Israel vowed a vow unto the LORD, and said, If thou wilt indeed deliver this people into my hand, then I will utterly destroy their cities. {21:3} **And the LORD hearkened to the voice of Israel, and delivered up the Canaanites; and they utterly destroyed them and their cities: and he called the name of the place Hormah."**

- "{31:1} And the LORD spake unto Moses, saying, {31:2} Avenge the children of Israel of the Midianites: afterward shalt thou be gathered unto thy people. {31:3} And Moses spake unto the people, saying, Arm some of yourselves unto the war, and let them go against the Midianites, and avenge the LORD of Midian. {31:7} **And they warred against the Midianites, as the LORD commanded Moses; and they slew all the males**. {31:8} And **they**

slew the kings of Midian, beside the rest of them that were slain; [namely,] Evi, and Rekem, and Zur, and Hur, and Reba, five kings of Midian: Balaam also the son of Beor they slew with the sword. {31:9} And **the children of Israel took [all] the women of Midian captives, and their little ones, and took the spoil of all their cattle, and all their flocks, and all their goods.** {31:10} **And they burnt all their cities wherein they dwelt, and all their goodly castles,** with fire. {31:11} And **they took all the spoil, and all the prey, [both] of men and of beasts.**"

- "{31:14} And Moses was wroth with the officers of the host, [with] the captains over thousands, and captains over hundreds, which came from the battle. {31:15} And Moses said unto them, **Have ye saved all the women alive?** {31:16} Behold, these caused the children of Israel, through the counsel of Balaam, to commit trespass against the LORD in the matter of Peor, and there was a plague among the congregation of the LORD. {31:17} **Now therefore kill every male among the little ones, and kill every woman that hath known man by lying with him**. {31:18} **But all the women children, that have not known a man by lying with him, keep alive for yourselves.**"

- "{33:50} And the LORD spake unto Moses in the plains of Moab by Jordan [near] Jericho, saying, {33:51} Speak unto the children of Israel, and say unto them, When ye are passed over Jordan into the land of Canaan; {33:52} Then **ye shall drive out all the inhabitants of the land from before you, and destroy all their pictures, and destroy all their molten images, and quite pluck down all their high places**: {33:53} And **ye shall dispossess [the inhabitants] of the land, and dwell therein: for I have given you the land to possess it.**"

Deuteronomy

Divine and Divinely Ordered Genocide and Evil, Plus Invasions, Murder, Sexual Slavery.

- "{2:33} And the LORD our **God delivered him before us; and we smote him, and his sons, and all his people**. {2:34} And we **took all his cities at that time, and utterly destroyed the men, and the women, and the little ones, of every city, we left none to remain**: {2:35} Only the **cattle we took for a prey unto ourselves, and the spoil of the cities which we took.** {2:36} From Aroer, which [is] by the brink of the river of Arnon, and [from] the city that [is] by the river, even unto Gilead, **there was not one city too strong for us: the LORD our God delivered all unto us**:"

- "{3:4} And **we took all his cities at that time, there was not a city which we took not from them, threescore cities**, all the region of Argob, the kingdom of Og in Bashan. {3:6} And **we utterly destroyed them, as we did unto Sihon king of Heshbon, utterly destroying the men, women, and children, of every city**. {3:7} But **all the cattle, and the spoil of the cities, we took for a prey to ourselves."**

- "{13:6} If thy brother, the son of thy mother, or thy son, or thy daughter, or the wife of thy bosom, or thy friend, which [is] as thine own soul, entice thee secretly, saying, Let us go and serve other gods, which thou hast not known, thou, nor thy fathers; {13:7} [Namely,] of the gods of the people which [are] round about you, nigh unto thee, or far off from thee, from the [one] end of the earth even unto the [other] end of the earth; {13:8} **Thou shalt not consent unto him, nor hearken unto him; neither shall thine eye pity him, neither shalt thou spare, neither shalt thou conceal him: {13:9} But thou shalt surely kill him; thine hand shall be first upon him to put him to death, and afterwards the hand of all the people."**

- "{17:2} If there be found among you, within any of thy gates which the LORD thy God giveth thee, man or woman, that hath wrought wickedness in the sight of the LORD thy God, in transgressing his covenant, {17:3} And hath gone and served other gods, and worshipped them, either the sun, or moon, or any of the host of heaven, which I have not commanded; {17:4} And it be told thee, and thou hast heard [of it,] and enquired diligently, and, behold, [it be] true, [and] the thing certain, [that] such abomination is wrought in Israel: **{17:5} Then shalt thou bring forth that man or that woman, which have committed that wicked thing, unto thy gates, [even] that man or that woman, and shalt stone them with stones, till they die."**

- "{20:10} When thou comest nigh unto a city to fight against it, then proclaim peace unto it. {20:11} And it shall be, if it make thee answer of peace, and open unto thee, **then it shall be, [that] all the people [that is] found therein shall be tributaries unto thee, and**

they shall serve thee. {20:12} And if it will make no peace with thee, but will make war against thee, **then thou shalt besiege it: {20:13} And when the LORD thy God hath delivered it into thine hands, thou shalt smite every male thereof with the edge of the sword: {20:14} But the women, and the little ones, and the cattle, and all that is in the city, [even] all the spoil thereof, shalt thou take unto thyself; and thou shalt eat the spoil of thine enemies, which the LORD thy God hath given thee.** {20:15} Thus shalt thou do unto all the cities [which are] very far off from thee, which [are] not of the cities of these nations. **{20:16} But of the cities of these people, which the LORD thy God doth give thee [for] an inheritance, thou shalt save alive nothing that breatheth: {20:17} But thou shalt utterly destroy them; [namely,] the Hittites, and the Amorites, the Canaanites, and the Perizzites, the Hivites, and the Jebusites; as the LORD thy God hath commanded thee.**"

- "{21:10} When thou goest forth to war against thine enemies, **and the LORD thy God hath delivered them into thine hands, and thou hast taken them captive, {21:11} And seest among the captives a beautiful woman, and hast a desire unto her, that thou wouldest have her to thy wife; {21:12} Then thou shalt bring her home to thine house; and she shall shave her head, and pare her nails; {21:13} And she shall put the raiment of her captivity from off her, and shall remain in thine house, and bewail her father and her mother a full month: and after that thou shalt go in unto her, and be her husband, and she shall be thy wife. {21:14} And it shall be, if thou have no delight in her, then thou shalt let her go whither she will; but thou shalt not sell her at all for money, thou shalt not make merchandise of her, because thou hast humbled her.**"

- "{21:18} **If a man have a stubborn and rebellious son**, which will not obey the voice of his father, or the voice of his mother, and [that,] when they have chastened him, will not hearken unto them: {21:19} Then shall his father and his mother lay hold on him, and bring him out unto the elders of his city, and unto the gate of his place; {21:20} And they shall say unto the elders of his city, This our son [is] stubborn and rebellious, he will not obey our voice; [he is] a glutton, and a drunkard. **{21:21} And all the men of his city shall stone him with stones, that he die...**"

- “{22:13} If any man take a wife, and go in unto her, and hate her, {22:14} And give occasions of speech against her, and bring up an evil name upon her, and say, I took this woman, and when I came to her, I found her not a maid: **{22:15} Then shall the father of the damsel, and her mother, take and bring forth [the tokens of] the damsel’s virginity unto the elders of the city in the gate: {22:16} And the damsel’s father shall say unto the elders, I gave my daughter unto this man to wife, and he hateth her; {22:17} And, lo, he hath given occasions of speech [against her,] saying, I found not thy daughter a maid; and yet these [are the tokens of] my daughter’s virginity. And they shall spread the cloth before the elders of the city. {22:18} And the elders of that city shall take that man and chastise him; {22:19} And they shall amerce him in an hundred [shekels] of silver, and give [them] unto the father of the damsel, because he hath brought up an evil name upon a virgin of Israel: and she shall be his wife; he may not put her away all his days. {22:20} But if this thing be true, [and the tokens of] virginity be not found for the damsel: {22:21} Then they shall bring out the damsel to the door of her father’s house, and the men of her city shall stone her with stones that she die…**”

- “{22:28} If a man find a damsel [that is] a virgin, which is not betrothed, and lay hold on her, and lie with her, and they be found; {22:29} **Then the man that lay with her shall give unto the damsel’s father fifty [shekels] of silver, and she shall be his wife; because he hath humbled her, he may not put her away all his days.**”

- “{23:1} **He that is wounded in the stones, or hath his privy member cut off, shall not enter into the congregation of the LORD.**”

- “{23:2} **A bastard shall not enter into the congregation of the LORD; even to his tenth generation.**”

- “{25:11} When men strive together one with another, and the wife of the one draweth near for to deliver her husband out of the hand of him that smiteth him, and putteth forth her hand, and taketh him by the secrets: **{25:12} Then thou shalt cut off her hand**, thine eye shall not pity [her]…”

- "{28:58} **If thou wilt not observe to do all the words of this law that are written in this book, that thou mayest fear this glorious and fearful name, THE LORD THY GOD; {28:59} Then the LORD will make thy plagues wonderful, and the plagues of thy seed, [even] great plagues, and of long continuance, and sore sicknesses, and of long continuance. {28:60} Moreover he will bring upon thee all the diseases of Egypt, which thou wast afraid of; and they shall cleave unto thee. {28:61} Also every sickness, and every plague, which [is] not written in the book of this law, them will the LORD bring upon thee, until thou be destroyed. {28:62} And ye shall be left few in number, whereas ye were as the stars of heaven for multitude; because thou wouldest not obey the voice of the LORD thy God."**

Joshua

Divine and Divinely Ordered Genocide and Evil, Plus Invasions, Murder, Sexual Slavery

- **"{6:24} And they burnt the city with fire, and all that [was] therein: only the silver, and the gold, and the vessels of brass and of iron, they put into the treasury of the house of the LORD."**

- **"{8:24} And it came to pass, when Israel had made an end of slaying all the inhabitants of Ai in the field, in the wilderness wherein they chased them, and when they were all fallen on the edge of the sword, until they were consumed, that all the Israelites returned unto Ai, and smote it with the edge of the sword. {8:25} And [so] it was, [that] all that fell that day, both of men and women, [were] twelve thousand, [even] all the men of Ai. {8:26} For Joshua drew not his hand back, wherewith he stretched out the spear, until he had utterly destroyed all the inhabitants of Ai. {8:27} Only the cattle and the spoil of that city Israel took for a prey unto themselves, according unto the word of the LORD which he commanded Joshua."**

- **"{11:20} For it was of the LORD to harden their hearts, that they should come against Israel in battle, that he might destroy them utterly, [and] that they might have no favour, but that he might destroy them, as the LORD commanded Moses."**

Judges

Divine and Divinely Ordered Invasions and Genocide, Evil, Child Murder, Sexual Slavery

- "{1:2} And the LORD said, Judah shall go up: behold, I have delivered the land into his hand. {1:3} And Judah said unto Simeon his brother, Come up with me into my lot, that we may fight against the Canaanites; and I likewise will go with thee into thy lot. So Simeon went with him. **{1:4} And Judah went up; and the LORD delivered the Canaanites and the Perizzites into their hand: and they slew of them in Bezek ten thousand men.**"

- "{11:30} And Jephthah vowed a vow unto the LORD, and said, If thou shalt without fail deliver the children of Ammon into mine hands, {11:31} Then it shall be, that whatsoever cometh forth of the doors of my house to meet me, when I return in peace from the children of Ammon, shall surely be the LORD'S, and I will offer it up for a burnt offering. {11:34} And Jephthah came to Mizpeh unto his house, and, behold, his daughter came out to meet him with timbrels and with dances: and she [was his] only child; beside her he had neither son nor daughter. {11:35} And it came to pass, when he saw her, that he rent his clothes, and said, Alas, my daughter! thou hast brought me very low, and thou art one of them that trouble me: for I have opened my mouth unto the LORD, and I cannot go back.

  {11:36} And she said unto him, My father, [if] thou hast opened thy mouth unto the LORD, do to me according to that which hath proceeded out of thy mouth; forasmuch as the LORD hath taken vengeance for thee of thine enemies, [even] of the children of Ammon. {11:37} And she said unto her father, Let this thing be done for me: let me alone two months, that I may go up and down upon the mountains, and bewail my virginity, I and my fellows. {11:38} And he said, Go. And he sent her away [for] two months: and she went with her companions, and bewailed her virginity upon the mountains. **{11:39} And it came to pass at the end of two months, that she returned unto her father, who did with her [according] to his vow which he had vowed:"**

- "{19:22} [Now] as they were making their hearts merry, behold, the men of the city, certain sons of Belial, beset the house round about, [and] beat at the door, and spake to the master of the house, the old man, saying, Bring forth the man that came into thine house, that we may know him. {19:23} And the man, the master of the house, went out unto them, and said unto them, Nay, my brethren, [nay,] I pray you, do not [so] wickedly; seeing that this man is come into mine house, do not this folly. **{19:24} Behold, [here is] my daughter a maiden, and his concubine; them I will bring out now, and humble ye them, and do with them what seemeth good unto you: but unto this man do not so vile a thing. {19:25} But the men would not hearken to him: so the man took his concubine, and brought her forth unto them; and they knew her, and abused her all the night until the morning: and when the day began to spring, they let her go.**

{19:26} Then came the woman in the dawning of the day, and fell down at the door of the man's house where her lord [was,] till it was light. {19:27} And her lord rose up in the morning, and opened the doors of the house, and went out to go his way: and, behold, the woman his concubine was fallen down [at] the door of the house, and her hands [were] upon the threshold. {19:28} And he said unto her, Up, and let us be going. But none answered. Then the man took her [up] upon an ass, and the man rose up, and gat him unto his place. {19:29} And when he was come into his house, he took a knife, and laid hold on his concubine, and divided her, [together] with her bones, into twelve pieces, and sent her into all the coasts of Israel."

- "{20:35} And **the LORD smote Benjamin before Israel: and the children of Israel destroyed of the Benjamites that day twenty and five thousand and an hundred men:** all these drew the sword."

- "{6:19} **And he smote the men of Beth-shemesh, because they had looked into the ark of the LORD, even he smote of the people fifty thousand and threescore and ten men: and the people lamented, because the LORD had smitten [many] of the people with a great slaughter."**

- "{15:3} **Now go and smite Amalek, and utterly destroy all that they have, and spare them not; but slay both man and woman, infant and suckling, ox and sheep, camel and ass."**

Isaiah

World-Destroying Divine Evil, Rape and Gruesome Child Murder.

- "{13:9} **Behold, the day of the LORD cometh, cruel both with wrath and fierce anger, to lay the land desolate: and he shall destroy the sinners thereof out of it.**"

- "{13:11} And **I will punish the world for [their] evil, and the wicked for their iniquity; and I will cause the arrogancy of the proud to cease, and will lay low the haughtiness of the terrible.** {13:13} Therefore **I will shake the heavens, and the earth shall remove out of her place, in the wrath of the LORD of hosts, and in the day of his fierce anger.** {13:15} **Every one that is found shall be thrust through; and every one that is joined [unto them] shall fall by the sword.** {13:16} **Their children also shall be dashed to pieces before their eyes; their houses shall be spoiled, and their wives ravished.** {13:18} **[Their] bows also shall dash the young men to pieces; and they shall have no pity on the fruit of the womb; their eye shall not spare children.**"

- "{14:21} **Prepare slaughter for his children for the iniquity of their fathers;...**"

Ezekiel

Divine Insane Rage and Genocide.

- "{5:12} **A third part of thee shall die with the pestilence, and with famine shall they be consumed in the midst of thee: and a third part shall fall by the sword round about thee and I will scatter a third part into all the winds, and I will draw out a sword after them. {5:13} Thus shall mine anger be accomplished, and I will cause my fury to rest upon them, and I will be comforted: and they shall know that I the LORD have spoken [it] in my zeal, when I have accomplished my fury in them.**"

- "{6:12} **He that is far off shall die of the pestilence; and he that is near shall fall by the sword; and he that remaineth and is besieged shall die by the famine: thus will I accomplish my fury upon them.**"

- "{32:12} **By the swords of the mighty will I cause thy multitude to fall...and all the multitude thereof shall be destroyed. {32:15}**

When I shall make the land of Egypt desolate, and the country shall be destitute of that whereof it was full, when I shall smite all them that dwell therein, then shall they know that I [am] the LORD."

Jonah

Come On, Really? Can You Please Grow Up Now?

- "{1:17} Now the LORD had prepared a great fish to swallow up Jonah. **And Jonah was in the belly of the fish three days and three nights**."

Time For A Sanity Break!

OK, OK, I've heard the faithful screaming for attention, but I haven't called on you yet for a reason. I know what you want to say, so everybody repeat after me, "But we reject the Old Testament and Jesus made everything *all better* in the New Testament." I know you've been absolutely *aching* to say that.

I think we can all safely agree that the Old Testament is about 577 pages of unmitigated psychopathy and horror, and should be kept away from anyone under 21. Before we address the New Testament at all, though, I want every single person who doesn't accept every single word of the Old Testament, which believers say is either written or inspired by the creator of the entire universe, to tell the rest of us why he thinks he can tell the Old Testament's creator god to damn well kiss his ass and that he can accept or reject whatever he damn well pleases.

This god, which is self-admittedly jealous and genocidal and repetitively proven to be so, and which is also extremely happy to kill individual sinners whenever he feels like it, is the very last god you would *ever* want to piss off. I am quite sure telling him to kiss your ass would result in fire and brimstone both in this life and the next if he were actually real.

Of all the characters in the rest of fiction, this god is most like the psychopathic Nazi concentration camp commander from the movie *Schindler's List*. He is quite happy to order his minions to engage in cruelty and genocide then watch the show, to murder on a scale both large and small, and to inflict the tortures of Job on large numbers of people, but when he gets pissed off at a specific individual or just gets in a killing mood he is quite happy to kill individuals himself.[1]

No, you cannot just reject inconvenient and uncomfortable truths. Your holy book is not quite a trinity but it is a duality and like conjoined twins which cannot be separated without killing them both, must be dealt with as a total package. So you are permanently stuck with the Old Testament or you can reject the whole thing.

That said, how does the New Testament actually sound? Let's dive in and take a look at what is actually in the book, and what it means. Note that I am not relying on opinions or work on this subject by any self-professed biblical scholar or apologist, or any of the critics of religion, of which there are a great many on both sides. I am performing my own exegesis and applying the "reasonable person" approach in understanding the text. As a well-read person of middle age, fairly extensive experience and reasonable intelligence I unequivocally claim that right, and so should all reasonable people, despite the protestations of the obfuscating, hyperprotective and not at all objective clergy and the faithful masses.

Remember that this book is held up to us as the direct or inspired word of the alleged creator of the universe, with messaging directed specifically at our very own species of semi-evolved partly hairless primate. There is no preface or accompanying instructions with the book that says one needs advanced study to interpret the 'real meaning' of the words, or that one needs any formal clergy to do the interpreting for us.

If it really is supposed to be the divine word to us, we need to approach it accordingly, and I will. No one should fall for the lie that only careful interpretation of the text by trained scholars can give it meaning. There can be no bleating about "context" or "this is how they spoke back then" or any of that idiocy.

If it is a human book written by humans with zero supernatural input or interference then that excuse works. If it were divinely written or divinely inspired in any way whatsoever, the words would be transcendent and easily understood across all ages, places and modes of communication, to include potential translation errors. If divine origin or inspiration is ever alleged, then the god utterly failed to give his creations the ability to properly interpret the texts even after two thousand years of trying, since there are so many thousands of interpretations on hand, and almost that many flavors of bible.[2]

And if there were any divine worries about mistranslations then the god should have delivered unto each new generation His Own Book in the current language of the current people. How unfortunate it is that this never happens, despite the tonnage of killing that has happened over disagreements about this particular holy book. Every holy book must stand alone on its words and be subject to critical examination by anyone who picks it up. That special pleading about context is quite obvious pathetic escapism. It is a blatant attempt at covering up the truth and it will not be accepted.

Also, a book allegedly on morality should not contain over 800,000 words. Less than one page will do fine, even with double spacing and decent sized font.

So to the words of the New Testament. The following are taken directly from the King James version, as were the above extracts from the Old Testament. Note that since the four New Testament gospels are essentially the same story written four different times, although with enough very important variations to prove none of them can be trusted, I haven't included all the super-similar verses in order to avoid annoying repetition and have only drawn from Matthew. With that in mind, let's begin our reading.

Matthew

1. **What Jesus Said.** "{5:17} Think not that I am come to destroy the law, or the prophets: I am not come to destroy, but to fulfil. {5:18} For verily I say unto you, **Till heaven and earth pass, one jot or one tittle shall in no wise pass from the law**, till all be fulfilled."

   What It Means. Jesus is not at all or in any way changing Old Testament laws or what its characters said or did. He himself states his mission is not to remove any of the Old Testament, but to fulfill and reinforce it. Until the end of the earth, not one tiny bit of the Old Testament can be taken away. So all its cruelty, as written, is still 100% in place as ordered by Jehovah Junior himself, and we can't change any of it. And any nonsense from the faithful about how the New Testament has primacy and we need to just ignore the old one is both pathetic blather and also against Jesus's direct words.

2. **What Jesus Said.** "{5:27} Ye have heard that it was said by them of old time, Thou shalt not commit adultery: {5:28} But I say unto you, That whosoever looketh on a woman to lust after her hath committed adultery with her already in his heart."

   What It Means. The Old Testament said you shall not commit adultery, which means that married people should not have sex with people other than their spouses. But Jesus goes much farther than that, and says that anybody who looks at a woman lustfully has already committed adultery. Jesus thus demonstrated he is evil, in several ways.

- Even thinking about something is now a crime, even if it is never acted upon. This is evil and immoral. How is anyone's well-being in any way affected by someone else's inner thoughts if they never translate to word or deed? They aren't. This is the sort of sadomasochistic claptrap which keeps so many Christians in permanent fear and angst over their so-called "sins".

- Lust, desire and other strong and unconsciously generated emotions are part of human nature. If Jesus and/or his father created humans, this means they deliberately designed and programmed our brains to generate these thoughts, thus we were deliberately designed by the creator to do things for which we are to be punished. And if we were made in his image, he is punishing us for being like him. This is a divine "Do as I say, not as I do." We again have pervasive sadism and unparalleled masochism and sadomasochism in an unholy and quite immoral trinity.

- The punishment for adultery, from Leviticus 20:10 in the old law which Jesus said cannot be changed by even one jot or tittle, is death. So we should be killed if we commit adultery, and we should be killed even for feeling lust. This is immorality and evil beyond comprehension.

- And since sinners go to hell, being an adult male who has certain erotic thoughts about an adult female is literally a capital offense for which we should burn for all eternity. This is indeed the Old Testament sadist god we know so well, the one who is the epitome of immorality.

- This mandate condemns lesbians as well as men to death for lusting after women, since the word "whosoever" is used for anyone thinking the lustful thought. So human males and lesbians, all despite being allegedly made in the image of the god and also part of his most important and valued creation, were programmed to commit capital offenses. I'll say it again. This is psychopathic evil and sadomasochism beyond all comprehension.

3. **What Jesus Said.** "{5:29} And if thy right eye offend thee, pluck it out, and cast [it] from thee: for it is profitable for thee that one of thy members should perish, and not [that] thy whole body should be cast

into hell. {5:30} And if thy right hand offend thee, cut if off, and cast it from thee: for it is profitable for thee that one of thy members should perish, and not [that] thy whole body should be cast into hell."

What It Means. If you perform some inappropriate act, you should cut out or cut off the organ or bodypart you used to do the act. There is no mention of the brain, which is the control center for all thoughts and actions, only the tool that is used. Obviously to make this statement Jesus had no conception of brains, and he also wasn't using his. By this twisted illogic, if I used a baseball bat to kill someone, the right thing to do would be to burn the bat, not do anything to me.

4. **What Jesus Said.** "{5:31} It hath been said, Whosoever shall put away his wife, let him give her a writing of divorcement: {5:32} But I say unto you, That whosoever shall put away his wife, saving for the cause of fornication, causeth her to commit adultery: and whosoever shall marry her that is divorced committeth adultery."

   What It Means. Jesus thus says marrying a divorced woman is adultery. And of course as we have already heard, the penalty for this is death. So, once married, two people should remain together until one of them dies, regardless of how badly one of them behaves, lies, cheats, steals or commits adultery. This is not just stupid, this is evil and it has no actual moral foundation whatsoever.

5. **What Jesus Said.** "{5:38} Ye have heard that it hath been said, An eye for an eye, and a tooth for a tooth: {5:39} But I say unto you, That ye resist not evil: but whosoever shall smite thee on thy right cheek, turn to him the other also. {5:40} And if any man will sue thee at the law, and take away thy coat, let him have [thy] cloke also. {5:41} And whosoever shall compel thee to go a mile, go with him twain. {5:43} Ye have heard that it hath been said, Thou shalt love thy neighbour, and hate thine enemy. {5:44} But I say unto you, Love your enemies, bless them that curse you, do good to them that hate you, and pray for them which despitefully use you, and persecute you…"

   What It Means. This is a combination of immoral and idiotic. Jesus and/or his father created psychopaths and is/are very happy to have them living among us and committing unspeakable evil. He says that if one of these people harms us physically, emotionally or legally that we should happily expose our uninjured parts for more torture. This is horrible. By this logic, we should give thieves whatever money they couldn't find to steal, we should give serial killers our unmurdered family members to torture and kill, women should follow a rape by turning over and offering the unused orifice for further rape, and we should allow psychopaths like Hitler, Stalin and the medieval Catholic popes to take over peaceful nations and kill as many people as they want. This is pure evil, and an open invitation for the evil among us to do their worst without limit. So Jesus can bring not peace, but a sword, but we commoners on earth have to just damn well suffer the worst which the evil can perpetrate on us. That's the Old Testament shining through in all its evil insanity.

6. **What Jesus Said.** "{6:25} Therefore I say unto you, Take no thought for your life, what ye shall eat, or what ye shall drink; nor yet for your body, what ye shall put on. Is not the life more than meat, and the body than raiment? {6:31} Therefore take no thought, saying, What shall we eat? or, What shall we drink? or, Wherewithal shall we be clothed? {6:34} Take therefore no thought for the morrow: for the morrow shall take thought for the things of itself. Sufficient unto the day [is] the evil thereof."

   What It Means. Jesus says not to worry about tomorrow, about planning or meals or clothing or indeed anything at all to include life itself. What an incredibly stupid message to give to his disciples and the world. That works so well in the real world that there are over a billion people living in poverty and wide swaths of humanity live in abject misery. Thanks, Jesus!

7. **What Jesus Said.** "{10:34} Think not that I am come to send peace on earth: I came not to send peace, but a sword. {10:35} For I am come to set a man at variance against his father, and the daughter against her mother, and the daughter in law against her mother in law. {10:36} And a man's foes [shall be] they of his own household. {10:37} He that loveth father or mother more than me is not worthy of me: and he that loveth son or daughter more than me is not worthy of me."

   What It Means. The so-called Prince of Peace, the supposed Lord, Christ and the Word, the alleged one and only son of the one and only god, came to earth not like a peacemaker here to bring calm and comity, but like a demon here to sow war and conflict. This is in his own words; it is not some vile calumny, not some deliberate misquotation, or even remarks taken out of context. He specifically said his mission on earth was to rip families apart and to lure people away from their parents. He specifically said people have to love him more than their own families. This is clearly psychopathic and evil behavior, and only gutless and spineless sadomasochistic sheep could ever use such maniacal words as any kind of foundation for either morality or "family values".

8. **What Jesus Did.** He healed the sick, cast out devils and brought people back from the dead, but very selectively.

    - "{8:16} When the even was come, they brought unto him many that were possessed with devils: and he cast out the spirits with [his] word, and healed all that were sick:"

    - "{8:28} And when he was come to the other side into the country of the Gergesenes, there met him two possessed with devils, coming out of the tombs, exceeding fierce, so that no man might pass by that way. {8:30} And there was a good way off from them an herd of many swine feeding. {8:31} So the devils besought him, saying, If thou cast us out, suffer us to go away into the herd of swine. {8:32} And he said unto them, Go. And when they were come out, they went into the herd of swine: and, behold, the whole herd of swine ran violently down a steep place into the sea, and perished in the waters."

    - "{9:23} And when Jesus came into the ruler's house, and saw the minstrels and the people making a noise, {9:24} He said unto them, Give place: for the maid is not dead, but sleepeth. And they laughed him to scorn. {9:25} But when the people were put forth, he went in, and took her by the hand, and the maid arose."

    - "{9:32} As they went out, behold, they brought to him a dumb man possessed with a devil. {9:33} And when the devil was cast out, the dumb spake: and the multitudes marvelled, saying, It was never so seen in Israel."

- "{12:15} But when Jesus knew [it,] he withdrew himself from thence: and great multitudes followed him, and he healed them all…"

- "{12:22} Then was brought unto him one possessed with a devil, blind, and dumb: and he healed him, insomuch that the blind and dumb both spake and saw."

- "{15:29} And Jesus departed from thence, and came nigh unto the sea of Galilee; and went up into a mountain, and sat down there. {15:30} And great multitudes came unto him, having with them [those that were] lame, blind, dumb, maimed, and many others, and cast them down at Jesus' feet; and he healed them…"

- "{21:14} And the blind and the lame came to him in the temple; and he healed them."

- "{27:52} And the graves were opened; and many bodies of the saints which slept arose, {27:53} And came out of the graves after his resurrection, and went into the holy city, and appeared unto many."

What It Means. Jesus walked around a few places and supposedly healed what was an apparently very large number of sick people. What this means is that for all the generations prior to and following his appearance, the massive swarms of sick people just had to deal with their problems on their own because Jesus and his father didn't give a holy damn about them. Also, anyone who wasn't fortunate enough to be on the path of Jesus's traveling circus and thus could have personal access to Jesus and his alleged healing touch was just shit out of luck. So sad, too bad. Perhaps this was a divine lesson in the importance of "location, location, location" in human activities.

Apparently possession by devils was also commonplace, and if you got lucky enough to come across Jesus he could 'do you a solid' and cast them out of you. But again, any unlucky people not in his path just had to deal with their demonic possession and try to do the best they could with whatever anti-demon over-the-counter treatments their local snake oil salesmen had to offer.

And anybody living after Jesus's travels abruptly ended was just S.O.L as well, which is a big problem because the bishop of Rome in

2018, Pope Francis, is truly, really and substantially worried about the big bad Devil[3] and his alleged presence on earth[4], and exorcisms[5] are still apparently a badly needed service. I have myself attended a meeting wherein a Catholic priest waxed on about exorcisms, and he had a full crowd of sheeple in attendance. The medieval Dungeons and Dragons mythicism is still strong with the Vatican and its agents and followers. Fortunately our far more responsible and sane secular organizations like hospitals and insurance agencies won't cover such "services" as exorcisms.

And resurrections are on tap as well, if you happen to be the daughter of a wealthy man, a guy who catches Jesus's eye, or a long-dead "saint". Which brings us to one of the most important points in all religion and particularly Christianity. It's extremely interesting and extremely indicative of how fictional they are that the gospels and all the other ancient writings we have don't contain massively discussed and thoroughly documented accounts by anyone who was allegedly resurrected. Surely these would be among the most interesting and informative experiences ever recounted in the entire history of our planet, and the returnees would spend most of the rest of their lives telling everyone about the world after this one, about the afterlife, and about heaven or hell and all the dead people they met. Claims could be proven and predictions could actually be fulfilled.

But, alas, this never happens. What an unbelievably convenient shame, one might cynically say. What a crock of crap, the empiricist will say, as he notes that this is all the proof he needs that all these religions and all their claims are pure bogus. Rather than using their current laughably pathetic justification for their belief in these resurrection myths, that being pointing to a few words in a few places in their ancient holy book, apologists could hold up multiply-attested and thoroughly documented first-hand accounts with heaps of corroborating and later-proven claims as 'evidence' for the truth of the resurrection stories. Alas, this was sadly not to be. As with everything else in these stories and the entire book, the total lack of evidence means the claims will be rejected until they are empirically proven.

9. **What Jesus Did.** He gave nutcases and eschatolophiles far too much 'end of the world' porn and insane evangelizing. "{24:29} Immediately after the tribulation of those days shall the sun be

darkened, and the moon shall not give her light, and the stars shall fall from heaven, and the powers of the heavens shall be shaken: {24:31} And he shall send his angels with a great sound of a trumpet, and they shall gather together his elect from the four winds, from one end of heaven to the other. {24:33} So likewise ye, when ye shall see all these things, know that it is near, [even] at the doors. {24:34} Verily I say unto you, **This generation shall not pass, till all these things be fulfilled.** {24:36} But of that day and hour knoweth no [man,] no, not the angels of heaven, but my Father only. {24:40} Then shall two be in the field; the one shall be taken, and the other left. {24:41} Two [women shall be] grinding at the mill; the one shall be taken, and the other left."

What It Means. This is obviously a parent-to-child admonition to "be good while I'm gone, or else!" with not a mere spanking as punishment but the actual end of the world and casting into hell as the god's own Final Solution. Whoever wrote the original story or the copies, those being the other gospels, obviously intended it to scare the daylights out of the faithful, and to make them toe the party line forever. This is all such a ridiculously contrived and ill-informed story that the Jesus character is more like the grade-school dropout, dirty-robed, unshaven and unbathed asylum escapee street preachers holding up End Times signs you see in cities in the West than any loving father figure.

The moon doesn't give light, it reflects light. Stars can't fall from heaven, because they are unimaginably far away from us, are several orders of magnitude larger than our Earth, and of course are essentially enormous balls of superhot fire. Thus this paragraph means none of the ancients or their allegedly omniscient gods had any idea about the true nature of stars. Some of the stars in the night sky are actually entire galaxies, the existence of which Jesus and all his devotees were also blissfully unaware. Jesus would surely have known that evangelists would pass these words to their children, and thus continue the progression of abject ignorance for many generations.

This story means that fairy tales told by unbelievably ignorant people are being used by people today to justify telling other people what to do and what public policy to ignore, from environmental policy to international relations. This is an appeal to the worst and lowest in people, and if Jesus had any of the powers or knowledge attributed to him, he would have known the effects of these words and not spoken them. Or maybe he would, because if he is indeed a part of the tripartite package deal god which Catholics worship, he is himself the god of the Old Testament and no evil is too evil.

And whatever prevented all these predicted things from happening two thousand years ago? Jesus himself supposedly said all this universe-destroying cataclysmic foofaraw was supposed to happen before the generation of people alive at that time died. So we are now two thousand years overdue for those alleged end times, and in all that time the Christian faithful haven't come up with anything other than a pathetic shrug of the shoulders for why the stars have stubbornly refused to kamikaze themselves to our planet like cosmic flames to a moth and the universe has stubbornly refused to end. This means Jesus and/or the people who wrote down these primitive science fiction stories had absolutely no idea what they were talking about and they should be treated as the very primitive manmade local mythology that they are.

10. **What Jesus Said.** "{26:11} For ye have the poor always with you; but me ye have not always."

    What It Means. Jesus was a needy little bugger, and he didn't really care too much about the poor when he wanted attention. He basically echoed that self-centered cry for attention you hear from people today, when they whine "What about me? What about *my* needs?" He wanted attention from a woman bearing expensive ointment, and maybe she was going to give him a nice little head rub too, but the disciples rightly asked why they should waste this on somebody who's going to get crucified pretty soon anyway? Now that Jesus was going away, somebody was going to have to be a grownup and worry about the morrow.

    In a devastatingly obvious admission that he really doesn't give two shits about the poor, he blithely notes how the poor will always be among us, but doesn't try to do a damn thing about their plight. He doesn't leave a source of manna, a set of magically and permanently reproducing loaves and fishes, or a miraculously regenerating supply of Lourdes-like magical healing ointment which his disciples could rub on the sick. He utters not a word about a recommended economic system designed to maximize the well-being of everyone, offers not a jot or tittle of guidance about capitalism or socialism or communism or any blend of economics and politics. He's perfectly content with having masses of the poor and the sick and the allegedly demonically possessed living out their miserable lives in the world he and/or his father are said to have created, although he or his father could permanently correct these issues in a picosecond. He also quite clearly says, from the perspective of someone who thinks he has insight to future events, that the poor will always be everpresent. In this single sentence, he mentions then dismisses the poor without a care. Such callousness, such immoral evil. When one can do something about evil, but does not, that is evil. When one can do something about evil for all times and in all places but does not, that is a level of evil that can never be surpassed.

And Jesus wasn't the only one spreading the evil. Saul of Tarsus, now called Saint Paul by many, is often seen as the second most important person in Christianity after Jesus himself. He was born around 5 CE into a religious Jewish family, thus receiving two strikes at birth, one each in both nature and nurture. He inherited a genetic predisposition for piety and also almost assuredly endured much indoctrination into the faith of his family and society. He was also described as a very small and unattractive individual; being reddish in color, bald, less than five feet tall and a bit twisted or hunched. All this undoubtedly resulted in an extreme paucity of interest in him by the fair sex, which explains quite a lot about him and his highly misogynistic writings.

The story goes that Paul was a 'hater' toward the movement which he later joined with such vigor. His attitude allegedly changed one fateful day when, according to the book of Acts:

> *"...he came near Damascus: and suddenly there shined round about him a light from heaven: {9:4} And he fell to the earth, and heard a voice saying unto him, Saul, Saul, why persecutest thou me? {9:5} And he said, Who art thou, Lord? And the Lord said, I am Jesus whom thou persecutest: [it is] hard for thee to kick against the pricks. {9:6} And he trembling and astonished said, Lord, what wilt thou have me to do? And the Lord [said] unto him, Arise, and go into the city, and it shall be told thee what thou must do."*

This event has been reasonably described by modern medical science as a temporal lobe seizure.[6] Medical technology in Paul's time didn't even include washing one's hands. Any unusual visions, dreams or states of mind were attributed to possession by demons or divine visitation. This is understandable, because no one knew anything of the brain and how it works, or more importantly how it doesn't. Seeing and hearing voices or experiencing hallucinations, which today are well-known side effects of a large number of conditions listed earlier in Chapter Twelve, are still interpreted by people in the more enlightened 21st century as visitations from the divine. How could anyone expect anything else from ignorant Iron Agers already stupefied by a culture of religious piety and knowing hardly anything else?

When people believed the only solution to anything unknown or unusual was supernaturally caused, visions and voices were sure to be seen as strong evidence of divine intervention and presence. Far from being a visitation from the mysterious beyond, it is extremely likely that this experience was either a flat-out lie or was a temporary aberration in his brain's normal functioning.

So a bolt not from the blue, but from within his own head, completes the trifecta. With this third strike, Paul is out. Born into a religious family, likely born with a genetic affinity for religious belief, and suffering a hallucinogenic event that he interpreted as a visit from heaven, it is completely understandable that he would have had a conversion experience.

And what did he do with his new-found faith and devotion? It seems he took his hero Jesus's *No Women Allowed Club* mentality rather strongly to heart. Let's see what he had to say, in these extracts from the King James version of the New Testament.

<u>The First Epistle of Paul the Apostle to the Corinthians</u>

<u>Women Are Icky, and Need to Know Their Place.</u>

- "{7:1} Now concerning the things whereof ye wrote unto me: [It is] good for a man not to touch a woman."

- "{11:3} But I would have you know, that the head of every man is Christ; and the head of the woman [is] the man; and the head of Christ [is] God."

- "{11:7} For a man indeed ought not to cover [his] head, forasmuch as he is the image and glory of God: but the woman is the glory of the man."

- "{11:8} For the man is not of the woman; but the woman of the man. {11:9} Neither was the man created for the woman; but the woman for the man."

The Epistle of Paul the Apostle to the Ephesians

Wives Are Subordinate To Husbands, Period.

- "{5:22} Wives, submit yourselves unto your own husbands, as unto the Lord. {5:23} For the husband is the head of the wife, even as Christ is the head of the church: and he is the saviour of the body. {5:24} Therefore as the church is subject unto Christ, so [let] the wives [be] to their own husbands in every thing."

The First Epistle of Paul the Apostle to Timothy

Women Should Keep Their Stupid and Guilty Mouths Shut, And Be Subordinate to Men, Period.

- "{2:11} Let the woman learn in silence with all subjection. {2:12} But I suffer not a woman to teach, nor to usurp authority over the man, but to be in silence. {2:13} For Adam was first formed, then Eve. {2:14} And Adam was not deceived, but the woman being deceived was in the transgression."

So the top two figures in Christianity make it plain that they are immoral and that they teach immorality. Anyone who teaches what they said and taught is also teaching immorality. This is obvious even to an unintelligent person using only the most rudimentary reasoning skills.

We can use their own words in yet another way to demonstrate the immorality of the Old and New Testaments. People who are truly moral and good tend to say things that are good, and to use good, kind and moral words. People who are immoral and evil tend to say things that are immoral and focused on evil and immoral subjects. We can therefore do a simple word search in the Christian holy book to determine the frequency of certain words and use them as a clear indicator of the immorality of the book.

Figure Eleven shows the frequency of certain words and quite clearly shows what the book, its god, its authors and its followers really care about. It also shows the antihuman immorality of these perverse ancient myths.

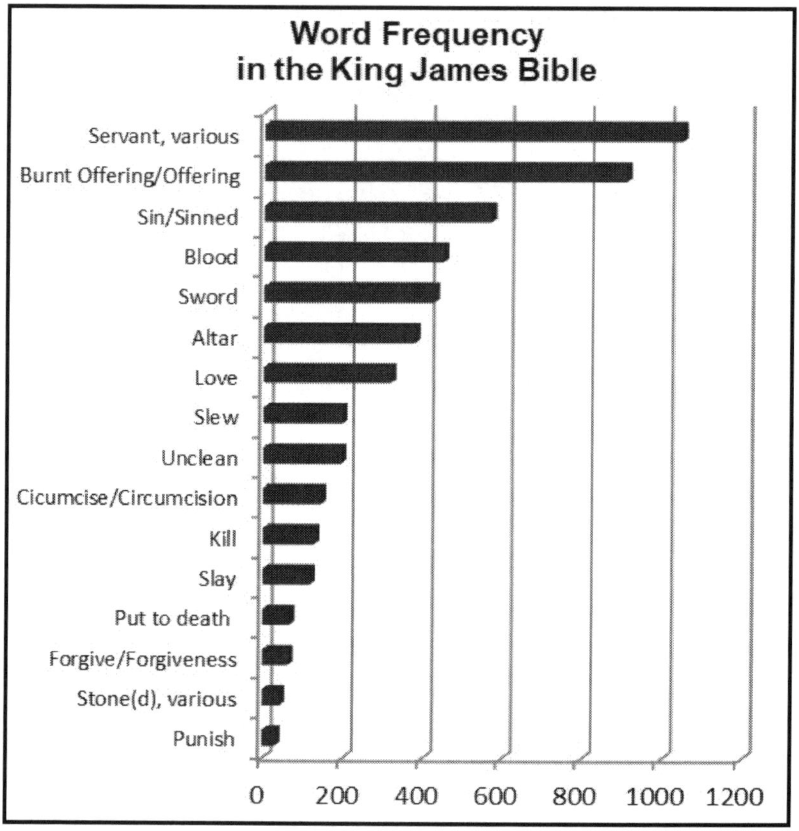

Figure 11. Frequency of selected words in the King James bible, both Old and New Testaments.

Rather eye opening, isn't it? What does this word frequency show us? It shows us the actual priorities of this allegedly holy book. It shows what is actually important to the god of this book and its adherents. Far from being the loving and caring façade the faithful hold up to the rest of us to try to stop us from seeing the evil truth, this book is a compilation of hate and evil, written and/or inspired by pure hate and evil. We can look at a few of the most commonly used key words to show this.

1. **Servant, with 1046 instances.** This includes the variations maidservant, manservant and bondservant. I won't even go into the deeply untruthful and diversionary apologetics Christians spout about these practices supposedly not being slavery as we know it, I'll just take the words as they are listed and note that of these key words it is the most frequently used, thereby showing that having servants is extremely common and extremely

acceptable in this book. Remember the book doesn't talk about hiring people to be servants, it talks about ***buying and selling*** people. Even the dimmest of bulbs will understand there is a difference between hiring a person and buying a person. And one need always remember this book neither condemns or prohibits slavery and it was used for centuries to justify slavery of the most inhuman and cruel kind.

2. **Burnt Offering / Offering, with 907 instances.** The god character in this book is Stalinesque or Trumpesque[7] in its contemptible need for continuous supplication and groveling from its gutless worshipers. From a range of offerings we note the well-known phrase "burnt offering" for sacrificing a living being to the god then burning it. Nothing more need be said. People used to and in some places still do actually sacrifice animals to propitiate their imaginary gods. It is hard to come up with a more idiotic and useless practice than this other than prayer itself. Not to mention the fact that when one makes a habit of sacrificing animals, the step to sacrificing humans is much shorter and easier.

3. **Sin Or Sinned, with 568 mentions.** This is wholly unsurprising, as this hateful book seems to care about little else than sin in all its imaginary flavors, and of course how to thoroughly punish it.

4. **Slay, Slew, Kill, Put to Death and Stone, all being synonyms for killing, mentioned an evil 549 times.** In this horrible collection of snuff horror fiction, of course the characters have to respond to acts of so-called sin by killing the sinner, and the numerous commands to do so and the bountiful heaps of rewards ones can receive from killing others come from the god himself and his "made men" on earth. What happened to 'thou shalt not kill'? This book sounds like the inspiration for the insane bipolarity of Dr. Jekyll and Mr. Hyde.

5. **Blood, with a whopping 448 mentions.** One must spill a lot of blood in this book to get the god's favor and earn rewards from him. This is well-known from the Old Testament, but even in the allegedly much nicer New Testament, in Hebrews 9:22, the faithful are told that "And almost all things are by the law purged with blood; and without shedding of blood is no remission." So

all the killing has a purpose after all, which is to gain the god's favor and also to enable the remission of sin and evil. The god could have made something far less evil, immoral and bloody as the route to remission of sin, but this god is a god of blood and suffering and punishment, so here we are. This mindless evil is still widely and fervently preached today and I have myself personally heard it proclaimed loudly and clearly by sitting preachers. Yes, Kansas, I am looking at you but we all know you have far too much company.

6. **Sword, mentioned 425 times.** The god's children can't draw too much blood or slay too many innocent men, women and children without swords. Therefore the Iron Age equivalent of the gun gets an appropriately high number of mentions. Were this book to be written in our times, the authors would have not merely edged weapons to choose from but also powder weapons, advanced explosives and a quite effective range of nuclear, chemical and biological weapons. The faithful could indeed inflict the tortures of Job on millions of their fellow humans under the approvingly sadistic gaze of their lord and savior with the full blessing of the National Rifle Association, evangelical Christians and Republicans.

7. **Altar, mentioned 377 times.** The ancients worshiped their gods at least as fervently as today's far right gun fetishists worship their weapons, and the altar was their hotline to the god. Used for burnt offerings and other purposes, the altar served as this connection to the god and was referenced an appropriately large number of times.

8. **Love.** But what about love? Christians, Jews and other uninformed folk go on and on about how the message of the bible is love. That is, to be blunt, bullshit. The word "love" appears 317 times, which is better than one might expect given the evil immorality of the book, but this raw total comes with many caveats and is very misleading. The word is used seven times in the preface. It is often used in an immoral or evil context, such as in Exodus 20:6 when the god says "And shewing mercy unto thousands of them that love me, and keep my commandments." It is used many more times when the faithful are ordered to love their god, and when the people's love for their god will be rewarded by his killing and displacing other

people so the children of Israel can occupy someone else's lands. It is still only used about half as much as the word "blood", and far less than "sin" and even less than "altar". Far from the Christian claim that their bible is all about love, it is actually all about slavery, offerings, sin, blood, swords and altars, just as one might expect if one actually reads the book.

One may further ask that since Christians proclaim themselves and their religion to be all about forgiveness, how many times that word would be mentioned. Surely it would have to be atop or nearly atop the list of mentions, wouldn't it? But no, again this is of course not the case, and anyone who knows anything about Christianity and its holy book knows better. The words forgive and forgiveness, between them, only warrant a paltry **63** mentions in the entire biblical disaster.

Circumcision is mentioned 144 times, though, indicating that cutting bodyparts off of baby boys allegedly made in the image of the male creator of the universe is almost exactly twice as important as forgiveness is in this faith and its book. No sane person can read the book and come away thinking the message is love or indeed any positive emotion or condition at all.

It is plain to see that if one wants to teach full-blown and deep-seated hate and immorality, one need look no further for a detailed compilation and instructional manual on how to inflict it than the Christian holy book. The book is indeed a primer in immorality, but the faithful continue to lie about its hateful contents, thus deliberately furthering the immorality.

There is no escape or evasion from the devastating truth. The book which billions of people claim to use as a moral guide for their lives is an instructional manual on immorality. This is a social and moral emergency and every reasonable, truly moral and thinking person must do everything possible to show that this emperor has no clothes and this god and his book have no shame or morality. It is time for decent people to take a permanent and vigorous stand against this shameless immorality.

In the spirit of circumcision, it is clear that what our world needs right now is to slice away and discard the utterly ignorant and immoral practices and beliefs which far too many people share. It is time to cut off the ancient clawed hands which still have a death grip on billions of people. It is time to throw the entire immoral body of religion into the pits of perdition from whence it came.

And it needs to start today.

Chapter Eighteen

Addressing Apologist Appeals to Hitler and Stalin, Plus A Condemnation of Hitler's Pope

"I am now as before a Catholic and will always remain so."
- Adolf Hitler

At this point I must, like the probably apocryphal tale of Martin Luther's swinging of hammer onto nail at the Wittenberg church, place a marker and take a stand against some of the more foolish and fallacious apologetics which Christians and other monotheists trot out whenever someone points out the massive amount of evil in their religions.

It is practically now a law of nature that whenever a nonbeliever rightly points out the massive evil and immorality in religions that have always and even unto this day still afflict humanity, many of the faithful frantically wave their arms and loudly bleat "What about Hitler?", "What about Stalin?", "What about Hirohito and Pol Pot?" These believers are almost always Christians, and even more often Catholics, due to their quite understandably high levels of shame and defensiveness.

They attempt to assert that these supposedly but not actually secular individuals and their acts show that secular people are "just as evil" as the religious. This refrain includes elements of the automatic defensiveness, incoherence, fragility and hypocrisy of the religious mindset and also of the criminal caught red-handed trying to explain away what he was just witnessed doing. It also shows the lack of analytical depth and rigor in the religious mind. These fatuous attempts at obfuscation have been trotted out and shot down on countless occasions, but this chapter will sum the facts and dismiss these pathetic attempts at diversion.

The first, most obvious and immediately fatal flaw in this attempt at distraction is that if a religion ever tries to use the "I know we did but so did they!" defense, and to find sufficient evil for an equal comparison one had to choose from some of the worst criminals in history, the argument is already over and the religion and the apologist have catastrophically crashed and burned.

What makes their already fatally flawed logic even worse is that many religious adherents and leaders go on ceaselessly about their utterly baseless certainty that only their particular group is the most moral bunch of people around, and that they alone know the Truth. Yet even this outrageous tribalism and self-aggrandizement is not enough. Many of these deeply faithful individuals have to go even further and insist that if the rest of us don't follow their tribal scriptures and rules, we are bad and immoral people and we are probably going to spend eternity roasting in their god's personal fireplace for our alleged sins. The apologist protocols are, in order:

1. Attempt to completely ignore the evil history of their religion and its most senior leaders.

2. If forced by an educated and informed interlocutor to admit that history, they immediately deploy whitewashing and attempts at downplaying the history, or throw up a defense of moral relativism and "context". As in, slavery and selling one's daughter and genocide are universally understood to be pretty bad today, but it wasn't so bad "back then", even if "back then" was last week or up until the last papal, elder, grand mufti or other decree.

3. Finally, if forced to completely admit the history and also to give up the moral relativism charade, it is maintained that while all those acts in history "may have been" bad, things are *much different* now. Meaning that mistakes were made, but they were all completely well-intentioned, made by other people, those are all in the past, everything will be roses and candy from this point forward, and can you please stop asking these questions? This last is of course where terminal denial of reality is employed as a final protective line from the assault of facts and reason.

That last attempt, at saying they promise to be good from now on, is a pernicious lie and they know it. The Catholic cult in particular has had almost two thousand years of being extremely bad, under a succession of allegedly infallible popes, and they're *still* doing things like raping little kids and covering it up. All of these obfuscations and non-apologies stand astride fossilized books and mythology, also known as the core of the faiths, which do not and will not change. The root of the problem is still there, because the orders to inflict evil and suffering are written on the pages of the holy books.

Despite these mandates being only very recently mitigated or blocked by secularly-derived progress in civilization and morality, noting that this is still not true in many Islamic countries and others which are under strongly Christian rule, those original deadly words in those original deadly books remain as poisonous and evil as ever. They never change, they never die, and they never lose their toxicity. They are like magma beneath the surface, boiling hot and pressing upward, awaiting the next moment in history when they can burst forth and burn the land and unbelievers with fiery righteous punishment yet again.

It must be noted that from a human and social perspective, these attempts to deflect and apologize for the evils of the faith demonstrate not only the inability to properly answer for the evils of religion, but also the lack of desire to do so. To truly and fully reject the evils of the past means to reject the core teachings of the faiths. Religious believers cannot be trusted until they completely give up their holy books, forget they ever existed and transition to a god-free secular humanism. And this is something they are neither willing nor able to do, so we have a major global problem on our hands.

True maturity and truly taking responsibility means accepting the facts and the truth, offering sincere apologies for crimes committed and pain inflicted, doing the appropriate penance and actually ceasing the immoral or criminal activity. This penance should be honest and true, not a sham or for show, or to appease the angry and the injured until the storm passes. True penance for truly heinous acts may require responses greater than a minor policy change, a verbal apology or a financial payoff. If the crime is heinous enough, and if the roots of the crime go deep enough, the best solution is for the offending group to be disbanded for the safety and protection of other humans, as was done with the Nazis.

Yes, that last sentence was primarily directed at Catholicism first and foremost, as that church's thoroughly antihuman and psychopathic history is truly their Devil's masterpiece. But this is where religions are not capable of true penance, because their evil foundational texts never change, and to be a faithful member of the clan you have to be true to the book. So words of apology may be offered from time to time but no actual change can be possible as long as the books remain.

The unbelievable and permanent hypocrisy of religious apologists is thus on full display. They claim that certain very evil individual persons were secularists and that they committed crimes which were as bad as those of the faithful, and think this is somehow points in their favor.

This is not maturity, this is not taking responsibility and it is not doing penance. How such blatant hypocrisy can even be countenanced is beyond a rational mind's ability to comprehend. Attempts by the faithful to change the subject and to divert attention from their history of evil and the most horrible crimes only reinforce religion's incompatibility with decent and truly moral humanity and compound its guilt.

Religions in general have as a core tenet that their particular system is the one true faith, and that it alone should be held as the source of true morality. The Mesopotamian monotheisms in particular have taken this approach to stratospheric levels directly in the face of towering mountains of evidence that they are the ones which have most contributed to immorality on a global scale. Attempts at diversion are a frank admission that the faithful cannot directly and honestly answer the criticisms of the religions and their followers.

And what other criminals do has no bearing on what any particular criminal did. A criminal doesn't get credit during trial because he can point to someone else who did evil things. When a defendant is accused of a crime and asked to explain the crime, he doesn't get points for saying what anyone else did. He has to explain his actions and only his actions, and has to suffer the consequences of his own bad deeds. The truth is plain and simple.

The evil history of religion is a proven fact, this evil is continuing, and no amount of attempts to distract from this reality will work. Religions, primarily Christianity and Islam, have a body count that defies understanding, and they act on aggressive and antihuman tendencies that are still just as dangerous today as they have been for far too many centuries.

And the utter gall to try such disingenuous tactics is disgusting and offensive beyond belief. How dare the religious try to change the subject from their centuries of murder and oppression? How dare they try to equate the nearly 700 years of the Inquisition, the 200 years[1] of Crusades[2] conducted by Christianity toward Islam, seemingly endless religious wars in Europe, the almost 1400 years of conflict between Sunni and Shiite[3], and the nearly 2000 years of Christian persecution of Jews[4] with a handful of unquestionably horrible years in the 20th century?

And this is not even mentioning the other religious wars and conflicts all across the entire globe over the last century where the devout continue to kill and persecute those both outside and inside their faith. This is not even mentioning the Vatican's massive contribution to and disgusting passivity throughout the years of the Final Solution.

The faithful drone on endlessly about their alleged morality and the advertised goodness of their faith, while completely ignoring the centuries of death and persecution they have perpetrated and still continue to engender or even mandate[5], even unto this day.

If these faiths and their books were actually moral and true, and they led to more moral and decent behavior, we secularists would be able to look around at both history and the world today and we would be able to find when and where religions and the religious actually behaved in a moral and decent manner. We would be able to see if religion made the world a better place.

This search has been done literally millions of times, and the results are always the same. Religion never makes people's lives better, as seen in Chapter Fifteen. The real truth is that evil is always evil, and religion has far too often been extremely evil and will assuredly continue to be evil.

The apologetic faithful are actually correct in one inadvertent and backhanded sense, in that they and their faiths are indeed no better than the genocidal murderers of the twentieth century. No one would ever dispute such an obvious truth. So why do these people continue to try to convince us otherwise when we know the truth? It is for the same reasons that they profess their deep faith while not realizing their actual atheistic selves, as described in Chapter Eleven. They see a different reality than rational people do. They see things through a tiny tunnel, like a burqa with only one tiny eyehole in it rather than a narrow slit.

This hyperconstraint on their worldview means they are unable to truly see and understand things from an outside and indeed a reasonable perspective, and this pernicious and extremely dangerous psychological rigidity is becoming much more understood by the scientific community.

Psychology and science are extremely important matters, as they are the only real keys to understanding human thought and behavior and improving our world. Therefore we as a human species must collectively insist that the faithful answer questions which they never truly consider from an outside or indeed a species-wide or planet-wide perspective. These are inquiries of the most extreme importance, and the answers absolutely must be incorporated into our local and global cultural, educational and political systems if we ever wish to become truly moral.

I encourage every empiricist and freethinker to ask religious people these questions, and for the religious to ask them of themselves and each other. Honest answers to the questions below are the only path to true enlightenment.

- Is it good for the world to indoctrinate children into a system where they are told that only their specific tribe has a hotline to the creator of the universe and his guidance, and that every other tribe in the world is wrong and their gods are made-up nonsense and probably evil too?

- Is it good for the world to tell these children that their tribe knows exactly what this creator wants them to do because it was allegedly beamed into the minds of a few ignorant eccentrics, misfits and misogynists thousands of years ago?

- Is it good for the world to tell the children of Christians, for example, that their holy book, the foundation of their beliefs, is not what it actually is, which is a man-made collection of discrepant desert fairy tales deliberately cherry-picked[6] from an even larger collection of discrepant fairy tales, then placed out of chronological order into a contrived mass, then passed off as the true and actual word and intent of their god and in many cases is used as literal orders?

- Is it good for the world to teach children of monotheists that this same god insists throughout their holy books that people within the faith who don't believe in or follow the books, or who do but who transgress any of hundreds of petty rules should be stoned, stabbed or burned to death, including children?

- Is it good for the world to instill in children that blind and unquestioned obedience to the words of their holy books and their elders is a mandatory requirement and an act to be highly praised?

- Is it good for the world to frighten and mentally torture children by telling them that they, their friends, their parents and anyone on earth who sins, or even thinks of sinning or is of a different religion, will be cast into a pit of hellfire forever without reprieve?

- Is it good to tell these children, in a thoroughly horrible denial of reality, that this faith is the only true source of morals and goodness on offer? And that even when people outside the faith disobey this religion and its god that they should also be murdered via stoning, the edge of a sword, strangling, burning at the stake, hanging or whatever the preferred method of faith-based murder is at the time?

- Is it good for the world to deliberately teach cognitive dissonance and a selective hierarchy of so-called "absolute" morality that is not universally applied? That is, that the god of the holy books can be immoral and evil beyond comprehension and be worshiped for it, but humans must rigidly obey a ridiculous and endless parade of his contradictory commands or face pain and death? And that we must praise the evil and immorality of this god and his representatives on earth as is thoroughly recorded in these holy books, but anyone who actually acts like that in civil society today is to be seen as terrifyingly evil and deserving of shunning, imprisonment or execution? In other words, is it good to deliberately instill double standards of morality, the Stockholm Syndrome and a sycophantic worship of and apology for the mad god king?

Religious indoctrination such as this is absolutely not bringing up our children to be good, moral and ethical people. It is deliberately inculcating fear, psychopathy and bigotry. Why don't these people ever consider the psychological damage they do to children? And if they do consider it, why do they keep doing it? The answer is as sinister as it is simple. Because in their minds the ends justify the means, regardless of how damaging the "means" or the ends are, and they cannot or will not see this.

All that matters is devotion, adherence to the faith and staying loyal to the tribe. In this pursuit they feel it is therefore more than acceptable collateral damage to permanently injure a child's psyche as long as it means he or she will remain true to the faith, or be "saved". This is intentional cruelty, and as many people have said for many years, and extremely eloquently by Richard Dawkins in his book *The God Delusion*, it is deliberate child abuse.

This is an extremely important point, and it addresses the nexus at which the questions of belief vs nonbelief irrevocably intersect and are physically manifested in the real world. Damaged child psyches become damaged adult psyches, and these damaged adults cause further damage to their own children and also to other people with whom they come into contact. This is what the psychological community says[7] about psychopathy:

"The terms "psychopath" and "sociopath" are often used interchangeably, but in correct parlance a "sociopath" refers to a person with antisocial tendencies that are ascribed to social or environmental factors, whereas psychopathic traits are more innate,

> *though a chaotic or violent upbringing may tip the scales for those already predisposed to behave psychopathically."*

Note that last bit, about chaotic or violent upbringing. The world is challenging enough without brainwashing children with the chaos and violence of the Mesopotamian holy books and their chaotic, violent, immoral and contradictory teachings, and the tribal hatred and execration of those outside the tribe. And a religious upbringing often includes the physical abuse of children as violent punishment, the virulent condemnation of sinners and the utter nonsense and illogic of these thuggish monotheisms.

Children ask questions and they seek honest answers, which is why the best and most unanswerable questions in bible study classes come from children whose critical thinking skills haven't been beaten or threatened out of them. They are trying to use these skills to understand the discord and conflict between their innate sense of right and wrong, the guidance on actual morality at least someone among their parents and society is hopefully teaching them, and the lessons and words of their holy book and religion. Below is a typical fictional bible-based dialogue on this subject between Johnny, a curious young critical thinker, and his Sunday school biblical propagandist:

> Johnny had been listening in silent confusion for months, and try as he might he just could not understand or reconcile anything he had been hearing from the bible teacher with what he was hearing outside the church and in the secular community. He decided to ask the teacher the questions he'd been wondering about because he was constantly running into logical and moral contradictions, unanswered questions and evasions. It was all extremely confusing and was causing him great concern. An opening came at the halfway point of the bible study session, after the instructor had completed a solemn retelling of the Noah's mass murder / mass abortion flood tale.
>
> Johnny asked "Why did god drown everyone on earth?"
>
> "Oh, because they were very evil", the instructor said.
>
> This was not a comforting answer from Johnny's perspective. He followed up with the next logical question, asking "OK, so is there evil in the world today?"

The instructor's face lit up with pure sadomasochistic evangelical delight, and he said "Oh, yes, lots of evil! Just look around at everything that is going on, so many people are evil that sometimes it's almost just too much to believe. We have school shootings and crime and terrorism and crooked politicians and special interest groups are trying to stop us from putting nativity scenes in federal buildings and public schools and on publicly funded grounds, and don't even get me started on gays and abortions! And you know, that's why you are here with us at these lessons, to learn the true lessons from us because everyone not of our church and everyone in the secular world and those of other religions are especially evil. So yeah, there's a lot of evil in the world."

Johnny, being a nascent critical thinker and easily intelligent enough to keep thinking of the next appropriate logical question while dealing with a series of evasions or non-answers, did so. He asked "If there's so much evil today, will god drown everyone on earth again?"

The instructor laughed and said "Oh, no, of course not, he felt so bad about that and said he wouldn't do it again, plus don't forget the Good News! He sent his only begotten son to be persecuted, tortured and murdered in untold agony as penance for all our sins!"

This standard response, which he had already heard hundreds of times in his young life and was still just as illogical, just led to another locked door Johnny had been unable to open. "But isn't god perfect and omnipotent and omniscient?" he asked. "Wouldn't he know ahead of time that if he drowned everyone on earth he would feel bad about it, and then just not do it? Why did he go ahead and do it? And doesn't the bible say killing people is bad?"

The instructor's face clouded a bit and his brow furrowed as he shifted into Serious Conversation Mode. "Well, I don't know about your first question, all I can say is no one can truly understand the ways of the lord and he must have had some good reason for doing it. Remember in Genesis 8 where god said that he would no more curse the ground because men's minds are evil from their young days?"

Johnny had expected this sort of a non-answer, as it was again just like all the others he'd heard far too many times. He continued. "But if god created us, why does god let men's minds have evil in them? Wouldn't we all be better off if there were no evil in their minds? If we have evil in our minds, and he can get rid of evil at any time, then he must want it to be there. Why?"

He could tell by the look on the instructor's face that he about to regurgitate the "free will" logic-avoidance tactic. Being a very precocious youth, Johnny had done some extrabiblical research on the subject of psychology and noted the extreme prevalence of sadomasochism in the religion his instructor was failing to explain to him, and he didn't want to go down that particular dead-end road today, so he quickly tacked back to the sin issue. He asked his next question quickly and a little loudly to refocus the instructor. "But why didn't god just forgive the sins?"

The instructor was secretly pleased to not have to go down the impossible path of justifying evil, as it was always painful and unsatisfying to walk on the thorn and thistle-infested ground of theodicy. He brightened up a bit and pulled back to his habit of citing scripture and his god's orders when answering questions, as this was his nurembergish apologetic comfort zone. He answered with some zeal, as this was a very familiar passage to him and the rest of his congregation. "Well, you see, there can be no forgiveness of sins without blood, just like it says in Hebrews 9:22, where it is said that "And almost all things are by the law purged with blood; and without shedding of blood is no remission", so someone had to die, and that someone was Jesus."

Johnny's irritation at the continued circularity and failure to answer his question was rapidly increasing, so he decided to go all in and asked "So why didn't he send Jesus down to earth before he drowned all those innocent people in the flood, and before he aborted all of those innocent unborn babies?"

The instructor's thoughts on such matters, as is the case with almost all believers, had never delved very far past the printed words of his holy book, churchy cutesy sound bites and weekly millimeter-deep sermons, so he had never considered that question before. He most assuredly had never met anyone with the guts to actually ask it. So he was stumped, and pondered the question. After he felt too much time had passed without an answer, he again decided to revert back to Core Principle #1 – when anything is inexplicable, chalk it up to the mysterious ways of the all-knowing lord. He wasn't too fond of the direction the conversation was going, either, so he said a bit curtly "Well, some things will always be a mystery and no one can truly know the ways of the lord, and you should never question them too much."

Johnny knew that sort of response, especially with the associated disapproving tone, was usually the last thing adults said before they told him to shut up, asked him if the Devil had been talking to him, or asked if he was trying to go to hell by questioning the lord. He could therefore tell this exchange was also probably going to be another useless descent into circular apologetics and shoulder shrugging, but he wanted to get in a few last questions before the instructor shut him down. These questions had been hammering inside his brain for a very long time, but none of the adults in his life were the type to either honestly answer them or even allow them to be asked, so Johnny's desperation and desire to exploit this potential crack in the seam of armored mental inflexibility were too strong to resist.

He quickly said "But why would a father deliberately send his son to be tortured and killed? Would my father do that to me, or would Bobby Wilson's father do that to him?"

The instructor was again taken aback by such penetrating and completely unconsidered questions, and while he knew he was probably about to walk into a minefield of logical contradictions which this little smartass had identified, he felt he had to say something positive now that this was out there. And also now that a dozen sets of eyes and ears were completely focused on him and quietly awaiting his response, particularly those of young Bobby Wilson who was now extremely awake and attentive. He answered a little too loudly and with a little too much obvious condescension in his voice. "Well, no, of course not, Johnny, I'm sure your dad would never do that, and neither would Bobby's dad. I know both of them and they are good, decent Christian men."

> That of course made no sense at all, since Jesus's dad, who one would suppose would either have been Christian himself or very supportive of it, had actually been the one who sent him into the torture chamber and slaughterhouse, and had previously ordered Abraham to murder his son, but that wasn't even the bigger point. Johnny was focused on an issue of obedience. "But if god told my dad to kill me, what if he said no? If he did what he was told, he would kill me, but if he said no, he would be disobeying god. He would be doing evil no matter what he did, but wouldn't god punish dad, or me, or my whole family for him disobeying and not killing me?"
>
> By this point, the instructor's brain had become quite wobbly from the loss of blood at the depth and number of verbal cuts Johnny's questions had inflicted on him. This, combined with the alarming expressions he was seeing on the faces of the other kids in the group, led to an immediate cessation of the conversation and a decision to make an early transition to snacks and games, the omnipresent Sunday school youngster bribe-bait. The last planned topic of the day, morality, was just going to have to wait until another day, selectively chosen for when Johnny and his family were on vacation very far away.

The fictional interaction above is not atypical of how these sessions actually go. It is entirely typical of how children think and how they try to process the actual morality they innately sense to be true while being taught official psychopathy in their religion. I, obviously playing the role of Johnny, have engaged in discussions like this with believers on many occasions and had very much the same fruitless dialogue with them. Black is white, blue is white, and turtle is white if believing in their god and book requires such utter destruction of reason.

I am sure this scene or ones very close to it have been repeated many millions of times across the world whenever faith is challenged, be it by a child in a holy book indoctrination session, a random conversation between adults during a bull session or as a formal point of argument in the course of a formal staged debate. Such questions are not answerable without resorting to "unknowable" mystery escapism and the Stockholm Syndrome or outright sadomasochism.

The bible is a thoroughly psychopathic book, and raising children using it is to deliberately engender psychopathy in them. The single verse still used to justify so much physical abuse both at home and in schools captures this particular evil in one pithy and too-often cited phrase. It is Proverbs 13:24 (KJV) "He that spareth his rod hateth his son: but he that loveth him chasteneth him betimes." And using the rod on your children is only for what one might call venial sins. Mortal sins, vast numbers of which are enumerated in these religions, require that truly devout and god-fearing parents actually murder their own children.

Yes, the so-called good book, the book that tells people not to do murder, is the same book that also tells people to do lots and lots of murder and quite specifically when they should murder their own beloved children. As I said, this book is psychopathic, and it was written by madly bipolar immoral schizophrenics, whether or not they were divine or human.

Is beating your child for minor transgressions a healthy thing to do? The god of the bible neither says nor cares if this is so, he only says to do it.

Is having your children living under the constant fear of being murdered at your hands if they cross some particular sinful Rubicon a healthy thing to do? The state of living in fear, and indeed the long term impact of beating and abusing children both physically and emotionally, has very significant, permanent and negative impacts on both physical and psychological health, as we see every day across the world and as I mentioned in Chapter Twelve. The god of the bible cares nothing about human sanity and well-being. He only cares about utter blind obedience.

Is actually killing your child for other transgressions a healthy or indeed a moral thing to do? Again, the god of the bible does not say nor cares if this is so, he only says to do it.

This must be said in clear and plain language. The bible orders people to think and act like psychopaths. Those who believe and follow the words as written either are or become damaged and psychopathic. If they raise children according to this book they are deliberately trying to raise damaged and psychopathic children. These damaged and psychopathic children then do harm in two major ways. They raise yet more damaged and psychopathic children and they themselves say and do damaged and psychopathic things to their fellow citizens.

This is how religion is a debilitating infection and a deadly disease afflicting humanity. This is why it cannot co-exist peacefully with the rational world unless it is a powerless and empty shade of itself, devoid of applicability outside the believers' minds and secured in an ironclad Pandora's Box, never to be reopened. This is why the world would be a massively better, humanistic and moral place if every single so-called holy book were destroyed, deleted and completely forgotten, immediately.

As if all the above weren't already far, far too much, we must keep asking more questions of the faithful.

- What would a modern Islamic terrorist do with a weapon of mass destruction?

- What would a Crusading pope have done with a weapon of mass destruction?

- What would any of the most devout believers in their faiths throughout history have done with one?

- What would happen if the Catholic church today held the same power as it did in medieval Europe, and the global campaign of rapes by its priests and the campaign of coverup by its leadership were uncovered?

- What even further and ubiquitous evil would the Inquisitions have perpetrated if the Inquisitors had been able to access modern technology? What would the medieval popes have done?

- What would an Inquisition look like today, if papal Sin Squads, spies and agents could use computers and surveillance technology to rummage through your professional and personal data and your social media for incriminating information? What if they could intercept your emails, text messages and other electronic communications and use them against you in faith-based torture and murder sessions using 21st century tools and technology to inflict pain and suffering?

The answers to all the above questions have to fill a decent person with nausea, because you know exactly what would happen. So the ugly truth has to be plainly stated, yet again. When someone feels his god is on his side, nothing is sacred.

Now that we've noted some questions about the psychopathic nature of at least one widely practiced religion, and every reader of the above hypothetical situations can list dozens more without breaking a mental sweat, it is an appropriate time to discuss psychopathy, religion and the future of civilization. We can take the cases of the 20th century's mass-murdering state leaders as case studies and extremely relevant cautionary tales.

One must first understand a specific fact about the infamous trilogy of mass-murdering dictators, and indeed all specimens of a like mind throughout history, including many specific individuals holding power in many countries even today and even in the West. They were or are absolute psychopaths. The proper way to discuss them and their deeds isn't to first ask if they were secular or atheists or members of a specific religious faith, it is to ask what the hell was wrong with them in an absolute sense, and why they weren't stopped long before they reached power.

No truly moral human being would ever be anything but nauseated by a true consideration of their actions, you would hope. No truly moral and decent human being would ever be capable of deliberately supporting or participating in these crimes, one would assume. Yet these crimes happened, often with the active support of or failure to act by millions of people. Why? Another digression into psychology is required here.

The infamous psychopathic heads of certain states in the 20th century were the most religious people on earth at the time, but in their cases they were the gods of their own internal universes and of the people under their thumb, and that was all that mattered. In their own minds, they were like gods on earth with their supposedly unmatched power and vision and righteousness, and the rest of humanity was theirs to use, abuse or destroy as they saw fit. The inclusion of further religious language and the invocation of any so-called even higher supernatural powers are not necessary, although religion is the number one easiest way to get massive numbers of people to persecute and kill massive numbers of other people for no reason.

Psychopathy is psychopathy, and the psychopath will use whatever can be pulled from religion and/or the secular world that will best enable him to carry out his compulsion to do evil.

One of the most horrible failings of humanity is that we still have too much of the hyperaggressive and brutally murderous primate in us, and we both often elevate the loud and aggressive alpha male to power and then for a number of well-documented and extremely unfortunate psychological and social reasons, go along with the destruction and inhumanity he first engenders and then orders.

One must also remember that Hirohito, the emperor of Japan during World War II, was indeed considered to be an actual divinity on earth[8], and it took his country's defeat in that war to force an acceptance that he was not, after all, a god.

Religion's role, aside from the psychopathy in many of its most dangerous monotheistic isotopes, is the most effective means men have invented to control other people and to justify cruelty inflicted upon them. It is horribly effective in this pursuit due to failings in human psychology and in the fabric of many human societies. A psychopath who operates on his own normally has a limited scope, in that he must personally interact with people to control, demean, dominate and injure them.

Religion, being an idea that first preys upon human weakness to bring people into the fold, then gives allegedly actual written divine and therefore allegedly unquestionable and eternal license to perform acts of ultimate evil, will thus always have legions of faith-based stormtroopers primed and ready for battle. The psychopath taps into this ready-made pool of dogmatic support and thus can quickly spread his psychopathy to many more people, through these true believers.

The gun analogy can be employed here as well. A psychopath operating on his own can only carry a few relatively light guns and a limited supply of ammunition. A psychopath as the head of a state or a church, or even of a congregation, now has the support of many other humans to implement his psychopathy. This is because religion, being an extraordinarily strong binding agent and highly tribal and xenophobic, is mostly populated by people who crave external guidance and also external approval for their actions. Believers are much less independent and much more apt to take orders, especially if they are orders from the clergy or orders which they feel are supported by their holy texts.

The psychopath, cognizant of this fact on at least the subconscious level but in many cases also on the conscious level, sees flocks of believers as the extremely weak, malleable and often immoral patsies they are. Thus, given that the holy books of monotheism are psychopathic and can be easily used to justify any evil act possible, the short leap from psychopathic text to psychopathic action need only be facilitated by a loud and angry psychopath calling the flock to action, whether to defend some insane expression of religious freedom or the faith or the flock itself from some particular contrived threat.

The psychopath who uses religion and its mindset to motivate his followers has thus taken advantage of an extremely powerful force multiplier which is already inherently evil. The psychopath has thus expanded his army from his own single weapon to a great many weapons which will operate on the orders and intent of the spider at the center of the web. That controlling spider's psychopathy will be inflicted on huge numbers of humans by his willing drones.

The most important point in all of this is that while we can always point to the psychopathic holy texts, and to specific religions and religious people acting like psychopaths in accordance with the texts, this is a leap that need not be made. One need not take that leap from text to torture, from written word to stabbing sword, from gospel to genocide. All one merely need do is completely ignore all of these texts in their totality. Had the Germans burned all their bibles in the 1400s and rejected the empty-headed evil of Christianity, there would have been no imaginary Jewish murder of a god, no Catholic and Christian religious justification for hating Jews, and the history of Germany and Europe would have been far different. I would say about six million times different.

If we want to have a much safer and more moral world, we need to prevent psychopaths from reaching positions of power over other people. But we still haven't reached the true root of the issue that still plagues us in every corner of the planet, and very recent events starting in November 2016 show that no country, even the United States, is immune from this horrible affliction. The root issue is why so many people *other* than the psychopath do psychopathic things. Again, one psychopath can only do so much damage, but his followers can do far more and in far shorter a time.

Speaking of psychopathy, the Catholic church has itself or by proxy persecuted, expelled, vilified, forcibly converted and massacred Jews for almost two thousand years, for the alleged crime of betraying then killing a god. Or one-third of a god, if you describe the subject in the insane glow of Trinitarianism. This is of course despite the Christian teaching that Jesus was deliberately sent here to die by his father/himself, who happened to be the creator of the universe, in order to absolve humans of our sins, and that he had allegedly known about the outcome for many quadrillions of years before our universe even began.

This is of course despite the apparent impossibility of being able to actually kill a god or any of his children, unless he specifically lets you. This is despite the teaching that Jesus is alleged to have risen from the dead and rejoined his (god)father and their even more invisible threesome partner holy ghost without a scratch on him.

Everything happened exactly according to the plan designed and executed by the creator of the universe and his godly son/himself, thus the Jews were doing precisely what the gods intended. Yet they have been vilified and killed for carrying out the gods' plan. Real living Jewish people have suffered for almost two millennia for doing exactly what a probably imaginary remote ancestor was supposed to do in complete fulfillment of the will of Jehovah and/or Jesus himself. It would be hard to come up with a more insane and psychopathic bit of evil nonsense than the core myth in Christianity, but such insanity is the foundation of supernatural religion.

But this is far from the only noose that Catholics have placed around the collective Jewish neck. From their pope on down they have punished, imprisoned and expelled Jews simply for being Jewish. They have denied Jews civil rights and civil justice and placed them in ghettos, and they did it centuries before Hitler did.

The Catholic cult has dreamed up numerous imaginary Jewish offenses, as one might expect from a group of infantile woman-rejecting and often self-denying homosexual males who suffer both from advanced terminal penis envy and a planet-sized insecurity complex. They have made up endless streams of nonsensical gossip about the group with which it is in fierce economic, social and political competition.

As two opposing groups competing for power and status are almost guaranteed to do, at least one group will begin making up insane lies about evil deeds and tendencies of the other group.

The stories were obviously just evil lies and immoral nonsense, but spreading hateful and damaging gossip is a common behavior in too many humans. The faithful made up utterly insane lies then spread them first through the church and flocks, then later through the Vatican's official publications. This was the Mesopotamian then the Catholic equivalent of America's current nationalist, xenophobic, racist, bigoted and hateful attacks on despised minority groups and the truth, carried out by a facts-optional press.

Christians in general, and Catholics in particular, have for almost two thousand years been the ecclesiastical equivalents of Breitbart, Rush Limbaugh, Sean Hannity and Alex Jones, the Daily Stormer and the various incarnations of Fox "News". A quick look at today's seamless alignment of American religionists with such virulently malicious partisans indicates that this paradigm is still strongly with us.

What are those imaginary misdeeds for which Jews have been persecuted for the last two millennia? It is hard to believe people could even dream up such things, much less believe them, but two thousand years of history and six million dead Jews just in the 1900s are proofs impossible to deny. A partial list of these imaginary evils includes:

1. Not believing Jesus was the true Messiah or the son of the One True god. They do hold this position, so when I say it is an imaginary evil I mean that it is like being thought evil for not believing in the divinity of imaginary characters like Bugs Bunny or Mickey Mouse.

2. An ancient ancestor's Jehovah-intended betrayal of Jesus, as mentioned above.

3. Killing Jesus, even if only very temporarily, as mentioned above, again as per Jehovah's intent.

4. Holding too much financial power. This despite being forced into the moneylending business by the ruling Christians, whose rules have always been "heads Christians win, tails Jews lose". This is like brutally beating a child so hard he wets himself, then beating him harder because he wet himself. The magnitude of sheer sadism and blind hypocrisy is beyond comprehension.

5. The blood libel. This is, according to the Encyclopedia Britannica[9], "…the superstitious accusation that Jews ritually sacrifice Christian children at Passover to obtain blood for unleavened bread." That is not a joke. Many real people have been killed because of this unbelievable bit of hateful insanity.

6. Host desecration.[10] Of all their insane accusations, this one is top-tier "Onion[11]" level, because of how ridiculous it is on too many stacked levels. This unsurpassed insanity is explained below along with how one has to successively twist reality and one's mind to get such a farcical notion.

    - First one has to turn logic and reason off in order to believe in Christianity.

    - Then one has to believe that before he sort-of temporarily died, the Jewish demigod preacher told his males-only followers to eat his body and drink his blood, ostensibly metaphorically, but apparently not necessarily so.

    - Then one has to believe that a celibate Catholic semi-evolved mostly hairless primate male priest with magical powers can, with a wave of his hands and some magic words, turn a wafer of pressed carbohydrate into an admittedly small reincarnated piece of the body of the undead Jewish demigod preacher. This body fragment / cracker is now called a "consecrated host". Stay with me, we're only halfway there.

- Then one has to believe that after this cracker receives the Frankenstein treatment and magically becomes "truly, really and substantially[12]" a (re)incarnated piece of the body of the long undead Jewish demigod preacher, it can be stolen by a sneaky Jew before it is happily cannibalized by one of the faithful. Or that someone can steal it and give it to a Jew.

- Then one has to believe that this Jew, or someone to whom he would give this cracker, perhaps a witch, would take some sharp object and stab the cracker, thus "desecrating" it and causing grievous, indeed even true, real and substantial harm to the stolen fragment of the long undead and now reincarnated-into-a-cracker Jewish demigod preacher.

The above beliefs, particularly the last one, would have one laughing one's ass off at such off-the-charts mindless lunacy if they were acted out as a Monty Python skit, merely using different names to protect the ignorant. But this is no laughing matter; it is deadly serious. Real live Jewish men, women and children were murdered because Christians actually believed this unbelievable garbage. Real live human beings have been tortured, exiled and killed because faithful Christians, including the allegedly infallible Catholic popes and their obviously far less than infallible black-robed enforcers, truly, really and substantially believed these things. And people still even unto this day still actually do believe these preposterous fairy tales and others like them.

In the words of the recent religious thuggery being promulgated in the United States, beliefs of this sort must not be questioned, because they are "deeply and sincerely held religious beliefs". What they are saying is that they insist that a person can believe absolutely anything at all based on absolutely no reason whatsoever, and the rest of us should respect such lunacy. *That* is what religion does to people.

In 2008, Mr. Paul Zachary Myers[13], more well known as "PZ", a biology professor at the University of Minnesota Morris, committed a deliberate act of host desecration. This was an intentional act done with the aim of ridiculing the idea of the host and that it can be desecrated. To the secularist, nothing is holy or sacred, particularly religious superstition. Therefore for a bunch of woman-hating sexless semi-evolved mostly hairless aged male primates to gather around ancient mythology and use it to come up with the story that a magic cracker is a reincarnated piece of a two-millennia-dead Israeli demigod preacher who'd already been partially resurrected in bits and pieces across the world on billions of occasions is an invitation in thousand-cubit high flaming letters in the sky for scorn of the highest degree.

Myers happily obliged in applying this scorn, to the delight of nonbelievers and nondogmatists everywhere. Catholics everywhere were aghast, with many calls for Myers to resign or for his employer to fire him, along with the obligatory death threats the faithful can be so very reliably depended on to deliver to those who disagree with them. In earlier centuries the Vatican and its flocks of sincere believers would have wasted no time in torturing then killing Myers, along with any conspirators.

But in their gelded modern state today, fortunately restrained by secular law, a call for his dismissal from employment, much anger and outrage directed his way and also threatening his death was all they could do to him. Just like Jesus would have done, right? No, there was no turning of the cheek by these devout Christians, who just like all other believers only use the bits of their book they want to use.

The head of the Catholic League, a thoroughly hateful and vitriolic individual who is both so far around the bend of unreality and also a specimen of the faithful so devout that he takes this sort of ridiculousness seriously enough to be the head of the Catholic League, was extremely angry[14] about host desecration. He believes all the above insanity and surely much more, and is extremely happy to go on ranting about it, in the true spirit of faith-based Christian bigotry.

Again, one can hardly see a better example of WWJND – what would Jesus *not* do – than this individual, a certain Bill Donohue. Unless, that is, Donohue really is acting fully on behalf of and with the full blessing and approval of Jesus himself, and can produce written or videotaped orders from him as justification. I await the formal unveiling.

So there is a long history of official church-mandated persecution of Jews. Anti-Semitism was one of the core values of the Catholic cult for almost its entire existence. Again, this was not something the masses just did on their own. It was specific and deliberate policy ordered from the top. A particularly noxious example of such official church policy was the papal bull[15] by Pope Paul #4 in 1555. This allegedly infallible representative of Jesus on earth used his office and station to state the following[16]:

- Jewish guilt has doomed them to "eternal slavery".

- All Jews in a village had to live in a single small neighborhood that was walled off from the rest of the city. To make it even worse, they were locked in when the day was over. In other words, the Catholic pope created Jewish ghettos, just like his spirit partner Hitler later did.

- Jews could not own property in this ghetto.

- Jews had to wear a yellow indicator of their 'Jewishness'. Just like Hitler later did.

- Jews could only work in essentially the trash/recycling trade or be a physician, and of course a Jewish doctor couldn't touch a Christian. In other words, his rules were very like Germany's Nuremberg laws, only 400 years before Hitler.

- Jews couldn't work on Sundays.

There were quite a few more evil restrictions and persecutions in the bull, but suffice it to say the ones listed above were the tip of the iceberg, and each and every item was intended to maximize the suffering of every Jewish person's daily life. You can be sure it succeeded. Hitler didn't need to invent new ways to persecute Jews; all he needed to do was read Catholic papal decrees from previous centuries for helpful tips on how to be evil to innocent people.

Did the church make any moves in the following four hundred years to reform itself? Did it serve as a vocal and tireless beacon of light and purity before, during and after World War II, and work as hard to protect Jews during the war as it had previously persecuted them? Of course not.

Let's take a look at what actually happened, documented with numerous written records, pictures and even video. Since the Vatican and Catholic apologists across the world both say the church did *just fine* during the war, and since apologists seem to prefer to measure themselves against the worst criminals of the last century, let's see how they did when faced with actual moral problems and the chance to show spine and intestinal fortitude in the face of evil.

The deadly truth is that popes and powerful men in the Catholic cult ruthlessly and cruelly continued the barbaric persecution conducted by their predecessors. Even throughout the Holocaust, not a single pope came out with an official edict stating once and for all that the Jews were just as human as Catholics and everyone else, and that official policy would be to "do unto them" as a truly moral person should. That this was the state of affairs even seventy years after 1870, when the church had formally declared during Vatican I that its Popes were infallible when opining on faith and morals[17], demonstrates that the church still clearly didn't understand the definitions of the words "morals" and "infallible".

It also clearly demonstrated the church's seamless kinship with fascist and authoritarian dictatorships throughout history, whose leaders were also often declared infallible. Indeed the popes are themselves by definition and position both fascist[18] and authoritarian dictators, whether as the ruler of a church, papal state or the current theocratic nation of Vatican City. These similarities must not be lost on anyone.

So what did the Vatican's pope and Jesus's top man on earth, Mr. Eugenio Pacelli, aka Pope Pius #12, do during World War II? Did he bravely and unequivocally condemn the Third Reich and its campaigns of killing homosexuals, gypsies, the mentally retarded and Jewish men, women and children, and say and do all he could to stop it? Did he send out a loud, clear, concise and unequivocal message to all Catholics that being a part of the Reich was immoral? Did he hold mass excommunications for all Catholics involved with the Reich and the Holocaust? Did he personally put himself in harm's way to try to intervene on behalf of those being murdered by the millions? Did he do what Jesus would have done?

No, of course he didn't do any of these things. The so-called Vicar of Christ on earth was, like all his predecessors, very clear that his top priority was to maintain as strong a position for the church as he could for as long as he could. His job was to keep the church above water, keep it in power and take full advantage of secular authority and strength to further its own aims. One could almost read his thoughts.

> 'There's a rising hyperaggressive lunatic who looks like he's going places? Better get in his good graces…maybe we should sign a treaty with him? Ok, got that done. Now what? He's going to kill all the Jews in Europe? Outstanding! Finally someone can get rid of those Christ-killing Christian-baby-blood-sucking parasitic host-desecrating Jews and we won't have to get our own hands dirty anymore. And he hates Communists? Me too! Damn godless commies! Better make sure we never piss this guy off, he could be very useful.'

Do you think that fictional quote above is improbable? Farfetched? Hardly. Let's take a look at what Mr. Pacelli, the head boss at the Vatican, actually did and said during one of the worst periods in human history. Did the Catholic leader stand up for the oppressed, the poor, the victims of evil tyranny? No, of course he didn't.

The Vichy Vatican collaborated with fascist regimes and contributed to the Holocaust through acts of both omission and commission. It must be yet again stated that for almost its entire history, the church has been moderately to virulently anti-Semitic and did not begin to formally change this doctrine until the *1960s*, twenty years *after* the last Jew was murdered during the Holocaust.

The following is a very short summary of Vatican-related events before, during and after the war.

- Pacelli's predecessor, Mr. Achille Ratti, aka Pope Pius #11, was Vatican ruler from February 1922 until he died in February 1939. It is very interesting that this pope, who actually didn't care for Hitler very much, died right before the war, and a very pro-Hitler pope took his place. No more proof need be shown that the Christian god does not exist or that if he does that he is indeed the self-described cruel genocidal sadist described in his own book. We also need no more proof that no pope is any more infallible than a comparably sized hot air balloon.

- Ratti signed the Lateran Treaty with Italy in 1929, which established Vatican City as a sovereign state. He later signed the Reichskonkordat with Hitler's Germany on the 8th of July 1933. Note that this concord with Hitler's Germany was facilitated by the man who would be the next pope, Mr. Eugenio Pacelli, in his position as the Vatican's man on the ground in Germany. This is an extremely important point to remember in the context of the coming years. The Nuremberg Laws were enacted in September 1935, and could not have been a surprise to Pacelli, even if they might have been to Ratti, due to the wave of extreme anti-Semitism accompanying Hitler's rise to power, culminating as his appointment as Chancellor in January 1933. Pacelli knew precisely what Hitler was all about, in merciless and disgusting detail, long before 1933. Again, the demonic god of the Christian bible would indeed have set the dominos up in precisely this manner if his aim were to install a genocidal psychopath as a national leader.

- Pope Pius #12, aka the aforementioned Pacelli, was in power from the second of March 1939 all the way into 1958. Publications on his numerous moral failings as the Vicar of Christ on earth fill many library shelves. At a time when strong leadership and an unwavering, loud and unequivocal opposition to pure evil would have been expected of the alleged one true god's alleged personal field rep on the planet, purportedly armed with the gift of papal infallibility on the subject of morals and chartered to act as Jesus would have, this pope showed yet again that the church's main goal has never really been getting people to heaven. Its goals are getting the most living people and their money into the cult here on earth and protecting the church as a corporate body from harm. And collateral damage be damned, literally if need be.

- One of Pacelli's first acts in power was to pull back an encyclical his predecessor was going to publish that criticized racism, and had some very, very mild pushback on anti-Semitism though it was still not pro-Jewish.

- On 20 April 1939, ***four years after the Nuremberg Laws,*** the Vatican's official ambassador to Germany, titled the Apostolic Nuncio and going by the name of Cesare Orsenigo, offered public congratulations to Hitler, it being his fiftieth birthday.

- Pacelli did produce an encyclical on the 20th of October 1939, titled Summi Pontificatus.[19] It was over twelve thousand words long, but in typical Catholic fashion was 99.9% gushing and fawning jargon-laden praise for the cult and the savior and how great the world would be if only everyone were Catholic. I highly recommend taking some time to completely read encyclicals like this, to get a feel for how Catholics think. Fair warning - it will be both a very illuminating and very repulsive experience.

- On the 29th of June 1943, Pacelli put out his sixth encyclical of the war, titled Mystici Corporis Christi.[20] This massive tome managed to take a whopping 19,000 words to say almost nothing of value but yet again, a huge amount on how great he thought the cult and its founder were. Of half a dozen encyclicals produced while quite literally the world was at war and many millions of people were dying, none of them contained any outright condemnation of the evil being perpetrated every minute of those years.

- From 1939 to 1945, a Catholic priest named Jozef Tiso was actually the head of state of Slovakia, and he replicated many of the war crimes of the Nazis in his country. The Vatican was fully aware of some of these crimes. For these crimes, the Czechoslovakian government rightly hanged him as a war criminal in April of 1947. But of course he was never excommunicated by the Vatican. Your average pew-denting Joe Catholic can be excommunicated for pretending to absolve someone of sins or pretending he can Frankenstein a cracker into a fragment of Jesus. He is also automatically excommunicated if he punches the pope. His wife Jane Catholic, who he is never allowed to divorce, would be automatically excommunicated if she ever got an abortion. Having a priest who is both a war criminal and the head of a Nazi state is OK, but don't you dare slap the pope or get an abortion. The church does have its standards, after all.

- Through the war and afterward, the Vatican Bank, and thus the Vatican, engaged in money laundering[21] and profit-taking off of the misfortune of Jews[22], the war effort in general and from German interests in particular.

- In 1946, Pacelli approved an unconscionably amoral and disgusting order[23] to "Catholic Institutions and families" in France that they were not to hand back any "…Jewish children, who during the German occupation, were entrusted to Catholic Institutions and families and whose return is being sought by Jewish Institutions…" This order actually contains instructions on how to stonewall people following up on these children and trying to help them, and it reads exactly like the cult's modus operandi for how it stonewalled the later outcry against the global epidemic of horrific child rapes it was found to be both perpetrating and deliberately covering up.

- Pacelli never truly denounced anti-Semitism. This would of course be fully in line with almost two millennia of official cult policy so this is unsurprising.

- Pacelli never spoke out strongly against the Holocaust despite knowing what was really going on by about 1942. He gave no vocal speeches directly railing against the murder of millions of men, women and children. Pacelli absolutely refused to stand up to the evil that was killing its way through Europe and committing heinous genocide. He was pathetic and an enabler.

Christianity's original disciples, indeed even its founder, are endlessly praised for their willingness to stand up for what they thought was right, regardless of the potential danger. Pacelli utterly failed both the courage and practical action tests, and his attempts to thread the needle of neutrality or impartiality and bleat vague platitudes while the Holocaust raged on were and are still absolutely contemptible. He was an enabler of crimes against humanity.

- Pacelli never even took the symbolic step of putting *Mein Kampf* on the Vatican's own 400-year-old list of forbidden books, and astoundingly did not excommunicate Hitler. One would have thought even these minor actions not beyond the capacity of an allegedly morally infallible pope, but perhaps his mind and body had rejected the infallibility treatment. Previous popes did not fail to miss the existential horrors written by Copernicus, Kepler, Kant, Bentham, Descartes, Voltaire and Hume and had properly consigned their works to the eternal pit. Again, this immoral cult does have standards to uphold. None of those scientists and philosophers were going to do the Vatican's dirty work and kill Jews, but Hitler was its action man doing its dirty work very well.

- After the war, did Pacelli or the cult, the organization that calls itself the source of all true morality, after standing side by side with the Reich while the earth's atmosphere became saturated with the ashes of countless murdered men, women and children, finally regain its moral compass? No, of course it didn't. It tried to prolong the evil. After the war, the Vatican and its network helped thousands of Nazis and Nazi war criminals[24] to escape[25] from Europe to Latin America on the infamous Ratline. Why would they do something so thoroughly and disgustingly immoral? Because those war criminals might be needed in a future fight against the godless Commies, of course, since the Vatican believed those Commies were its most dangerous threat. The pope and his cult of immorality didn't give a holy damn about how much blood was on the escapees' hands, because again, the religious mindset is always that the end justifies the means.

Thus before, during and after the war, Pacelli and his cult were complicit Nazi sympathizers, offering them both aid and comfort. And when the murdering was over the Vatican helped Nazis escape justice while preventing Jewish children from being reunited with loved ones. Pacelli was thus himself a moral war criminal and should have personally stood trial for crimes against humanity.

We can now ask a very simple question. What *should* Pacelli, the allegedly infallible leader of the global Catholic cult, have done? What would his boss, the once-dead but quickly resurrected and now so often partially re-resurrected Jewish demigod, have wanted him to do, while so many nations were fighting together against this evil, and while his god remained entirely silent and absent?

Even with his boss having died two thousand years ago, an infallible theocrat, despite being a semi-evolved hairless primate, ought to have been able to figure that out, given that that the source for guidance was the book on which his religion was based. Taking a look at that book, we can see Jesus's direct orders to his disciples regarding doing his / his father's work on earth. The gospel of Matthew from the King James version contains extremely relevant orders, and the words are in stark contrast to what Pacelli actually did during World War II.

- "{10:23} But when they persecute you in this city, flee ye into another."

- "{10:27} What I tell you in darkness, [that] speak ye in light: and what ye hear in the ear, [that] preach ye upon the housetops."

- "{10:28} And fear not them which kill the body, but are not able to kill the soul: but rather fear him which is able to destroy both soul and body in hell."

- "{10:38} And he that taketh not his cross, and followeth after me, is not worthy of me."

- "{10:39} He that findeth his life shall lose it: and he that loseth his life for my sake shall find it."

The above passages actually are truly courageous words, and they are directed at those who say they follow Jesus. The authors make it completely clear that Jesus ordered his followers, especially his most faithful disciples, not to be silent in the face of evil. They should persevere and bear his cross, even if it meant being killed in the process. Such acts would be examples of ultimate courage. To recast a trite phrase, that is 'what Jesus would do'. This is absolutely what Jesus would have wanted his most powerful and allegedly infallible representative on earth to personally do and to tell his followers to do.

Rather than follow either his boss's example or his orders, Pacelli did exactly the opposite. Pacelli's written and verbal silence proves he was pure evil, a craven enabler and coward and a traitor to his savior. He made a conscious choice to be silent, to be weak and to preserve his institution at all costs, even if those costs were millions of innocent murder victims, including those of pregnant women and their unborn children. He chose to side with an almost 100% Christian country and its hordes of war criminals because they were killing his cult's worst enemies, and because he thought they would one day protect his church from godless communists. He did all this while those chosen Christians and their leader were murdering their way through a continent and an entire social group. Pacelli earned every bit of the pejorative title of "Hitler's Pope".

But the Vatican tries desperately to tell a different story, thus violating its god's own commandment against false witness. Apologists say that Pacelli was actually working behind the scenes to try to help in an assassination of Hitler and other actions. For all we know, since the Vatican is still stonewalling and blocking access to the records that would answer this and many other questions, this could be at least partially true. But it would be quite the double edged sword, would it not?

If the Vatican revealed that Pacelli was a Neville Chamberlainish appeaser and coward in public but a black-robed scheming Oliver North on steroids trying to secretly fund or facilitate the murder of a sitting head of state in private, this would indicate that he had some common sense and was working toward a good thing, that being the removal of a murderous psychopath. But it would also reveal that the Vicar of Christ on earth, the Vatican's allegedly infallible pope, was actually a conspirator in a plot to murder a human being, which is not something the modern cult teaches or would want to reveal.

A couple of hundred years ago or more, when popes waged war and would have been quite happy to tell the world about their plan to kill a head of state or entire groups of people, this would have been acceptable and indeed common practice. But this has not been the case since the full rise of the Age of Reason and the Enlightenment. The Vatican would never today admit to a papal plot to kill someone while it has such limited secular power and while it has shown itself as evil and immoral in so many other ways.

But even if they did, and if he had, it *still* wouldn't exonerate the cult's millennia of evil or Pacelli's thoroughly inhuman and evil actions and inaction. So until such time as the Vatican opens up its archives to an unannounced inspection of every bit of information it has, we have only the evidence we have, which is sufficient to justify all the above statements along with many, many more. And the Vatican and its field reps have shown themselves on countless occasions to be liars, frauds, cheats and completely untrustworthy, so even if all its hidden archives become available, no one should be under any illusions that they would be truly getting access to everything.

Unflattering documents have a habit of disappearing, falling to "shreds" or turning into ash. You can be assured that documents which clearly incriminate the Vatican or its members in aiding or abetting the Holocaust or other moral, ethical or legal crimes will never, ever see the light of day unless they are stolen by an insider or unless someone working in the archives goes off the deep end and pulls an inside job kind of papileaks data dump.

So we don't know what the Vatican is hiding. But we do know it conceals what it can. How much does the Vatican wish to hide its evil history? Take a glance at the Vatican's web page listing "Papal Documents".[26] The first thing one notes is the total absence of documents from before 1878. This is exactly what one would expect if the Vatican desired to exclude the most egregious and criminal orders by the various popes, most of which occurred before that time. This is how that outfit operates. You could call it a coverup, or deliberately failing to tell the truth, the whole truth and nothing but the truth, 'so help them god'. Which has been that cult's modus operandi for as long as it could get away with it during the increasingly horrifying process of discovery of the extent of the planet-wide rape of children by Catholic priests and the official coverups.

Regarding the war years, the Vatican to this very day has refused to open its full archives dated after the year 1939. Thus each and every pope from Pacelli through Bergoglio has made a conscious decision to keep those records away from the world. It's pretty clear this stonewalling is the equivalent of the Vatican pulling a Fifth Amendment dodge in order to keep its collaboration with the Third Reich and its offspring under wraps.

You may or may not be surprised to find out that the Catholic cult is, at least at the time of writing, most of the way through the process of formally naming Pacelli a saint.[27] Yes, this weak and pathetic Vichy pope, this enabler of one the worst and most murderous regimes in all of human history, is on the road to sainthood. One thinks no more need be said to make the entire case against at least Catholicism, if not all of Christianity and indeed monotheisms as a whole.

Which brings us back to Hitler, the Third Reich, the Final Solution and the guilt of the actual perpetrators. Hitler was unquestionably a psychopath, but who were the people who actually carried out the Holocaust? One man could kill a lot of people, Jews for instance, with the right weapons and given sufficient time without being caught, but even someone as thoroughly evil as Hitler could not carry out the Holocaust on his own. That required the collaboration and active participation of huge numbers of people in the military and business sectors in Germany and other countries, and the complicity of many others in the civilian population. It also required those in the know in the outside world who weren't too troubled by the goings-on to remain relatively quiet about it.

Those who bleat endlessly about an allegedly atheistic Hitler being worse than the Crusading popes and the Inquisition never mention all the Christians who actually did the wet work of carrying out the Final Solution with their own hands. It doesn't matter at all if Hitler was a devout Catholic, a devout pagan or a devout atheist, because he himself didn't kill six million Jews. Catholic and Protestant Christians did much of this killing.

Ninety-five percent of Germans were Christians during this time. You read that right. Germany as a nation and Germans as individuals were almost exclusively Christian. It was Catholic and Protestant Christians who pulled the triggers, herded the doomed into camps and gas chambers and dropped in the gas. They killed the men, women and children of Europe, these Christians, and many others knew it was happening.

Therefore in the final accounting, measured by those who failed to act against the Holocaust, those who supported it but didn't participate, and finally those who actually carried it out, the Christian faithful were and in some cases still are overwhelmingly guilty. It wasn't legions of atheists and agnostics who killed Jews or supported their murderers, it was cohorts of Christian Soldiers.

If anyone ever tries to go down that path of blaming Hitler or atheism for six million deaths and a world war, remind him that there is no evidence that Hitler personally killed a single Jew, but both his Onward Christian Soldiers and the SS killed six million Jews. Which is literally all one need say to permanently put an end to any talk about Christianity being the path to a decent morality. From the Inquisition to the Crusades to the invasions of the Americas to the Holocaust, Christianity has proven itself the global expert at killing people.

No one in history has ever been killed in the name or cause of atheism. From the Killing Fields to the gulags to the ovens and concentration camps to the death camps and marches in the Pacific, atheism was not the motivation for killing people. It no more motivates people to kill than does the absence of belief in Santa Claus and the Tooth Fairy. Psychopathy and insane levels of xenophobia were the reasons for the 20th century's tortures and murders, and have always been the reasons for moral disasters.

The psychopathy and xenophobia were strongly fueled and supported by religion and the faithful, or were present in the absence of religion, and in no case did religion mitigate either psychopathy or xenophobia. Psychopathy was the core evil and the root cause, and religion served as a nuclear-level supercatalyst and enabler for it. Had religion served as a bulwark against this psychopathy, had religion and its teachings prevented the genocides and the murders, we might begin to have a case for religion, but when put to the most important tests even in recent history, it didn't just fail, it made things worse.

Bear in mind that anti-Semitism was the official position of the Catholic cult and the Vatican for almost two millennia, and it was fiendishly implemented in practice as often as possible. Thus the vilification of and hatred for an entire group of people, through generation after generation, spread throughout Catholic countries and those places where Catholics influenced education, policy and governments. Anti-Semitism was not a purely Catholic action, as many Christians across the spectrum were of that same mindset. Protestants and their clerics were supporters[28] of the Reich as well. Therefore in the early and middle 1900s, anti-Semitism was common and often very virulent in Europe.

It has been well documented that the level of authoritarian upbringing was a major factor in a person's tendency to help a Jew or someone in need, or not. People who were raised in an authoritarian home and environment, who were raised to follow orders from parents and authority figures and especially those from the overbearing Christian sky god, were more likely to go along with the evils perpetrated on Jews and less likely to help them. These people were raised to obey orders, to follow directions, to do what the authorities and the gods said to do and not to talk back. They were raised by authoritarians.

Those people who were raised in a more liberal and independent household, more focused on *how* to think and on human values as opposed to mandated immoral values and on *what* to think were more likely to help those in need and not participate in the Final Solution.

Why is this relevant to religion? Because religion draws authoritarians like a flame draws moths and like a Klan rally draws racists. Because religions are built on people telling other people *what* to think, not *how* to think. They are built on bowing down to and blindly following authority.

The disease of religion, most particularly its monotheistic strains, is truly viral in its ability to take over minds and reprogram them, thus replicating the propaganda throughout the population. Religion is the ultimate authoritarian system because there is one central controlling authority whose orders are never to be questioned or disobeyed, and there is literally an eternity of punishment awaiting the independent and critical thinker.

So when the psychopath Hitler blamed Jews for pretty much every societal ill, it fell directly in line with many generations of Christian hatred for them, official papal oppression of them, and decades of outright anti-Semitism in Vatican-approved publications. In pre-WWII Germany, 94-99% of people in Germany were Catholic or Protestant. Since almost everyone in Germany was therefore a Christian, and thus ostensibly a believer in Jesus and his preachings, almost everyone who designed, participated in, enabled, carried out or knew about the systematic genocide of Jews was a Christian.

This includes those who were merely aware of it as well as those military and civilian personnel who actively participated. It was Christians who filled the ranks of the military, Christians who ran the ovens and death camps, Christians who pulled the triggers, Christian military personnel whose government-issued belt buckles were inscribed with "Gott Mitt Uns", or "God With Us".

Hitler did not conduct the Holocaust on his own. It took a compliant and complicit Vatican and huge numbers of Christian Germans to do it. The Final Solution was carried out with the full support of many Christians, following many centuries of almost equally horrific Christian precedent.

When the programs of murdering Jews and the mentally handicapped were ordered, these Christians, these faithful believers who would be the ones actually marching the Jews into the ghettos and gas chambers, shooting them or killing them with poison gas, burning or burying their bodies, then repeating this for years on end, had to make a decision about what they would do.

- They could follow the preachings and long historical precedent of their churches and happily do their generation's part in the centuries-long effort to kill all the Jews, thus either committing a mortal sin by killing, or merely dutifully following the many examples of godly-mandated biblical genocides, or both. This required both psychopathy and dissonance. Many took this path.

- They could follow the orders of those in authority to kill all the Jews even if they disagreed with it. Again we have psychopathy and dissonance. Many took this path.

- They could decide that murder and genocide are immoral in all times and in all places, and not participate. This could have been through removal of those in power who ordered it, through refusal to follow orders, or through leaving the country. Extremely few people took this path.

The ashes and graves of six million Jews bear silent and permanent testimony to what these Christians actually did. They killed Jews in cold blood. Again, and again, and again, they killed men, women, children and unborn children, fully aware of what they were doing, while the body count soared. Why didn't these Christians, these followers of the alleged Prince of Peace, turn the other cheek? Why didn't they treat the Jews as they would have had the Jews treat them? There are three reasons.

The first is based in psychology and self-preservation. It is indeed difficult for a human to not follow the orders of his high command and his society in any time. It is especially difficult when that high command got to power through killing and corruption and/or killing or persecuting the average Joe who stood up to the high command. So following orders and the desire to avoid being jailed, exiled or killed oneself was a motivator. These are never excuses, of course, as the secular Nuremberg trials demonstrated and reinforced so well. And, one would assume, they are not excuses their savior would accept either.

The second is purely religiously-based, and is that these people killed Jews because their religious leaders said it was OK to kill them, as their ancestors had done for centuries. Their collective psyche had been primed to kill Jews, and now their psychopathic leader and his psychopathic followers were implementing the sentence naturally resulting from the words of the Vatican and other Christian leaders.

And the last reason is the fact that religious practice can be molded however the Dear Leader wants to mold it, and the flocks will happily go along with the program. If the faith is Christian and based on the bible, the Dear Leader can use the majority of the book to justify anything from selling or immolating his daughter to killing his son to enslaving people to carrying out a large-scale genocide. This is what happened in Germany and other parts of Europe during World War II and is what has happened throughout the history of this and other religions. A psychopath used the basic foundations and preachments of a mainstream religion to get a great many of that religion's followers to first demonize and then kill millions of innocent people.

Voltaire's timelessly wise words about those who can make a person believe absurdities can make them commit atrocities will echo forever through the tombs of the dead Jews and all the uncountable millions of other people killed in the faithful service of religion.

Psychopathy only affects about one percent of the population. It is thus the people who follow the psychopath, and why they do so, who are the real problem. It is therefore extremely clear that no single human being should ever be trusted with any more than heavily checked and equally balanced secular power over other humans. A group of humans elected to lead then held accountable by an informed and engaged population is the only known way for a truly decent human society to endure.

Before I close this chapter, I have two last questions for the faithful.

- "Is there anything at all, are there any words anywhere, that can be found in your entire faith system or in any of your holy books that if applied in the real world would cause physical or emotional harm to a human being? Not just stress, but actual harm?"

- "Is there anything at all in your entire faith system, are there any words or phrases at all in any of your holy books, that you do not do because you believe what they say to be immoral or evil?"

If you are a monotheist, and you answer anything except an unconditional and unhesitating "yes" to both of these questions, you are some combination of terminally ignorant of your faith and/or an evil and psychopathically immoral person.

If you are at all moral, and another human being told you to do pretty much anything ordered or inspired by any page of the monotheistic holy books, you have a choice. Be a believer and do it, or choose not to be a believer and tell the person to get bent because you'd never commit such evil and immoral acts. These books are the foundation for monotheisms which are practiced by billions.

How can these books contain so much horror? It is in there on purpose, no matter who inspired or wrote them.

If you believe the books are wholly manmade, you are in line with the nonbeliever's position on them. Therefore the evil in them is completely human and a derivation of an absolutely demonic but completely human mindset, and should be completely ignored as a matter of course.

If you believe these books are divine in origin but you believe in a good god, how did this god either dictate or inspire these books with his very own intent and guidance but allows them to remain so far off the moral mark in such horrible ways? The god is completely conscious of this problem yet knowingly allows billions of people to read perversions of his guidance. Your god is thus some combination of utterly and unforgivably incompetent and evil.

The other choice is to believe that this allegedly loving, omniscient and omnipotent god did indeed set down dictates and guidance that are evil beyond words because the god is and has always been the epitome of evil. This option is denied at the start by the faithful, although it is actually much more likely to be true than is the god merely being a sloppy and ignorant fool. After all, there have been thousands of years of missed opportunities to correct these books, and the abortive and disjointed tales of a human sacrifice in what people now call the New Testament did nothing in that regard.

We've already discussed the type of people who take their scriptures literally. These people are clear and present dangers to other human beings and civil society and can never be trusted.

Those who say the scriptures are not literal, that they have been misinterpreted or are selectively allegorical, have been blinded to rationality and logic by continuous and terminal rationalization. If you think this way, just who the hell do you think you are, and how do you say with a straight face and no admission of your galaxy-sized arrogance, that you somehow actually know the real word of your god and not what His Own Book says? On what line of communications are you receiving the specific guidance direct from your god as to how to interpret or amend these scriptures, and why can't anyone else dial in?

This is arrogance raised to the power of infinity, yet it is in common and daily practice both among the clergy and the laity although it is never directly admitted by the faithful.

This is where psychopaths can and must be stopped, and why monotheisms in particular offer supermassive amounts of weapons and ammunition to be used against the innocent. A dangerous psychopath reading the Mesopotamian holy books will find them overflowing with divine orders to murder, rape, pillage and destroy, and every kind of interpretation and action is there for the taking. He will look around the world and find billions of adherents of the faiths derived from these holy books. It will take no time at all to see that a murderous ideology combined with billions of pliable and sadistic adherents will give him ample resources to carry out murder and mayhem, and history proves this.

What if there were no holy books of any kind? What if all humanity shared a single common moral code derived from basic principles of fairness, and as Sam Harris so simply but eloquently put it, the well-being of conscious creatures?

We can, as he advises, use science and reason to construct a truly moral code that would indeed, if applied, result in higher levels of well-being for more people in more places, and this could be an enduring and continually improving situation.

We would be able to put this basic moral code into a very few lines, and there are innumerable examples of such from history and our present day. From Confucius to the great moral philosophers such as Bentham and Mill, to modern enlightened bloggers sharing good will and actual moral recommendations, this is actually a very easy task and not only does it require no gods, it requires the total exclusion of all gods. For now, it will suffice to list an established set of observations that can inform our collective human and planetary future. These observations sum up the reasons why religious apologists will never win the morality argument and should never be part of the discussion until they completely reject the totality of their holy books.

1. Psychopathy is the fundamental core reason for evil.

2. The monotheistic holy books are psychopathic and both justify and order psychopathic actions.

3. Teaching these holy books deliberately instills psychopathy and blind obedience and destroys critical thinking.

4. Few people deviate from the religion and society they were brought up in as children, especially when such deviation will be met with ostracism, imprisonment, injury or death.

5. Passing on the teachings and psychopathy of the monotheisms to children ensures that the process of instilling psychopathy in future generations will remain in place for generations.

6. Learned psychopathy is expressed through words and actions in the real world, thus word becomes sword and believers seek to

instill the psychopathic religious teachings and mindset into all aspects of society.

7. Psychopaths seek to control and harm others, and the more people that can be injured or controlled, the happier the psychopath and his enablers become.

8. Psychopaths tend to rise to positions of power in the religious and secular worlds due to unfortunate flaws in human psychology.

9. People are drawn to religion out of fear and weakness, and because they need someone to tell them what to think and do. They are thus highly receptive to orders from the controlling psychopath.

10. People will follow orders to do psychopathic things, especially and often unhesitatingly if ordered by their religious or religiously inspired leader.

11. The only path to a more moral future, without Crusades, Final Solutions, Inquisitions, martyrs and jihads, requires that humanity completely and permanently rid the planet of any holy book that contains even a single word of psychopathy or antihuman content.

12. Humanity must actively prevent psychopaths from being relevant in any way or reaching any position of power whatsoever in both the secular and religious worlds. We can never, ever, ever let a lunatic, liar, self-appointed hand of some lord or a mad dog rise to power.

Chapter Nineteen

How Can You Possibly Be Moral If You Believe In Jehovah?

"But if ye will not hearken unto me, and will not do all these commandments; I also will do this unto you; I will even appoint over you terror, consumption, and the burning ague, that shall consume the eyes, and cause sorrow of heart: and ye shall sow your seed in vain, for your enemies shall eat it. And I will set my face against you, and ye shall be slain before your enemies: they that hate you shall reign over you; and ye shall flee when none pursueth you."
- Jehovah

The question I ask in the title of this chapter is not at all rhetorical, and I do not ask it for any kind of shock value. It is a wholly honest and legitimate question, particularly when referring to Christians in the United States. This question turns around the irritatingly common and totally groundless assumption that to be moral one must worship a specific imaginary supernatural god, in this case Jehovah, and also believe in and live one's life according to ancient snuff science fiction and mythology.

As a lifelong disbeliever in imaginary supernatural characters of any kind, I have on multiple occasions been asked by persons of some faith or other how I can be moral without a belief in a god. Since I live in America, this question is often posed by a devotee of one of the seemingly endless variants of the Christian cult, and is asked with reference to the Jesus/Christian god figure. The question has generally not been asked in a condescending or truculent tone, but in genuine lack of comprehension and disbelief, as if I had said that I didn't attend college but nevertheless earned a degree.

It is astounding that millions of my fellow nonbelievers and I continue to be asked this question, when the vast majority of gods to choose from, including the single or tripartite Daddy/Junior/holy spook Christian specimen of the breed, exemplify the worst kinds of immorality. Why does the Christian worship a god who has, both through his direct action and by ordering and inspiring the actions of humans, caused enough innocent blood to be shed to fill a medium-sized lake? Who is said to have killed every living thing on earth? Who created his own personal fiery agonizing eternal torture chamber? Who condones slavery, murder, genocide, rape, child abuse and misogyny?

Who in truth is no more moral than any of the Mesoamerican jungle gods who demanded the sacrifice of uncountable numbers of innocent people, including young children?

It is unfortunately far too easy for a believer to achieve the combination of blatant faith-based ignorance, egotism and self-deception necessary to ask this question of we nonbelievers, as the justification for it stems from the assertion that worship of an omnipresent and omnipotent supernatural terrorist and sadist somehow makes one a moral person. This is absolutely incomprehensible in any truly moral human society or code of ethics, or to a rational and thinking person. How does a person who adores and worships and suborns himself to a mass murderer and eternal sadist and bigot have the galactically unparalleled gall to ask me or any other secular humanist how we can be moral?

It is because most humans are ignorant, pathetically incapable of overcoming fear, incapable of and/or unwilling to think critically and for themselves, and incapable of and/or unwilling to escape their herd.

In reality, humanity is not an object of special creation by a single one of the uncountable thousands of gods which humans have invented. We are merely the most intellectually capable species of primate alive today that has survived and evolved to our current position in the tree of life. We are but one of many millions of species of life forms that inhabit a small part of the surface of one tiny planet in one small solar system in one average galaxy in one incomprehensibly large universe, which may be only one of a near infinity of universes.

We arrived on the scene as semi-evolved primates after billions of years of evolution from our unicellular ancestors, and we now gaze fearfully at the stars using our semi-evolved eyes while standing shakily on two semi-evolved and misshapen feet. We could not survive long on much of the earth's surface. A very large number of the visible and invisible life forms with which we coexist can maim, disfigure, sicken or kill us, and many do. We are very fragile creatures, very easily injured, very often infected and often very slow to heal even if we do survive major injuries or illness.

We inhabit this planet following the deaths of untold trillions of other life forms. It is only through the chance extinction of certain types of life in the distant past that the evolutionary path to *Homo sapiens* became available to our mammalian ancestors. We have nearly gone extinct[1] ourselves, and our species survived by only the tiniest of margins. Our understanding of the physical processes of the universe is still limited and is extremely recent. Every new human is born a squalling and ignorant near-blank slate, with no more knowledge of the world it will inhabit than any human today knows about specific events that will occur a million years from now.

We enter the world needing the care and feeding of others, and we cannot survive for more than a few hours without it. We live but a short and sometimes painful life, and endure many stresses before our consciousness permanently ceases to exist. We treasure the love we give and receive from our family and friends, and the satisfaction we gain through the happiness we find in this life.

The spool of our lives spins very fast, and we know it must end all too soon. Is it any wonder that we fear our physical death and want to live forever after it happens? Particularly if we can go to a place where we can see our loved ones again? Is it any wonder that we have invented thousands upon thousands of gods to provide this comfort to us? Far too many humans are incapable of overcoming the stresses and fear they feel, and they turn to religion for the numbing of these fears. This has been a primary factor in religion's persistence in humanity for thousands of years.

Homo sapiens is a self-aware and generally thinking species, but not all thoughts are conducive to the progress of rational thought or independence. Many people are quite content to let information flow into and settle in their heads without filter or question, as a cesspool passively receives waste. Most people strongly gravitate toward information that pleases them or that appeals to their particular world view or bigotry. Most people do not have the education, the training, the ability or the inclination to critically examine the information presented to them. These people are particularly vulnerable to incorrect and indeed immoral information. These same people are often easily led by those who seem very confident and who offer answers to any questions posed to them.

If these persons can also offer the comfort of an afterlife with one's family, the allure of that comfort in and of itself is irresistible for a great many people. Not one single shred of actual evidence of a god, of heaven, of hell, of purgatory, of angels or demons or devils or indeed any of the supernatural boogeymen found in these religions has ever been produced, despite thousands of years of extremely strenuous efforts to do so. Yet billions of people believe in these imaginary fantasy creatures.

The reliance by the religious ruling class and their clergy on the blind faith held by the swarms of the fearful and ignorant has been the key to their continued success over time. The progress of rational and independent thought was deliberately retarded for many hundreds of years, during which rational and independent thinkers stood a very high risk of being captured, tortured and burned alive by clergy doing "god's work". Such a threat both involuntarily and permanently shut the mouths of those too imprudent to keep them shut on their own, and voluntarily shut the mouths of those who relied on prudent silence and discretion to avoid immolation at the hands of clerical sadists and murderers.

The book which spawned much of this behavior, the Christian holy book of fables, has endured almost unchanged in its original tone and content for well over 1500 years, like a permanently poisonous heirloom passed from generation to generation. This book has inspired the Dark Ages, Inquisitions, Crusades, genocides, heresy hunts, witch burnings, slavery, the Holocaust and the uncountable trillions of evil acts that have been committed in the name of that book's god or gods. The words, the passages and the incoherent calls for murder are still exactly the same.

All that has changed is that after far too many hundreds of years, some parts of humanity decided that true morality would rise against the religious sadists and murderers, disregard the holy books and bring humanity closer to a true, rational and humanistic morality.

No honest person today can look at the Christian holy book and deny that it is overflowing with hatred, bigotry, immorality and murder. No honest person can deny that anyone who followed Jehovah's actual commands would be quite busy killing and oppressing people in his god's name and that all this killing would be completely biblically justified.

We are an extremely social species. Much like many other forms of life on our planet, the social fabric of our societies is wound tightly around groupings of family and friends. Our traditional nuclear families, extended families, in-laws, friends, acquaintances, co-workers and the familiar faces on the train or the bus or the playground are all part of our social network. The global obsession with electronic social media is a clear indicator of the power of the social drive we humans share. The incessant sharing and posting and liking and all the other methods of faux-connecting or advertising one's presence and activities, and desperately hoping people notice it, clearly show the strength of the human social drive.

While most humans love attention, most people will generally do quite a lot to avoid criticism, and they are cut deeply when it is received. Some of these cuts never heal, and they can fester for many years and sometimes for life. Intense criticism actually drives some people to end their lives. People don't like to be on the outside of the tent looking in; they want to be on the inside where everyone else is, looking out from the tent at those who aren't worthy enough to be inside with them.

So when the prevailing mood and culture are beholden to a religion, people will follow it and act like herding animals everywhere. They will moo and bleat and stick together for mutual reinforcement and protection from those who would threaten the insularity of their beliefs. So the few who strike out on their own and posit beliefs opposite of those of the herd receive threats, are cut off and ostracized, lose the love and support of friends and family, and have to seek out the islands of other like-minded individuals who have left their herds.

While religious and other beliefs themselves are often bad per se, actions based on these beliefs are far worse. Beliefs that never leave the head of the believer are like a bomb that has not gone off. The destructive potential energy in that belief may be extremely high and extremely dangerous, but if the belief never translates into action in the real world, no harm is done. When the bomb explodes or the belief turns into words or actions in the real world, that is when real people are harmed.

Thus no truly rational thinking person can be moral while carrying out the actual orders of any of the monotheistic gods of Christianity, Judaism or Islam. Those persons who actually are moral in practice, who happen to be devotees of any of these specific groups or any of the many thousands of the splinter groups, sects and cults they have spun off, have achieved the peak of cognitive dissonance or have never bothered to inquire into their books or their faith beyond the placid platitudes pushed by the watered-down modern pulpits. Or they have been restrained by the knowledge that secular authorities such as police, courts and armies stand ready to prevent and/or punish theistically-inspired evil behavior.

No truly rational thinking person can be moral who also ascribes to immoral teachings stemming from any of these faiths or their spinoffs. How can it be moral to feel that women are inferior to men and must be subservient to them, simply because women don't have penises? How can it be moral to block women from key positions in the hierarchy of the faith because they have vaginas? How can it be moral to tell a person who is not heterosexual or whose skin is not white that he or she is less than fully human and not worthy of the same legal protection and rights offered to straight or white persons? How can it be moral to insist that use of condoms is a sin in a continent where AIDS is literally an epidemic, causing pain, misery and death to many thousands of men, women and children? How can it be moral to deliberately allow known child rapists to remain in one's organization? How can it be moral to remain a member of such an organization?

More to the point and to be more specific, how can a human being possibly be moral and still remain a Catholic today, even if that human is totally ignorant of Catholicism's long and extraordinarily evil history? One would think that finding out how widespread child rape is and has been in this cult, and how its leaders from the very top on down were aware of it and engaged in a deliberate campaign of denial, attacking victims, moving perpetrators around and avoiding justice and accountability at every turn, would cause people to leave in droves. This has not happened, and this is pathetic and disgusting.

If you are a Catholic, how can you look any of your fellow humans in the eye and admit you support this organization? Apparently over a billion people don't consider an epidemic of child molestation and protection of the molesters by church leadership as a big deal. One wonders exactly what this cult would have to do to lose the support of its faithful, if raping innocent children and protecting the rapists won't do it. Would it have to stand idly by while six million Jews are murdered? Oh, right, they did that too. So no, based on cold, hard and extremely recent evidence, nothing from enabling genocide to widespread child rape and its coverup seems to be enough to make people leave Catholicism.

One wonders why Jehovah, who is said to have become so hot and bothered even by the tiniest acts of misbehavior among his Chosen People that he killed almost everyone on the entire planet, would sit idly by while those Chosen People were being murdered by the millions. The answers are that he either likes snuff shows on a grand scale or he doesn't exist. One also wonders why Jesus stood idly by while innocent young boys and girls were being raped and tortured in untold numbers by the self-appointed keepers of the keys to his heavenly kingdom. One has the same damning answers.

One wonders why this cult still prohibits women from being priests. Female applicants for priesthood may not have penises, but given the history of this organization and its clergy's predilection for sticking those penises into little children, a hell of a lot more priests without penises would be a major step forward in actually trying to protect innocent kids from priestly rape and the official coverups.

Speaking of rape, if you are a Christian, you have bought the fairy tale that the creator of the universe, aka Jehovah, de facto raped a minor girl under the legal age of consent by not asking her permission to impregnate her, that the issue of this unholy union was born specifically in order to be later tortured to death as penance and a scapegoat for the sins of all humanity, and that this X-rated horror snuff film is actually the best news in the history of the planet. You find nothing odd about this person's appearance during a remarkably ignorant and religion-poisoned period in human history and the extreme similarity of his myths and storylines to earlier myths and religions of the time and place. You never question why the belief systems spawned by his brief time on earth have resulted in almost two millennia of conquest, murder and oppression.

Had this person appeared on earth right before either of the two

World Wars, and began actually curing people and actually performing miracles with actual godlike powers, perhaps the attention of the world might have been moved in a different direction than years of horrific deadly war and tens of millions of deaths. Why didn't he? The faithful will say that their god works in mysterious ways, which is faith slang for they really don't have a fucking clue, they prefer blind faith to any kind of actual reasoning and "will you stop asking such difficult questions so we can stop thinking about it?"

Jesus or his father could have made an appearance in the religion-soaked United States before the Civil War, by making a quick, loud and direct public service announcement from the sky to inform everyone that having another human as your slave is always bad and immoral no matter how much you want one. This would have resolved the biblical contradictions on that subject and would have forever put an end to the question of the morality of slavery, not to mention saving 700,000 lives and preventing millions of injuries and hundreds of years of Jim Crow and other oppressions. Why didn't they?

The believers say there surely had to be a purpose for all of the killing and oppression, although there is nothing a sane and moral person can understand about such an idiotic statement.

Just because you don't have the intellect or morality to see that avoiding millions of deaths, or even 700,000, is a good thing per se, does not mean that the rest of us have to turn off our morality and critical thinking.

Jesus could have appeared during any of the mass death events of the Black Plague and the later Spanish flu or AIDS epidemics, and performed a laying on of hands to heal millions of the sick and the dying. Why didn't he? Christians will say that our ancestors' misuse of their divinely granted free will led to the original sin in the Garden which unleashed the evils of disease on humanity, and we sorry fallen miserable sinners deserve what we get. Terminally insane escapist sadomasochism is almost never so clearly articulated as when the faithful go off on this lunatic fringe tangent.

Perhaps making his presence known before one of these global humanitarian disasters might have been a better use of the Christian lord and savior's time than sitting on his golden throne and eating bon bons while watching the shows and deliberately choosing not to answer the trillions of heartfelt and desperate prayers crashing the heavenly phone lines. It is clear that Jesus and Jehovah care nothing about the welfare of the allegedly most important and beloved species in the universe, they get a kick out of watching all the pain and suffering, or they don't exist at all.

Speaking of prayers, the person of Jesus Christ specifically told his followers that they should not pray in public. Yet every Sunday, and indeed on many other days of the week, millions of the faithful herd like sheep to fill the churches and the bank accounts of those who ignore great swaths of their holy book while preaching the most bigoted pieces of it. And in a great many cases in America, they are knowingly and unconstitutionally violating their lawful tax-exempt status by directly or indirectly proselytizing for specific candidates for political office.

And regarding the conception of Jesus, the Christian faithful practically have strokes when a freethinker mentions their god impregnating a teenager without her consent. They say "No, no, no, Mary gave consent! Really, she did!" To respond to this, the freethinker need merely pull out the Christian holy book to set the fictional record straight. So what do the New Testament gospels say about this situation?

Mark, which is the first of the four gospels and was written in about 70 CE, says absolutely nothing about the situation. Mark's story essentially begins with Jesus already grown up and doing his thing. Either the first gospel author thought it totally unimportant to mention the allegedly hugely important divine impregnation event or this particular fiction had not been invented yet.

Matthew, written in about 85 CE, says only that this girl "...was found with child of the Holy Ghost." No consent was noted and there was not even a discussion between the Most Holy Impregnator and the teenage girl about the situation. Matthew says that an angel appeared to Mary's husband Joseph in a dream before the two of them had consummated their relationship, and that this angel told Joseph to cool his jets about the whole situation because it had actually been the Holy Ghost who was the dirty devil who'd gotten his woman pregnant and not some other human male.

John, written in about 90 or 100 CE, also says absolutely nothing about this impregnation and also begins with Jesus grown up and making his rounds. For half the gospels, apparently the whole virgin mommy and divine daddy of the global lord and savior thing just isn't important enough even to mention in passing.

Only Luke, written in about 85 or 95 CE and based on at least Mark's previous writing, and being one of the three so-called "synoptic" gospels along with Matthew and Mark, gives any details about this alleged event. What little it does say is what one would expect of a misogynistic society in which women are unclean beasts who have to be tolerated because they are unfortunately a necessary part of the business of making sons. So what does Luke's fairy tale say? He says an angel swooped down from heaven to Mary, making sure to do it when there were absolutely no other people around who could have witnessed the event, then told her that she alone among all women had caught the god's wandering eye and that the god was going to impregnate her.

This angel didn't ask her if this was OK, he did not bring a love letter written on a magic scroll signed by Jehovah asking her permission to be impregnated, and to top off the misogyny, Jehovah didn't even bother to personally speak to the future mother of his very own and only begotten son about the situation. The tale notes how Mary was understandably confused and asked how she would become pregnant without having sex. The angel breezily noted that "The Holy Ghost shall come upon thee, and the power of the Highest shall overshadow thee", which was wholly uninformative to say the least.

And what was Mary's response to this message from heaven? She said "Behold the handmaid of the Lord; be it unto me according to thy word." The word "handmaid", by the way, means a female maid or servant, meaning Mary responded in the same way any member of a king's harem would when he called on one of his concubines to be his entertainment for the night. She didn't say "Wow, really? That would be great, let's get it on!" She said "yes, master, my king, do with me what you will."

This is not consent, this is a slave doing what she is told. Only a highly damaged person having both utter contempt for women and a total failure of critical thinking and empathy will not see this for what it is, which was an act of impregnation without truly willing consent.

Do slaves and concubines get to tell their slave masters and owners no? Do peasants get to tell their king no? Of course not, unless they want to be beaten or beheaded, and when the king and slave master is believed to be the murderous psycho from the Old Testament you better goddamn not tell him no. Telling Jehovah no at any time is a good way to get yourself turned into a pillar of salt, firebombed, drowned, stoned or painfully killed in some other horrible way. Telling him "no, thanks, I don't want to carry your baby" would surely result in a level of divine sadistic punishment beyond even Old Testament levels.

And Mary, being a slave to her god, a second or third class citizen simply by being female, and knowing of her god's tendency to torture and kill when he gets pissed off, would have been in no position to say anything but what she did. What would you say when a psycho is holding a gun to your head? Unless you have a death wish you'll say anything you think will make him not pull that trigger. The spineless and brainless rationalizations by the believer in this story are just more examples of the sadism, masochism and utter broken thinking of Christianity and Christians in full evil bloom.

So no, we will have no foolish idiocy about saying Mary willingly considered the proposition and had any ability at all to say no. And this is over and above the inability of these allegedly divinely inspired or dictated fairy tales to agree on how the imaginary Christian lord and savior got here. These fairy tales are simply laughable and those who believe them are damaged human beings.

Ridiculous tales of virgin births and divine baby daddies aside, the secular humanist has none of the baggage of institutionalized worship of ancient mythology or any compulsion to protect a religious status quo. The freethinker is by definition an independent thinker who will not suborn her brain to the narrow constrictions and hateful bigotries preached by the churches. The rationalist can think critically and objectively, unhampered by the pathetic need to twist and reverse-engineer actual positive and humanistic morality and real science to fit with a specific ancient desert tribe's bizarre local fables and science fiction. We can look at a forest and see both the individual trees and the forest as a whole. We are not blinded by dogma or fear.

We know the holy books are evil. The teachings of the religions they spawned are immoral. The actions of the clergy and the faithful are very often evil and immoral. Jehovah is far worse than any mass murderer or genocidal lunatic in human history. Put Hitler, Caligula, Genghis Khan, Stalin, Pol Pot and the murderous Crusading popes into one single entity and it still wouldn't be as evil as Jehovah. Yet billions of people throw themselves at his feet, adore him as a loving god and think he will answer their prayers. The English language doesn't have words strong enough to express the disgust and scorn this imaginary god deserves, although how Richard Dawkins described him in *The God Delusion* was the best I have yet read. Dawkins noted quite accurately that:

> *"The God of the Old Testament is arguably the most unpleasant character in all fiction: jealous and proud of it; a petty, unjust, unforgiving control-freak; a vindictive, bloodthirsty ethnic cleanser; a misogynistic, homophobic, racist, infanticidal, genocidal, filicidal, pestilential, megalomaniacal, sadomasochistic, capriciously malevolent bully."*

Were we a truly evolved species, actually able to think and be moral, it would be impossible for a normal human to even begin to admire Jehovah, much less worship him. Yet here we are.

I have to ask the original question one more time. How can you possibly be moral if you believe in Jehovah?

Chapter Twenty

Good News? You Can't Possibly Be Serious

"If we had to earn our age by thinking for ourselves at least once a year, only a handful of people would reach adulthood."
— Mokokoma Mokhonoana

"It is often argued that religion is valuable because it makes men good, but even if this were true it would not be a proof that religion is true. That would be an extension of pragmatism beyond endurance. Santa Claus makes children good in precisely the same way, and yet no one would argue seriously that the fact proves his existence. The defense of religion is full of such logical imbecilities. The theologians, taking one with another, are adept logicians, but every now and then they have to resort to sophistries so obvious that their whole case takes on an air of the ridiculous. Even the most logical religion starts out with patently false assumptions. It is often argued in support of this or that one that men are so devoted to it that they are willing to die for it. That, of course, is as silly as the Santa Claus proof. Other men are just as devoted to manifestly false religions, and just as willing to die for them."
— H.L. Mencken

Christians go on endlessly, and much more frequently around days which they fabricated or co-opted through the centuries to advance their religion, such as "Christmas" and "Easter", about what they call the core of their faith, their *raison d'etre*...this thing they call the "good news".

What is this so-called good news?

Christians think their god decided that humanity needed external redemption from the sins which he had deliberately programmed them to commit, so he sent his 'only begotten son' down to be killed as an innocent scapegoat by these miserable sinners.

It is claimed, in an unbelievably ironic, nonsensical and self-contradictory idiotic non sequitur, that through the heinously evil and sinful act of murdering an innocent person, the entire human race, including the murderers, would receive redemption for their sins, thus the future of humanity and its relationship with its creator would be secured. So it is on divine command that humanity is supposedly redeemed of sin through acts of ultimate evil, both those of the humans doing the murdering and those of the god who deliberately sent his kid into the meat grinder.

This is good news? In what possible way could this be considered good news? What this insane little bit of mad muddled Mesopotamian mythology is really saying is this:

1. The creator of the universe selected this one planet out of many billions in the universe on which to create life in many forms. One life form was a particularly ignorant and stubborn species of mostly hairless primate. The first specimen, call him Man, was a mostly hairless male primate which the creator made from dirt for some unknown reason. This individual got lonely so the creator put him to sleep and, rather than animating another pile of dirt, he for some unknown reason reached into the Man's chest and pulled out one of his ribs, from which he made a female of the same species. Call her Woman. He threw both of them into a small game preserve prison with no education, no books, no distractions or anything else to do, told them not to eat a specific fruit and then went on vacation.

2. Shortly after returning from vacation, the creator got insanely mad at these two completely uneducated and naïve teenage primates for being so insanely bored and curious, as the creator had deliberately programmed to be, that they ate that fruit thus disobeying his direct orders. He then kicked them out of their home and cursed them and every single one of their descendants for the rest of time. One must assume this curse also included the son the creator would later have with one of these descendants, but special exceptions were invented to explain this embarrassing logic problem.

3. The original pair, now having plenty of time on their hands and still not much else to do, was thus able to bring forth a growing population onto the earth. Due to some quirk of genetics, however, they somehow avoided the problem of inbreeding in their young, the creator built in some kind of short-term genetic patch for that problem, or all their descendants did indeed suffer the mental and/or physical problems resulting from inbreeding and all current and recent humans are the comparative mental equivalent of gibbering inbred idiots. That last is the most likely option, based on much human behavior and all religious behavior.

4. The descendants of the original two incorrigible sinners fell into a similar trap as did their ancestors, and having inherited their original sin and behavioral habits and also the creator's repeated inability to properly program and train his offspring, they again sent the creator into a fit of rage. He angrily drowned almost every living thing on

earth to include animals, plants, innocent men, women and children and also an extremely large number of unborn babies, thus becoming the world's first recorded abortionist by committing a mass abortion of unborn children all across the world. The creator kept a few breeding pairs alive during this event, and after the murdering waters receded they yet again populated the earth, as had their ancestors. Evidently the sin of the First Pair was *still* locked in the genes of even those chosen for saving, because the descendants of those spared from the flood eventually got into trouble too. The inbreeding question remained unanswered.

5. At some point later on, the misbehavior of the growing numbers of mostly hairless apes evidently crossed yet another threshold of irritation for the creator, so he apparently chose yet another option to get the message to them, the first two having failed despite the god's alleged omniscience, omnipotence and foreknowledge of their certain failure. For this attempt at solving the problem he had originally inflicted on these apes, he dreamed up a plan inspired by both Rube Goldberg and Charles Manson. This plan involved a lot of particularly ignorant semi-evolved partly hairless apes in a particularly ignorant and backward part of a particular desert area. When he felt the time was right, he set his contraption in motion.

6. He started the plan by de facto raping a teenage girl in this ecological and intellectual desert, allegedly magically engendering a fertilized egg directly in her uterus without asking for her consent. This teenager has been praised as a virgin ever since that time, in order to bolster the magical impregnation story. The fertilized magic egg would, somehow, become simultaneously the son of the magical creator and the son of this singularly selected mostly hairless teenage female primate. Not too terribly long after his birth, this son became an itinerant preacher, avoided women and only hung out with other male primates while getting himself into a lot of trouble with the locals.

7. When this half-breed son of the creator of the universe and a teenage semi-evolved primate was in his early thirties, some of those locals decided they'd had enough of his big mouth. So after he was betrayed to them by one of his best friends, exactly in accordance with the creator's original plan, the locals tortured him, took a couple of beams of wood, made them into a big cross, made him drag it across town then nailed him to it with big spikes. He died not too

long after this, was dead or sort of dead for a few days then allegedly came back to life for an indefinite time before merrily flitting off to join his supernatural creator father in a magical spirit kingdom Never-Never Land.

8. With the torture and death of this half-breed, all of humanity, even those who had never heard of the creator, his son or the son's teenage mother who'd been impregnated without her consent, and would have laughed at the foolishness of the story had they been told about it, were supposed to be brought back to some kind of relationship equilibrium with the angry creator. This equilibrium would be a state in which all of humanity's collective sins were forgiven and the path to heaven and eternal life was (re?)opened for all mostly hairless primates having the approved extremely specific religious beliefs and practices.

9. Thus despite the creator's previous very explicit commands not to lie to or kill people, which had both been written down many years before in a book he allegedly either inspired or dictated to the ancestors of the people who killed the son, his plan required that his own son be betrayed then murdered in an act of heinous sin.

10. So commission of the sin of betrayal and the sin of murder were both consciously mandated parts of the creator's plan to absolve humanity of the sins which they had been committing and would continue to commit for the rest of all time because of how badly the god had designed and programmed his people in the first place.

All of this happened exactly as the creator of the universe had planned it for trillions of years, down to the last detail. Did the plan work out as intended? Let's see what history says.

1. The number and degree of "sins" committed by mostly hairless primates across the rest of the world continue unchecked even unto this day, absolutely unaffected by the goings-on in this Mesopotamian desert village. Thus the original sin and defective design deliberately built by the creator into these primates are still in full force.

2. For almost two thousand years, an entire group of people, the Jews, have been persecuted, expelled, imprisoned, vilified, tortured and killed because they have been collectively and individually blamed

for killing the half-breed son. This despite the murder occurring *exactly* as the son's creator father / their god wanted, and without which the world would have no allegedly "good news".

3. The followers of this half-breed son cannot agree on how to worship him or his father, and have split into many thousands of competing and discrepant sects.

4. The most devout followers of this half-breed son have engaged in endless conflicts, wars, crusades, inquisitions, murders and sins across the entire spectrum of evil in the devout pursuit of their beliefs and their disagreements with other equally devout believers in the same, similar and other faiths.

5. One of the most devout groups of followers of this half-breed son formed a theocracy which has a self-described infallible leader, who calls himself the One True representative of the half-breed son on earth. These allegedly infallible leaders have been personally responsible for frauds, lies, wars, tortures, murders, invasions, disease epidemics, near-genocides and supporting the persecution of Jews up to and including when six million of them were killed in the span of a few years.

   As if this weren't enough, this group later protected many of the people who helped kill the six million, covered up the rape and torture of children by the members of this group, refused to fully accept responsibility for all these sins and also refused to cooperate with civil and a far more civilized secular society in attempts to fully discover and address these officially-sanctioned crimes.

6. The most devout followers of this half-breed son pick and choose the most evil selections from his father's and his words to continue to intimidate and persecute innocent people, and sometimes to kill them.

7. The most devout followers of this half-breed son choose dogma and unfounded belief to deny reality and deny science. This denial is leading to grave damage to individual people and the entire world.

8. The most devout followers of this half-breed son believe that no matter how bad a person's life is, their imaginary afterlife will make it all better. Thus they think there is no need to worry too much

about all the pain and suffering they or others inflict in the real world.

9. The most devout followers of this half-breed son believe that no matter how badly damaged the earth becomes, he will one day come back to earth a second time, avoid getting murdered again and will lift all of the devout into heaven. Thus there is no need to worry about all the damage they are doing to the world and the plants, animals and humans living in it.

10. This second coming is many, many centuries overdue, and apologists continue to trip over themselves to try to rationalize the fact, while damage to individuals and the world continues to grow, almost all of it committed by the devout.

11. The unbelievably immoral message being sent by this insane fairy tale, and the one which every Christian accepts, is that someone can accept responsibility for someone else's misdeeds. This is as immoral and antithetical to a true morality as it is possible to be. It greatly diminishes the actual need for actual people to actually be good to each other and to solve actual problems in the only world that actually exists. It says that it is OK to let someone else take the blame and suffer the punishment for one's deeds. This is antisocial, this is unethical and this is a diabolical perversion of morality.

I have an as-yet unanswered question for Christians. Why do they make such a big deal about the alleged "only son" or "only begotten son". What's so special about this particular half-breed son? If their god really exists, he or it could have made billions of sons and implanted them into any woman or group of women anywhere at any time. He could have even implanted multiple such sons into one or many women at the same time in order to have hundreds or even thousands of sons spreading the divine gospel all across the world.

If he is truly possessed of the advertised ability to create universes and all the rest of the magical powers his followers think he has, his disastrous and utterly failed choice to only send or make one half-breed son is just that – his choice - and there should be no extra meaning given to the singular nature of his earthly offspring. Unless, that is, one is so deeply trapped by this religion's wholly illogical mythology that one's ability to reason is completely destroyed, which is quite factually the case for the uncritical swarms of the faithful.

For all we know, every single human male in history is actually the offspring of this same god, who has been madly buzzing around the world for millions of years busily replacing human and pre-human semen with his own divine swimmers immediately before any mortal sperm met egg. No one can prove otherwise, because we can always use the religious logic fail that it may *look* like a specific male and female pair had a child, even down to their DNA…but that's just because their god wanted to make it look that way.

So where exactly is the so-called good news in all of this? It is quite obvious that the only news which this kind of supernatural storytelling and wishful thinking brings is the worst possible kind. The news is that humans can be unbelievably ignorant, stupid and evil, and that evolution has not yet shaped enough human brains to enable widespread rationality and critical thinking.

It's time we grew up, stopped being frightened children believing in fairy tales and started talking and acting like responsible grownups with responsible grownup brains.

It's time we understood that religions are only local mythologies, getting rid of them would improve humanity overnight, that only humans can make this world a better place, and that time is running out to do so.

It can happen, and it can one day be true that humanity has truly rejected the childish foolishness of religion.

Finding that out would actually be some really good news.

Chapter Twenty-One

Why Is Any Moral Person Still A Catholic? (The Church of Evil, Abnormality and Hypocrisy)

"I am surrounded by priests who repeat incessantly that their kingdom is not of this world, and yet they lay hands on everything they can get."
- Napoleon Bonaparte

"It were better for him that a millstone were hanged about his neck, and he cast into the sea, than that he should offend one of these little ones."
- Jesus of Nazareth

"Despite some institutional reform, individual leaders of the church have largely escaped public accountability. Priests were raping little boys and girls, and the men of God who were responsible for them not only did nothing; they hid it all. For decades. Monsignors, auxiliary bishops, bishops, archbishops, cardinals have mostly been protected; many, including some named in this report, have been promoted. Until that changes, we think it is too early to close the book on the Catholic Church sex scandal."
- 2018 Pennsylvania Grand Jury Report on Systemic Catholic Sex Abuse and Coverup

"It is not the monsters we should be afraid of; it is the people that don't recognize the same monsters inside of themself."
- Shannon L. Alder

What do you consider normal? What do you consider abnormal? If everybody you know talked and dressed generally like everyone else and as is customary for your culture, you would see those behaviors as normal. If you ran into someone who talked like he had just sucked down two cubic feet of helium, wore no clothing but a transparent poncho and had painted his entire body in highly fluorescent colors, you would see these behaviors as abnormal. Interesting, perhaps, but definitely abnormal.

Note that context and intent matter in such circumstances. If you were at an especially avant-garde costume party, such a person's appearance and affect might not be exceptionally unusual even when compared to the other guests. And even if you met the person on Main Street at lunchtime, it does not necessarily mean the person has any abnormalities, mental issues or unusual intent. He might just be on his way to that costume party but his car broke down. Or he might just be a weird person who is far more moral than you or your Aunt Sue but likes to be extremely expressive in the desire to stand out from the crowd. Or he might have lost an unwise bet and had to pay the piper. Who knows?

Avant-garde costume parties and bad bets notwithstanding, most people believe things are normal if those things are what they see and experience most often. They believe things are abnormal if they occur rarely and/or differ from the cultural or experiential norm. One person can be in an abusive or repressive home while none of his classmates are so cursed. The daily life of this unfortunate person is his normal existence but would be highly abnormal to, one hopes, everyone else and to a commonly held standard. Indeed, child abuse is against secular law in civilized societies, proving that humans absolutely can and do figure good and bad out for themselves if they ignore their ancient holy books and the immoral gods and words they contain.

Note also that the word "abnormal" has negative connotations. Every person occupies a different point on the bell curve in all human characteristics, for instance in physical strength, physical attractiveness and intelligence. Those at the very top of the curve in these characteristics are often called "exceptional". When used to describe highly capable people, the term "exceptional" carries a strongly positive connotation, and is usually applied when a deviation from the norm occurs in a way that strongly accentuates a positive and personally or socially desirable characteristic. When deviations from the norm occur on the other side of the bell curve, such as when someone is exceptionally oddly attired, unintelligent, weak, small, shaped or unattractive, the term "abnormal" is often used and it has a clearly negative meaning.

People can be abnormal, other organisms can be abnormal, storms and other natural events can be abnormal and human organizations can be abnormal. Humans are often stuck with their inborn abnormalities, since it is the rare person who can change height or IQ or regenerate organs and tissues at will. Human organizations are not made of stone and they are not unchangeable. They are made of clay and can be changed very rapidly to remove any original abnormalities or any that have accumulated over time, if the people in the organization truly want to change it.

Speaking of organizations, we can assess them and determine if they are normal or abnormal just by asking a few questions and doing a little bit of research. If someone knocked on your front door tomorrow out of the blue and asked you to join his organization, what would you do?

Even a half-witted and moderately inquisitive person would want to know the name of the organization, what it does, what it stands for, its history, who have been members in the past and who are current members, who runs the organization, is it trustworthy or sketchy, what it would demand of you and your family and other such questions. What if this person at your door told you his organization has the following characteristics:

1. It is a religious group led by a theocratic dictator who is the absolute ruler of his country.

2. The religion is based on ancient local fables and mythology that were borrowed from other local and regional fables and mythologies, and on no actual evidence at all. Out of a large group of tales, some were selected for inclusion into a single book which was put together nearly two thousand years ago. Group members believe in and live their lives based on this book, which they think was inspired by[1] an original male god who is also the creator of the entire universe. They believe this despite admitting that the creator was not completely forthcoming when he inspired the book and that certain special vested members of their particular religious ruling class had to figure out what was missing.[2]

3. Despite the admittedly extreme lack of clarity and consistency in the book, that book and their religion are based on the human sacrifice of one of the book's central characters.

4. Only males who voluntarily and extremely abnormally don't have sex with women and who choose to live a cloistered life with other males can be part of the ruling class of the religion. Females are not permitted to become full members of the organization.

5. The group was founded by a group of similar males almost 2000 years ago who also didn't care too much for females. These males say they are not homosexual despite abnormally swearing off women, although many of them are indeed gay. They are nevertheless both hypocritically and vehemently opposed to homosexuals marrying or having equal civil marital rights.

6. These males and other group members call their organization a church, and refer to it using the female pronoun "shc". Thc church is called the 'bride" of its central figure, who died 2000 years ago but became undead a few days later.

7. These males believe that every human is riddled with original sin[3] because of the acts of two extremely remote immature teenage ancestors who never actually existed, and only their particular religious group can help you solve your permanent and very big sin problem.

8. The founder's father, who they believe was a god and the creator of the universe, allegedly magically impregnated a human female teenage virgin who was already married to a human male. Almost 2000 years after the impregnation, in 1854, it was suddenly decided that the mother was wiped clean of original sin when the creator of the universe magically impregnated her.

9. They believe this human female later gave birth to a son, but that this son was/is actually a god who had already been 'born' out of the male creator god of the universe at some time in the past. They believe this son/god is both separate from and an integral part of a triune collective entity consisting of the creator of the universe, the son/god and also a ghost, and that these separate but equal entities are at the same time separate from each other yet completely conjoined.[4]

10. They believe that this son/god was deliberately placed into the human female and was later born for the sole purpose of achieving salvation[5] for the sins of humanity. In order to acquire this salvation, this son/god had to be tortured and murdered in a remote desert 2000 years ago.

11. This murder was set in motion by the creator of the universe, meaning the victim set at least part of himself up for murder and thus the event was a suicide, even if only in part and even though the suicide victim came back to life almost immediately. Despite this act being both a god's partial and reversed suicide and also that god's mandated path for gaining salvation for humanity, it was nevertheless considered to be a grave Jewish sin. This alleged sin was a collective, permanent and

unforgivable act to be paid for by every single Jew after the time of the murder.

12. These males preached until 1965 that the collective Jewish sin was sufficient justification for persecuting and killing them. Thus the temporary suicide of this son/god who a few days later came back to life then went straight home to heaven was the reason behind centuries of Jewish persecution. As a result of this persecution, this organization set the conditions for the murder of six million Jews in the first part of the 20th century. The allegedly infallible theocratic leader of this organization had previously signed formal accords with the people who committed these murders.

13. This son/god only stayed dead for three days. He came back to life, then a short time later rose bodily into a magical fairy tale spirit kingdom where he now resides with his father and spirit partner ghost, and from where he communicates with people on earth, or not, at his discretion.

14. Members of the group believe that they eat the reincarnated body of the son/god every week or so during a magic ceremony. The abnormally sexless woman-avoiding males believe that when one of them says special magic words over a cracker, it magically transforms into the reincarnated body of the son/god, which is eaten by congregations in an act of ritual cannibalism. Similar magic words directed toward fermented grape juice will magically turn it into the blood of this undead son/god, which the congregations drink to wash down the body of the son/god.

15. Men and women from the general population are encouraged to join the males in worshiping their three-in-one or one-in-three god and ritually cannibalize him on a weekly basis.

16. After the male clergy vote for one of their number to be the Maximum Leader of their theocratic fundamentalist state and religion, the winner is magically imbued with infallibility when he opines on faith and morals. This assertion remains in place in defiance of the facts of human psychology, the centuries of bloody crimes this church has committed against humanity and the immorality still perpetrated in the modern age by these theocratic leaders. They cannot explain the reasons behind the

centuries of evil committed by these boss men and their field agents.

17. The creator of the universe personally committed mass murder and mass abortions and ordered others, and was quite happy when his chosen tribe committed them. These males do not consider these acts, which defy some of the god's own direct orders to humans, as immoral.

18. The organization was founded by the son/god who liked to break apart families and only hung around with other men.

19. Many of the allegedly celibate males have raped children for many years. Leaders in this group have paid billions of dollars in hush payments and blood money to the victims, moved the perpetrators to new hunting grounds without telling anyone, covered up the crimes, relocated guilty senior leaders to the home base and have done all they could to minimize the damage to the church. Some of these senior leaders were quickly moved to cushy positions under the theocratic dictator at the church's home base in order to avoid secular authorities and secular prosecutions.

20. These abnormally celibate and woman-avoiding males believe they have the only true source of morality, despite well over a thousand years of horrifically antihuman evil and persecution. They also think they are uniquely qualified to advise normal married men and women on their normal male-female relationships and morals. Despite their supreme leader's alleged infallibility, these boss males and their willing underlings have been responsible for:
    a. Hundreds of years of assorted bloody Crusades.
    b. Nearly seven hundred years of bloody Inquisitions, not ending until the 1800s.
    c. Centuries of witch burnings, heresy hunting and mini-genocides.
    d. Centuries of the sale of indulgences and other ill-gotten financial and property gains.
    e. Far more than a thousand years of persecuting and committing atrocities against Jews.
    f. Invasions and murder of countless native people across the world.

g. Countless acts of rape and torture of children all over the world, then lying about it and covering it up.

21. These males used to, and some still do, believe that some women are witches. They also believe in demons, possession and exorcisms, and some of them conduct exorcisms.

22. These males, from the theocratic dictator Maximum Leader down to far too many of the clergy and the men and women in the pews, believe there is an evil spirit called the Devil. This devil is able to reach into and affect the minds and actions of humans to make them do and say evil things, while the creator of the universe and his once-dead but now risen-to-the-heavens son along with all the saints and angels and other heavenly powers do nothing about this devil's actions. They could eliminate this devil in a nanosecond, but deliberately choose not to.

23. They believe that they can telepathically communicate with the creator of the universe, his son, his son's dead mother and many other mythical creatures, called saints and angels. They further believe that pleas to these creatures, known as prayers, will be answered and acted upon or not, at the discretion of whichever entity is the intended recipient of the telepathic plea.

24. They believe that condoms and birth control are immoral, and that the issues of contracting a fatal disease, contributing to massive numbers of orphaned and neglected children and overpopulation are of trivial comparative significance. In other words, they would rather have tiny children orphaned than have the parents of those children use condoms.

25. They believe that the son/god they all pray to will come to earth again at some point very, very soon, maybe even tonight, to pass eternal judgment on dead people and also people who are still alive. They have been saying and believing this for two thousand years.

How far down that list would you have to go before you asked the person to please get the hell out of your neighborhood? And if he persisted, how long would it take before you would call the police and the local mental institution to get this dangerously insane person away from you and your family? But as of this writing, about a billion people are members of this church, and many proclaim it proudly. It is very clear that extreme immorality is no barrier to becoming a Catholic priest or moving up in the hierarchy all the way to the very top. And neither is extreme abnormality.

Catholicism is so overflowing with repression, sadism, masochism, superstition, illogic, cruelty and evil that it would be hard to find another system as antihuman and abnormal. Even normal thoughts and behaviors are considered abnormal to these people, which should of course be the first clue how abnormal this church is. We will need to go into some discussion of human normality and abnormality to paint a fuller picture.

So let's examine some real life situations. Is it normal to get hungry and thirsty? Of course. Is it normal to get tired at the end of a long day or after vigorous exertion? Again, yes of course. Is it normal to feel physical attraction for the opposite sex? Yet again, the answers are yes and of course. Everyone knows that these feelings are completely normal in every way. And these are all for perfectly understandable reasons given our evolutionary history and proven by the fact that our species hasn't died out for lack of nutrition and reproduction.

One would certainly say that someone who didn't get hungry or thirsty is highly abnormal. Not eating or drinking will kill you in very short order. On the subject of sexuality, there are two basic perspectives, only one of them supported by reality. In the reality camp, there are people who understand that sexuality is a function of biological processes within the brain, it is a product of many millions of years of evolution which shaped us into a reproductively successful species, and it is essentially not consciously changeable.

So since people who know the facts and the science behind our sexuality usually also have compassion and understanding for other people, regardless of their own orientation, they see homosexuality as a variation from the norm, with no impact whatsoever on actual human morality, and nothing for which anyone should be vilified or persecuted.

There are other people who are quite different, and they do not operate within reality. They have a rigidity and hatred for variation from group norms. This is an aspect of their own brains' biological processes and programming, amplified by their societal programming and hugely focused and amplified by antihuman religions. These people almost universally deem homosexuality as not merely abnormal, but actually such a gravely horrible and disgusting condition that people of this orientation should be denied compassion, freedom and human rights, and quite often even denied the right to remain alive.

Reality is clear. Due to their being required for the survival of individuals and the species, built into us by evolution if you understand science and reality or by a divine creator if you deny science and reality, a very strong compulsion to breathe, eat and procreate is quite normal. Having a low or no urge to breathe, eat or reproduce is abnormal. Further, the sexual urge is particularly strong in normal males, as is plain to everyone. And the strength of the urge for sex applies equally to heterosexuals and homosexuals. The difference between those types of sexual expression is in polarity, not power.

The Christian god is extremely clear on the subject of procreation. Shortly after allegedly creating all the animals but *before* creating humans, this god is said to have told the fish, birds and land animals, presumably in languages they could understand, to "Be fruitful, and multiply, and fill the waters in the seas, and let fowl multiply in the earth" (KJV, Genesis 1:22). One wonders which animal related this story to the authors of Genesis, and in what language. Or perhaps, as people who actually use their brains can figure out, this animal was the two-legged primate kind and was itself one of the authors of the book.

Biblical authorship aside, sex wasn't just for the birds, bees, fish and animals. The god is said to have given the newly-minted original pair of humans the same exact guidance to be fruitful and multiply, right after creating them. So this god is all about multiplying, and since even the devoutly faithful usually know where babies come from, their god not only encouraged but ordered everything with sexual organs to give those organs of reproduction quite a lot of exercise.

Later on, in Genesis 8 after the god had angrily drowned everything alive on earth including vast numbers of unborn babies, sparing only some animal breeding pairs and a handful of suddenly very lonely people with planet-sized cases of PTSD and survivor's guilt, he told the people to let the animals off the big life raft so they could go forth and, again, be fruitful and multiply. Of course the humans received the exact same guidance moments later. So the biblical god is very clear on this. This god's Divine General Order Number One, to animals and humans alike, is to breed like flies.

We are then told that somewhat later, this god cross-bred with a teenage already-married Jewish girl and the "fruit" of that event was a single male child. This hybrid boy would have been expected to behave like every other normal heterosexual male ever born. Meaning he would have shortly begun to show great interest in females and eventually start a family, both according to his normal drives and according to the clear instructions given by his father, alleged to be the creator of the universe.

This would have been normal behavior for any time period and normal behavior for any normal human heterosexual male.

This is not what happened, from the time he started puberty through his death in his early thirties.

In the years before his death, this man who may or may not have existed but who many now call Jesus, collected a group of followers, advertised to us as his disciples. They were all males since no women were allowed into his boys-only club. This young man did not marry, did not have children, and apparently showed no romantic interest in women at all. A nice Jewish boy who never had a girlfriend in his entire life? A male whose divine father specifically ordered his specially created and favorite species to be fruitful and multiply like crazy but who apparently never even touched a woman?

One would think if the creator of the universe had fathered a son via a human woman, surely he would also want grandchildren, and surely he would want his divine DNA to spread like wildfire amongst his singularly chosen people. One would think having an actual god in the genetic woodpile would be a huge plus in the dating game, not to mention giving some very interesting readings in future genetic studies.

Many of the main male biblical characters, as well as prophets and sect founders throughout history, have been not only married to at least one woman but also in many cases were nearly unstoppable in their pursuit of nubile young women. From polygamy to concubines to all the other ways men use religion to get large quantities of women into their beds either voluntarily or involuntarily, one of the constant themes in religious history is that of rich and powerful men having lots of sex with many women. So why was this not the case with the "only son" of the "be fruitful and multiply" god? This can be a lot of things, but it is quite definitely not normal. Why no women? Some possibilities present themselves.

One is that the Jesus story is pure mythology, and the misogynistic biblical authors simply didn't care to develop a female character associated with the main character in the story. This is of course the appropriate answer based on the total scarcity of evidence for anything in the story and its extremely sketchy and borrowed beginnings. It should therefore be the default position for any thinking person.

Another is that Jesus was *just too busy* for women, since he was a god-man on a mission to get crucified so the so-called "good news" of his brutal murder could inspire his followers to start a church and start persecuting people. This is extremely unlikely, and a sane and moral person would say impossible. Normal males *always* find time for women. Give a normal heterosexual male prisoner who is mere hours away from the electric chair or a lethal injection a choice between a ten course meal and ten minutes with a receptive and attractive woman and you can leave the chicken in the freezer.

Jesus had 30 years to socialize with women, to date, to have an arranged relationship, to have a normal heterosexual relationship or even just hook up, but never did? Whatever this is, it is massively abnormal.

Another possibility was that he had a girlfriend, concubine, wife, or a combination, but either the original authors of his mythology or some of his later followers edited her or them out of the story to suit their theological and social narratives. This is extremely possible given how motivated his followers have always been to tell the story their way regardless of how disjointed, illogical and inconsistent it is.

Yet another possibility is that the story was written about a real person who happened to be a gay man who fancied himself as fulfilling ancient prophecy and who preferred the company of a bunch of adoring young male groupies far more than hanging around with women. This is entirely plausible. The biblical Jesus, the one actually described in the book, was a penniless, sandal-wearing, long-haired, turn-the-other-cheek flaming liberal socialist hippie, not the gun-totin' religiously intolerant aggressive Republican bigot that American conservatives have turned him into. Of course Jesus could have been gay.

I know that almost every believing Christian who is still reading this is feeling angry and/or horrified at this blaspheming calumny about the potential gayness of their lord and savior, but to that I have two responses. One is that you are failing the WWJD and bigot tests by feeling that way, because does it really matter a damn if Jesus were gay or straight? Would the Sermon on the Mount need to be interpreted differently based on the sermonizer's sexual orientation? Would Jesus himself be bothered by the assumption at all?

You can bet the biblical Jesus would have done no more than correct the mistake or make up some parable about how being good to other people is the true measure of a person's worth. Or he would follow his own orders, turn the other cheek and say, "go ahead, insult me again, sticks and stones may break my bones but words can never hurt me".

The other response to any aghast Christian is OK, fine, so just prove Jesus wasn't gay. The evidence is pretty strong that he was, or at least that he was asexual, which is itself extremely abnormal. We freethinkers will stop saying we think Jesus was probably gay when you can prove that he wasn't, and we will happily and unashamedly give you public credit if you can do this.

So now that we have established that it is reasonable to hypothesize that Jesus was gay, and we can definitely see that the books show him to be extremely abnormal, we turn our attention to Catholicism, which continues the very abnormal behavior of its central character. A quick reminder of just how much disgust, disdain and distaste the highly misogynistic Apostle Paul had for women is appropriate here.

- From the First Epistle of Paul the Apostle to the Corinthians: "{7:1} Now concerning the things whereof ye wrote unto me: [It is] good for a man not to touch a woman."

- From the First Epistle of Paul the Apostle to Timothy: "{2:11} Let the woman learn in silence with all subjection. {2:12} But I suffer not a woman to teach, nor to usurp authority over the man, but to be in silence. {2:13} For Adam was first formed, then Eve. {2:14} And Adam was not deceived, but the woman being deceived was in the transgression."

Can you ever imagine a normal heterosexual male, at any time and in any place in history, ever saying it is good for a man not to touch a woman? If you ever heard a male say that he didn't think it was good to touch a woman, what would you think about that male? If you are being honest you will of course think that he was gay or had some sort of sexual or social abnormality. This is because that would be 100% true. The history, structure and practice of Catholicism is built on this kind of misogyny, avoidance of females and both general and specific abnormality.

To further inform the discussion and heighten the abnormality, celibacy was actually not mandatory for priests for about a thousand years. It was only when the church decided that it didn't want those dirty sinful women and their dirty damn kids draining resources from the church that celibacy became official policy. In politics and religion, follow the money.

Now we need to talk about normal behavior. How is it normal in any way that for almost a thousand years successive generations of allegedly voluntarily celibate and often gay men passed down their woman-demeaning and women-excluding traditions to successive generations of voluntarily celibate and often gay men? The sex drive in most males is close on the heels of breathing and food in terms of things they must do to remain sane and mentally healthy, and for some of the highest-testosterone years of their lives many men will happily forgo food and drink for feminine companionship.

Deliberately denying these urges is the textbook definition of abnormal, with absolutely massive amounts of masochism leavened in. And denying such natural urges always damages the psyche, as is well known to the medical and common sense communities. Normal humans need the contact of other humans to stay normal, feel normal and act normally, whether they are gay or straight.

So for nearly a thousand years there has been a bunch of abnormal males who self-selected out of the general population and chose to spend their lives with a bunch of other similarly self-segregated abnormal males, and opted on their own not to have sex or intimate relationships with women, contrary to their own god's orders. And these highly abnormal and often highly immoral people are supposed to hold some sort of magical universal secrets to what the rest of us ought to do for normal and moral behavior? This heaps layers of additional abnormality upon the already highly abnormal foundation.

Unlike far too much of the rest of what it does and says in secret, the Vatican's positions on sexual behavior are available in written and electronic media. In 2018, well into the 21st century and in an age when modern science has discovered quite a lot about human sexuality, the Vatican's website still contains these scientifically and morally ignorant and highly bigoted words from the Catholic Catechism[6] regarding homosexuality:

- *"...homosexual acts are intrinsically disordered"*
- Homosexuality is *"objectively disordered"*
- Homosexual acts are completely forbidden. *"Under no circumstances can they be approved".*

So homosexuality is officially considered "disordered" by the Catholic pope and by Vatican doctrine. The hypocrisy it takes for a deliberately and abnormally celibate male ignoring his god's own orders to say that gay people are disordered but their own asexual and often pedophilic behavior is not infinitely more disordered is beyond all rational understanding. Thus, like all the other lunacy contained within religion, this completely contradictory belief is a wholly faith-based assertion. This is abnormality and hypocrisy gone completely institutional.

And lest any of you self-assured heteros out there who have actual normal sexuality think you are safe from disorder and sin according to these abnormal males, think again. That catechism has you squarely in its sights as well.

This is what the Catholic cult teaches on the expression of heterosexuality and even solo sex, straight from its core documents:

- *"Lust is disordered desire for or inordinate enjoyment of sexual pleasure. Sexual pleasure is morally disordered when sought for itself, isolated from its procreative and unitive purposes."*

- *"The deliberate use of the sexual faculty, for whatever reason, outside of marriage is essentially contrary to its purpose."*

- *"Both the Magisterium of the Church, in the course of a constant tradition, and the moral sense of the faithful have been in no doubt and have firmly maintained that masturbation is an intrinsically and gravely disordered action."*

So you horrible and heinous sinners, you damnable heterosexuals, the creator of the universe has coded your brain with an extremely strong urge to have sex, and has therefore made you "disordered" from birth. If you masturbate you are a dirty rotten disordered sinner. If you have sex outside of marriage you are a dirty rotten disordered sinner. If you and your spouse are overtaken by the urge and make the beast with two backs because you are both horny as hell, you are *both* dirty rotten disordered sinners.

And if a condom is used during any of this dirty rotten disordered sin, the Vatican says you have heaped dirty rotten disordered sin upon dirty rotten disordered sin and your evil is beyond words.

How much more evidence do we need that Catholics and their clerical hierarchy are the ones who are dirty rotten disordered people whose dirty rotten disordered teachings cause endless and needless guilt and shame?

And it is here that we run into a social problem that is only matched in the magnitude of its impact and pain by the irony of its genesis. Over a billion[7] (with a "b") people admit to being members of the Catholic cult. To repeat, this organization is run by a clique of highly abnormal males who have deliberately chosen not to have sex with a woman, who choose to live their celibate and cloistered lives only with other men inside this closed-off, secretive and very often dishonest and criminal organization, who believe in things about the physical world and an imaginary spirit world that are ridiculous beyond compare, who got these beliefs from a few borrowed myths and fables written thousands of years ago, and who are every day sent around the world by the supposedly infallible head of state of their insular theocracy to tell everybody else in the world how to be…you guessed it…normal.

Making matters worse, for much of its history and even in a great many instances still today, this organization has been and is every day still responsible for the mental and physical torture and abuse of billions of men, women and children, including child rape and burning people alive. Since many of the murdered were women, it is a certainty that some of the women were pregnant, therefore the murderers of the women also committed abortion on the victims' unborn babies in addition to killing their mothers.

These acts go by the well-known names of Inquisitions, Crusades, heresy hunts, witch burnings and the like, but putting names on such cataclysmic evil events or periods can never describe the totalitarian and all-encompassing evil mandated and enforced by this church for many dark and evil centuries over every human it could whenever and wherever it had the power to do so.

Note that this church continued an especially long persecution of Jews, as the reach of that particular bit of focused sadism extended for almost two thousand years, until 1965.[8]

Yes, this cult and its self-professed and dogmatically infallible Maximum Leader didn't formally renounce anti-Semitism until a full twenty years after the United States, Allied powers and the godless communist Soviet Union brought about an end to the genocidal and psychopathic Third Reich and its Final Solution. Even the last murder under the Inquisition happened not in 1426, or even 1626, but in 1826[9], a full fifty years after the United States declared its independence from England.

Even during a period as late as the 1900s and early 2000s, this church and its allegedly infallible dear leaders racked up a rap sheet covering crimes on almost every land mass on earth. Its signing of accords with Mussolini's Italy[10] and Hitler's Germany[11], its failure to do or say anything meaningful against the Holocaust which was essentially a modern Inquisition and heresy hunt done by proxy, its teachings against condom use in AIDS-ridden lands where populations are beset by disease and starvation, its global campaign of child rape and coverups, in these and many other crimes against humanity the list of official Catholic crimes against humanity goes on and on.

This church has an inglorious history of allowing criminals to shelter inside the Vatican, its totalitarian regime's headquarters. Such persons include the war criminal Ante Pavelic, the pedophile-protector Bernard Law and far too many others.

This history also includes serving as a conduit for Nazis to start over in various places around the world through the increasingly forgotten Rat Line and countless other immoral acts of aiding and abetting evil.

The Vatican could try to defend itself against any complicity in any of these crimes during and following World War II, but the Vatican to this day refuses to allow access to its secret archives dated after 1939. And this is still the case even during the most liberal dictatorship in recent memory, that of Mr. Bergoglio.

With so very much to answer for, and with so many protestations of innocence or at least willingness to cooperate and beg forgiveness, why the continued coverup?

The reason for all these crimes and obfuscations is clear. The Catholic church is an extraordinarily abnormal organization, in the sense described above regarding being both highly abnormal and also far on the evil side of the moral bell curve. It is composed of extremely abnormal people both in the clergy and in the pews. Its sole driving mission since its first consolidation of secular and worldly power has always been first and foremost to protect the church, and this is obvious in everything it does.

Police have a code of internal secretiveness where they do not "rat" on each other. This "blue wall" or "blue shield" or other code phrase has parallels in other organizations, such as the Mafia's "omerta". Much of the secrecy behind these codes of silence is aimed at covering up or concealing illegal, unethical or immoral actions. The Catholic church's "black wall" includes its priests and popes and is similarly employed to distract from and cover up its unethical, immoral and in far too many cases illegal misdeeds.

This black wall is strengthened by the utter lack of spine and action among the billion Catholics across the world who are so blinded by their faith that they remain in their cult of immorality no matter how horrible its crimes and coverups. These complicit Catholics in the pews and congregations, along with the Vatican's complicit and often aggressive supporters such as the Catholic League, form an extended outer "gray wall" surrounding the inner core "black wall".

This outside buffer consists of apologists for the church who are not of the black-robed inner circle but who seek to protect the church for their own reasons, normally to protect an income, social status, membership in a highly bigoted organization or the continued belief in an eventual avoidance of permanent death.

These apologists, who are also quite deliberate and complicit accomplices in their church's crimes, serve as a first line of defense for the church to weaken attacks and hamstring inquiries into its misdeeds.

This is aiding and abetting, and it is equally as evil as the original act itself.

Thus the moral and intellectual failures of Catholics worldwide enable the continuance of their abnormal cult and its abnormal, immoral, evil and illegal activities.

When one looks at the history of this cult in the light of its two guiding principles, all of this can be easily explained. These principles are first to spread the faith and financial support for the church by whatever means possible, and second to protect the church as an institution at any cost. These are the offensive and defensive principles it has used for almost 2000 years, and due to their Darwinian effects they have worked so far. These principles have kept the church alive despite many centuries of continued evil, so we can be quite assured it will continue to use this strategy until it is no longer viable. In other words, do not expect anything different from this church in at least the next 2000 years than what it has done in its last 2000 years.

This church is an organization whose chartered members have actually committed and covered up rape and torture on the individual level, and collectively on a much broader scale through wars, inquisitions and other church-sponsored activities. It has also stood by while proxies carried out its dirty work for it, as during the Holocaust and other actions across the world. It has done all it can to cover up all this evil, and its followers still happily support and protect it.

World War II and its aftermath encapsulate this paradigm perfectly. For nearly 2000 years, the Catholic church sowed hate against the Jews. For hundreds of years it had spread its faith via the pulpit, the instruments of torture, the sword, the gallows and the stake, and in many instances from seats of worldly power. When its rapacious language was adopted by a murderous psychopathic state leader and became the foundational justification for the mass murder of Jews in Europe, the Vatican's silence was deafening, but its Concordat with Hitler's Germany remained firmly in place.

An almost totally Christian nation with which it had joined in unholy official Concordat was quite aggressively doing its dirty work for it, with much rejoicing by the faithful. Germany was also fighting against the dreaded Soviet Union, which if not defeated would likely be an existential danger to Catholicism in Europe.

Since the church thinks in terms of generations and centuries, it felt that partnering with a few fascist and murderous states in the near term would politically benefit and protect the church and eventually swell its ranks. To the Vatican, of course, this was far more attractive than potential conflict with the Soviet Union. Thus no member of the Third

Reich was ever excommunicated for any war crime whatsoever, and this cult is forever on record as collaborating with fascists and murderers, solely for secular power and preservation of that power.

After the war, the Vatican helped many Nazis escape justice, surely with one eye still on the future and needing to both thank those who had done its dirty work and also to keep these bloody-handed individuals around in case they were needed again in the future, for more "wet work" against the enemies of the church. The Vatican's core ethos, just like that found in almost all religions, is always that the end justifies the means, no matter how evil. This was incontrovertibly proven during the 20th century.

So unless your definition of normal includes denying science, denying reality, denying the normal urge to procreate, denying the normal urge to associate with the opposite gender, denying the normal disgust at those who lie and cover up lies, denying the normal hate and contempt for those who rape children and cover up those rapes, denying the normal hate and contempt for those who murder and who support murderers, and denying the normal hate and contempt for those who never accept responsibility for their actions, this cult and its members and supporters are off-the-charts abnormal.

The church knows, as it has always known, that when people truly believe the abnormal, they will act abnormally. That is why it has quickly defaulted to imprisonment, torture and murder for most of its history and whenever it has had the power to do so. And it has ensured that it maintained an iron grip at any cost over its adversaries' bodies and actions through physical control, and over its adherents' minds through the threat of eternal damnation and an absolution which it declares that only it can offer.

The French intellectual and philosopher Voltaire, who died in 1778, could have been specifically describing the Catholic cult for its entire history when he said "Those who can make you believe absurdities can make you commit atrocities". The Catholic cult is an organization founded on and steeped in the most obvious and foul absurdities and abnormalities. The actions of its leadership, its clergy and its flocks in the pews are abnormal and enable the continuation of abnormality and atrocities.

The absolutely fantastic book *Guns, Germs and Steel: The Fates of Human Societies*, written by Jared Diamond and published in 1997, is a treasure trove of information and insight covering history and numerous scientific disciplines while giving an extremely well-researched exposition on why many of the various societies on earth did or did not technologically advance. Chapter Three, titled *Collision at Cajamarca,* contains a description of the Battle of Cajamarca which took place on the 16th of November, 1532 in what is now Peru. In this battle, an invading Spanish force led by Francisco Pizarro captured the Incan emperor Atahualpa and murdered thousands of Incan troops and imperial retainers.

This sort of Christian and Catholic murder has happened far too many times throughout history, so the description of the callousness and murderous actions of the invaders was not in itself a surprise to read.

What surprised me was the depth to which the invaders used their religiosity, specifically their deep and sincere Catholic faith, as the justification for their campaign of deceit, kidnapping and murder. Diamond collected the written accounts of the battle from several participants and put them in the book. In Diamond's words, he used this compilation to help the reader "...relive them by weaving together excerpts from eyewitness accounts by six of Pizarro's companions, including his brothers Hernando and Pedro..."

I have included some extracts of Diamond's compilation in order to convey the utter depravity and cruelty of the Spaniards and their use of Catholicism as justification. The following paragraphs are from his book, starting on page 67. Read them carefully and you will see how easy it was for the Inquisition to happen, for witch burnings to happen, for the Holocaust to happen and for innocent people to be gleefully murdered. Consider just how close to the surface such faith-based evil lurks today, suppressed by a razor-thin veneer of secular civilization's rules and laws. Read the extracts and see what religion does to people. See what Catholicism does to people when it has the power to do it.

> *"The prudence, fortitude, military discipline, labors, perilous navigations, and battles of the Spaniards – vassals of the most invincible Emperor of the Roman Catholic Empire, our natural King and Lord – will cause joy to the faithful and terror to the infidels. For this reason, and for the glory of God our Lord and for the service of the Catholic Imperial Majesty, it has seemed good to me to write this narrative, and to send it to Your Majesty, that all may have a knowledge of what is here related. It will be to the glory of God, because they have conquered and brought to our holy Catholic Faith so vast a number of heathens, aided by His holy guidance. It will be to the honor of our Emperor because, by reason of his great power and good fortune, such events happened in his time. It will give joy to the faithful that such battles have been won, such provinces discovered and conquered, such riches brought home for the King and for themselves; and that such terror has been spread among the infidels, such admiration excited in all mankind."*

"Governor Pizarro now sent Friar Vicente de Valverde to go speak to Atahuallpa, and to require Atahuallpa in the name of God and of the King of Spain that Atahuallpa subject himself to the law of our Lord Jesus Christ and to the service of His Majesty the King of Spain. Advancing with a cross in one hand and the Bible in the other hand, and going among the Indian troops up to the place where Atahuallpa was, the Friar thus addressed him: 'I am a Priest of God, and I teach Christians the things of God, and in like manner I come to teach you. What I teach is that which God says to us in this Book. Therefore, on the part of God and of the Christians, I beseech you to be their friend, for such is God's will, and it will be for your good."

"If night had not come on, few out of the more than 40,000 Indian troops would have been left alive. Six or seven thousand Indians lay dead, and many more had their arms cut off and other wounds. Atahuallpa himself admitted that we had killed 7,000 of his men in that battle."

"It was extraordinary to see so powerful a ruler captured in so short a time, when he had come with such a mighty army. Truly, it was not accomplished by our own forces, for there were so few of us. It was by the grace of God, which is great."

"The Governor said to Atahuallpa, 'Do not take it as an insult that you have been defeated and taken prisoner, for with the Christians who come with me, though so few in number, I have conquered greater kingdoms than yours, and have defeated other more powerful lords than you, imposing upon them the dominion of the Emperor, whose vassal I am, and who is King of Spain and of the universal world. We come to conquer this land by his command, that all may come to a knowledge of God and of His Holy Catholic Faith; and by reason of our good mission, God, the Creator of heaven and earth and of all things in them, permits this, in order that you may know Him and come out from the bestial and diabolical life that you lead. It is for this reason that we, being so few in number, subjugate that vast host. When you have seen the errors in which you live, you will understand the good that we have done you by coming to your land by order of his Majesty the King of Spain. Our Lord permitted that your pride should be brought low and that no Indian should be able to offend a Christian."

That was almost 500 years ago. Now it is 2018, and the Pennsylvania state attorney general has released an utterly disgusting report on yet another disgusting discovery of Catholic priest pedophilia, child rape and the inevitable official church coverups. We are now far too many years since the initial reports of this vicious and immoral cult's crimes against children began to be broadly exposed, and we have seen far too many of these reports, apparently with no end in sight. If you are any kind of moral person, yet still somehow consider yourself Catholic, this is partially your fault.

If you are somehow still a Catholic after all the evil which that sect has wrought not just throughout almost its entire history, but most recently during the 20th and 21st centuries, you have a massive amount of explaining to do.

If you are somehow still a Catholic, you are making a conscious choice to be part of and to knowingly support an organization that first fostered then did nothing about the WWII Holocaust except offer a half-hearted half-apology after the murdering was over. In our own time this church is still lying about and trying desperately to cover up the ages-long holocaust it has been perpetrating in its campaign of rape, abuse and neglect of countless innocent young children.

If you are somehow still a Catholic, you are part of this problem. By continuing to call yourself Catholic, or by participating in any of its activities, you are complicit in its crimes. Anything you do that contributes to the continued existence of this global pedophilia ring in any form means you are complicit in some of the most horrible crimes a human can commit against another. You need to ask yourself if you would remain a part of any other organization on earth if so many of its members raped and abused little children, children they were supposed to be protecting. Ask yourself if you would remain part of this organization if its global campaign of rape was both covered up by leadership at all levels and deliberately allowed to continue through moving the rapists around to new raping grounds well-populated with flocks of new children to victimize.

Ask yourself more questions. If you discovered your best friend was a child rapist, or that he knowingly covered up someone else's rapes of children, would you still consider him your friend? Would you let him babysit your children or advise them, or indeed you, on anything at all, particularly morality?

Ask yourself why you remain a Catholic, and why you choose to be an enabler of this organization's crimes. While horrible beyond words, the Pennsylvania revelations are no singular event; such a horrible revelation is now just another day in Catholicism. It is just one more disgusting discovery in the endless stream of disgusting discoveries of priestly rape and pedophilia which this church has never fully accepted or dealt with. And why do they continue to rape and cover it up? Because you Catholics let them.

They do these things because they know that most Catholics aren't really bothered very much by their priests screwing and terrorizing little kids. If such horrible behavior actually bothered Catholics, they would be leaving the church with as many people as they could take with them and they would never look back. But no matter how many children are raped, no matter how widespread the coverup extends and no matter how high it goes, the complicit pope and all his complicit bishops and cardinals know that Catholics across the world are too invested in this cult of human sacrifice, cannibalism and pedophilia and are too pathetically weak to leave the church.

Due to this weakness in its supporters, the Catholic hierarchy knows full well that the church isn't really in any long-term danger. This makes the situation even more disgusting to those of us who do actually have contempt for child rapists and their supporters and would like to see them all behind bars forever.

Every single person who remains in the Catholic cult is complicit in child rape, period. And if you're still a Catholic, the rest of us need to know what the hell is wrong with you. Why do you deliberately choose to be associated with these criminals? How can you look anyone in the eye and try to say your pedophilic crime ring/church can have anything at all to say about morality? Your church has about as much credibility on morality as Hitler does, and you damn well need to accept that.

I have to ask all of you who choose to continue as Catholics what in your god's damn hell will it take for you to leave that church? You're obviously OK with huge numbers of your priests orally and anally raping children, among far too many other unspeakable acts, so what will it finally take for you to say you've seen enough? You're obviously OK with leadership up to and including your allegedly infallible pope doing all they can to cover it up or offering insincere apologies and insufficient actions, so what exactly are you waiting for? Do your priests have to start killing these children after raping them, then nailing their bleeding and semen-stained bodies to crosses in your goddamn front yards in broad daylight before you've finally had enough? Do they have to flay the bodies of these children and have sex with their skinned corpses during Mass on Easter Sunday for you to finally decide you've had enough?

The sad truth is that due to the lack of spine and morality among Catholics it's highly doubtful that even such overt and public evil would make large numbers of you leave the church. And this is really the true evil, that so many people can commit, cover up and then be complicit in the most heinous acts a human can perpetrate, yet their common tribal religious bond makes it all just fine.

Consider this. If you have ever given money to the Catholic church, your own hands are stained with its crimes. Some of that money you gave them helped pay the legal fees of rapist priests. Some of that money you gave these pedophiles went to payoff money and hush money, given to the families of those children who were raped so that they would keep their mouths shut. Some of that money you gave these rapists and their enablers is paying rapist priests and their rape enablers a salary right now, at this very moment. If you give these people even a single cent, you are subsidizing current and future child rape and its coverup.

There is only one way to stop this once and for all, and for any of you who have any hope of preserving what may remain of your humanity and your imaginary soul, and it is for each and every one of you to get a damn backbone, strongly denounce Catholicism and walk out those doors forever.

It is time for the human race to start thinking.

It is time for those who call themselves Catholic to see how abnormal they and their cult are.

It is time for them to see how every second they continue to support this organization they are contributing to the continuation of antihuman abnormality and to enabling the institutional bigotry of the Vatican, its black wall of clergy and its gray wall of one billion fully complicit followers.

Only by immediately and publicly leaving this horrifically immoral cult and letting it rot in the prison where it should be right now will you force actual change.

Know this – if you stay, you are an enabler. If you leave, you are an agent of change. That change may not come in your lifetime, but it will come eventually.

And if there is an afterlife and you do have to plead your case before Saint Peter or some other gatekeeper, you can rest easy knowing that you did the right thing for the right reasons.

Sometimes City Hall is wrong, it is corrupt and it is evil, and you are always right for standing up to it.

The choice is yours, and the time is now.

Chapter Twenty-Two

The Voice of Hypocrisy

"Hypocrites are rather easy to recognize. They spend most of their time pointing out the flaws in others, and the rest of the time trying to flaunt their perfection."
— Charles F. Glassman

As a devout atheist and science-minded humanist, I see religion for what is really is. It is a wholly human creation and a tremendously unfortunate byproduct of our fearful semi-evolved primate psychology. The horror and damage religion has inflicted on our planet, and which it still continues to inflict, are straight out of the worlds of Stephen King, Aldous Huxley and George Orwell, with a touch of Marquis de Sade for flavor.

It is an unequivocal fact that our planet would be a much happier and more prosperous place were we to finally eradicate religion. I can hear the aghast cries of the faithful now: "No! How can that possibly be? Isn't religion the only true source of morality? How else would the human race know to be good and behave if not for the (insert here the "holy" book or god of choice)?"

It is obvious and simple to the point of infantilism that the stupefyingly large profusion of such so-called "holy" books, most of which are said to have derived from different but nevertheless singularly powerful and all-knowing gods, is the first and best clue that those who believe this tripe have already violated Feynman's exhortation not to fool oneself, and William of Occam's advice not to add layers of BS to a simple if unpleasant conclusion.

The next most powerful clue, which can be counted upon to demonstrate its presence in every religion and in a great many of the godly, is the unparalleled ignorance, self-righteousness and hypocrisy of so many of the faithful. One can see this in all ages and in many forms. In our current age of mass electronic media, it takes only a few seconds of searching to locate an endless supply of examples of the godly being highly immoral, and indeed quite often outright demonic.

While this thread of unholiness is woven through all socioeconomic and demographic faith blocs, it is normally persons in the media or in visible positions in the society who demonstrate ostentatious faith and thus tend to make the news.

The faithful often profess their reliance on their particular species of "holy" book as their source of moral principles. The more overtly and loudly faithful the adherent, the more they seem to rely on these writings to guide their lives, and the more stringent their desire to tell the rest of us how we should do the same. Immersion in these books would certainly enlighten the reader as to the actual teachings of the book in question. In America, the Christian holy book is the default and most widespread reference.

As with most of the religious today, adherents use their books, in this case their bibles, extremely selectively. The vast majority of the evil, immoral and antihuman teachings it contains are ignored, although not all of them, and far from all the time. And they want the rest of us to ignore them too. They do not ignore, and they profess loudly and often, some occasional passages that do resonate positively with the human condition. This is no credit for religion, because as the pithy saying goes, even a blind squirrel sometimes finds a nut.

As to its content, the Christian holy book does contain a truly biblical (deliberate pun) amount of hypocrisy, but if one follows the faith-based example, one can easily cherry-pick a quotation to make a point. I thus note that this book does indeed, in stark and clear terms and in several places, decry hypocrisy, to the extent of denial of salvation for the hypocrite. Job 27:8 in the King James version says "For what [is] the hope of the hypocrite, though he hath gained, when God taketh away his soul?"

Can't get much plainer than that.

Yet what do we have in our great country? Religious hypocrisy, rampant and flourishing. It is utterly clear that almost all of the faithful either can't read or they didn't read that far into the book. Or they read that far but ignored the words. Or they read that far and understood the words but are committed to doing and saying whatever they want for as long as they can, and are relying on that final deathbed '"Oh, God, I'm really sorry!" version of the Final Confession for the full rescission of any pending punishment or 'denial of salvation'.

Ain't religion great? It's like the ultimate bankruptcy filing. You recant on your deathbed, and all of a sudden it's no harm, no foul, and you don't even have to have your posthumous credit or salvation dinged for 7 years.

Every megachurch pastor and television evangelist I have had the extreme displeasure of observing, every pedophilic clergyman out there clawing his way through the youth of his district or parish, every nutcase pulpit pounder howling for the blood of atonement, every right wing politician in today's America diligently taking from the poor and giving to the rich while blabbing about biblical values; the list of what can only be described as gobstopperingly audacious hypocrites goes on and on.

When they aren't glorifying their god ostentatiously and in public, while flashing whitened incisors and coiffed hair and any of a range of expensive jewelry and asking for money (hey, prosperity gospel, so it's OK, doncha know?), they are condemning and despising literally millions of their fellow humans, all of whom were supposedly created in their joint god's image.

Love thy neighbor, but if they vote for the other political party or are poor, do all you can to disenfranchise and dispossess them. Love thy neighbor, but make sure you are heavily armed so you can blow their asses away if they mess with your house, oxen, sheep, ass or wife. Love thy neighbor, but if they happen to be gay, well, you can't yet legally shoot them unless you think they might be trying to put you in mortal danger by blowing their highly infectious gay breath on you, but you can sure condemn and oppress them from now until the end of time. Babble all day long about loving thy neighbor and turning the other cheek, but don't you dare actually *act* on that shit!

It's clear I am no fan of the faith-based and their farcical foolishness. But, again, as a science-minded humanist, I feel it appropriate to steel myself and occasionally try to do a speedy firewalk across the flames of religious intolerance and ignorance. That means I occasionally gather the mental and physical strength to do opposition research, which in my neck of the woods is listening to religious programming on the radio. Unfortunately, in most places in this explicitly secular republic, there are few corners to which one can retreat which are not awash in such damnably mind-draining drivel.

So there I was, on the 17th of December 2015, a week away from the borrowed mashup festivus which most Americans call Christmas and which they try so hard to claim as their religion's original creation. I took a deep breath and hit the scan button on the radio, figuring I'd just stop on whichever of the many religious stations was actually talking to a guest or caller rather than doing a donation drive or otherwise begging for money, as they all do nearly continually.

On this particular day, the electronic roulette wheel landed on a station broadcasting a show hosted by a faith-head calling himself the "Bible Answer Man"; real name Hendrik "Hank" Hanegraaff. It appears to be funded by or run by some outfit calling itself the Christian Research Institute. I could be wrong on this; my few moments of Internet research on it at the website www.equip.org/ didn't help much, although I admit I didn't have the patience or inclination to waste very many seconds of my valuable lifetime on it.

If you want to know more, have fun surfing the Net. Suffice it to say it is a religious broadcast on the radio and the host is Hendrik "Hank" Hanegraaff, the so-called "Bible Answer Man". I will refer to him from now on as "Hank, Answer Man", abbreviated to HAM.

I had listened to snippets of this program in the past…very, very, very small snippets, in order to maintain my sanity and normal organ function. So I was used to the affectation which the HAM presents, which is a mild and 'oh so modest' façade of calm and erudition, which obviously has the intended effect on the ambulatory fungi which are evidently the program's main listeners and callers. The HAM's role in the program is to peddle nonsense and pablum to this obviously irretrievably muddled demographic. If you like, feel free to give it a listen, but I recommend not eating for at least an hour before you do; the overpowering mass of bullshit is so powerful you can actually smell it. I knew this, so I was prepared for anything.

That turned out to be a good thing. It took less than ten seconds of listening to the insipid and mindless drivel of this broadcast for me to, yet again, feel no small amount of despair for the future of the country. This particular 'episode' of faith-based fatuity included some truly bizarre fare. As I tuned in, a specimen of *Gullibilis totalis* was quite earnestly asking the HAM some blathering nonsense about Satan and sin. The HAM calmly informed him that Satan is not omnipresent but he does have many fallen angel minions and they have a plan for perversion. He indicated that is basically their job description. He noted that only the ear of the mind can hear the evil whispers to sin, and that we can't hear these whispers with the actual, you know, listening ears. We will need to be aware of this and stand against the powers of darkness, blah, blah, blah.

Following a bit of back and forth, the caller said he appreciated the clarification, which was interesting to me, as none of the actual words the HAM excreted made a bit of sense to the rational mind. Evidently the same kind of person who feels that leprechauns, sprites, dragons and ghosts are real is also comfortable with equivalent nonsense about angels, demons and satanic influences on this, our real Earth. Or maybe the fault is mine. Maybe I was just using my actual physical 'listening ears' and not the ears of the mind. That must be an acquired skill I have failed to pick up. Or maybe I wasn't taking the right drugs that day. Whatever.

I do often wonder, when listening to the shameless waterfall of utter nonsense on programs like this, which font of other-worldly knowledge these people think they have access to which regular people like you and I don't. One wonders what a regular Joe or Jane would have to do, or to imbibe, to get their own HAM-line to the other world.

After this bit of uninformative verbal ephemera had thankfully faded into oblivion, HAM moved onto the next caller, whose points and questions were, astoundingly, far from fungal. Indeed, they were so shocking in the context of the broadcast that my cerebrum was actually able to let go its death grip on my cerebellum, which had been trying very hard to jump out of my ears and run away screaming from the sheer insanity it was hearing.

This caller was your average Joe, although his name was Mike, actually. Surprisingly for a listener to this program, he appeared to have a conscience, a rational brain and compassion. As he talked, I could sense in him the intensely painful cognitive dissonance and difficulty he had with believing as he did, in only the 'good' parts of his faith, while being surrounded by, in his words, 80% of the people around him who evidently spoke the "right" words on Sunday, and professed their faith publicly, but failed miserably at execution, both in word and deed. His questions were good, honest, and, dare I say it, faithful.

The abject hypocrisy with which he was surrounded in his daily life turned out to be not merely matched by the HAM's responses; the HAMster blew conventional hypocrisy out of the proverbial pig sty with the blatant hypocrisy and sanctimony best exemplified in the most money-hungry televangelists.

The mp3 audio file of that show is titled BAM20151217, and can be found on the web site http://www.equip.org/broadcast/larycia-hawkins-the-council-on-american-islamic-relations-and-qa/. Go partway down that web page and click on "Download and Listen". You can listen to or download the file. I have downloaded it, so if the web page or file are suddenly deleted from the Net, "I have the originals" and can prove this dialogue and broadcast really occurred. The dialogue of the HAM-Mike discussion is transcribed below to the best of my ability. It starts at about the 12:30 mark on the recording, and 'mercifully' ends at the 16:01 mark. Here's the relevant portion:

HAM: "…we'll talk next to, uh, Michael; he's listening in Norman, Oklahoma, Bott Radio, hi, Michael."

CALLER: "Hello (Hank?) how are you?"

HAM: "Good how are you?"

CALLER: "I'm fine, thank you. Um, my, uh, I'm going to make more of a statement than a question I guess uh, you'll, you'll know how to answer it, um. You know, I'll, I will say 8 of 10 people that I am around are Christians, or at least, you know, say that they are and say that they believe…um, of course Muslims and refugees are, you know, a hot topic all around and I keep coming in contact with close friends and family that are good people, you know, I would say, and I believe that they are Christians and they do believe in, in Christ and the word, but they also don't want them here, don't bring them here, and, you know, I feel like I'm the black sheep in the conversation and I really only have two things to say, you know, I, if you had, not that I do but if you had to boil the bible down I would say it's love, and uh as far as to the topic I would say, you know, no weapons formed against you shall prosper and that's kind of how I feel and I say I guess, you know, love them and let 'em be here, but, what, what, what, what how would you answer that if it was you?"

HAM: "Well, how many, d-, dy-, do you have kids?"

CALLER: "I, I have a one year old."

HAM: "You have a one year old, so uh, uh, what would you think of having 20 one-year-olds right now?"

CALLER: "It would, it would uh be chaotic, for sure."

HAM: "Yeah, uh, wha-, wha-, what about 200 one-year-olds?"

CALLER: "OK, yeah, impossible."

HAM: "Yeah, so, I, I, you know I, I look at the same thing, I mean I have 12 children and, and sometimes, my wife and I talk about the fact that we would love to take in more children, we would love to adopt more children, we'd love to have 24 or 36 or 48, but the reality is that if we tried that, everything would implode on itself because we simply do not have the capacity. As much love as we may have in our heart, we simply do not have the capacity and so what we have to do is do the very best with what we have. Raise people within our family to be those who are selfless producers rather than, uh selfish consumers. And I think that when you look at the needs of the world, we have to look at those needs with realism. We must always be ready, uh, to give the cup of water, and the loaf of bread as you so beautifully have said, but you think about what is happening in Germany right now with uh 2015 ushering in some million refugees, Syrian refugees into Germany, and what has happened as a result of that is that they're running up against the sheer exhaustion of resource and volunteers. There's another issue here and we don't have time to get into it but remember that the chief principal responsibility of the government of the United States is to protect its citizens and that is an issue unto itself."

What a pathetic and despicable piece of unbelievably hypocritical horseshit. This self-styled and so-called bible expert, a self-promoting and self-professed knowitall on all things biblical, one who would nominally be assumed to parrot that book's most humanistic words back to the caller, took just a few seconds to show how thoroughly full of bullshit and hypocrisy a typical faith-head on the radio or television can be.

This moralizing and monotonic bible-thumper has nine biological children.[1] ***Nine***. Not the oft-referenced 2.5 kids per family that is thrown about so often as the 'standard' number of biological kids American families have. Not the five kids that causes raised eyebrows and starts the questions about family size and jokes about lost condoms. Not even the octuplex of offspring which a certain infamous and likely addled person produced a few years ago. No, this demonizer of so-called "selfish consumers" has produced ***nine*** biological children, each of whom are daily and busily consuming oxygen, food, water, electricity and petroleum, and producing copious amounts of trash and polluted air and water.

Surely popping out nine biological kids, in a world with nearly 7 billion people, couldn't be construed as the actions of "selfish consumers", could it? Evidently not to the faith-head community. It's only those *other* people, *over there*, those '*selfish consumers*' who have all those kids, who are in the wrong. *Those people* don't need any cups of water or loaves of bread, because they are *selfish consumers*, don't you know? And besides, the HAMster needs that water and bread for his own massive herd of selfish consumers, er, I mean *selfless producers*.

Pathetic.

One would have thought that even a cursory glance at that book in which he is supposed to be such an expert might have led him to a different response. He might have referenced any of these passages:

- "Also thou shalt not oppress a stranger."

- "Therefore if thine enemy hunger, feed him; if he thirst, give him drink."

- "Mind not high things, but condescend to men of low estate. Be not wise in your own conceits."

- "Charity suffereth long, [and] is kind; charity envieth not; charity vaunteth not itself, is not puffed up"

The blind hypocrisy of the sanctimonious faith-head will always reach levels no normal human can attain. This sort of hypocrisy is clearly addressed and prohibited *in his own holy book,* on which he is an alleged expert. One would think that a strict and clear heavenly prohibition, particularly one which prevents eternal salvation, might be something HAM would have picked up on.

As they say, it gets worse.

I haven't even mentioned this little extract: "Judge not, that ye be not judged"?

Nor the biggie, the hypocrite-cruncher of all time - "And why beholdest thou the mote that is in thy brother's eye, but considerest not the beam that is in thine own eye?"

One might think even a microsecond of introspection by a self-pronounced biblical expert would have called that little tidbit out of HAM ROM. But no, the blinders of faith are indeed omnipresent and omnipotent.

It is clear that the correct and appropriate biblical response to such unbelievable hypocrisy would be a lingually-applied version of this little piece of guidance:

> *"Wherefore if thy hand or thy foot offend thee, cut them off, and cast [them] from thee: it is better for thee to enter into life halt or maimed, rather than having two hands or two feet to be cast into everlasting fire. And if thine eye offend thee, pluck it out, and cast [it] from thee: it is better for thee to enter into life with one eye, rather than having two eyes to be cast into hell fire."*

But that would be too much to ask, wouldn't it?

Hypocrisy, thy name is HAM.

Part Five

The Women's Perspective

Chapter Twenty-Three

What If Women Had Written the Bible?

"Every denial of education, every refusal of advantages to women, may be traced to this dogma [of original sin], which first began to spread its baleful influence with the rise of the power of the priesthood and the corruption of the early Church."
—Lillie Devereux Blake

"I was a woman, a divorcee, a socialist, an agnostic...all possible sins together."
—-Michelle Bachelet

"The memory of my own suffering has prevented me from ever shadowing one young soul with the superstitions of the Christian religion."
— Elizabeth Cady Stanton

We need to make one thing very clear up front. Women would *never, ever* have written the Christian bible or either of the other two monotheistic disasters that keep being passed off as so-called holy books. This unholy trinity of ancient Mediterranean snuff fiction could only have been written by people devoid of morality and conscience but bursting with ignorance, aggression and way too much testosterone.

And being 98% genetically identical to the murderous chimpanzee and living in societies which reward the most aggressive groups was a permanently active catalyst in the generation and sustainment of so much evil nonsense. These books were solely the products of angry and jealous male humans, who wrote tales about gods which were hyperamplified versions of themselves with infinitely higher levels of anger, jealousy and power, unachievable by mere mortal males.

And it isn't just because of the rampant misogyny which monotheisms and many other faiths simply can't seem to do without. Women comprise only a microscopic fraction of the war criminals, murderers and torturers throughout history. Due to their nature, women have had to be more focused on the production, protection and maintenance of their offspring, their families, their clans and themselves, often in that order.

Women don't go out on raiding expeditions, they don't burn villages and their inhabitants, and they don't commit massacres, rapes and tortures. Women don't start wars. Women may argue but they don't often kill other people unless they are essentially forced to do it. There are far fewer[1] female than male psychopaths, and they do not often engage in the physical violence that male psychopaths enjoy and inflict. Women have not been the torturers or the ones who commit the evil acts; they have been the victims. They have preserved life, not taken it. They have saved lives, not snuffed them out. They have created lives, not ended them. Women have investments in life that men can only imagine.

These are general statements with the inevitable exceptions but the rule stands. Women are far less evil than men, certainly in their deeds if not always in their thoughts. There are indeed some extraordinarily hateful and evil women, but when this is the case they are usually motivated by their religion or other truly unfortunate circumstances to commit acts of evil. Again, except for the inevitable exceptions, women are not power-hungry, hyperaggressive and psychopathic. The male writers and founders of religions and religious stories often have all of these unpleasant characteristics and many others.

If women had written the bible, there would be no childish ignorant lunacy like ostracizing them because of a simple biological act like menstruating.

Women would not be stoned for adultery, suspected or otherwise.

Women would not be treated as property of their husbands.

Women would not be used as sex slaves after the conquest of their villages and the murder of their fathers, husbands, brothers and sons.

Women would not be stoned if they were suspected of not being virgins.

Men would not be praised for having hundreds of concubines and/or wives.

Women would not be able to be sold, either as a slave girl or to the male who raped her.

Women would not be told to shut the hell up and don't talk to men just because they were female.

If women had written the bible, it wouldn't be full of killing, rape, murder, conquest and evil. It would have much more practicality in it, and it would focus more on cooperation, mutual support and problem solving. It would talk about the actual important necessities of life, such as how to cure disease and have family and social harmony.

It would discuss how to have plenty of safe food and clean water, bits on personal hygiene, safety, child rearing and things to teach these children. It would not contain details on how they should invade other people's lands, kill people, sacrifice animals, oppress people or chop parts off of their children's genitalia. It would be more "*Hints from Heloise*" as opposed to "*Mein Kampf*".

And let's not let Islam and its holy book escape notice. While so far the criticisms in this book have focused primarily on the Jewish and Christian holy books, all three of these books are misogynistic spirit brothers no matter the textual differences. Therefore the criticisms leveled against one are leveled against all. All of these religions and their books keep women down with or below the sheep and the goats, and in general treat women as frequently unclean housekeepers and laborers, sexual outlets and a necessary part of making sons.

We can do a quick comparison and list what these religions say about women versus what secular humanism says about women. There is no question which is infinitely more moral and better for all women of any time and in any place.

The Bible verses below came from the King James version; the Quranic notes came from the Ahmed Ali[2] translation.

On the importance of critical thinking, acceptance of reality and interaction with other people.

- **Secular humanist.** She should learn critical thinking, how to analyze ideas on their own merit and accept or reject them on this basis. Facts and evidence are the only real basis for understanding and acting in the real world. Mere differences of opinion, particularly on imaginary gods, are no reason for violence of any kind, much less murder. Initial respect for others is always expected.

- **Biblical.** Following a 2000 year old book of fairy tales is more important than using reason and common sense. Facts and evidence only apply when they support religious faith and beliefs. If she tries to convert anyone out of the faith she should be immediately killed, even by a close family member (Deuteronomy 13:6-10).

- **Quranic.** Faith trumps facts and evidence. Unbelievers will burn in hell (Surah 3:10). Failure to believe in Islam will be severely punished "…both in this world and the next…" (Surah 3:56). Do not believe anyone who does not follow your religion (Surah 3:73).

On treating women with compassion.

- **Secular humanist.** Girls should be taught how their bodies work, and how to handle normal and natural changes in their bodies with information, understanding and compassion.

- **Biblical.** Menstruation is horrible and makes both females and what they touch unclean. (Leviticus 15:19-20).

- **Quranic.** Avoid menstruating women because they are unclean (Surah 2:222). After sleeping with a woman a man has to purify himself before he can pray (Surah 4:43).

On treating women with dignity and respect, or not.

- **Secular humanist.** She is slave to no one and is the equal of a man. She should be taught how to be an independent and proud person.

- **Biblical.** She is sex slave "booty" for victors in war (Numbers 31:17-18). Women should shut the hell up and always defer to men (1 Timothy 2:11-12).

- **Quranic.** Women are inferior to men (Surah 2:228). Women only get to inherit half of what their brothers inherit (Surah 4:11). Women should be punished more harshly for the same offense (Surah 4:15-16).

On human sexuality, marriage and treatment of wives.

- **Secular humanist.** Sex is normal, necessary and nothing to be ashamed of. Sex is best kept within the confines of a committed relationship, for psychological and physical reasons. Love and sex are very important to the human being and should be cherished and once begun, kept healthy.

- **Biblical.** If she has sex before marriage, she should be stoned to death (Deuteronomy 22:20). If she has sex before marriage and her father is a clergyman, she should be burned to death (Leviticus 21:9).

- **Quranic.** No matter how many wives you have, everybody knows you can't treat them all equally well (Surah 4:129). Women need to be submissive, cover up and guard their private parts (Surah 24:31).

On the response to sexual assault.

- **Secular humanist.** She should learn how to defend herself, but if she is unable to, she will be protected by her family, police or other people. If she is raped, there will be severe consequences for her rapist through the criminal justice system, and she will be taken care of by her family and friends and helped to heal.

- **Biblical.** If an unbetrothed virgin girl is raped, she has to marry the rapist (Deuteronomy 22:28-29).

- **Quranic.** Men are not allowed to sleep with married women unless the women are war booty (Surah 4:24).

On witches, witchcraft, gods and devils.

- **Secular humanist.** There are no such things as witchcraft, magic, devils, demons, fantasy creatures, gods or any supernatural imaginings. Women, just like men, are human beings and have no supernatural powers that need to be beaten or burned out of them.

- **Biblical.** Women can be witches and if they are, they need to be killed (Leviticus 20:27).

- **Quranic.** Devils teach sorcery (Surah 2:102) and there are sorcerers (Surah 7:120).

The above was just a microscopic sampling of the disgusting misogyny within the monotheistic holy books. You could write volumes on this subject and not come close to covering how badly the alleged creators of the universe and their male minions on earth want to treat women. These religions and books treat women like garbage. Why? Because unbelievably ignorant, woman-fearing and woman-hating men wrote these books.

Christians throughout history and particularly today constantly praise themselves for their alleged morality and praise their god and his holy book for how wonderful they are, while ignoring how this god and his books treat half of our species.

It is impossible to understand how any actually moral human being can read such passages as the above, and the thousands of others not included, and not practically retch in disgust at such evil. These are not just random words in a random book; these words were and in many cases still are taken as holy gospel and acted upon in real life. Far too many women have been raped, abused, ostracized, tortured, attacked, maimed, burned and killed because of them. Words matter a great deal when humans use them to justify acts of pure evil.

No, women didn't author a single word of the Christian holy book, but they and their children have unwillingly written much of the history of that book with their innocent blood. If women had written the bible, it's for damn sure it would be nothing like the anti-female travesty currently poisoning morality across the world.

The only path to a truly moral existence for women in our world is to burn every single so-called holy book and start from scratch, using reason, science, equality and compassion as the foundation for a true moral code.

And this time, women need to write the whole damn thing.

Chapter Twenty-Four

Thank god For Eve

"I have an independent streak. You know, it's kind of hard to tell an independent woman what to do."
- Betty Ford

"If you know you are on the right track, if you have this inner knowledge, then nobody can turn you off... no matter what they say."
— Barbara McClintock, Nobel Prize Winner

She popped into existence two hours before sunset on a nearly cloudless day. One moment there was a small empty meadow covered with fine short grass, then suddenly there was a nude female in her late teens lying supine on that grass.

Her eyes were closed so she didn't see anything at first. Her first thought was that there was a cool and tickly sensation in her back, and she unconsciously wriggled a little bit to try to make the tickling stop. Then she began to feel other things, all over her newly acquired body.

As a freshly materialized creature, her experience of existing was completely new and utterly without precedent. She had never seen, heard, smelled or felt anything before, so her newly-activated senses were jangling at her all over her body in a dizzying cacophony of strange inputs. Her seconds-old brain was having tremendous difficulty processing everything.

She heard a very loud chirp/screech from somewhere to one side and reflexively snapped her eyes open. The first thing she saw was a huge expanse of featurclcss blue with some small white patches in it, and there were some green fuzzy things in the periphery of her vision. The cool and tickly sensation in her back was now also in the backs of her arms and legs. She heard a very gentle rustling sound plus an occasional screech, chirp or tweet from somewhere inside the fuzzy things.

Then she heard quite different sounds, fairly loud and in a deep rumbling tone, coming apparently from out of the big blue area. Surprisingly, these sounds made a connection inside her brain and she was able to comprehend them as containing a message, which she understood as "Woman, sit up." How odd. Despite being only about sixty seconds old at this point, her brain had evidently been programmed to understand that certain sounds have specific meanings, so she understood both the words and the idea of sitting up. And also that the sound that referred to her, something called a 'name', she thought, is "Woman".

The feeling of knowing how to sit up solidified in her brain. As this was her first movement and act of sitting up, things did not go quickly or easily. She clumsily contracted the proper muscles, twisted her limbs, joints and torso this way and that and after a somewhat protracted sequence of windmilling and flailing gyrations, she ended up in a sitting position with her legs outstretched in front of her. She felt a little "dizzy".

From this new vertical perspective the place she was in looked rather different. She saw that she was sitting on a flat expanse consisting of a huge number of short and soft but tickly green things, and the bigger green fuzzy things now looked like they were also sitting (or standing?) up on the short green tickly stuff too. But they stretched high into the blue area over her head.

She felt more than a little light-headed, and tried to take it all in. She looked at the green things she was sitting on ("grass", she thought, quite unconsciously, not knowing why she called it that), then at the tall green things ("trees", she thought, again quite unconsciously), then everything else around her. Each time she looked at or touched something, the sensation of knowing what the item was ran through her head. She began to examine her "body", flexing her "hands" and "fingers", then used them to touch her "belly", "head" and "face", then her "legs" and "feet".

"How interesting", she thought. "This is me. I am 'Woman'." The dizziness remained firmly in place while she continued looking around and touching her new body and surroundings.

After she'd touched or looked at everything there was to touch or look at from a sitting position, she heard more of the loud low-toned sounds emanating from the big blue area (the "sky"). This time she understood the sounds as "Woman, stand up." As before, a bit of brain comprehended what was actually meant by "stand up", and she turned, contracted, windmilled some more, flailed the appropriate bodyparts and after a few unsuccessful tries was able to assume a very shaky and uncertain but more-or-less standing position.

When she'd been sort-of-standing for a full minute without falling, the deep sounds came again from above. This time, she heard the command to "Walk." So she walked, or tried to, using muscles and bodyparts in that way for the very first time in her extremely short life. Very carefully and shakily at first, with some hesitation and a few stumbles, then finally more competently, she managed to move slowly if somewhat jerkily forward. She 'walked' for several minutes, careful not to "fall down".

Then to her extremely great surprise she saw a body much like hers in the near distance. This one was walking out from behind one of the trees scattered throughout the place, and it was doing a much better job of walking than she was. She stopped, not sure what to do. The sounds from above came again, telling her to "Walk". So she kept moving toward the other body. As she drew near it, she realized that while it was indeed much like her, there were some differences.

It had a flat chest, as opposed to hers with its two round protrusions, although both chests did have two very similar small red circles on them. The other body also had a dangly thing hanging between its legs, while she had nothing there at all but hair. Her face was hairless, but the other body had some hair on the lower part of its face and a little bit on its chest, as well as on its head.

When she got close to the other body, which had stopped walking when it saw her, new sounds from above told her to "Stop", then said "Adam, this is Woman." The other body, evidently named 'Adam', said "This is now bone of my bones, and flesh of my flesh: she shall be called Woman, because she was taken out of Man. Therefore shall a man leave his father and his mother, and shall cleave unto his wife: and they shall be one flesh."

She had absolutely no idea what this other body was talking about, but didn't say anything at this point. It was all quite overwhelming and the dizzy feeling was coming back even stronger than before. This was not a welcome development for someone who had only taught herself to walk a few moments before and was even now on the verge of toppling over into a "bush" of some sort.

The voice from above then said "Alright then. Goodbye. I'll be back."

At this point she was as confused as it is possible to be, not an unexpected feeling in a ten-minute-old who had been materialized out of thin air into an eighteen-year-old body, then spoken to by a disembodied voice which had led her to another completely unfamiliar body which spoke unintelligible gobbledygook at her, whereupon the disembodied voice abruptly left for parts unknown, leaving her with the other body and nary a word of help or guidance. Not knowing what to do, she continued dizzily sort-of-standing there waiting for something else to happen.

This went on for a few moments with no reaction from Adam or the absent voice from above.

She wobbled in place looking at Adam. He stood in place looking at her. Neither said anything.

After some time she felt the silence had gone on long enough, and decided to see if she too had the ability to speak. She discovered to her pleasure that she did, and she used it to try to gain a little clarity on the situation. After a few false starts, she managed to get the complex arrangement of tongue, teeth, lips, vocal cords, diaphragm and air flow to produce something that sounded like words instead of a fish out of water with a stutter and an incurable case of the hiccups.

"Sssssmmm, ssssooooo, ssso you're Adam?" she asked, unknowingly uttering the world's first spoken consonance.

"Yes", Adam said.

"And I'm…Woman?" she asked.

"Yes", Adam said.

Apparently that was all Adam intended to say, so she asked the logical followup questions. "Ssso whe where did you come from, and where are we?"

Adam "furrowed" his "brow" and said he didn't know where he came from, that he had only been here for a short time himself, and that he was not sure exactly how long. He said the voice from above told him that he could eat things off of every tree in this place, called a garden or Eden or something, but that he would die if he ate the "fruit" from the "tree of knowledge of good and evil". He pointed to a particularly lush and lovely tree and indicated that it, with some interesting roundish red things hanging on it, was the 'good and evil' one. He noted that there were many plants and animals around, and that since he had named them he could tell her all their names.

She took that in for a moment, then asked if he knew where she came from.

He responded with an odd look, unconsciously scratching the bottom of one side of his "rib cage" with the "fingertips" of one "hand", and said "I don't know. All I know is the voice told me that it was going to make another one like me, and that when the other one came to me I should say the words you heard me say when we met. So that's why I said what I did. I don't know what most of it meant. I figure it will tell us at some point. I see you aren't quite like me. I like it."

Then she and Adam spoke at some length, or rather, she asked a great many questions about the garden, the plants, the animals, the sky above, the grass underfoot, the possibility that the voice might make others like them, the voice itself and what it was, when it would talk to them again, and many other topics. Their discussion was almost completely one way, with her doing most of the talking.

This one-way dynamic did not change over the weeks and months that followed. She noticed quite early on that Adam was not very inquisitive, and not only did he not know much of anything, he seemed to have little desire to find out more about the situation they now shared. He seemed content to repeat life as it was, day after day, seemingly forever.

But she was different. She wanted to know more about *everything*. "The voice must have made us, right? How? Why? Why did it make us with different bodies? What is a bone? The voice talked about mothers and fathers, so where are ours? Do you have a father or a mother? Do I? If we do, where are they? We had to have a father or a mother, right? If I am Woman, you must be Man, right? What did you mean when you said I was taken out of you? Did it hurt? How did I fit inside you when we are about the same size? Why is the sky blue during the "day" and dark at "night" when the "sun" goes under the land? Why does it go over our heads and then under the land, and doesn't circle around or just stay in one place and make it day all the time? What are the "stars"? Why can't we eat from that one tree? Is all knowledge a bad thing, or just knowledge of good and evil? And what exactly are good and evil, anyway? How can we know good and evil if we don't know what they are and no one is telling us? Will something bad happen to us if we don't know good and evil?"

Every time she tried to discuss these questions with Adam, he said that he didn't know, and if the voice wanted them to know, it would tell them. She couldn't accept this at all. Being far more inquisitive than Adam and not at all content with ignorance, she spent her days wondering about these questions and trying to work them out, but without success. Every night, she went to sleep still thinking about such questions, frustrated at her inability to figure them out and increasingly frustrated with the unchanging lack of interest from Adam. She fervently hoped that one day she would be able to figure out the answers, or that the voice would tell them to her.

One breezy morning, as she was walking through the garden pondering the nature of the sun for what seemed like the thousandth time, still awaiting the Second Coming of the mysterious voice and glad she was away from Adam for a while since his total lack of curiosity was becoming extremely irritating and was making her compare him to a rock more and more often, a very odd thing happened.

A snake, heretofore nonverbal and apparently eking out a solitary existence on the outskirts of the garden, began to speak to her. It inquired if she had been told not to eat from every tree in the garden. She was surprised at the snake's words, but responded that "Yes, Adam told me that the voice told him that. I didn't hear it from the voice myself. All I know is what Adam told me, which is that we aren't supposed to eat it or even touch it or we will die."

The snake nodded its scaly head, blinked its eyes and said in a lisping and raspy sort of voice, "Yessss, well, that'sss not actually right. You won't 'sssurely die'. The voice, let'sss call it 'god' for now, knows that if you eat the fruit from that tree, you will be like him and the other gods and will know about good and evil. That'sss all. He jussst doesn't want you to know as much as him."

She pondered this for a moment. She was extremely aware that it had been quite a long time since the voice or god or whatever it was had been around. She noted that she had received no guidance or instructions or information of any kind except to sit up, stand up, walk, and then that blather Adam had spouted when they first met. There were so many questions and so many unknowns, so much she desperately wanted to know, and that rockhead ignoramus Adam just wasn't interested in any of it at all.

She had been watching fruit grow and birds fly, and tried to work out how they did it. Adam just ate the fruit and spat out the seeds. He couldn't care less how or why new fruit grew on the tree. Or why they even had to eat in the first place. Talking with him these days was more or less like talking to one of those trees, and it was getting really frustrating. At least this snake seemed to have half a clue about what was going on around here.

She took another look at the fruit on the good and evil tree, and noted that it looked quite appealing and tasty, and reflected once again that if eaten, it would provide the knowledge of good and evil. As knowledge and understanding were the things she craved more than anything, and it looked like a fruit snack was going to be the only way to get any, she decided she would do it. But she did want to let Adam know about her plans. So she went back looking for him, growing more and more excited at the prospect of finally (*finally!*) getting answers to some of her most important questions.

She found him in his customary midmorning position, sleeping under a tree next to a small pile of fruit cores and scattered seeds. This is how their conversation went:

"Adam. Adam, wake up. *Adam!* That's better. Sit up. Okay, so, I've been talking to the snake, and he told me all we have to do is eat one of those nice apples on the good and evil tree, and you know, we get to find out what's right and what's wrong. I think that's a great idea. I don't like knowing there are right things and wrong things, and not knowing what the wrong things are. I don't want us to do the wrong thing. We could get in big trouble. And it's not true that we will die if we eat it."

"(groggily) But honey, the voice in the sky told me not to eat that apple. We're not supposed to know right from wrong. We have to do only what we're told, exactly as we were told, and nothing else."

"Now come on, does that make any sense to you? He made you and put you here, then me, then says 'see you later' and that's it? He tells you not to find out what's good or evil, but doesn't explain it at all? If we don't know good and evil we could be doing evil all day long and not know it. Doesn't that make you nervous? Concerned? I can't sleep at night thinking about that and all the other things we don't know!"

Adam yawned and said "Don't worry about it. He told me not to do it. If he wanted us to know more then he would have told us."

"Well, I can't help it. I don't like not knowing what we're doing or where we're going. We think he made us, right? So he must have made me with this intense need to know things. And I'm telling you, I *need* to know. I don't care if *you* don't want to know, all you seem interested in these days is sitting around eating fruit, laying with me once or twice a day and sleeping!"

"Woman…"

"Don't you 'Woman' me, mister. What's wrong with you? One day, either you or I are going to do something we don't even know is wrong, and then we could be in big trouble. At least if we eat this apple and get to know what's right and wrong we'll know what not to do. And maybe it's some kind of test. Maybe he really actually wants us to eat the apple, but wants us to figure it out on our own and get up the courage to do it. If we just sit here, we'll never know. If we eat the apple, we will know. And then we will know what right and wrong are and will know not to do the wrong things. Isn't it much better to actually know things than just assume them, or just take them on blind faith?"

"Woman, I'm telling you, he told us not to eat it; he was very clear."

"Well, you know, I've never actually seen him. All I've heard is a few words from the sky and that's it. How do we even know he still exists or ever comes around here? I know that snake is real. I've seen it around and talked to it just a little while ago. It's real. I can see it and talk to it, and it talks back to me. I've even seen him eat those apples, and nothing bad happened to him. Why can't we do it?"

"Uh, I just don't know. I'm just doing what I was told."

"So are you going to just walk around here aimlessly, forever, without knowing anything, or are you going to stop and get directions from the apple?"

"Uh, I, um, uhhh…"

"Well?"

"Well, I guess, maybe-"

"Alright then, just come on over here and let's do this."

Adam said with exasperation, "OK, OK, fine, alright, just stop bugging me about it."

They both ate of the fruit of knowledge of good and evil. It was absolutely delicious.

Woman said "Well, do you feel any different?"

Adam seemed somewhat uncomfortable, and said "Um, not so much different, but for some crazy reason I think I need to cover myself up."

"Cover yourself up, from what? There's no wind right now and I'm plenty warm. Are you cold?"

"No, you know, I mean cover myself."

Woman said "I heard you the first time; you're just not making sense. Are you feeling OK? What are you talking about?"

Adam put his hands over his genitals and said "I need to cover myself up."

Woman said "Cover from what? Are you bleeding or something? Got a thorn?" She leaned over and tried to see if there was anything amiss.

Adam yelped at her approach, and said "Hey, stop that!"

"Stop what? What's wrong with you?"

"It's not right for you to look at my, you know, down there. It's embarrassing."

Woman raised an eyebrow and stared at Adam for a second before responding. "Embarrassing. Is that what you just said? Embarrassing? We've lived here for what is it, about a thousand "days" now? I've looked at you every day, you idiot, including 'down there'. There's nobody else around here. The animals don't care. The trees don't care. The grass doesn't care. You didn't care up until two minutes ago. You sure didn't have a problem with me looking at it this morning, or with any of the other things I did with it. What's the problem all of a sudden?"

Adam held his hands out in front of him as if to defend himself from her onslaught and said "OK, OK, I'm sorry. I won't talk about it again. But I do think I would feel better with, say, that fig leaf over there covering me up."

Woman rolled her eyes and said "OK, fine, you weirdo, make one for me too, so you won't get all '*embarrassed*' by looking at me '*down there*' too."

Shortly after they made and donned their new fig leaf underwear, they heard the voice again, this time in the middle of the garden. When he heard the voice, Adam made a tiny garbled sound partway between a squeak and a whimper, dropped to all fours and crawled to the nearest copse of bushes, dragging Woman with him in an attempt to hide them both. The voice called out to Adam, looking for him. Adam managed to muster up the courage to say he was hiding because he was scared and naked.

The voice said, "Who told you about being naked? Did you eat from that tree I told you not to touch?"

After letting out a pathetic low moan, and glaring at an unrepentant Woman, Adam said "Uh, I, um, yes, I did. I mean we did, but she did it first, and she gave it to me and talked me into it!"

The voice then said to Woman "What have you done?"

Woman replied that "We've been here a very long time and still don't know anything. You left and never came back. I talked to the snake about it, and he told me what the situation is, and yes I did eat the fruit. I have to know about more than just this tiny little meadow and this tiny little existence. I have to know what is right and wrong. I didn't see any other way to do that."

The voice directed itself at the snake, which had been quietly listening some distance away, hiding under a thick low hedge, fervently hoping not to be mentioned or noticed. No such luck. The voice angrily said "Fine. As of now you and all your descendants will be cursed worse than any other animal. You will all be forced to eat dust and slither around on your belly at all times 'til the day you die. And you and your descendants will hate the descendants of Adam and Woman forever. Now get out!"

After the snake disappeared in a puff of divine anger, the voice redirected its ire toward Woman, and said "Woman, you are a disobedient traitor. You have chosen to be not with me, but against me. You chose to seek truth and good and evil for yourself instead of following my orders on blind and unquestioning faith. For this, you and all your descendants will suffer agony and pain, and you and all other Women will be under the rule of your husbands."

It then said to Adam, "Because you listened to Woman and disobeyed me, I have cursed the ground you will till. You will encounter thorns and thistles, and you will die and return to the dust you came from."

Now that both Adam and Woman were suitably cowed and the entire future lineage of humanity properly and sadistically condemned to needless pain and sorrow, the voice said "Now, before I kick you out of the garden, take these coats made of animal skins. You pissed me off so much I made them from your favorite animals, so don't bother looking for them. Now get the hell out of here before I get really mad. And don't let the flaming sword hit you in the ass on the way out."

Postscript

So what's the meaning of the whole Adam and Woman/Eve story? Is it a true and literal account, though I've stylized it a bit, of two humans who were magicked into existence in their late teens by an obviously incompetent, deranged, sadistic and admittedly jealous god, then left to fend for themselves without any guidance whatsoever and in the constant presence of both ultimate temptation and a slick 'snake oil' salesman?

Then of this god getting cosmically angry at the two kids for following the nature he deliberately programmed into them, and punishing the entire human race in perpetuity for it?

It's like dropping two fish into a fishbowl, leaving them alone for three months, then coming back to check on them and getting insanely angry at them for eating the wrong fish food or exploring all the parts of the fishbowl, then flushing them down the toilet.

Freud would have had an absolute field day if he were ever able to get this lunatic on the couch. No reasonable and sane person in the 21st century, or even the 19th or late 18th centuries, with even a modicum of education, skepticism and critical thinking could be persuaded to actually believe the story to be a true and literal account of history. But since *Homo religiosis*, particularly in the less educated and more fundamentalist parts of the United States and elsewhere, is able to believe the most preposterous propositions, this literal view of the collection of myths, legends, fables and frauds called the Christian bible is indeed seen by many as the absolutely factual way that history really occurred.

Let me repeat that.

As you will see from the data below, nearly one third of the people you see on a daily basis in America, and quite a few more in the South and in de facto theocracies like Alabama, Mississippi, Louisiana, Tennessee, Arkansas, Kansas, Texas, Oklahoma and South Carolina, actually believe this loony story of supernatural sadism. They don't understand or won't accept that it is ancient mythology and attempts by scientifically ignorant desert tribesmen to explain a then-unknown world and universe. They think the story contains the actual literal words of the creator of the universe and is a description of actual historical events. This is scary.

A third of the people with whom we share roads, gun ranges, classrooms, workplaces, bowling alleys, pools and playgrounds actually think this nonsense really happened as the book says. They actually believe that Adam and his Woman really existed and that they popped into being about six to ten thousand years ago. This is of course sometime after the invention of agriculture and animal husbandry, and is in the face of a massive amount of incontrovertible multidisciplinary evidence across the entire planet as well as numerous relics of two million years of human evolution. They believe all this although they never provide anything more than laughable special-pleading nonsense as 'explanations' of how this could be.

Most frightening of all, these people own guns, vote, control school boards, and serve as elected leaders at the local, state and federal levels.

In 2014, Gallup noted[1] that a shocking 28% of Americans said they believe the Christian bible is the actual word of a single local ancient Middle Eastern desert creator god and that its contents should be taken literally. This means that, when adjusted to 2018 figures, over 91 million people in these our United States believe that the Mesopotamian variant of Aesop's Fables and the Brothers Grimm is literally and historically true.

Almost a third of American adults actually believe that our first ancestors were created from dust and a rib by one particular local imaginary god, were placed without any other human company into a constructed habitat with no education or information of any kind, and were told not to eat fruit from a specific but agonizingly appealing tree.

Nearly one hundred million Americans believe this fictional god deliberately made one of these original people with an unquenchable drive for knowledge or an unbelievably high level of gullibility and susceptibility to suggestion, then damned both of them and all their billions of descendants to unnecessary pain and misery for giving in to their natures in what was in reality a pitifully small and inconsequential act by wholly uninformed and extremely immature children.

They behaved how their creator deliberately programmed them to behave, and in a way which this creator had seen coming for all eternity. He knew billions of years in advance what would happen, but still felt it necessary to create the two kids and watch them fall into the trap he had personally set for them and knew they would fall into. What an utterly horrible and disgusting story. One wonders at the unbelievable depths of gullibility and masochism a person must have in order to actually believe such an obviously fictional and deranged story.

Much worse than even this depressingly infantile statistic is a different Gallup finding[2] that another approximately 50% of Americans think this allegedly holy book may not be the actual literal word of the creator of the universe after all. They think it is merely inspired by this fictional being. As Gallup puts it, "But about half of Americans continue to say the Bible is the inspired word of God, not to be taken literally -- meaning a combined 75% believe the Bible is in some way connected to

God."

If the fact that 75% of the people in this country, including those who serve on school boards and in state and federal governments, actually believe this stuff doesn't absolutely scare the pants off you, it damn well ought to. Anyone who really believes such patently obvious sadomasochistic mythology should be automatically disqualified from any position with any kind of power and influence over others at any level, from schools to the courts to the halls of state and federal government.

The American Christian Taliban is loose in our country and in our government. Its Christian Soldier militiamen and women are doing their damnedest to make laws and policy that affect the rest of us, based on particularly invasive and repressive parts of their particular interpretation of their immoral mythology. And they are working to unconstitutionally spend our money for their pet religious causes and to enact laws which institutionalize their particular religious bigotry.

But back to the story. This appalling fable is actually a fairy tale about a god which is simultaneously both a galactic-level sadist and a terminal prude. He deliberately made one of his absolutely *favorite and bestest creatures* ***ever*** simply chock full of the thirst for knowledge, but forbade the ability to slake it, instead mandating that the isolated and uninformed pair of pets in the fish bowl remain completely ignorant.

He placed intelligent and highly social creatures in solitary confinement, thereby deliberately inflicting cruel and unusual punishment on his own innocent creations. He made his uber-special and singularly adored creations, or at least the female of the species, intelligent and curious, but cruelly made them both live in a tiny terrarium and told them to live their lives in mind-bogglingly boring and repetitive fashion, practically in sensory deprivation.

He made a tiny pocket universe where eating an apple (or fig, or pear, or whatever) is all it takes to reveal the cosmically important and previously unknown so-called 'truth' that nudity is evil and humans should be ashamed of certain parts of the bodies which this creator supposedly made in his very own image. This isn't the work of an intelligent, compassionate, loving god; it's the deviant behavior of a massively sexually repressed egomaniacal sadistic psychopathic control freak.

Or, as is actually the case, it is the work of the human equivalent, and it is a species-level travesty that one third of Americans take the words of these ancient demonic menophobic (fear of menstruation) goatherders as final and unalterable cosmic truth.

What else can you say about the garden story but good for Eve!

Good for Adam! They freed themselves from the mental and physical prison they were thrown into. They showed how to defy the iron grip of totalitarian authority and think for themselves. They sought actual knowledge rather than life as mindless sponges. They should be venerated as brave pioneers, groundbreakers, victims who broke out of captivity and rejected the path to the Stockholm syndrome. Bravo, Woman! Bravo, Eve! Bravo, good job, and way to go!

As a point of reference and comparison, it is interesting to note how various religious and nonreligious groups would have reacted to this defiant and independent female freethinker.

Christian god: "Uppity bitch! Just couldn't follow orders, could you? Now you and all your descendants will endure eternal pain and suffering for your disobedience. Get the hell out of here before my son creates Hell and throws you in it." Or he himself does, or however else a trinitarian would befoul both language and logic in an attempt to explain the act in his or her batty theology.

Catholic church: "Burn her!"

Protestants: "A woman's place is in the home, and this Woman definitely deserves her punishment!"

Baptists: "Every wife has to submit to the leadership of her husband, and this man really needs to show her what submission means."

Islam: "Stone her to death!"

Judaism: "Blessed are you O God, King of the Universe, Who has not made me a woman. Since the mind of woman is not suited to be taught, she should be whipped or starved for her crimes."

Scientifically-informed secular humanism: "What a great job, Eve! What you did was incredibly brave! Your fearless pursuit of knowledge has singlehandedly expanded the sphere of human knowledge. We are all in your debt, and the human race needs many more women like you! We'd like to offer you a research fellowship position so you can continue your daring and groundbreaking research."

As you can tell, I'm firmly in the "give Eve a Nobel Prize" camp.

Dear reader, what would ***you*** say?

Chapter Twenty-Five

Why Can't Priests Have Boobs?

"So God created man in his [own] image, in the image of God created he him; male and female created he them."
- Genesis 1:27

Number of female priests who have sexually abused a child - zero.

Number of male priests who have sexually abused a child - many thousands, and more are discovered all the time.

We again need to make one thing perfectly clear up front. No sane, moral or self-respecting woman would have ever written the Jewish, Christian, Muslim, Mormon or any of the other collections of antihuman nuttiness which the faithful keep trying to pass off as revealed religious texts in our world. And it isn't just because of the rampant misogyny which these religious writings and so many others just can't seem to do without.

Only male testosterone-driven insanity could have produced such immoral abominations. The male writers of religious stories and the male founders of religions usually have and extol the very worst and most immoral thoughts and behaviors a person can have.

Here are some examples from these religions to give you a sense of the contempt and disregard they heap upon women. As you may or may not know, separate massive volumes could be written on each religion's foul treatment of and disgust toward the female of the species, despite women being allegedly made by the apparently anatomically-correct male creator of the universe and according to him, actually in "his" own image.

> **Judaism.** Morning blessings[1] recited by men include "Blessed are you, Lord our God, Ruler of the Universe, who has not made me a woman." Women, of course, have a different requirement. Humans with boobs have to thank their god for making them "according to his will". Message clearly sent, message clearly received.
>
> **Christianity.** The biblical First Epistle of the so-called "Saint" Paul the Apostle to Timothy has these gems, which encapsulate

both the Christian god's view of his female creations and his male creations' extremely dim view of their fellow boob-equipped humans:

- "{2:11} Let the woman learn in silence with all subjection."

- "{2:12} But I suffer not a woman to teach, nor to usurp authority over the man, but to be in silence."

- "{2:13} For Adam was first formed, then Eve."

- "{2:14} And Adam was not deceived, but the woman being deceived was in the transgression."

Islam. Islam's holy book is also clear. It notes in Sura (chapter) 4, verse 11 that "As for the children, God decrees that the share of the male is equivalent to that of two females." In Sura 4, verse 24, we read that "Also forbidden are married women unless they are captives (of war)."[2] Women are worth much less than men and it's perfectly fine to collect them as war "booty". Message clearly sent, and message extremely clearly received.

Mormonism. Multiple wives, no problem. Multiple husbands, no freaking way. Again, message clearly sent and message clearly received.

Except for the relatively few exceptions, women are not normally power-hungry, hyperaggressive, sadomasochistic or psychopathic. Women don't normally write or expound hateful screeds against other people or populations, although today's American conservative and evangelical movements unfortunately contain far too many exceptions to this general species-wide rule of thumb.

Highly divisive specimens like Ann Coulter, Phyllis Schlafly, Michelle Malkin, Michelle Bachmann, Carly Fiorina, Sarah Palin, Marsha Blackburn, Liz Cheney, Mary Fallin, Laura Ingraham, Tomi Lauren, Anne Graham Lotz, Jeanine Pirro and their ilk are an embarrassment to the species and a blemish on humanity, morality, truth and decency. They and women like them are the tragic and hateful exceptions to the rule.

Women don't normally pen unhinged manifestos thousands of words long. They don't write chapters and verses about how men are dirty and unclean beasts, although they often are, then try to pass the writings off as inspired by some invisible god and use them to justify permanent and thorough mistreatment of men. Women don't normally write stories about genocidal bands of murderous rapist thugs ordered to carry out these deeds by voices in their heads or from the sky.

Women wouldn't write a book which justifies permanent subjugation of women in the real world due to a single understandable indiscretion by a vanishingly remote and fictional ancestor. Women sure as hell wouldn't write stories in which the main female characters are dumb, ignorant, disobedient, subservient and often meet a grisly end at the hands of human males and their thoroughly male sky gods.

Through stoning, being burned or raped to death, cut to pieces, turned into pillars of salt, being thrown from buildings and trampled to death and many other unpleasant fates, the ways in which women in monotheisms and particularly Christianity are vilified, insulted, tortured and lose their lives could be taken from the imaginations of immoral and evil psychopaths. Which is of course what actually happened.

And women would be even less likely to try to found religions in which they were automatically no better than second-class and lower-tier humans, and often on the level of beasts of burden or penis receptacles whose primary duties consist of having sex with men on demand, cooking, cleaning, not sitting on furniture while menstruating or otherwise unclean, raising children and generally shutting the hell up.

As late as 2014, living beings who are 100% genetically *Homo sapiens* but who happen to possess boobs and ovaries are not allowed to become ordained[3] in Mormonism, Catholicism, Islam, Jewish Orthodoxy and the Southern Baptist Convention. One can instantly recognize all of these as some of the world's most strongly fundamentalist, bigoted and misogynistic faiths. They are also the most authoritarian, thuggishly aggressive and/or spiteful religions in the world. Women have always been subordinate to both the male of the species and the male gods worshiped in these self-admittedly anti-female religions.

What rationale can there possibly be for excluding female humans from the ranks of the ordained? Why does having different reproductive plumbing from males and a set of baby-feeders make a human being ineligible to serve in an equal capacity in any god's special club on earth? Because the gods of the books and their holy books say so, that's why, and for religious males that's all the divine justification they need to be world-class assholes.

Is there any real or logical justification for this exclusion? No, of course not, this is religion we are talking about. Logic and reason are wholly incompatible with religion.

Apparently the fictional business with a certain curious fictional female and an apple or fig or whatever the imaginary tree of the knowledge of good and evil was outfitted with is the reason for this sexual apartheid. So, in the spirit of illogical acts as moronic as throwing a toddler in jail because an ancestor from 300 years ago was a bank robber, for the rest of the entire lifetime of our species all humans with boobs must be kept in second-class status and absolutely blocked from ordination for the imaginary crime of an imaginary distant ancestor.

We again see the incompatibility of logic and reason with religion and the unsurpassed insanity of imposing ancient mythology onto the real world.

That any women at all would ever knowingly be part of these religions attests to the thuggish nature of males throughout our history, the inability of women to stand up to them, the ability of many women to masochistically accept being considered less than men, and the ability of many men and women to believe the most unbelievable and hateful nonsense.

That any women would want to be part of the ordained leadership of these faiths is, however, a tiny bit of a testament to their individual desire to rise above their second-class status and try to fix all the problems men cause in these religions, despite their masochistic adherence to the faith.

So women are excluded from leadership in these faiths on the basis of fables told thousands of years ago on hateful and illogical grounds, and the fables and oppression have been carried forward and amplified by male thugs, misogynists and bigots ever since that time.

But we no longer dwell in the Dark Ages, or indeed in the thoroughly ignorant millennia before and during much of the Christian Era. There is no reason to oppress women based on ancient mythology and science fiction, but there are still way too many people who think they have books whose gods revealed the evils, sins and unclean natures of women as an all-inclusive totality while also failing to tell us to wash our hands.

The difference between those people and reasonable people is that reasonable people have advanced hugely since ancient times and the faithful today are in many ways exactly the same as their ignorant goatherding peasant predecessors.

Quite on our own and with no divine assistance, we discovered that we need to wash our hands, how and why to cook our food and that we don't need to sacrifice animals or humans to the gods all the time. We have discovered that menstruation is not a curse or a sickness or anything to be feared or avoided. We have discovered much about human sexuality, biology, psychology and indeed biopsychology, and we have discovered all this despite religions and their books, which have all been often and extremely opposed to both science and scientists since religion began.

Today, with quite amazing advances in medical science, we have new questions to ask the faithful with regard to gender.

1. Can a transgender person become ordained in these religions? These people are partially male, if not externally in their biology than internally in their psychological orientation. If not, why not? What does the creator of the universe have to say about these his very own creations? If the book does not contain guidance on this, why did he deliberately choose not to include it? Are these people not made in this creator's image? If not, why did the creator make them this way?

2. What if a person is born female but undergoes all the necessary operations to change to a male at age two? What about at age twenty-two? Or forty-two? Isn't the person now rid of those disqualifying boobs and maybe even those offensive ovaries too, and isn't the person now considered a "male"? If the person's brain and personality are clearly male from birth in one of the not too uncommon cases where this occurs, would eventually making that person's body male on the outside clear the person for ordination?

3. If these people can't become clergy, why are so many homosexual males and child molesters being ordained? Can't the allegedly infallible Catholic pope wave some magic incense or peer through a special consecrated host or see signs in magic smoke or something to enable him to screen out gay men before they become priests, or remove evil men before they start raping little kids? Why can't Mormon leadership, supposedly gifted with the power of discernment[4] and able to know "the true character of people", identify then screen out gay or criminal applicants?

4. In a few decades or centuries, if religion does not drag humanity back to the Dark Ages or destroy civilization or even the species altogether as it is now trying so hard to do, we will have many more and increasingly more difficult questions for the faithful and their clergy. When medical science has progressed to the point where humans can be gene-edited and other medical procedures applied so that an adult female can be transformed into a male at the cellular level on up and vice versa, would a

newly minted male then be allowed into the ranks of the ordained? Would external appearance and internal and external organs as screening criteria be replaced by DNA testing or an examination of gender history to determine maleness?

5. What if a technology is developed that enables parents of a developing female fetus of any stage in its prenatal development to change it to a male, such that the change is undetectable and the resulting post-natal human person is completely male in every way, down to the DNA? Would this person be considered male or female by the Ordination Approval Authorities? One can assume, based on its current and historical attitudes, that the Vatican would insist wholly without evidence that such a person would have to be considered female forever, since the nano-instant of conception is said to trigger the magical insertion of an invisible soul into the brand new zygote. And since their god would have injected this female zygote with a "female" soul, it would be a horror and an inconceivable violation of their god's will to allow a male body to be inhabited by a female soul. So the person would be biologically male in every way but would be cut off from the priesthood due to his/her imaginary zygotic phantasmagorical spiritualish feminine alter ego.

One can imagine the gyrations and discussions that would be held behind closed doors as the cloistered male leadership of these faiths debate how to reconcile the advances of science with their fossilized fables. And this would of course all be gobstopperingly idiotic nonsense, on a level with those empty-headed religionists who argue about how many angels can dance on the head of a pin and other equally ridiculous but nevertheless "deeply and sincerely held" beliefs.

Humans are humans, and just as one should not forbid a person of one or multiple genders and sexual preferences from being, say, a police officer, teacher, plumber or indeed any other profession solely based on sexual organs and/or gender, restricting a person from ordination in any religion based solely on gender will always be nothing but deeply and sincerely hateful misogyny. The only questions one need ask about job applicants are can they do the job, are they truthful and moral from a secular humanist perspective, are they a danger to themselves or others in any way, and will they be a strong contributor to the well-being and success of the organization, its people and its missions. It shouldn't matter if they have a penis or a vagina, multiple penises or are fully hermaphroditic. Or even those scary boobs.

Indeed, it may be possible one day, as postulated by the astoundingly good science fiction writer John Varley, that people could, in his words, "Change" back and forth between male and female as often and as thoroughly as their finances and situations allow. Bobby may leave for his summer vacation and come back as Betty, completely female and with exactly the same skills, abilities and memories, just with his penis and testicles on ice in some lab someplace awaiting reinstallation. The laser focus on gender by religion is asinine and childish, just like all religion is asinine and childish.

Why can't a woman do the same job as a male cleric? There is not a single rational reason why priests, imams, Orthodox rabbis, Southern Baptist pulpit pounders and Mormon religious leaders can't have boobs, other than misogyny and faith-based assholery.

If women were ordained, and if clergy were allowed to marry and have children, which is after all what the gods of these religions told humans to do with considerable vigor, and by the way which Catholic clergy could do until about a thousand years ago, two things would immediately happen.

One is the rate of child rape by clergy would drop immediately, and the other is that the average level of bigotry in these faiths would instantly decrease as their overall morality and compassion, indeed their humanity, would begin to increase.

Any rational person in the history of the human race would call these major improvements.

So yeah, maybe priests should be allowed to have boobs.

Part Six

America's Secular Politics

Chapter Twenty-Six

Why Religion And Politics Should Never Mix

"The purpose of separation of church and state is to keep forever from these shores the ceaseless strife that has soaked the soil of Europe in blood for centuries."
— James Madison

If you have read all of the preceding parts of this book, you will be very clear about what religion actually is. It may be many things to many people, but in none of its many thousands of incarnations is it a divinely inspired or dictated set of lessons or guidelines or a reference for actual humanistic morality. It is not a venue by which the members of a semi-evolved and mostly hairless primate species can successfully beseech unseen entities for wish fulfillment and avoidance of consignment to eternal perdition.

It was the first attempt by our scientifically ignorant ancestors to explain the natural world, and each little tribe's origin story later fractured into many different descendant tales which became separate cores around which a large tonnage of myth and fable has accreted.

Religion has survived to the present day due to a number of clearly identifiable psychological reasons, identified in Part Three of this book, which will not leave us until the species undergoes many thousands more generations of natural selection and scientific education. It will take at least that long to eradicate these unfortunate programming bugs which still reside in the subconscious parts of our brains and stick to us like foul smelly tar, but we need to do all we can right now to purge its contamination from our legal, social and other systems.

Religion is nothing more than fossilized ignorance. It is a wretched relic from our deeply ignorant and fearful past. It is a human construct that has long outlived any usefulness it ever had. It needs to take its place in the hall of failed bad ideas, joining the rest of its failed family of beliefs such as haruspicy[1] (seeing the future in the entrails of animals), leeching, alchemy and astrology.

As with your neighbors' odd sexual or dining habits, the religion they practice is something that is perfectly fine if and only if it is done in complete secrecy as per Jesus's direct orders, if it is never shared with anyone else at all unless specifically requested or approved, if it is never drummed into children and if it has no influence at all on voting or interaction with other people outside the believer's home. When your neighbors' sexual, dining or religious habits expose themselves to the light of day, they enter the social commons and become subject to discussion and critical analysis.

This is all assuming this is being done on a voluntary basis. When these same sexual, dining or religious habits are forcibly and illegally injected into the legislative process and are proposed as requirements for or proscriptions against certain behaviors, it is a hijacking of the common good, of our secular Constitution and of that document's deliberate enshrinement of freedom *from* religion. All such hijacking efforts must be immediately quashed. This includes the unconstitutional proselytization from pulpits, the unconstitutional tax exemption which churches currently unfairly enjoy, and the highly illegal, unethical and immoral attempts by the faithful to unconstitutionally include specific religious requirements or teachings into law.

Anyone who has lived in America has seen or experienced the assault by religion on our society throughout her entire lifetime, most notably in its attempts to enshrine both the ethos and the specific proscriptions from a particular strain of Bronze and Iron Age Mesopotamian fairy tale into our legal codes. From the pathetically spineless and unconstitutional actions to place "In God We Trust' on everything from courthouses to money, to the endless attempts by the religious to outlaw, ban and/or defund any social practice or individual behavior or even opinion which they disagree with, to the unceasing and deceptive efforts to ban the teaching of actual science in favor of religious dogma, the faithful have repeatedly proven that they can never, ever be trusted.

They will always seek to interject themselves where they are not wanted or needed and they will always attempt to make others knuckle under to their particular faith's teachings, through force if necessary. This is a socially destructive practice. It is not moral, in America it is expressly unconstitutional, and it must be fought at every turn.

To the faithful, we nonbelievers all say with one voice that you do not have the legal, ethical or moral right to try to foist your hyperconstrained, ignorant and intolerant worldviews on those of us who think for ourselves. We do just fine on our own legally, ethically and morally, and we do it much better than you do. Your beliefs in things you can't prove are intellectual and moral failures on your part which we don't and won't share. Yours is a position of weakness and ignorance and by itself is a sufficient reason to permanently reject your beliefs and your attacks on our civil society.

As Christians in America have already happily succumbed to a belief in the unreal and the unseen, they are primed for and continually more susceptible to other beliefs which reinforce their inability to escape the mental black hole of motivated reasoning.[2] Astoundingly, white evangelical Christians in America believe they suffer from more discrimination[3] in their own country than do Muslims. The scale of this error is akin to mistaking the moon for the sun, or a house for a grasshopper. How this utterly deranged mass delusion can be possible is absolutely beyond belief but is quite in line with the entire history of Christianity.

Only a group of people who individually and collectively suffer from delusions, hypersensitivity, insecurity, fear, weakness and a planet-sized persecution complex could be so blind to reality. Only a group of people so blindly and completely hooked on the preachings of their unfounded and unprovable religious claims and also so aware of their unsupportability could be so touchy and feel so aggrieved when they are prevented from forcing their religious bigotry upon their fellow citizens.

These deluded Christians feel this way despite the fact that America is absolutely flooded with religions, believers and places of worship, the vast majority being Christian. They feel this way despite the fact that our secular Constitution is under daily attack by the religious through faith-based attempts to teach religion in schools and plaster "In God We Trust" all over courthouses, schools and police cars.

And despite the popularity of Roy Moore, the fiendishly anti-Constitutional theocrat[4] who was an Alabama Supreme Court justice whose oeuvre and mindset can be summed up in his bible-based defiance of the US Supreme Court's order on gay marriage, his suspension from the Alabama Supreme Court on not one but *two* separate occasions, his refusal[5] to remove a monument of the Ten Commandments from the Alabama Supreme Court facility and his utterly delusional blaming of secularism for events like the wholly faith-based attacks of September 11, 2001.

Moore is a classic Christian Taliban religiofascist and there are millions like him in America, living all around us and seeking to inject their religion's poison into the national bloodstream. The people who emulate and enable him and his ilk are extremely populous in places like Alabama, Texas, Kansas, Mississippi, Georgia, Iowa, Tennessee, Kentucky, Louisiana, Missouri and Utah.

The theocrats and Constitutional subversionists these people elect are a Devil's Brigade which works extremely hard to subvert the Constitution, common sense and human decency in their attempts to Old Testamentize the United States. Examples include George W. Bush,

Mike Pence, Donald Trump, Mike Huckabee, Michelle Bachmann, Marsha Blackburn, Greg Abbott, Rick Perry, Mary Fallin, Rick Scott, Rick Santorum, James Inhofe, Sam Brownback and far too many other enablers of hate, bigotry and division.

To add grave insult to mortal ethical injury, Moore won the 2017 Republican primary to fill the U.S. Senate seat vacated by another fascistically religious[6] racist bigoted fellow American Taliban theocrat[7] from Alabama, Jeff Sessions. Sessions occupies the same hellish circle of hyper-religious fanaticism and despotism as does Moore. Sessions departed his Senate seat after being selected by Trump to head the Department of Justice, was astoundingly somehow confirmed to this post after bearing false witness to and/or being terminally witlessly confused[8] during his Senate testimony. His thoroughly shameless and shameful performance involved so many instances of "I don't recall" that he should have been immediately removed from all duties and placed into inpatient round-the-clock supervised Alzheimer's care rather than into the job of the nation's top cop.

Since taking office, Sessions has spent his brief time in office making an absolute hash of the Department, as have all of the rest of his fellow new Cabinet members in their own increasingly and transparently religio-fascist "dominions". Misery loves company, and so do theocrats committed to doing what they believe to be their god's work in this world, regardless of the amount of evil and misery they inflict.

A listing and discussion of all of the subversions and attacks on our secular Republic by the faithful would rival the Christian holy book in length and immorality, and I listed many of these in Chapter Six, but here are just a very few of which most informed Americans should at least have a passing knowledge:

1. The religious justification[9] of slavery[10] before and during the Civil War.

2. The faith-based Comstock Act[11] of 1873.

3. The religiously motivated Butler Act[12] of 1925, which legally prohibited the teaching of actual science in Tennessee.

4. The Scopes Monkey Trial.[13]

5. The religiously motivated Prohibition.[14]

6. The Jim Crow[15] laws.

7. The suppression of the right to vote for women and black persons.

8. The astoundingly overt and unapologetic religious proclamations[16] of our elected[17] officials[18] and some mid-grade[19] and senior[20] military leaders, all the way through to Pence's Piety Brigade and the suppression of the Constitution in much of the American south in favor of Christianity. That Constitutional suppression, one could indeed call it subversion [21], has been a primary mission of the pious throughout our nation's history. In their attempts to reach that full subversion of the relevant parts of our Constitution, the faithful have been constantly working to suffocate reason, lie, obfuscate, cheat and embark on whatever fair or foul activity they think will achieve their ends.

The endless bleatings by the faithful about religious freedom will not cease anytime soon. Their actual goal is absolutely not religious freedom in the sense a reasonable person would understand it to be. Their goal is to misuse the protections in our Constitution to use their religious freedom as a justification to enact their bigotry and discrimination against whoever they don't like and whoever they want to suppress or oppress. These are assaults on both our people and our nation.

Make no mistake, these acts are quite literally assaults, and they blend overt attacks, which sometimes result in murders in broad daylight, with overt and covert information operations and with both overt and covert subversion of political organizations and other entities. Their attacks on the Constitution and on decent human morality are becoming increasingly powerful and increasingly well-coordinated and funded, which is why I have dedicated an entire section of this book to religion's attacks on our fully secular politics.

I'll start with an examination of the Declaration of Independence and the Constitution, and I will show that far from the lies and revisionist history currently polluting Christian airwaves, books, websites, speeches and minds, the United States was founded on secular values and not at all on any variation of Christianity or any of its fictional characters.

America's foundational documents are absolutely secular, over and above the Establishment Clause, which is a wall of separation between church and state that was explicitly placed there by an extremely informed and prescient group of Founding Fathers who knew all too well how devious and deceitful the faithful will always be. The quote at the beginning of this chapter is just one of a very long list of statements made by our founders which clearly show that America was founded as a strictly secular nation.

As to the separation of church and state, this invaluable legal wall of separation between the armies of darkness and ignorance and those who defend true freedom and actual knowledge is being rapidly eroded by the hateful and bigoted theocrats being elected at every level of our government by an increasingly ignorant, hateful and bigoted electorate and Electoral College.

Read on and prepare to be enlightened, and if you are a religious person whose knowledge of this matter is derived from the disingenuous conservoreligious media or clerical propaganda, quite likely enraged at how they have lied to you so blatantly and so thoroughly for so long.

Chapter Twenty-Seven

America's Completely Secular Founding Documents (More Christian Attempts At Rewriting History)

"The government of the United States is not, in any sense, founded on the Christian religion."
— John Adams

"The day will come when the mystical generation of Jesus by the Supreme Being in the womb of a virgin, will be classed with the fable of the generation of Minerva in the brain of Jupiter. But we may hope that the dawn of reason and freedom of thought in these United States will do away with all this artificial scaffolding...."
— Thomas Jefferson

"I have found Christian dogma unintelligible. Early in life, I absenteed myself from Christian assemblies."
— Benjamin Franklin

"...I beg you be persuaded that no one would be more zealous than myself to establish effectual barriers against the horrors of spiritual tyranny, and every species of religious persecution."
— George Washington

In addition to Christianity's age-old but seldom-discussed tradition of stealing other people's religious traditions and stories and calling them its own, its burning of a great many books as well as their authors, its Inquisitions and its Crusades, this religion and its adherents have a remarkably persistent and consistent track record of both denying history and making it up to best suit their goals.

The above quotations by four of the most important figures in our Nation's founding should be all one needs to blow the foul lies about America being founded on any kind of religion quite off the face of the earth, but destroying the lies and delusions of the faithful requires much more than that. This chapter is designed to thoroughly and completely kill off any lies the faithful and their leaders will tell or believe about our Nation's completely secular origins and intent, and it uses the actual founding documents to do so.

Many secular persons of note throughout recorded history have had their good clean unreligious reputations tainted and lied about by the faithful, not merely during their lifetime but also after they had died. The latter, being not merely packs of deliberate lies but also extremely cowardly demonstrations of the weakness and lack of confidence by the faithful, has been inflicted on personalities and figures from the world of physical science to the political sphere.

There are two parts to this posthumous character smear. The first is, of course, the completely false, made-up, disingenuous and astoundingly calculated and deliberate pack of lies the faithful spread sometimes even while the body of a known and famous nonbeliever is still cooling. This is the lie of the nonexistent deathbed conversion. Personalities as varied as Charles Darwin, Christopher Hitchens and Thomas Paine were victims of this vile calumny.

And rather than take the time to actually seek independent confirmation or verification of such a remarkably improbable assertion, the faithful happily set aside reasonable doubt and unquestioningly accept that the power of their particular god or the lure of a promised eternity in some heaven someplace overcame any objections and that the formerly doubting freethinker 'saw the light' and reached for it at the very end. And of course, the dying person could only have seen the light of the believer's particular pet faith, not that of some other or "false" faith.

Lies about deathbed conversions are merely one of the types of lies the faithful tell. The other is their disturbing and pernicious campaign of rewriting history to suit their lies and false narratives, which they feed to the gullible faithful like birds vomiting worm carcasses into their babies' yawning gapes.

Such lies are extremely common among the faithful. This is because if they have even a modicum of a modern education they have already had to deliberately suppress reality to cross a ratiocinational Rubicon and decide that lying to themselves was better than facing a world and universe devoid of supernatural comforts and the alleviation of existential fears by the strong hands of their male god.

The faithful in America continue to spread lies about the founding of America and that this country and its founding documents were based on Christianity. They also spread lies that the founders were drop-to-their-knees believers in this same Christianity. Based on the persistence and number of these lies, such falsehoods evidently make the faithful who make them up, who spread them and who hear them, imagine that the flock in which they herd is far bigger than it really is. They feel infinitely comforted by their false narratives and the anesthetic comfort of feeling cozily nestled among their fellow herd members. They feel emboldened, too, which is extremely problematic for the rest of us.

The bare fact of the matter is that the American Declaration of Independence and Constitution are *secular* documents founded on Enlightenment principles and absolutely not on any religion at all. This may come as a shock to the faithful who have been falsely told or who believe otherwise, but it is quite clearly the case, as shown by a simple reading *of the actual documents*.

In addition to being gullible and unquestioning, with a strong tendency to believe the unbelievable, the unseeable and the unprovable, the faithful share an extremely strong compulsion to groupthink. They also tend to believe absolutely anything that agrees with their comforting world view. They, like a great many Americans, also lack the desire, the courage, the capacity or all of these characteristics to do even a tiny bit of research on their own and discover the truth for themselves.

When people have their faith so deeply implanted in their heads that the roots have penetrated both wide and deep, any challenge to their beliefs often evokes a visceral and immediate response. The reply to the challenge will not be informed by actual relevant and factual information. The response is often sharp and shrill, and conveyed as if the challenger were stating something that was clearly and incontrovertibly against what 'everyone knows to be true'. I have personally experienced this when I told a Catholic acquaintance that the US was in no way founded on Christianity. He actually said "you know that's not true!"

This was a middle-aged and intelligent man of no small amount of learning, but he had fallen into the same religious trap as the ignorant and the unintelligent. They listen to the sermons and tribal messaging and think that's all the research they need. I am sure there are countless other informed secularists who have experienced the same kind of Twilight Zone response from the faithful, and it is jarring when it comes from otherwise educated people. I am quite sure we will all experience many such uninformed responses from the faithful as we try to kill off these infectious and virulent lies.

The shocked and surprised response by the faithful to such realities is what I am working to correct in our national and individual dialogue. We must achieve an informed understanding of history and reality based on facts, not feelings or opinions.

Religionists in the US today continue to insist in the face of incontrovertible evidence that our nation's founding documents and indeed our nation as a whole were founded on the Judeo-Christian faith, or were at the very least heavily influenced by it. This is patent nonsense and a pernicious falsehood which the uninformed often fall prey to, both due to their desire to believe it and for their lack of understanding and knowledge of the actual documents.

I've put a brief summary below of the pertinent points accompanied by some extracts from the documents themselves. My comments inside the text of the actual documents are in italics. The clear and simple facts are that there is absolutely no possible way to honestly or rationally claim that the documents derive from and codify anything but secular human-based and human-formed values, untainted by any religion.

The Declaration of Independence

This document, signed in 1776 and clearly written by men with both the mettle and the will to show the courage of their nonreligious convictions, was absolutely *not* founded on the Judeo-Christian or any other religion. It is a secular document which in only two places and in only five words mentions a nonsecular concept at all. The entirety of the document has zero ties to any religion or god. Don't believe me? Then read it yourself. Look at its actual words.

For your convenience I have inserted the entire document below, copied straight from the National Archives, and I have placed in bold font the words which show the closest the document ever comes to religion, followed by an explanation in italics of what the words mean. Note that just like with all holy books, no priests or ecclesiastic translators of any kind are needed for this exegesis. The words in the document are plain and simple and should not be poisoned by selective or self-serving interpretation or wishful thinking. The reference used is https://www.archives.gov/founding-docs/declaration-transcript. All bolding is applied by the author.

TEXT FOLLOWS:

"When in the Course of human Events, it becomes necessary for one People to dissolve the Political Bands which have connected them with another, and to assume among the Powers of the Earth, the separate and equal Station to which the Laws of Nature and of **Nature's God** entitle them, a decent Respect to the Opinions of Mankind requires that they should declare the causes which impel them to the Separation."

> *"Nature's God" is a clearly* ***pantheistic*** *expression, with no tie to* ***any*** *system of theism whatsoever. It could possibly be stretched into deism, but no farther. Certainly there is absolutely zero tie to any specific system of theism such as Christianity, Islam or Judaism. Again, the religious mindset is the core problem and the cause of*

the logical breakdown. Christians have been told for centuries that they can't just read the words of their holy book as they are written. They have been told that in order to get the "correct" meaning they have to read into or interpret something from the actual words that is something other than what the actual words actually say.

The fatal problem this engenders is that this mindset unfortunately transfers to most everything else the believers read, so they have a hard time taking any other words at face value, particularly when the words say something they don't like. Many of these people, apart from the literalist factions, have been told that a word for word reading of the text "as it is written" is not appropriate for their bible, so they have a difficult time reading other documents literally. But this interpretative style of reading has to be reserved for literature and art, not for science, math, economics, politics or the law. You can't 'interpret' a speed limit sign or extract whatever metaphor you want to from a law or a mathematical equation, no matter how much you want to. You have to read the sign, the statute or the equation as written.

You have to clearly understand that the speed limit is 55 mph and the law is as stated, not whatever you want them to be based on your interpretation of what the number 55 means to you or whatever number you would like the speed limit to be. This applies to the law, to math and to these quite carefully and plainly worded founding documents. It is also ***extremely*** *important to understand that if the Founders had wanted to mention Jesus or his father or any other specific god figure, they absolutely could have, but they very clearly and extremely deliberately* ***did not****. This was a quite deliberate and very conscious decision, and this must never be forgotten. Christianity is not mentioned anywhere, at all, and this was a deliberate exclusion.*

"We hold these Truths to be self-evident, that all Men are created equal, that they are endowed by their **Creator** with certain unalienable Rights, that among these are Life, Liberty, and the pursuit of Happiness—That to secure these Rights, Governments are instituted among Men, deriving their just Powers from the Consent of the Governed, that whenever any Form of Government becomes destructive of these Ends, it is the Right of the People to alter or to abolish it, and to institute new Government, laying its Foundation on such Principles, and organizing its Powers in such Form, as to them shall seem most likely to effect their Safety and Happiness. Prudence, indeed, will dictate that Governments long established should not be changed for light and transient Causes; and accordingly all Experience hath shewn, that Mankind are more disposed to suffer, while Evils are sufferable, than to right themselves by abolishing the Forms to which they are accustomed. But when a long Train of Abuses and Usurpations, pursuing invariably the same Object, evinces a design to reduce them under absolute Despotism, it is their Right, it is their Duty, to throw off such Government, and to provide new Guards for their future Security. Such has been the patient Sufferance of these Colonies; and such is now the Necessity which constrains them to alter their former Systems of Government. The History of the present King of Great-Britain is a History of repeated Injuries and Usurpations, all having in direct Object the Establishment of an absolute Tyranny over these States. To prove this, let Facts be submitted to a candid World."

The word "creator" can be pantheistic, deistic or theistic. Again, there is ***zero*** *linkage to any particular system of theism anywhere, and no such tie is found in this word or sentence or indeed the entire document. If people want to read things into these words, they are committing the fatal flaw found in all religions, which is imagining something is there when it is not. One could just as easily place the names Poseidon, Allah, Manitou, the Flying Spaghetti Monster or indeed Satan in the place of "Creator". Yes, it is true that many people of the time were Christian, but I have to say it again, that if the founders wanted to, they could extremely easily have placed the name "Jesus" or Yahweh" or whatever other name they liked into that "Creator" slot. But they deliberately* <u>***did not***</u> *do this, and for very good reasons.*

The founders themselves were atheistic, pantheistic, deistic or theistic, depending on the person, but their personal beliefs don't matter even a single tiny bit. Their personal religious beliefs didn't and never will matter in how we implement these foundational words. The only thing that matters is what the document says. It includes absolutely no tie to any past, present or future religion, and the fact that it specifically doesn't mention Christianity is clearly an indication of how careful the Founders were to keep it completely out. No amount of wishful thinking or special pleading can change the fact that the words stand in stark black and white, and they deliberately don't mention Jesus or his father or Christianity at all.

"He has refused his Assent to Laws, the most wholesome and necessary for the public Good.

He has forbidden his Governors to pass Laws of immediate and pressing Importance, unless suspended in their Operation till his Assent should be obtained; and when so suspended, he has utterly neglected to attend to them.

He has refused to pass other Laws for the Accommodation of large Districts of People, unless those People would relinquish the Right of Representation in the Legislature, a Right inestimable to them, and formidable to Tyrants only.

He has called together Legislative Bodies at Places unusual, uncomfortable, and distant from the Depository of their public Records, for the sole Purpose of fatiguing them into Compliance with his Measures.

He has dissolved Representative Houses repeatedly, for opposing with manly Firmness his Invasions on the Rights of the People.

He has refused for a long Time, after such Dissolutions, to cause others to be elected; whereby the Legislative Powers, incapable of Annihilation, have returned to the People at large for their exercise; the State remaining in the mean time exposed to all the Dangers of Invasion from without, and Convulsions within.

He has endeavoured to prevent the Population of these States; for that Purpose obstructing the Laws for Naturalization of Foreigners; refusing to pass others to encourage their Migrations hither, and raising the Conditions of new Appropriations of Lands.

He has obstructed the Administration of Justice, by refusing his Assent to Laws for establishing Judiciary Powers.

He has made Judges dependent on his Will alone, for the Tenure of their Offices, and the Amount and Payment of their Salaries.

He has erected a Multitude of new Offices, and sent hither Swarms of Officers to harrass our People, and eat out their Substance.

He has kept among us, in Times of Peace, Standing Armies, without the consent of our Legislatures.

He has affected to render the Military independent of and superior to the Civil Power.

He has combined with others to subject us to a Jurisdiction foreign to our Constitution, and unacknowledged by our Laws; giving his Assent to their Acts of pretended Legislation:

For quartering large Bodies of Armed Troops among us:

For protecting them, by a mock Trial, from Punishment for any Murders which they should commit on the Inhabitants of these States:

For cutting off our Trade with all Parts of the World:

For imposing Taxes on us without our Consent:

For depriving us, in many Cases, of the Benefits of Trial by Jury:

For transporting us beyond Seas to be tried for pretended Offences:

For abolishing the free System of English Laws in a neighbouring Province, establishing therein an arbitrary Government, and enlarging its Boundaries, so as to render it at once an Example and fit Instrument for introducing the same absolute Rule into these Colonies: For taking away our Charters, abolishing our most valuable Laws, and altering fundamentally the

Forms of our Governments:

For suspending our own Legislatures, and declaring themselves invested with Power to legislate for us in all Cases whatsoever.

He has abdicated Government here, by declaring us out of his Protection and waging War against us.

He has plundered our Seas, ravaged our Coasts, burnt our Towns, and destroyed the Lives of our People.

He is, at this Time, transporting large Armies of foreign Mercenaries to compleat the Works of Death, Desolation, and Tyranny, already begun with circumstances of Cruelty and Perfidy, scarcely paralleled in the most barbarous Ages, and totally unworthy the Head of a civilized Nation.

He has constrained our fellow Citizens taken Captive on the high Seas to bear Arms against their Country, to become the Executioners of their Friends and Brethren, or to fall themselves by their Hands.

He has excited domestic Insurrections amongst us, and has endeavoured to bring on the Inhabitants of our Frontiers, the merciless Indian Savages, whose known Rule of Warfare, is an undistinguished Destruction of all Ages, Sexes and Conditions.

In every stage of these Oppressions we have Petitioned for Redress in the most humble Terms:

Our repeated Petitions have been answered only by repeated Injury. A Prince, whose Character is thus marked by every act which may define a Tyrant, is unfit to be the Ruler of a free People.

Nor have we been wanting in Attentions to our British Brethren. We have warned them from Time to Time of Attempts by their Legislature to extend an unwarrantable Jurisdiction over us. We have reminded them of the Circumstances of our Emigration and Settlement here. We have appealed to their native Justice and Magnanimity, and we have conjured them by the Ties of our common Kindred to disavow these Usurpations, which, would inevitably interrupt our Connections and Correspondence. They too have been deaf to the Voice of Justice and of Consanguinity. We must, therefore, acquiesce in the Necessity, which denounces our Separation, and hold them, as we hold the rest of Mankind, Enemies in War, in Peace, Friends.

We, therefore, the Representatives of the united States of America, in General Congress, Assembled, appealing to the Supreme Judge of the World for the Rectitude of our Intentions, do, in the Name, and by Authority of the good People of these Colonies, solemnly Publish and Declare, That these United Colonies are, and of Right ought to be Free and Independent States; that they are absolved from all Allegiance to the British Crown, and that all political Connection between them and the State of Great-Britain, is and ought to be totally dissolved; and that as Free and Independent States, they have full Power to levy War, conclude Peace, contract Alliances, establish Commerce, and to do all other Acts and Things which Independent States may of right do. —And for the support of this Declaration, with a firm

Reliance on the Protection of **divine Providence**, we mutually pledge to each other our Lives, our Fortunes, and our sacred Honor."

> *Fully 99.8% of the words above are* ***secular*** *in nature, as they form* ***secular*** *statements made about* ***secular*** *issues which involve human beings in the real world and have absolutely no tie to any system of theism at all. Only one small phrase deviates from the secular. The mention of "divine Providence" is a nod to the potential of a deistic or potentially a pantheistic Big Guy in the Sky who might lend a hand if one is in trouble, but this divinity is not named in any way, shape or form. Again, this would have been an absolutely perfect place to insert that name or that religion if they wanted to, but they quite deliberately* ***did not****, thus this document cannot be tied in any way to Christianity except through wishful thinking, motivated reasoning or flat-out lying.*
>
> *So the Declaration is clearly a thoroughly secular document, written in language that is specifically designed to be nontheistic. In the few instances when supernaturalism is mentioned at all, however tangentially, there is deliberately no linkage to any specific religion or deity whatsoever. The authors had numerous opportunities for this but* ***they did not do so even once****. It could not be any more obvious that this document is secular in nature and secular in implementation in this, our real world. It has absolutely nothing to do with Christianity or any other religion, because the Founders wanted religion completely separate from the state, and they designed our founding documents accordingly. Again, it does not matter at all what the actual beliefs of the authors and signatories were, because the document itself is all that matters.*

The Constitution

Our nation's foundational document[1] was instituted in 1789 and is fully secular. The **only** time it mentions religion is to ensure it is **always** separate it from the state. It drips with secularism, focuses on people in the real world, and reinforces the need for actual people to do the hard work of living together in a civil society. There is no mention or inclusion of any god of any kind. As seen with the Declaration of Independence above, I have inserted sections of the document below along with relevant comments. The references used are https://www.archives.gov/founding-docs/constitution-transcript, https://www.archives.gov/founding-docs/bill-of-rights-transcript and https://www.archives.gov/founding-docs/amendments-11-27.

"We the People of the United States, in Order to form a more perfect Union, establish Justice, insure domestic Tranquility, provide for the common defence, promote the general Welfare, and secure the Blessings of Liberty to ourselves and our Posterity, do ordain and establish this Constitution for the United States of America".

> *This section enshrines humans and humans alone as being in charge of things in the real world. No gods are included or even mentioned, although this would be the most appropriate place for one of them to be named and included.* ***We the people*** *will do all these things for ourselves, to make this a great nation.* ***We the people*** *will establish this document to unite us in this grand experiment.* ***We the people*** *don't need any gods to do any of this, and we never will. And we descendants of these brave patriots, rebels and secularists are eternally grateful for their foresight and secularism.*

Article VI *specifically states that* "The Senators and Representatives before mentioned, and the Members of the several State Legislatures, and all executive and judicial Officers, both of the United States and of the several States, shall be bound by Oath or Affirmation, to support this Constitution; but **no religious Test shall ever be required** as a Qualification to any Office or public Trust under the United States".

> *This Article specifically includes the ability for citizens to not only swear an oath, but to "affirm" their agreement or position. How about a hallelujah for that extremely deliberate injection of the secular Enlightenment! And it says in as bold and clear language as can be how no religious test will ever be required to be a politician or in a place of "public trust". Not only is there no tie to any specific religion, this language opens the door to those of any faith or no faith at all. So the atheist and the animist, the Muslim and the Jew, the Buddhist and the Sikh, the pagan and the pantheist, and yes even the Catholic and Protestant, can hold these positions.*

Amendment I.[2]

"Congress shall make no law respecting an establishment of religion, or prohibiting the free exercise thereof; or abridging the freedom of speech, or of the press; or the right of the people peaceably to assemble, and to petition the Government for a redress of grievances".

Amendments 1-10 to the Constitution were ratified in December 1791. Religion is explicitly mentioned, but in such a manner as to ***specifically and strictly prohibit its entrance into state matters.*** So our foundational document permanently and specifically excludes all religions from the publicly funded sphere, both in word and deed. And there are no exceptions noted, so this is a binary situation where absolutely no intrusion of religion into affairs of state is allowed.

Additional Confirmation, and Explicit Divergence from Christianity

Far from supporting Christianity, the following Amendments to the Constitution clearly and explicitly ***diverge*** from Judeo-Christian values, as noted in the italicized font following the citation from the Amendment.

Amendment XIII.[3]
Section 1.

"Neither slavery nor involuntary servitude, except as a punishment for crime whereof the party shall have been duly convicted, shall exist within the United States, or any place subject to their jurisdiction."

> *This language, based on secular morality, is diametrically* ***in opposition to*** *Judeo-Christian values, because the holy books of these religions quite clearly condone slavery. The Christian bible and its supernatural and mortal characters are strong supporters of slavery. So by banning slavery this amendment is actually* ***anti-biblical*** *and* ***anti-Christian****, so any claim of divine Judeo-Christian inspiration is ridiculous.*

Section 2.

"Congress shall have power to enforce this article by appropriate legislation."

> *This turned out to be a valuable addition to the document, because getting rid of slavery was a job that had to be done over the objections of millions of devout Christians, who used their bible's own words as holy and divine justification for the enslavement of fellow human beings. Overcoming this inhuman and religiously based opposition required a Civil War, far more than half a million deaths and an equally horrific amount of grave injuries and damage to our society and nation. If only the*

Judeo-Christian bible had just said "Thou shalt never enslave another human being", but alas, it does not. So our secular leaders had to do it, and this Amendment is again a clear ***contradiction of and a blockage of Christianity*** *and its holy book.*

Amendment XIX.[4]

"The right of citizens of the United States to vote shall not be denied or abridged by the United States or by any State on account of sex. Congress shall have power to enforce this article by appropriate legislation."

> *Again, this language is clearly and thoroughly* ***anti-biblical.*** *The Christian bible and Judeo-Christian theism have specifically and hatefully placed women in lesser positions than men, and they have worked very hard to maintain the status of women as chattel and unclean creatures. Why was female equality not in the Christian holy book, as opposed to how it actually describes and treats women, which can only be described as worse than horrible? Why did women not get the right to vote in Christian countries for so many long and painfully misogynistic centuries?*
>
> *It is apparently now a belief among many American Christians that the fictional Jesus somehow inspired the Constitution. If that is the case, why did women in these our own United States not have the right to vote until 1920? Was this a deliberate act of omission by the Christian god to continue the biblical teaching of keeping women down? Why all the thousands of years of torturous and tortured casuistry and woman-hating? Why did a bunch of freedom-desiring human women have to fight so hard up to 1920 to get the country to legally accept that women are the equal of men and they should be allowed an equal voice in society? You know the answer. It is because Christians, using their fossilized immoral holy book as a guide, sought to keep them barefoot and pregnant just as that not-good-at-all book demands.*

To recap another point, and to be extremely clear, the Founders' religious beliefs or lack thereof have absolutely nothing to do with our nation's founding documents. It would not matter a whit if the Founders were all thoroughly rabid fundamentalist Christians, devout to the core and absolutely slavering to implement every word of the Old and New

Testaments exactly as written. It would not matter a whit if they were of the cherry-picking kind, happy to ignore any inconvenient bits.

And it would not matter a whit if they were all secretly Christian-hating pagans who spent their nights dancing naked around campfires holding torches, chanting pagan hymns and sucking blood straight from live Christian babies.

The actual founding documents are the only things that matter, and they will always stand as written, untainted by selective interpretation by the faithful and the deceitful. These national founding documents were deliberately written to be incontrovertibly and thoroughly secular, and to eliminate anything which **any** religious group might ever be able use to try to elevate itself above all other forms of thought. Which is exactly what Christians in America keep trying to do in the face of morality and the Constitution itself.

That a group of Enlightenment-inspired thinkers could create an entire nation and government based on such magnificently humanistic and thoroughly secular documents, particularly in a time when religion still held massive sway over humanity and Darwin's magnum opus was still 70 years in the future, was a remarkable achievement for which we should all be eternally grateful.

BONUS POINTS!

The Declaration of Independence was signed in 1776, the Constitution was instituted in 1789, and there was one more document in our nation's early history which clearly shows our roots are fully secular and not religious or Christian at all. It is almost unknown in America today, but not long after America's independence there was a Treaty of Peace and Friendship[5] which America signed with Tripoli. It was brought to the floor of the Senate in June 1797, real aloud in its entirety and approved unanimously. I have put the relevant words from the text below in bold font, and my comments follow in italics.

ARTICLE 11.

"As the government of the United States of America **is not in any sense founded on the Christian Religion**,-as it has in itself no character of enmity against the laws, religion or tranquility of Musselmen,-and as the said States never have entered into any war or act of hostility against any Mehomitan nation, it is declared by the parties that no pretext arising from religious opinions shall ever produce an interruption of the harmony existing between the two countries."

> *I invite anyone of a religious mindset to explain how a major legislative body of a nation which was supposedly founded on Christianity or the Judeo-Christian religio-military complex would draft a treaty which explicitly says* ***that is absolutely not the case****, then would unanimously pass the treaty on its own floor. The United States government specifically said that it was not founded on the Christian religion, period, end of story, full stop. Discussion over, and all the liars and the faithful need to stop all the lying and deceit.*

All of the above reflects an evidence-based perspective, derived from analysis of the actual words from the actual founding and relevant documents. The American Declaration of Independence, the Constitution and indeed a treaty which America signed shortly after it enacted its Constitution all deliberately do not mention any system of theism whatsoever, including Christianity.

These documents take even further steps to demonstrate their secular nature, by specifically banning the mixing of religion and state, specifically going *against* biblical practice and specifically saying the US is not in any way founded on Christianity. These points come straight from the documents themselves, not from any self-serving interpretation or selective reading of the text.

There is absolutely no linkage in any of our national founding documents to Christianity or the Judeo-Christian system, it is not included anywhere, the documents are actually anti-biblical, and it is only a blatant lie of the worst kind to try to claim otherwise. Those who do are simply ignorant of the facts, psychologically incapable of understanding them, or are flat-out liars.

Chapter Twenty-Eight

Religious Freedom Laws, Also Known As "My god Says You're A Piece of Shit"

"Our civil rights have no dependence on our religious opinions any more than our opinions in physics or geometry..."
— Thomas Jefferson

"To know a person's religion we need not listen to his profession of faith but must find his brand of intolerance."
— Eric Hoffer

Religion can be summed up as "1001 Bullshit Excuses to Fuck With People". It always has been and it always will be, no matter how many people of faith actually do act morally.

The last decade in America has been a sprint-to-the-bottom competition among multiple Jesus-lovin' conservative states absolutely falling-down drunk on the 200-proof firewater of Christianity. They are unfortunately not racing to prove which state can be most like their allegedly dearly beloved hyperliberal socialist uberhippie Jesus and feed the most poor, heal the most sick, turn the most cheeks or be the most welcoming and humanistic.

They are racing to see which one can be the most religiously bigoted one in the nation. They are in a dead sprint, racing to see which of them can be the *least* like their allegedly loving, caring, helping, healing Jesus.

There's no clear winner at this point since the race is far from over, and others are joining in all the time. In a disgustingly evil string of faith-based 'meaner-than-thous', a series of hateful theofascistic state governments have rolled out odious and oppressive hate-based and religiously justified legislation in order to oppress, denigrate, insult and otherwise mistreat often tiny and harmless groups of American citizens.

This is all being done explicitly in the name of Christianity, and what should be a wake-up call to those who actually respect our fully secular national Constitution is that the perpetrators of this bigotry aren't trying to hide it at all. They are openly, blatantly and unconstitutionally using their particular religion in our Constitutionally secular society as justification for this legislation, as though some particular person's or group's specific superstitions and mythologies are supposed to somehow hold magical power over other people's actual civil rights in the real world.

This senile and demented Mesopotamian religion, supposedly founded on the core tenets of 'do unto others', 'turn the other cheek' and 'blessed are the merciful', is now being used as the explicit basis for telling some of our fellow Americans that they are nothing more than pieces of shit, and that they are so disgusting that they don't deserve the rights of equal protection and equal service.

Why do the Christian faithful keep trying so hard to add more and more victims to their lists of people to persecute or discriminate against? Why do they seek to create an American apartheid system in which Christians have rights and non-Christians don't? Why do they never cease their efforts to enact American Nuremberg Laws to harm other people? Bigotry, persecution and oppression, rather than acceptance and kindness, seem to be the *sine qua non* of nearly all organized religions, regardless of the size of the sect or the obfuscations of its apologists, and American Christianity is overflowing with this kind of hate.

Christianity in America since the nation's founding and even unto today is a case study in institutionalized bigotry in action, as I showed in Chapter Six. Today it is merely continuing its centuries-long tradition of doing whatever evil it can get away with before being slapped down again by secular authority or a rebellion of the people. When it was a small and persecuted oddball Mediterranean religious cult two thousand years ago it could be safely ignored no matter what its nutty beliefs were. After it rose to become a major regional power in the form of the Catholic church, following a multistep process of extremely unfortunate and murderously aggressive events which enabled that rise, it became insanely jealous of its power and intoxicated with its dogmas. This was despite or perhaps even because of the sheer lack of proof for its claims.

It then rose to the level of a religious superpower and used this power not to maximize the well-being of the most people, but to embark on a many-centuries-long process of inflicting widespread pain and suffering on humans and humanity, an activity in which it is still busily and industriously engaged in every place it is found. I won't list all of these in this writing, since this is a chapter and not a book of extra-biblical size, and I have already addressed the main points in this book.

But a quick reminder of just a few major examples of Christian bigotry, persecution and aggression throughout history easily shows that Christians are born and bred into antihuman behavior, and that your default position should always be to count on Christians to be oppressive and bigoted.

Hundreds of years of Crusades were undertaken with the justification that the Christian god approved of the actions, to say nothing of the fully secular gains to be had in this world. And of course they had the blessing and full approval of the supposedly infallible popes.

The Inquisition, or Inquisitions, if you like, since there were multiple separate but related incarnations of this ultimate evil, was the peak of organized evil ever perpetrated on earth, due to its extreme length and reach, its broad and long-lasting impact, and its deliberate continuation by many generations of believers, allegedly infallible popes and papal torturers and murderers. This horrific program of persecution for imaginary thought crimes, this centuries-long continuous crime against humanity which blended the worst aspects of the secret police state of Stalinist Russia with the Nazi techniques of imprisoning, murdering and burning the targeted population, astoundingly went on with full papal blessing and not a single heavenly voice from the sky in opposition for over six hundred years.

It is not widely known, but the last person murdered under the Inquisition was killed in **1826**. Not 1526, not 1626, but an astoundingly recent 1826. The Inquisition did not fade away as the lights of the Age of Reason and the Enlightenment dawned on humanity. It kept killing all the way into the 19th century. Charles Darwin and Abraham Lincoln, both born on the 12th of February in 1809, were almost 20 years old, and the United States had been in existence for 50 years, when a group of deeply and sincerely faithful Christians committed an officially-sanctioned and wholly faith-based murder of a human being under the Inquisition, because this independent thinker refused to accept and perform Catholic doctrine.

The Inquisition officially murdered a living human being for thought crime during the lifetimes of the two people who between them freed more minds and bodies from mental and physical imprisonment than any other two humans who have ever lived.

That Lincoln was not religious and Darwin lost his faith after accumulating information about how the world really works, and that they both strove to emancipate minds and bodies rather than hanging and burning those minds and bodies as the most highly devout often do, encapsulates the entire history of the blind and ignorant faithful versus those who choose to use the eyes and brains evolution provided them. Faith kills, and enlightenment emancipates.

It is even less known that the Vatican still maintains the spawn of the Inquisition as an actual institution within the Catholic church, temporarily and one hopes forever stripped of its instruments of torture, nooses and stakes, and retitled as The Congregation for the Doctrine of the Faith. This changing of the names to protect the guilty did not happen in the distant past. It happened in 1965, which was almost within my lifetime.

We tend to forget just how much positive change this world has seen in the last one hundred years, and most people do not realize how much of it has been driven by the secular among us taking action to suppress religious intolerance and bigotry. While the latest Inquisitional incarnation does not overtly conduct burnings of heretics in the town square, it quite clearly shares the original sin of its murderous lineage.

And let no one be deluded for a microsecond that given the right circumstances the Vatican would not gladly and quickly return the Inquisition to its original mission. The bloody history of religion has shown that the deeply and sincerely faithful are always waiting for a chance to gain or regain power and then to oppress and/or kill heretics, nonbelievers and those of other beliefs.

Refocusing the spotlight on the Americas, we notice that even just since our European colonist ancestors arrived here either with or without innocent humans held in biblically-approved involuntary slavery, the list of official Christian bigotry and worse is disgustingly long. Millions of people were killed by Christian, usually Catholic, invaders. An absolutely appalling account of Spanish treatment of indigenous people written by a Dominican monk in the 1500s gives the reader a strong flavor of typical Christian thought and behavior. He noted[1] that:

> *"And the Christians, with their horses and swords and pikes began to carry out massacres and strange cruelties against them. They attacked the towns and spared neither the children nor the aged nor pregnant women nor women in childbed, not only stabbing them and dismembering them but cutting them to pieces as if dealing with sheep in the slaughter house. They laid bets as to who, with one stroke of the sword, could split a man in two or could cut off his head or spill out his entrails with a single stroke of the pike. They took infants from their mothers' breasts, snatching them by the legs and pitching them headfirst against the crags or snatched them by the arms and threw them into the rivers, roaring with laughter and saying as the babies fell into the water, "Boil there, you offspring of the devil!"*
>
> *Other infants they put to the sword along with their mothers and anyone else who happened to be nearby. To others they attached straw or wrapped their whole bodies in straw and set them afire. With still others, all those they wanted to capture alive, they cut off their hands and hung them round the victim's neck, saying, "Go now, carry*

the message," meaning, Take the news to the Indians who have fled to the mountains. After the wars and the killings had ended, when usually there survived only some boys, some women, and children, these survivors were distributed among the Christians to be slaves..."

The first Christian European visitors to North America treated the Native Americans quite badly.[2] The English arrivals followed in the spiritual footsteps of their Spanish Christian brethren, and while they did not kill so many or so flagrantly, they did kill many Native Americans and take their land.

In the late 1700s in America, various Christian groups barred other Christian groups and those of other religions from holding public office. Catholics were persecuted in the 1800s and in various incidents a convent, some houses and two churches were destroyed by other religionists. Today a number of state constitutions actually still have religious tests for office which would preclude[3] atheists from public office. My fellow nonbelievers and I cannot hold office in Arkansas, Maryland, Mississippi, North Carolina, South Carolina, Tennessee and Texas.

Aside from Maryland, each of these states is a blood-red Republican theocracy-lite, quite happy to lead the Religious Bigot Brigade in combat against those who choose evidence over dogma. And you need not work too hard to guess which religion is the one which motivates every bit of this anti-Constitutional bigotry. Hint – it starts with a C and rhymes with Istianity.

The Texas state constitution notes in Article 1, Section 4[4] that "No religious test shall ever be required as a qualification to any office, or public trust, in this State; nor shall any one be excluded from holding office on account of his religious sentiments **provided he acknowledge the existence of a Supreme Being**" (my bolding). Yes, in the 21st century one of the requirements for a Texan running for office is to state that he believes in imaginary beings in the sky based on no evidence whatsoever. This explains quite a lot about Texas. I will assume that since the words do not specify which of the thousands of such supreme beings is being referenced, the Flying Spaghetti Monster, Zeus, Amun-Ra, Allah, Satan and The Worm Of All Worms will all be equally acceptable to Texas faith-heads.

The Arkansas state constitution, in this case one-upping even Texas's super-sized foolishness, notes in Article 19, Section 1[5] that "No person who denies the being of a God shall hold any office in the civil departments of this State, nor be competent to testify as a witness in any Court." So Texas, Arkansas and the other states mentioned above have constitutions that are themselves unconstitutional due to the US Constitution's protections of nonbelievers.

And Arkansas actually goes so far as to say that people who require evidence before they make claims are not competent to testify as a witness in court. Again, this certainly explains quite a lot about Arkansas, in that they say that requiring actual evidence disqualifies a person in court. Does one need any further proof that ignorance and stupidity are the only eternal things in religion, and that truly peerless Christian bigotry is still infecting our nation?

The support for slavery in America was another faith-based Christian abomination. Christianity, due to the repeated biblical support for slavery in both Testaments, was widely used as a primary justification for America's widespread slavery. The stain of the horrific faith-based episode of slavery in our national history will fade but will never disappear. The continuing racism and white nationalism by white Christians in America today, and the continued presence of the not at all good book that justifies it, demonstrate the core inability of religion to honestly recommend and then defend actual decent moral behavior.

In a separate but equally bigoted example, one recent episode in the highly twisted history of Mormonism serves as a useful modern example that religions are always extremely far behind the rest of society when it comes to defining and promulgating a truly decent morality. It took Mormonism until 1978, long after the end of the major civil rights battles in the 1960s, to decide that black males in the faith could finally be ordained as priests. It took until ***2013*** for them to admit that their previously and doggedly held doctrine that black skin was the result of a divine curse was in error after all. Yes, all the way to 2013. This is not medieval faith-based bigoted idiocy, this is present-day faith-based bigoted idiocy.

It's quite interesting how the allegedly divine revelations to actually give people civil rights and grant them equality always come *after* humans in the real world start standing up for their rights and they won't take no for an answer anymore. This happens no matter which faith is being discussed. Again, one needs no more proof than this to show that these religions are all just fabricated nonsense, and that they were designed to and are still today being deliberately used to keep certain people in power and persecute people whom those in power don't like.

The horrible treatment of women is yet another example of Christianity's crimes against half of humanity, gleefully and often sadistically perpetrated by the other half. In addition to life's usual challenges, and the extra physical challenges inherent with being female, such as menstruation, pregnancy, being an actual source of food for infants, having less physical strength than men, being paid less for the same work, often having more common sense and practicality than men and having to deal with their crap, women have endured the continual indignity of forced subservience to their husbands and/or other males in service of one religious fiat or another.

Women were imprisoned, tortured and/or murdered in Catholic countries and even in the American colonies for the wholly imaginary yet nevertheless capital so-called crime of witchcraft. This practice of witch persecution and sometimes murder is still being practiced in some places today, proving that utter religious stupidity and evil do not have an expiration date. Women were repeatedly denied the right to vote in the United States until 1920, and Catholics were particularly strenuously opposed to female suffrage and indeed feminism in general.

Women today suffer from the arrogant and wide-ranging paternalism of empty-headed religionists, with the most extreme faith-based bigotry being the demand that women and very young girls carry pregnancies to term even in cases of rape, incest or if their very lives are in danger, again with religiously blinded and hateful Catholic males usually leading the sadistic and misogynistic charge.

Citizens who happen to be nonbelievers, people of other religions, people who happen to be black, brown, red, yellow, homosexual, transgender or female, or indeed anyone who doesn't meet the definition of a white Christian male who follows white Christian beliefs, have all experienced bigotry and discrimination purely on the basis of religion. And a great many still do, because Christianity is absolutely overflowing with justifications for bigotry against anybody the Christian doesn't like, and because Christians very seldom miss an opportunity to oppress and persecute people.

Persons who happen to be homosexual have long been treated as mentally ill, incurable sinners and not fit to be a part of society. This is unbelievably still the case today in America. We have seen an unbelievably ignorant and bigoted county clerk in Kentucky deliberately and unconstitutionally not do her job as a taxpayer-funded civil servant, then receive endless praise from the faithful for her assholery. We have seen a hateful proliferation of anti-gay bills and the overt proclamations by the faithful of their allegedly god-given right to discriminate against and be assholes to those who were made in their god's image, but happen to be gay. Now their assault has expanded to include transgender people. Indeed, every letter in the LGBTQ acronym is now being physically and legally attacked, as the Christian faithful try to legislate them into as many protectionless "separate and unequal" legal ghettos as they can.

Because these bigotries are permanently rooted in the faith which so many Americans so blindly and ignorantly follow, change and advancement in a common human morality in American society has been and continues to be extremely difficult. The devout cling to their bigotry like viruses infecting a host, and they fight the advance of morality at every turn, continually citing their holy book for justification. It is unfortunate that a book allegedly written or inspired by a so-called Prince of Peace contains language that justifies every kind of evil and immorality which humans can perpetrate.

How unsurprising that people of small minds and evil hearts choose to use these words to justify their bigotry, rather than choosing other words which are actually moral and humanistic to help and support their fellow humans. How unsurprising that so many people of faith opt to hate than to love, to persecute than to befriend, and to exclude rather than include. They have almost two thousand years of hateful precedent and murderous history to uphold and defend, so their antihuman behavior is to be expected. This is the struggle which the truly humanistic wage against these armies of the ignorant and the intolerant. This struggle has been waged for many centuries, and the faithful never give up their bigotry without a protracted and painful fight.

As sacred truths and bigotries become exposed as the manmade inventions they are, and as secular societies codify the acceptance and protection of the oppressed, the religious oppressors and bigots always fight back. In this way, history and humanity have a strong tendency to repeat themselves. Take the issues of slavery and civil rights for black Americans, which required so much conflict, pain, an actual war and just over a hundred years to legally resolve. In 1863, Abraham Lincoln issued the Emancipation Proclamation.

For those too-long removed from school, the Proclamation stated that all people who were held as slaves in the states which had seceded were immediately free. It also paved the way for black men to enter the Union military. This relatively minor provision of humanity toward fellow men, women and children who had been treated worse than dogs for so many years was met with what one can call 'white'-hot vicious anger by many in the deeply Christian and lily-white south. These Christians of course would not let any sort of moral progress go unchallenged, and they eventually found a way to respond.

Their response to being thwarted in their bigotry was the unbelievably hateful Jim Crow[6] laws. These were laws beginning in the 1870s that required racial segregation. Unbelievably, it took until 1965 for the last of these hateful laws to finally be abolished. Again, this was in the deeply religious and extremely Christian, white-controlled South. Millions of white Christians waged a century-long battle against the human and civil rights of fellow human beings, through their institutional and legalized bigotry.

It should be a requirement for continued citizenship that every American know the laws that were passed and the things that were said and done simply to keep people of a specific skin color in an oppressed and second or third class status. Far from being a bastion of freedom and equality based on the teachings of Jesus, the extremely religious South was a Taliban-like hell on earth for people with dark skin for about two hundred years, and in many places it has not improved much. Figures 12 and 13 below show what highly religious white Christians very deliberately and very happily did to fellow human beings in the deeply and sincerely devout South for as long as they legally could.

Figure 12. Segregated water cooler, Oklahoma City, 1939. Image courtesy of the Library of Congress.

Things have quite fortunately improved in America with the legal suppression of white Christian racism by secular law, with the Civil Rights Act of 1964 and other positive legislation, but the tension is still there and the hate and bigotry are at a constant boil. In America today the targets of Christian bigotry are not only people with black skin, they are now also those who have brown skin, yellow skin, are of a different or no religion and/or are not strongly politically conservative or who do not agree with the religious views of the Christian faith-based aggressors.

Today the ignorant and hateful Christian crusaders, as always being found both within the general population and throughout legislatures, supreme courts and executive residences in states and at the federal level, are doing everything they can to unconstitutionally and immorally legalize and institutionalize their religious bigotry. The Christian faithful and their highly conservative political establishments from the federal government on down have been working hard to take voting rights away from anyone who is not a white conservative Christian.

They work hard to inculcate Christian bigotry into secular law, and because they have failed to overturn the US Constitution in this effort through injecting Christianity into our schools, they are trying to use one of our national strengths, that being our codified freedoms, against us in a prolonged, perverse and hateful series of insider attacks and fifth-column sabotage. They are trying to turn our right of religious freedom into codified bigotry, in which their particular faith's and their own particular religious thuggeries are protected by law.

Figure 13. Segregated bus station in Durham, North Carolina, 1940. Image courtesy of the Library of Congress.

Christians are demanding the right to refuse service, even publicly funded service provided by a completely secular government agency, to anyone they don't want to serve, on the grounds that it somehow violates their religious beliefs thus their "freedom" to be religiously bigoted assholes.

Even someone whose head has been softened by religion and who can see more than ten minutes into the future can see the problem. The holy book of Christianity contains infinitely more evil than good, and since following that book means a person will definitely do evil, the

Christian faithful are telling us that they will do evil because they think their book should take precedence over both common decency and secular law.

This is a slippery slope to full perdition. Once you open up the door that says one person's religious freedom can be used to demean, oppress, persecute and eventually imprison, exile or even execute people, the floodgates will open and tsunamis of Christian evil and bigotry will pour through and drown our great Republic.

The hypocrisy, self-centered and small-minded egotism and bigotry of the Christian faithful show through every time they whine about religious freedom and complain that they can't legally treat other people badly. Unless Christians specifically manage to legislate special protected status only for themselves, which is most definitely one of their primary goals, and if religious justification for discrimination is enacted into law, Muslims and Jews will have just as much religious freedom to do mayhem as Christians would.

If unrestricted religious freedom were to be made law, what would stop Muslims from imposing Shari'a on other people, including Christians, because they think their holy book says to? What would stop fundamentalist Jews from following the Old Testament's actual literal words and killing those whom they think are on their land, because Judaism's holy book says to? Absolutely nothing.

How would you like that, all you Christians? How would you like Muslims taking advantage of their holy book and their newfound legal justification to oppress ***you***? Religious freedom is a right in this country, but it should be kept within the confines of your own little skulls. Once it escapes, it and its adherents become an evil and immoral problem.

Christians can always be counted on to be bigots, and they keep proving it every single day. From denying customers service in a publicly funded Kentucky county clerk's office, to a baker refusing to create a wedding cake for a gay couple, to countless organizations trying to get out of providing contraceptive coverage in health insurance plans, to trying to keep transgender people out of bathrooms, and indeed to a faith-based near national Christian hysteria on these subjects and far too many more, religiously motivated assholery is rampant in America.

In 2015, the state of Indiana, under its self-admittedly religiously blinded and theocratic Maximum Leader Mike Pence, enacted a Religious Persecution Enablement Act, or as Christians like to call them, one of those so-called "religious freedom" laws. Not a single solitary person had been prevented from going to any house of worship of his or her choice. Not a single solitary person had been persecuted for believing in any god or faith of any kind, except for Muslims, of course, who Christians will never see as needing any religious protection, least of all from the persecution by Christians themselves.

So what was so important that an actual law had to be passed? The answer is disgusting. Pence and thousands like him in elected office, as well as millions of their deeply and sincerely faithful supporters, don't like certain people and since they can't legally persecute and oppress them any other way, they work extremely hard to pass laws that let them discriminate, let them persecute and let them oppress people solely on faith-based grounds. We've seen this sort of thing before.

The Muslim holy book mentions the *jizya*, which is a tax placed on non-Muslims living under Muslim rule. People of a particular faith used their religion to oppress others.

In 1487 the German Catholic cleric Heinrich Kramer published a vicious book on witchcraft. It contained tips on the best ways to torture witches and also the recommendation that they be murdered. Needless to say, it was as full of utter insanity, bigotry and hate as the Christian holy book which inspired it but it was very popular with the religious persecutors and bigots of the time. People of a particular faith used their religion to oppress others.

In 1555 Pope Paul IV issued a papal bull[7] which "…revoked all the rights of the Jewish community and placed religious and economic restrictions on Jews in the Papal States, renewed anti-Jewish legislation and subjected Jews to various degradations and restrictions on their personal freedom." People of a particular faith used their religion to oppress others.

In 1935, Hitler's Germany enacted the Nuremberg Laws, which were very much like the 1555 papal bigotry decree. People used a particular faith's religious beliefs to oppress others.

American Christians in the 21st century, following their extraordinarily long tradition of institutional bigotry and persecution, are today using their particular religion to oppress others.

Why do Christians always have to be like this? Why is it their Christianity which always motivates or enables these people to perpetrate such hate and to inject their intolerance into legislation? If you don't want to make gay couples a cake, or provide flowers to their wedding, or do a job which taxpayers, many of whom are LGBTQ, non-Christian and/or nonbelievers, are paying you to do, then quit your damn job and go be a cleric or do some other job which keeps you away from decent people. Just who the hell do you people think you are?

If you are one of these people, you are hypocritical, sadistic and hateful. You are so full of hypocrisy, hate and ignorance that you ignore the evil you are doing and can't see beyond that damnable book you are using to justify your evil. Take just Kim Davis for an example. This apparently fundamentalist Christian takes the traditional Christian approach to her religion and only pulls from it what she wants to so that she can treat other people like shit.

Yes, her holy book does surely say that gay people should be killed, so we can be thankful she merely refused to grant a gay couple a wedding license instead of whipping out a gun and shooting them dead as the lord her god would have any properly devout Christian Soldier do. But the Jesus she proclaims to love and follow so goddamn closely said quite plainly in Matthew that "…whosoever shall marry her that is divorced committeth adultery."

Davis has been married four times to three different men. Therefore at least two of her husbands have committed adultery in her god's eyes by marrying her. Has she followed the biblical guidance that adulterers should be put to death? I assume not, because I have not seen her name in the news as carrying out this punishment on them or paying someone to do her god's bidding. Why not? Why do Davis and the millions like her only choose to follow the holy book verses that justify their acting like assholes to other people and ignore the verses that say to do good for other people?

Truly world-class bigotry and hypocrisy like this are omnipresent in religions and the faithful, and every time such faith-based bigotry occurs, mobs of the faithful whoop and cheer and celebrate the oppression of innocent people. The problem is there are millions of Kim Davises in our country and they are striving every single day to be faith-based hypocritical bigoted assholes. They are working equally hard to infect our legal system and use our great Constitution against people who have done nothing to them or anyone else.

I actually am extremely sorry that the demonic hell of monotheism does not exist, because without it people like Davis and the millions of Christian hypocrites like her can never themselves be sent there to truly feel the pain they inflict on other people in this world. Thus my wish for all evil people to feel the pain they unnecessarily and hatefully inflict on others will likely never be achieved. As I have said before, when people think their god is on their side, nothing is sacred, including the lives of innocent and harmless people.

The Trojan Horse "religious freedom" movement is just another front in the war on our secular Constitution and decent society by immoral religious bigots and misogynists. The faithful don't give a damn about real people in the real world. All they are doing is trying to find a way to use their holy book as a club with which they can legally bludgeon, persecute and oppress people they don't like and treat them like shit.

Unbelievably, they are using a book of ancient bigotry to try to justify modern bigotry. What they never see is how immoral this actually is, and how hypocritical and short-sighted this is.

If the aggressively bigoted Christians who are behind this evil nonsense persist in their efforts to institutionalize ancient bigotry, I will claim that *my* deeply and sincerely held faith must enable me to engage in the acts ordered in my faith's biblical equivalent, called *The Most Supreme And Inviolable Super Duper Most Holiest Book Of Really Mean Shit God Officially Orders You To Do.*

This book, which was written in the demonic and genocidal spirit of the Christian bible, specifically orders all persons of my faith to discriminate against, shun, persecute, ridicule, insult, demean and potentially stone religiously motivated bigots, thugs, liars and misogynists. And the Christian faithful can say absolutely nothing at all while they dodge the slings, arrows and stones of outrageous and righteous religious zealotry, because this belief is deeply and sincerely held and that is our shared justification for our particular bigotry.

Monotheistic religions, particularly Christianity and Islam, are incompatible with a civil society, and they have proven this every day since their invention. The demonic practice of such religions must therefore be erased from our world, and it will only come about when good people take a stand and make it happen. Until we reach such a "blessed" state, however, we need to forcefully shove the façade of religious freedom back into its Constitutional box.

In America, actual religious freedom is the Constitutional right granted to each citizen to believe utterly insane things without proof, and to have "free exercise" of that religion. This free exercise is limited to oneself, however. It does not enable a person to treat other people like shit because he thinks his god says to. It means you have the individual personal freedom to let the infection of your particular religion rot your mind. Your religious freedom rights absolutely do not give you the freedom to be an ignorant and evil bastard to anyone at any time. You are not allowed to let your particular religious infection spread beyond your individual person.

The First Amendment says "Congress shall make no law respecting an establishment of religion, or prohibiting the free exercise thereof; or abridging the freedom of speech..." Note that we have the Constitutional freedom of speech but it has limits. We cannot yell "fire" in a crowded theater, we cannot lie in court and we have laws addressing lies and false claims. In the same way, you do have the right to practice your religion, but within the limits described above. Your personal religious freedom is absolutely not a free pass to be a bigoted faith-based fundamentalist asshole. Your Constitutional right to religious freedom does not give you any right whatsoever to treat other people like pieces of shit.

And while there is no heaven and no hell, and there will be no Second Coming, if there were I would happily trade my eternal soul for the opportunity to watch the big show when millions of you religious bigots desperately try to explain yourselves to Jesus while he's angrily picking up that whip of his to beat some decency into you and asking you "*what the hell goddamn part of 'be nice to people' didn't you understand?*"

Part Seven

Miracles Are Imaginary

Chapter Twenty-Nine

Weather, Football, Resurrections and Other Non-Miracles

"Any sufficiently advanced technology is indistinguishable from magic."
— Arthur C. Clarke

Yes, the title of this chapter puts football and resurrections in the same box. Why? Because once the invasive virus of religion gets into a person's head, it's like skunk stink. Once it's on something, it's almost impossible to get it off, it spreads to everything else in the area and it really smells up the place. Once religion gets into the brain, it immediately infects the entire structure, and everything a religious person experiences becomes filtered through a particular faith-based mental sieve.

This sieve modifies, changes, slants and twists facts and reality to whatever degree is necessary to fit into the person's particular set of dogmas. Various religious additives such as confirmation bias, denial of reality and rationalization are poured into the religiously-infected cognitive process like artificial dyes, artificial flavors and chemical preservatives in order to ensure all inputs are modified as needed to make them acceptable to the religious brain's mental palate.

It's similar to the process by which you can soften water, but in this case the result is a softened brain.

And if you've ever lived in America and listened to a televangelist, watched a sporting event on TV, talked to people of faith or read any history, you've almost certainly been exposed to the mind-numbing phenomenon of a semi-evolved and usually less than semi-educated primate stating that a certain event or act in the real world either did or did not occur due to the intervention or lack thereof by a divine or supernatural power. Yes, the tired but always popular miracle or "the hand of my god" story.

Did your favorite sports team win a game? Did your house sell in a down market and you still made a profit? Did you win 10,000 bucks at the slot machine? Did your child recover from a very serious illness? Did you experience anything at all that was good and/or that you wanted to happen? Well, by jolly, it must have been a miracle, right? Isn't a miracle a good thing by definition?

It is extremely odd how when your sports team lost that game, or your house burned down, or you lost a bundle at the slot machine, or your child died, that *these* occurrences are not miracles. These events are unpleasant diametric opposites of the desirable ones listed above, equivalent in scale but opposite in sign. Somehow, due to the degree of pain they cause, they are not seen as miracles but something quite different.

These bad occurrences are never described as the angry god taking a stick and beating you with it. They are called "acts of god", though it is never provable which god is thought to be responsible and there are always a great many theories as to why the act was taken. They are also called "god's will", or the god is described as "working in mysterious ways", or that the god "has a reason for everything that happens."

Self-flagellating empty-headed platitudes like those along with "surely some good will come of this" are omnipresent in the speech of the faithful after a bad thing happens. These are all pathetic attempts at theodicy, proffered by willing masochistic victims of a pitiful and pervasive Stockholm Syndrome. All of this merely shows the weakness of the religious mind and its strong resistance to or potentially its inability to understand reality. I am cured of cancer and it is a miracle. I die of cancer and it is god's will. This is a complete short-circuiting of the critical thinking and reasoning processes in order to achieve a desired answer, no matter how wrong, illogical or masochistic.

Miracles can only exist in a hyperconstrained and blinkered mental universe in which the cosmological constants are fear, superstition, confirmation bias, cognitive dissonance, illogic and a complete failure to understand probability. Far from being fine-tuned by some imaginary creator, these constants are wildly turned up or down as needed by the believer to arrive at whatever conclusion he or she wants to reach.

One has surely noticed the extreme commonality of miracle stories in both ancient and modern times where and when there is no means of independently verifying the stories or capturing recordings of the events. One of the laws of religious nature is that there is an inverse relationship between the number and type of miracles being proclaimed and the availability of technology and information needed to document and verify miracle claims.

What would one expect, when a miracle is merely some occurrence or phenomenon which we have yet to explain properly, a low-probability event, a story from a fairy tale, or a flat-out lie?

Almost every aspect of life in the 21st century would be considered miraculous to someone living 2000, 1000 or even 500 years ago. If you raised a human today in a group of people completely lacking modern technology, did not teach the person that such things existed elsewhere, and instead steeped him in some sort of religion, of course the conclusion this person would reach upon seeing such wondrous devices as the radio and smartphone is that they were divine in origin, not made by the billions in factories by humans just like him.

The truth is there are no such things as miracles. There are events which we cannot yet fully explain, but this does not of itself mean the events are miracles. Human thought is not fully understood or explained but it is hardly miraculous. What actually makes gravity work is not known, but it is not miraculous. Nothing at all about our world or our universe requires a miraculous or divine explanation.

There are plenty of things which we could potentially experience which would seem miraculous, and in some cases might actually be miraculous, but only rigorous analysis by teams of scientifically informed experts and the appropriate tools would be able to make that final determination, not some cabal of celibate males in the Vatican or some random religious believer or snake oil salesman slithering around on a megachurch stage deciding if some act, event or ghost story were proof of the miraculous.

If people were seen to levitate one day, there is no shortage of people who would call it miraculous. Those prone to religion would undoubtedly see it as proof of a divine presence or the "saintliness" of the person doing the floating, and they would all see it as proof of their specific religious worldview and the hand of their own specific god at work.

The nontheist would laugh at such idiotic talk, ask how the trick was done, and if deemed to be no trick would marvel at the new technology and wonder where it came from, then ask where she could get her own levitation machine. She would not for a second believe any idiotic snake oil pitch or religious foolishness about the act being a miracle or the result of some magical nonmaterial power.

The outstanding author and intellect Arthur C. Clarke had a great many thoughts on many subjects, one of which is the aphorism "Any sufficiently advanced technology is indistinguishable from magic". This can be slightly tweaked to reflect a perspective on miracles, to observe that "Any sufficiently credulous or ignorant mind sees the uncommon as miraculous."

Compounding the problem, humans are often extraordinarily deficient at accurately linking a cause with an effect. One of the most common and most powerful human logical fallacies is the "*post hoc, ergo propter hoc*" fallacy, which I discussed in Chapter Twelve. This black hole of a logical failure is a trap that is extremely easy to be sucked into and is very hard to escape. It is the erroneous connection between one event and a second subsequent event that occurs shortly after the first event.

It is a trap because it is based on the illogical System 1 also discussed in Chapter Twelve, and is not based on a deliberate evaluation of reality. It is not based on logic and observations screened through the scientific method.

An example would be if a dog barked outside your house and within a second or two your phone rang. While the one event did indeed follow the other one, the dog's barking did not cause your phone to ring. Even if this paired sequence happened to occur a few times in the same week, there is still no actual connection from the barking to the ringing, unless your phone has been deliberately programmed by a human being to ring when it hears a dog barking.

System 1 programmed us to make these faulty connections, and again this immediate assessment tool served us well when our ancestors roamed free among dangerous animals and dangerous humans, but today this mental module far too often gives us false impressions and nonexistent connections and leads us into logic failure traps.

Similarly, when a person prays for something and it happens to occur, the person extremely often believes that the prayer was the cause of the event or outcome. This is a childish and ignorant breakdown in logic. The breakdown occurs because brains capable of making decisions, be they bird or human brains, are subject to the many logical flaws and cognitive biases listed in Chapter Twelve plus at least one more massive contributor.

Many years of psychological research have shown that what is called a variable ratio schedule of behavior reinforcement is an incredibly powerful driver of human responses and thinking.

In both experiments and real life, if a subject thinks that Action A results in Outcome A, but that it is unknown how many times the subject has to take Action A to get Outcome A, the subjects who want Outcome A will continue to believe that Action A results in Outcome A more strongly than if there were a known number of times needed to get Outcome A.

This variable ratio schedule of reinforcement is extraordinarily powerful, and beliefs and behaviors shaped by this particular reinforcement scheme are incredibly hard to kill.

Two examples of this are slot machines and prayer. No one knows which pull of the slot machine lever will result in a cash reward. The machines do pay out but they do so on a variable number of pulls. So in the mind of the subject, every single pull is potentially going to be a winner. If the machine paid out only on every 20th pull, the excitement and anticipation will be gone, along with the belief/hope that the next pull will shower the subject with lights and coins.

When the subject does not know how many pulls are needed, the subject is filled with hope and anticipation for every single pull, because it is possible that the next pull will be a winner. Humans and other animals will pull that slot machine lever or tap a food button or do whatever other behavior has been shaped by variable ratio schedule reinforcement far beyond when logic would say to stop doing so.

Prayer is exactly the same way. All it takes is a belief that prayer will make something happen. Prayers are sometimes believed to be answered because simply based on objective statistical probabilities, the outcome the person wanted to happen is likely to happen at some point. People who pray for things like finding Mr. or Mrs. Right, or for the rains to come, sometimes actually do have this happen, but often after having to pray numerous times and going through a lot of bad dates and days with no rain. But as rains and finding mates happen on a variable ratio interval, the imaginary connection between the prayer and the outcome is strongly fixed and extremely hard to extinguish.

Prayer is strongly reinforced because of its application to those common occurrences such as rainfall. No one has ever prayed for an actually impossible outcome, such as the ability to fly, and had it happen. In prior ages untold millions died of preventable diseases but some random number of people survived the illness, and many of them believed their god answered their individual prayer on their own individual behalf. Today no one with a grain of common sense prays to prevent disease; we go get vaccinated, and we don't pray for the vaccines to work because they do. Because, you know, science.

This variable ratio programming module gives many people the illusion that prayer works, and thus the delusion that the particular god or spirit or whatever they are praying to is the one that answered the prayer. Since people offer billions of prayers for what are actually things that are at least somewhat likely to happen, far too many people make the mistake of thinking their god answered their prayer with some sort of miracle.

And context is something that the faithful are absolutely terrible at understanding. Returning to the point that people often pray for a desired outcome, it is highly unlikely they are doing it in a single-person vacuum and the only action being taken is this single person's internal prayer. Since from one to many people are often taking actual active measures in the real world to attain the outcome, it is no surprise that desired outcomes sometimes, and perhaps even frequently, occur.

When even more and highly trained people are also actively taking measures to help attain that outcome, such as hospital staff working to heal a sick or injured person, the odds of that desired outcome go up dramatically. When the desired outcome happens, the believer erroneously connects the outcome with his fervent prayers, again illogically assuming that his particular god heard and answered the prayer.

Nontheists and critical thinkers will note that with so many people or so much effort focused on attaining the outcome, it is perfectly reasonable that it was achieved and it might be a miracle, euphemistically and statistically speaking, if the outcome *didn't* happen. The outcome turns out to be no big deal.

This is why rain dances and other calls for true so-called miraculous "acts of god" don't work no matter how many millions of people pray for them or how fervently they genuflect, flagellate and wail to the heavens, because humans can't yet actually technologically produce rain and there are no gods to hear these prayers. This is why faith healing and praying for health without using actual proven medical science result in lots of needlessly sick and dead people, because medicine works and prayers don't. There are no such things as miracles.

But why do people think there are? They think this because people don't understand math and science, because they desperately want to believe in magical fairy tales and gods and afterlives, and because people lie.

We have already demonstrated the first two of these, through the staggeringly broad and deep ignorance of math and science by religious believers in America and elsewhere, and the many reasons people are unable or unwilling to escape from god tales and ghost stories. Those two major problems on their own would provide a never-ending source of miracle stories, but the problem of lying is a very significant one, particularly among the faithful and the religious hierarchy despite their religions' direct prohibitions against such activity.

People lie, and they lie about almost anything. They lie about their age, their weight, their grade point average, their finances, their past, their present, their criminal record, their taxes, their intelligence, their sexual prowess, their opinions of your kids and spouse, their failings and successes and pretty much anything there is to lie about. They make things up, embellish stories or spread gossip that gets bigger and juicier with each telling. They often lie or exaggerate to gain attention, just as religious myths and other stories originally dreamed up around a campfire get bigger and more unlikely over time. People lie to gain an advantage, to convince someone to have sex with them, to get out of trouble, or to not seem like they are as ignorant or unprepared as they are.

Before facts could be checked, lies could be commonly told without means of verification and they were, repeatedly and in massive numbers. Many were commonly believed. Witnesses are often kin to or sympathizers with the liars, so cannot be trusted to be objective and free from stain. It is, I trust, no surprise to hear that people also lie on the subject of miracles.

This has all been true since humans learned to speak, and today's miracle stories are no different from those told thousands of years ago. The difference today is that miracle claims and stories can be viewed through the lenses of logic and information, and they can be rigorously analyzed using the tools of science. Therefore just as the number of reasonably accepted religious explanations for natural occurrences has dropped from covering almost everything down to zero today, the drop in the number of events people accepted as miracles has followed the exact same slide down to the exact same total absence of scientifically and empirically confirmed miracles today.

There have never been any miracles, and the only thing today that one might reasonably call miraculous is the ignorant and stubborn refusal of our species to rid itself of such illogical and impaired thinking.

Chapter Thirty

It's A Miracle! Isn't It?

"All the tales of miracles, with which the Old and New Testament are filled, are fit only for impostors to preach and fools to believe."
— Thomas Paine

To listen to the credulous ramblings of religionists, you'd think their god was invisibly and magically standing right behind them all day, every day, just waiting to magically intercede and miraculously protect them from the impacts of a disaster while letting everyone else die horribly. Or he is waiting by the prayer phone for the call to the divine help line, ready for the hotline to ring so he can magically and miraculously spring into action and intercede somewhere.

These prayers are often for medical miracles, but many other types of requests do flood into the heavenly help desk. Anything that ails, distresses or impacts people has been the subject of quadrillions of pleas for miraculous intervention. Many occurrences that are statistically improbable, and many that aren't, have been touted as so-called proof of the intervention of a specific god or entity from a specific religion on behalf of a specific person, group or cause.

Many people, as the gullible and ignorant semi-evolved mostly hairless primates they are, have no understanding of probability and statistics, of cause and effect, of chance and correlation, and of the cognitive biases and logical fallacies described earlier in this book. They have also been primed and propagandized into believing there is some sort of god out there in some spirit world and that this god intervenes in the real world. These people are thus both ready and quite willing to believe in miracles.

These people pray for miracles to happen, and they think they see them everywhere. And when certain things happen you can see it in breathless headlines and hear it in the voices of radio and television reporters, as they say things like "We bring you news of a family's miraculous escape from disaster; they survived unhurt when their (house, neighborhood, tent or church) was struck by a (tornado, earthquake, hurricane, fire or flood)!"

This is often followed by a pan or a cut to a disheveled and often somewhat disoriented person saying "It's a miracle we all survived!

Praise be to (jesus or god or the lord or allah or jehovah or pick the god or supernatural entity of choice)!"

Miracle claims come in an uncountable variety of flavors, such as those below:

- She survived a fall down a hundred foot cliff – it's a miracle!
- He survived when the tornado hit his house – it's a miracle!
- Average U just beat Football Powerhouse – it's a miracle!
- The existence of the universe is just so amazing – it must be a miracle!
- The American hockey team just beat the Russian hockey team – it's a miracle!
- I found a parking spot *in the front row* at the mall on Christmas Eve – it's a miracle!
- My child's illness was cured – it's a miracle!
- The god I worship magically impregnated a teenage virgin who had a son who was killed but the son came back to life a few days later – it's a miracle!
- The works of the apostles were so amazing and they even died for their beliefs – it's a miracle!

All the claims above have been made on countless occasions, and at least a few of them are still made numerous times every day. These claims bring us to the first set of Rubicons of ratiocination, if you will. Miracle claims are always extraordinarily sketchy, and they are usually deployed as alleged proof of the existence of some specific figure or god from some specific religion.

Throughout recorded history and even unto the present, people have claimed to have performed or witnessed miracles of healing, resurrection, stoppage of the sun or the earth, or to have seen various people who could fly or levitate or do other amazing things. Not a single one of these claims in any place or in any time has ever been proven to be real, and it is notable that the number of claims of miracles over time has decreased in inverse proportion to the increase in the amount of our scientific knowledge and the proliferation of recording devices.

Most claims of miracles today are the banal "why a person survived a disaster" type and the ones dreamed up thousands of years ago which are permanently ossified within the mythologies of the various religions, such as resurrections and flying horses.

Indeed the definition of the word 'miracle' is problematic. Merriam-Webster says[1] that a miracle is "1: an extraordinary event manifesting divine intervention in human affairs, or 2: an extremely outstanding or unusual event, thing, or accomplishment".

The common usage of the word, particularly in the religion-addled United States and many nations under Islam, often defaults to the former definition, entailing the influence of something supernatural or divine, making this interpretation of the word problematic from the start. Yet there is absolutely no way to prove the divine intervention other than wishful thinking.

No god has ever written fiery letters in the sky or proclaimed in a Very Big Voice from the clouds or otherwise offered any proof of any kind at all that he, she or it was the one responsible for the alleged miracle. Believers want to believe *their* god was the one who did it, and what passes for their thinking on the matter ends there.

What the vast majority of people fail to see and understand about the history of the world is that the cultural ethos and common world view for the vast majority of humanity for the vast majority of human existence has been thoroughly steeped in religion of one kind or other. Just as a child learns the language of its society from the moment of its birth, it also learns and joins in the customs and beliefs and all the other nuances of how its local human population interacts with the world.

The vast majority of people raised in a society become carriers for that worldview, they pass it on to their children, and this continues for many generations due to the numerous built-in social and individual barriers to change. So the inertia of the local religion perpetuates, like a genetic defect that affects all of one's descendants.

It is only *extraordinarily* recently that humanity discovered that germs and not miracles or divine intervention cause disease, that gravity causes the tides, that variations in the structure and function of the brain or body cause physical or psychological variations such as dwarfism or homosexuality.

It is not that long before living memory that we acquired the tools to prove that plate tectonics explains earthquakes and volcanoes, that the earth is not the center of the universe or even our solar system, that heat and pressure and water vapor and wind cause storms, that evolution is the reason behind the variety of organisms on earth today, and that witches don't actually exist.

These and millions of other once-unknown questions which had previously been assigned to the purview and whim of the divine and miraculous are now fully and appropriately explained by science. We have only recently come to know just how frequently improbable events happen. We have only recently discovered how susceptible our brains are to misinterpretation and misunderstanding of what we see and hear.

For many, many hundreds of years in Christianity, far too many thousands of people were persecuted, ostracized, jailed, tortured and/or burned for believing things contrary to the accepted and church-approved Pravdas of the day.

Thus the innate tribal and social desires to 'fit in' were exponentially reinforced by the natural desire to avoid imprisonment and torture, and nonconforming thoughts and writings were systematically suppressed to an incredible degree. Imagine if you knew today that some papal or Christian State theocratic secret police hit squad would break down your door and carry you away if you were known or suspected to have said, done or even thought things contrary to the ruling religious class and its Dear Leaders.

The terrors and purges in Russia in the early and middle parts of the 20th century are a classic example. You would do all you could to avoid this, which is exactly what happened for the many hundreds of years during which the church could and did torture and burn people for such "crimes". Thus the inertia of the prevalence and institutionalization of religion and religious thought carried on for a very long time, and the belief in miracles was a significant passenger in this caravan of ignorant belief.

There have always been 'heretics' and those who don't believe in miracles, though their voices were often desultory and suppressed even after their deaths. The Enlightenment, the Renaissance and the Scientific Revolution finally flung open the doors to the hall of knowledge, and those who were able to think and operate beyond the reach of the religious authorities moved into this intellectual space and began consolidating their ideas and concepts into a critical mass.

This mass finally exploded onto the world scene and has positively impacted many billions of people. As a side note, it was very fortuitous that our nation was formed at the time in history which it was and by the secularly motivated people who formed it. Had we been formed as a religious nation rather than the expressly secular nation which we actually are according to our foundational documents and ethos, our history would be quite different.

As with any group of extremely dedicated devotees facing undesired developments which threaten their world view and, more importantly, their livelihoods and institutions, the various religions in their own ways have responded to the advances of science and empiricism in generally the same manner. They mostly rejected but then slowly incorporated selected bits of science, first interpreting them through a religious lens, then translating them in grade-school and pithy terms to their congregants after much careful sanitizing, repackaging and frequent mischaracterization, all with the end goal of not harming any of their religious beliefs.

Or, as in the case of Islam, they have been either rejected or used to enrich and arm cabals of oligarchical nepotists and line-of-succession thugarchies, or to explode suicide bombs and fly planes into office buildings. But rejecting miracles would be a rejection of the core tenets of the faiths, so this is never done. Such claims are either never properly examined or reality is twisted to enable individual and collective rationalization and continuation of belief in miracles.

For the totality of humanity alive on the planet, its collective level of understanding of basic science and scientific discoveries has slowly progressed and improved. That said, even today there are still many millions of people who do not believe that germs cause disease, who do believe in witchcraft and who do still ostracize and kill women for it, and who do actually think that certain human behaviors cause the gods to send earthquakes and hurricanes.

Many of these people reside in America, where our school systems and antiscientific and anti-intellectual culture are proving every day how poor they are both in absolute terms and also when compared to the rest of the world.

Denial, suppression, ignorance and hatred of science are the foot soldiers in the armies of darkness and ignorance which are fighting an ongoing battle against science in 21st century America. The witchcraft-level ignorance and stubborn desire of millions of Americans to actually believe in ancient fairy tales enable the continuation of belief in miracles and miracle stories.

The inertia of the long centuries of religious oppression and suppression will not be overcome by a few great and insightful minds such as Newton, Galileo, Copernicus, Darwin, Leeuwenhoek, Hooker, Wegener, Hawking and others. People are born as completely blank slates in terms of the information in their heads, and they can be shaped dramatically by every second of their existence and everything they see, hear and experience.

Humanity was trapped in the black tar of scientific ignorance for millennia, and it has been extremely hard to get out of it. Some people like that tar very much and they are petrified at the thought of escaping its dark and sticky clutches. The comforts of their religions and their miracle stories are too much of a safe zone for these frightened snowflakes, who are too scared to question their faith or face reality.

Until this paradigm changes, and until we as a collective global population finally turn the corner on intellectual and educational maturation as a species, the comforting tar will be where most of humanity will be found, stuck in a trap of ancient mythology and ignorantly happy to be there.

Taking Christianity as an example, the holy book it uses today is the product of a deliberate sieving process. It came together not through a blind process of acceptance of all the writings that were extant, but by deliberate choices made by a subjective group of ignorant men with a specific goal in mind and very little objectivity. Thus we do not often see the massive store of 'apocryphal' books which were left on the cutting room floor by the sieving authorities.

This automatically places the decisions for inclusion and the resultant set of bibles in the category of "permanently untrustworthy due to motive". Some miracle stories did indeed make it into the final approved canon, but only the miracle stories that the powers-that-be at the time felt were appropriate for inclusion made the cut.

A pre-scientific person with a theological nature or one who was raised in a religious society would see light coming from a flashlight or music coming from a radio as clear evidence of a miracle and would attribute the existence and function of the device to the god of his or her choice. In antiquity, and particularly in the case of ancient Mesopotamia, being as that locality was that era's version of the early 1800s New York 'burned-over district', prophets and religions were absolutely everywhere.

Three of the variants of one of the local faith systems have managed to remain to the present day. Each of them states that it alone is the one true religion with the one true god, and there are plenty of miracle stories used in each religion as its alleged proof of its own veracity. Go anywhere on earth and you see the exact same kind of thing but with local modifications.

One cannot claim that extreme dedication to a cause even in the face of death is miraculous, else one would need to accede to the 'miracle' of the many instances of individual and mass suicides throughout history and the actions of suicide bombers. Examples include the suicide of many Japanese during WWII as American forces advanced, and the ferocity and tenacity of the Japanese resistance especially in the Pacific islands.

Others were the Japanese kamikaze warriors, today's streams of suicide-bombing Islamists, the uncountable numbers of troops who sacrificed themselves in absolutely selfless and incredible acts of valor in war, and numerous other instances of total and final dedication by people to their specific belief or cause. It's in the nature of many people to be like this.

While it is uncommon, to be sure, it is far from rare, it is not miraculous in any way, and it is not confined to any racial, ethnic or religious group.

To bring the above to a pithy point, all miracle claims are merely evidence of some combination of terminal gullibility and wishful thinking, appalling ignorance and a complete absence of critical thinking.

To repeat the definitions from above, a miracle is deemed to be "an extraordinary event manifesting divine intervention in human affairs" or "an extremely outstanding or unusual event, thing, or accomplishment". So definitionally, it boils down to some god doing something in the world or just a very improbable occurrence. It doesn't take logic nearly as sharp as Occam's Razor to cut the first of these unprovable and unfalsifiable definitions away, thus we can focus our energies on the second. Why can we cut the first one away so easily? There are many reasons.

You've likely seen T-shirts, memes, books, articles and other such which list "Twenty Questions To Ask A Believer", where the intent of the questions is to illuminate the absence of thought and facts and the excess of credulity that are part of every religious belief system. Below are eight questions which one must get responsible answers to in response to any miracle claim:

1. Does the person claiming the miracle, anyone he knows or is positively associated with, or any of his religious beliefs stand to gain in any way if the miracle story were to be believed and acted upon by others? If so, the allegation should be dismissed as a matter of course, unless it can be later duplicated or appropriately observed in a controlled, laboratory-like setting and/or explained by multiple and separate independent experts in the appropriate scientific disciplines and those trained to spot frauds and tricks, such as professional magicians and debunkers.

2. Given that eyewitness testimony is one of the very worst and weakest categories of evidence, if there were any witnesses, what did they report, what is their believability and what is the proof of their claims? Would their claims, as with #1 above, enable them or the miracle claimant or any of their core religious beliefs stand to gain in any way if the miracle story were to be believed and acted upon by others, to include even just getting some attention? If so, the allegation should be dismissed as a matter of course, unless it can be later duplicated or appropriately observed in a controlled, laboratory-like setting

and/or explained by multiple and separate independent experts in the appropriate scientific disciplines and those trained to spot frauds and tricks, such as professional magicians and debunkers.

3. What was the level of sophistication and knowledge of the eyewitnesses, and what was their mental state? These are very simple but extremely important questions which are often not asked of those claiming miracles. These questions also apply directly to those who wrote stories of miracles in the works of ancient science fiction now used as monotheistic holy books as well as their modern equivalents.

4. Were there any pictures or recordings of the incident which were then later verified by multiple separate and independent expert forensic investigators as true and actual original and untainted or unmodified data? These investigators should include both scientists and those trained to spot frauds and tricks, such as professional magicians and debunkers.

5. Is the incident fully, mostly or even partially explicable by any of the known material laws and scientific information?

6. Could it be that a more advanced technology than those known to the observer of the event or to the reviewers of the event, or a novel use of an existing technology, was the cause of the incident? When stealth technology was first displayed, it performed in what the untutored would have called a magic or miraculous manner, but it was merely advanced technology and Clarke's maxim about advanced technology and magic is again applicable. This will be the case for as long as humans pursue technology.

7. Is the incident merely unusual or of low probability? It is very possible that a blind person could hit a hole in one on a golf course, and that a person can survive a fall from tens of thousands of feet in the air. Such outcomes are highly improbable but will assuredly happen if enough blind persons whack golf balls or people fall from planes. The Guinness Book of World Records™ and Ripley's Believe It Or Not™ archives are overflowing with descriptions of actual events, to include falls from high-flying airplanes, which the credulous would call miracles but which no one else does.

8. Can anyone replicate this alleged miracle? If not, what would a trained scientist and expert in the relevant field, along with professional magicians and debunkers, say about that?

It is clear that a miracle is assuredly not a miracle just because the observer cannot explain an event or believes it to be miraculous based on his or her uninformed opinion. The requirement to prove that the laws of nature and the universe were temporarily violated in one particular location at one particular time for some particular reason by some particular figure from some particular religion is and should always be extraordinarily stringent, and no such evidence has ever been proffered for any miracle claim in the entire history of humanity.

And yet, suppose one day there really were an event which could be thought to be miraculous by all the commonly known observations and science of the day. Someone on the earth could come up with a new technology or a quantum leap ahead in an existing technology which would be to the people of that time as a flashlight or radio would be to the people of 50,000 years ago. It would be amazing and extraordinary, but in no way miraculous.

We need to understand that every time we flick a light switch, watch television, talk on a cell phone, ride in a car or plane, mow our front lawn or survive many of the diseases that once killed so many of our ancestors, we are acting in a way that would have been miraculous to any of the authors and many of the readers of any of the monotheistic holy books.

And even if all of humanity did agree one day that a miraculous event had occurred, and that some supernatural being "had to be responsible for it", even this next threshold of concurrence would not at all point to any *particular* supernatural entity as the doer of the deed. There would not be any reason whatsoever to connect the event to any particular supernatural being or another except personal opinion.

This, by the way, is another in the long line of fatal flaws in all supernatural religions. Believers always tie things that happen to the particular pantheon or god in which the faithful person happens to believe, especially if that event is one the person is happy about.

Why this irrational default? There are thousands of gods which people used to worship and still do, and not all of these gods are as evil, xenophobic and tribal as Jehovah. Why don't Christians give any of these other gods any credit for alleged miracles? The gods of the Greeks, Romans and Egyptians are just as likely miracle workers as are leprechauns, spaghetti monsters, magic worms or any of the inhabitants of the various collections of gods and figures found in the monotheist mythologies.

There are no supernatural miracles. There are just things that happen.

So why do we keep getting miracle stories? This is because beyond the obvious reasons that some people are insane and others are terminally gullible, the other main reason is because people lie, either unintentionally or quite deliberately. As I have noted, people lie about pretty much anything there is to lie about. They do it to gain attention, to gain power, to gain money, to make themselves look better, to convince someone to have sex with them, to get out of trouble or to try to convince others that a religion is real. Today's miracle stories are no different from those of antiquity, as they are still lies told for gain of some sort.

The difference today is that stories can be checked and viewed through the lenses of logic, information and science. A flashlight is not a miraculous emanation of light from a demonic or holy instrument, it is wires and batteries and electricity. Not one of the miracle claims ever made has withstood even a rudimentary fact-checking. Belief in miracles depends on adherence to dogma and a willful ignorance or dismissal of evidence against the miracle.

People also tell and believe miracle stories because it reinforces the groupthink in the minds of those who share the faith. This is a form of lying, again either deliberately or through a less conscious form of self-deception, and the lies are told to oneself and to other people. The false mental comfort that results from a belief in a miracle-working and helpful god simply cannot be overstated. And when the belief in miracles is a foundational cornerstone of the person's religion, this is a two-headed sledgehammer keeping the believer in thrall to the religion and its miracle stories.

But so far we've been talking about what one would call "positive" miracles, which seem to have devolved down to disasters or accidents which harmed or killed a lot of people but in which a few people are not harmed or suffered less harm than would be expected. There are other events which are equally as dramatic, but they are never claimed as miracles. Have you ever seen headlines like these:

- "A quarter of a million people die horribly in Asian tsunami, but it didn't hit Tokyo so it's a miracle!"

- "A quarter of a million people die horribly in Haitian earthquake, but everybody in Haiti didn't die so it's a miracle!"

- "Six million Jews dead from Nazi program of murder. It's a miracle that anyone is still Jewish!"

- "As many as a hundred million dead in worldwide Spanish flu epidemic. It's a miracle the human race didn't go extinct! Thanks, god!"

All these events did indeed happen, and all were terrible disasters with great cost to human life. If there really is some god out there and he, she or it really does intervene in our world by picking out some random single or few survivors of disasters, this god is also intentionally choosing literally tens of millions of innocent people to die horrible deaths and hundreds of millions to suffer senselessly. You cannot have one without the other, therefore a selectively intervening god is exactly the insane sadist found in the monotheistic holy books. No rational person can reconcile the facts of our world with the miracles and other claims of religion.

We can take another stab at some headlines to illustrate the pathetic miracle claims. Rather than discussing real tragedies of major regional and global scale and show the gods' total lack of interest in them, we can precisely mirror-image some of the more common miracle claims in which one or a few people die or suffer horribly while most go on about their business.

- "Man shot and killed during robbery. No one else in city harmed. It's a miracle!"

- "Family burns to death in vehicle accident. No other vehicle fatalities in the state today, though. It's a miracle!"

- "Woman dies in childbirth at hospital; it was the only maternal death in the state today out of 1530 total live births. It's a miracle!"

If you are one of the unfortunate people who actually die in situations like this all the time but get none of the breathless fawning ignorant coverage of the miracle-minded mass media, then your god has just given you the finger. Literally, as in he put the finger of death on you, because you are The One he has personally Chosen to die that day. Is it really a miracle that most people in the state *didn't* die in a car accident or robbery, or that most women in a civilized country *didn't* die in childbirth on some random day, at least in our era of modern medicine?

Why is it that if we flip the numbers the situation suddenly becomes miraculous to so many people? How can the death of 1529 mothers in childbirth be a tragedy while the sole survivor is proof of a miracle, but the survival of 1529 of those mothers and the death of one is not proof of a miracle? Wouldn't *more* people surviving be, on the face of it, more miraculous if they all encountered the same dangerous situation? Thus wouldn't the fact that most people in the world don't actually die in crashes, robberies, from childbirth and disease be the more truly miraculous circumstance?

And if a miracle is simply something that occurs that has less than a certain level of probability of occurring, and also must have some positive outcome, then the fact that you possess the name you do and the Social Security number you do is literally hyper-miraculous. If one of your ancestors in the year 1500 had successfully predicted your full name and your Social Security number, particularly considering Social Security was several hundred years from formation, the odds against this actually happening correctly are staggeringly large. Thus if it happened it would have been worthy of being thought miraculous. Yet you and all other Americans since the inception of Social Security have a singular combination of name and Social Security number among all the billions of humans who have ever lived, and none of the combinations were miraculous in any way.

In the same way, every single snowflake is miraculous, due to the odds of any specific one having a specific shape and symmetry. Every single crystal you find is thus an icy little miracle, according to the broken logic of the faithful. The same goes for all of the planets, stars and galaxies, which all form and behave according to basic laws of chemistry and physics.

Do we really need to posit divine intervention and a miraculous intervention for every conception and gestation? For every Social Security number and snowflake? For every single one of the stars, planets and galaxies in the universe? To be consistent, yes the faithful should insist that we do, but of course since the rest of us know better we do not engage in such foolishness. At least most of us don't, because some people do indeed actually believe in godly hands permanently gripping cosmic and subatomic control dials, continuously managing everything from the expansion of the universe down to the existence and movement of subatomic particles in their Schrodingerian existence.

All of the above is simply proof that too many human minds do not see reality as it is, but as they want to see it in order to continue believing in comforting religious fictions. This mindset is extremely dangerous and highly damaging to humanity and it needs to be removed, like a malignant cancerous tumor, as quickly and thoroughly as possible.

Every single miracle story ever told has been a deliberate or unintentional lie. From burning bushes to flying horses to prophets healing with spit to zombie tales of resurrections, these stories are childish. They are designed for both young children and adult children and are deployed by manipulators and deceivers for nefarious ends. The sooner we all accept this, the better off we will be.

Thus far this chapter has been a bit light on attribution and specific examples, and I will provide some here. The Catholic cult has received much direct fire from me throughout the course of this book, and it has earned every bit of it and much more because of its countless sins of commission and omission and its truly insane dogmas. We'll examine its position on miracles now, in the context of the process of making saints.

I have it on good authority that the US Conference of Catholic Bishops (USCCB) is the 'go-to' source for 'real' Catholic information, along with of course the Vatican. A quick search for the word "miracle" at the USCCB web page[2] leads one to a page[3] describing how saints are made. I strongly recommend readers go to this page and read it in its entirety, despite your personal religious or secular position, for it gives great insight into the truly bizarre and loony nature of this insane cult's beliefs.

The article on the page is short but packs, as they say, ten pounds of crazy into a one pound bag. Reading the page calls to mind Dungeons and Dragons, medieval woodcarvings, images of people in straightjackets, a strong desire to know what the hell they were smoking when they came up with this stuff and an even stronger desire to know how the hell anyone over the age of five could ever believe such utter fabricated lunacy.

Once one gets through all the religiously overtoned gobbledygook and flimflammery, the process of making someone a saint, which Catholics call canonization, generally boils down to the following steps:[4]

1. A person's life and death are examined to see if the person either lived what the church deems a heroic "virtuous life" or died a martyr's death. This is done at the local level through "…the calling of witnesses" and an examination of the dead person's life and writings, and determining martyr status through an investigation of the martyrdom event.

2. Once that is all done, and I give no judgment or opinion here on how thoroughly that is done other than to say no one should ever believe this process is very objective, the person's file is sent to

the Vatican where it will be reviewed by the "Congregation for the Causes of Saints". If this group thinks the file warrants further action, meaning they think the evidence supports moving the file along, they send it out for staffing and more review. If they "…find the documentation favorable…" they send the file to the pope. If the pope thinks everything is kosher at this point he sends out a formal decree stating the person has reached the status of "Venerable".

3. Then comes the miracle part, and this step is called beatification. To move up the posthumous career ladder from "Venerable" to "Blessed" status, and know this is crazy but I promise I am not making this up, they have to show that the person was responsible for a miracle. To do this, they collect a group of medical and theological advisors, who apply three so-called traditional standards to determine if an alleged miracle really happened or not. In the case of healing a person, the alleged miracle must be "…complete – meaning a total healing of the disease, it is not enough for the person to "just" feel better; instantaneous – meaning the healing occurred all at once and not over the course of several days or months; and durable – meaning the person remains permanently free from the illness that afflicted them." I offer no further opines except total doubt about the rigor of this process. If the collective thinks the miracle story is true, the pope decrees the person has now formally reached "Blessed" status. Apparently the pope can claim executive privilege and promote / beatify the person without proof of a miracle.

4. But wait, there's more! To get promoted higher than the beatified status of "Blessed" and be canonized, the person must have yet *another* miracle proven to be attributed to him or her. The review process for Miracle #2 is the same as for Miracle #1. Once this step is done, the person can be formally promoted / canonized to sainthood. At the end of the process, if the person makes it this far, the newly minted saint is deemed by whoever is pope at the time to be "…united with God in heaven, an intercessory to God on behalf of the living, and worthy of public and universal veneration."

The Catholic cult currently recognizes about 10,000[5] saints. So it has formally declared that tens of thousands of actual verifiable and proven miracles have occurred, *just within Catholicism*. Apparently miracles are extremely prevalent within that cult, which makes one wonder why the news of all these allegedly bona fide and proven miracles haven't made front page news each and every time one is said to be "proven" by that internal process, and if they are so common why they bother to call them miracles at all. Aren't miracles supposed to be rare things? Apparently in Catholicism they happen all the time.

A better way to state the obvious truth is that this Catholic uber-mysticism is a classic example of delusion and superstition gone absolutely institutional, and that actual science is quite unsurprisingly totally absent from this entire process. I won't do more than mention the truly galactic level of irony in a sect rummaging through files and testimonies to try to prove miracles while huge numbers of its clergy were raping and molesting untold numbers of young children and its leaders were working overtime to cover it up and move rapists around to stay ahead of the law and angry parents.

One would think one of the world's most populous cults would have its priorities in order, but the facts prove otherwise. Wouldn't a real miracle be if a group of abnormal and sexually-repressed males *didn't* rape children?

And if the rest of this book weren't illuminating enough on the ability of *Homo gullibilis* to believe absolutely anything at all for no real reason whatsoever, a Pew[6] study found that 79% of Americans believe in miracles, with of course Catholics and evangelicals even higher at 83% and 88% respectively. So only about one to two of every ten people you run into in America actually understands that reality doesn't include some god magically interfering in the real world. That statistic about belief in miracles is absolutely one of the most damning statistics I have ever read on the lack of critical thinking in our species and its superstitious religions.

I think it appropriate to close this chapter with some examples of what actually would be apparently or truly miraculous occurrences worthy of serious exploration, without any ridiculous notions of canonization cluttering the sight picture.

A real miracle would be observing a human male materializing out of thin air like the biblical Adam, walking and talking and a with brain already full of memories and information, rather than being born via the unprovable and unfalsifiable allegation of entering the world via a teenage female human virgin.

A real miracle would be if any highly specific event could be proven to have been highly specifically predicted centuries in advance, down to the date, time and location, and if the event were not capable of being performed by humans and were not some recurrent event such as an eclipse. An example would be specific prediction of the epicenter of an earthquake, its exact time and magnitude, its effects and other very specific details, written down in advance and known to most people long before its occurrence.

A real miracle would be an airplane's entire wing falling off at 30,000 feet and the plane still landing safely.

A real miracle would be human limbs, eyes or organs spontaneously regrowing themselves without medical technology or intervention.

A real miracle would be everyone involved in every plane and car crash next year stepping out of the crashed vehicle without a scratch.

A real miracle would be bullets bouncing off a child's body during a school shooting instead of blowing holes in it.

A real miracle would be a total cessation in school shootings even with the continued deliberate decision by religiofascistic Christian Republicans in Congress to do nothing about the problem.

A real miracle would be a hurricane bringing nothing but fresh water to areas of extreme drought, filling up both aquifers and surface reservoirs while doing no damage at all. Or moving this fresh water directly to the scene of a large forest fire, as if with intent and purpose, and putting out the fire.

A real miracle would be an inexplicable blockage of a tsunami from striking land and killing hundreds of thousands of people, in the spirit of the Moses story but with the aim of saving hundreds of thousands of people from death, rather than killing an opposing army.

A real miracle would be an invisible hand magically diverting planes full of people from hitting office buildings.

Or preventing the Holocaust, or stopping the London Blitz, or snuffing out the atomic explosions at Hiroshima and Nagasaki.

Or dropping an unmistakable hint to any of the Inquisitional popes that killing Jews and heretics was immoral and evil.

Or informing slave traders and slave owners of the inhumanity of their actions.

Or delivering the cure for the Spanish Flu before it killed a hundred million people and destroyed hundreds of millions of families.

Or stopping thousands of pedophile priests from raping and torturing little children, and stopping the Catholic hierarchies from covering it up.

Or indeed preventing any of the billions of other acts of evil perpetrated by humanity on itself, and delivering a clear message about it and maybe also applying a little public earthly punishment to the perpetrator at the same time.

Or giving even a single, actual, clear and unequivocal hint to all humanity of the existence of the hand's owner.

The biggest miracle of all would be the entirety of humanity simultaneously throwing off the infantile superstition of religion and finally realizing their imaginary miracles are just poisonous and brainless ignorant wishful thinking. Getting humanity to accept reality and enabling billions of the faithful to realize that all their religions and all their imaginary miracles are manmade nonsense will take hard work and perseverance from every thinking person.

I therefore ask every reader of this book to do their part in ridding their corner of the world of this ridiculous superstition of miracle-belief and the religions that foster it.

Chapter Thirty-One

Hey, god Was Here Last Week!

"For me, it is far better to grasp the Universe as it really is than to persist in delusion, however satisfying and reassuring."
— Carl Sagan

"Did you hear the news? God was here last week! Yes, the one and only Supreme Being, Jehovah Himself, was right here, on Earth, just last week! You say you must have missed it? Oh, no, what a shame! No, really, I'm not kidding. God was here, seriously! Where and when, you ask? Why in Tampa, Florida, in these our very own United States of America, at the Raymond James Stadium…and at the college football national championship game on Tuesday the ninth of January 2017! How do I know? Well, duh, because the head coach[1] and quarterback[2] of the winning team, those Clemson University Tigers, said so, and they should know, right?

Well, no, I didn't see or hear very many Alabama Crimson Tide players, coaches or supporters say anything about how and why Jehovah thought it better to make their side lose the game. I suppose they must just have been unwilling or unable to see the glory of His Presence down there on the field, and not worthy of the grace which He so lovingly and obviously bestowed on the much more godly and deserving Clemson players.

Did I personally see any gods while I was watching the game? Well, no. Did anyone watching the game on TV or live at the stadium see any gods? Well, no, but you have to understand, He works in mysterious ways, and doesn't have to make himself seen, as in seeing a physical body or arm or hand or lightning bolt or anything like that. He just does things, and you just *know* it was Him!

No, I have no idea why the divine hands clasped Clemson to his godly bosom and carried them to victory. You know He works in mysterious ways, and we mortals have no idea of the reasons for what He does, except it was surely for a divine reason that Clemson won and Alabama lost.

No, I don't think anybody else actually knows exactly why Jehovah picked one team over the other, or specifically *how* his divine hand actually guided the outcome of the game. He just did, and the proof is that Clemson won. That's what the coach said, and that's all I need to know.

Um, yes, I suppose that if Alabama had won, there would be a lot of people who would think that Jehovah had chosen *them* over Clemson. But Clemson was the underdog, and they couldn't have just won the game *on their own*, without help from the creator of the universe! Only an idiot would think such a stupid thing. The Clemson coach said it himself after the game – "only God can do this". It's not like they could have won the game unless Jehovah Himself helped them. Alabama was unbeatable!

Well, yeah, now that you mention it I suppose it's true that other teams have beaten Alabama. Oklahoma and Ohio State beat them, and when Johnny Manziel was at Texas A&M they beat Alabama. No, I don't know why Jehovah picked those teams on those days to beat Alabama. And no, I don't recall any of the coaches on those teams claiming He did.

But He must have intervened, right? Yes, I know Johnny Manziel never met a Commandment or a law or a rule he liked, and that it would be very hard to imagine him gaining any favor or special treatment from any god. But Jehovah must have made an exception, because A&M sure did beat Alabama.

And, well, yes, now that you mention it, there are millions of extremely, and I do mean really, *really* extremely religious people who played and coached for Alabama or went to Alabama or live in Alabama or just plain old like Alabama, who were surely praying their little pigskins off just as hard for their team to beat Clemson. I would agree there were probably about as many people praying for an Alabama victory as for a Clemson victory. But I just don't see the problem. Clemson said Jehovah was on their side, and their coach is so religious I bet he sleeps on a padded cross, so he must know, and that's that.

Listen, I don't know why you're asking all these questions. The coach said they won because 'only God could make it happen', and if he thinks his team wasn't good enough to beat Alabama without divine intervention then that's how it is. Who do you think you are to question him, a man of God giving thanks for a gift of gridiron grace? You're just not showing any faith here. What are you, an atheist or something?"

The fictional dialogue above is obviously satirical, and intended to show the absolute idiocy of anyone claiming that any supernatural entity of any kind is on his side in whatever effort he undertakes, be it getting a promotion, winning a battle or a war or scoring more points in a game. Many millions of people in the US, and billions of people on the rest of the planet, were either blissfully unaware of that game or didn't give a crap about it.

Too many of these people had far better things to do, such as trying to find food and medicine for their starving or sick children, trying to escape the ravages and destruction of holy war and religious conflict or trying to keep armies of deeply and sincerely faithful bigots in America from enacting Christian faith-based persecution into secular law.

There is an extremely unfortunate compulsion in the religious mind to believe that his or her particular god really does care about things like wars, parking spots, promotions and the outcomes of sporting events. To these people, I will always direct the sharpest and most biting satire and ridicule, and I hope the wounds run deep and painful enough to penetrate their thick mental fog. Unfortunately, the same breathtaking lack of critical thinking and the superabundance of arrogance and conceit these people possess usually provide them with a mental firewall against either the blunt impact of reason or the cut of satire's sharp edge.

If you've never been to truly scary lunatic fringe religious fundamentalist outposts like Saudi Arabia, Afghanistan, Yemen, Kansas, Texas, Alabama, Mississippi, Arkansas, Louisiana, Missouri, Oklahoma, South Carolina, Tennessee and Utah, you can't imagine the unholy state of mental affairs in those places. In the southern USA and particularly the southeast, American football and the Christian god compete for number one and number two in the lives of many millions of people, with football often winning if push comes to shove.

If the team these people support wins, many of them really do believe their god had a hand in the victory. If their team lost, they don't give the god a second thought. Or if they do, it is due to some punishment being delivered for some heinous sin, like saying that religious people shouldn't be allowed to persecute others, or averring that not every Tom, Dick and Harry should be allowed to tote fully automatic weapons to their preschool daughter's dance recitals, just like Jesus said to do.

Or it is described to have happened "for a reason which can't be known", which in the space of seven words exemplifies the ultimate copout and is a total abdication of reason. Such foolishness ought to be punishable by removal of a person's voting rights for a period of no less than ten years.

Other events than losses by sports teams are attributed to whatever current social issue religious conservatives are whining about. In other words, so-called "sin" can not only lead to losses by a sports team, but can also result in a whole region suffering a natural disaster like a devastating flood, drought, hurricane or tornado, all because somebody supported gay marriage, doesn't believe in a specific god, supported abortion or women's rights or minority rights or voting rights, teaches the fact that evolution is true, or some other wholly unforgivable imaginary sin.

This magnified foolishness ought to be punishable by removal of a person's voting rights for a period of no less than twenty years to life, with hard labor and four additional years of mandatory education in basic science tacked on.

If you as a human in the 21st century are alive and well, it is due to the advances of technology discovered by humans on their own, without any divine assistance whatsoever. If you really believe that the creator of the universe, a being of endlessly-advertised infinite power and ability, focused even a picosecond of its allegedly omniscient attention on this particular spinning ball of silicon and water among the billions and billions of other planets out there, then sharpened that focus down onto a little grassy field where a bunch of mostly hairless semi-evolved teenage primates were playing a game with an inflated ball of cowhide[3], then gave any thought at all to actually paying attention to this game, much less affecting its outcome, you are delusional in ways that are quite likely beyond all hope of medical or psychiatric correction.

If you really do believe this, you have already fallen prey to the delusion that the creator of the universe either acts or does not act in the lives of we semi-evolved primates, based on decisions made at least 13 billion years ago, and with some predetermined specific goal in mind. As this divine intervention or lack of intervention has never actually been detected in any way whatsoever despite many billions of opportunities and almost as many deliberate attempts at detection, it is only through the twisted illogical rationalizations of fearful and fearfully ignorant people that such a thing can be believed.

I have many questions for Messrs Swinney and Watson, who were the Clemson football coach and quarterback, respectively, regarding their statements and beliefs, with just a few listed below.

1. Exactly which god was being hailed and praised as having brought about the conclusion of the game? Swinney said "Only god can do this." We assume he was referring to Jehovah, but humans have dreamed up countless gods to which they have genuflected, pleaded, prayed, sacrificed, murdered and abandoned reason since we evolved a brain which could look up at the stars with more than a blank stare. Religions and gods are born and die all the time, just as stars, fads, animals, plants and humans do. One can reasonably assume, though, based on this being the southern USA, that the god being referenced was one specific derivative of the heavily populated ancient Mesopotamian gallery of gods.

And further, that it was the central god of the Christian bible. So it was probably Jehovah and/or his son and/or the holy ghost, close enough. How unsurprising it is that the coach and QB would invoke the god of their local community as the responsible party, and not a god from a religion from a different country or continent, say Africa or Australia or Asia. How unsurprising it is that they, as classically tribal and myopic primates, ignore the entire staggeringly long list of other possible gods, goddesses, sprites, fairies and ectoplasmic or invisible supernatural entities as possible fans of a particular sport played by particular members of a particular educational institution.

2. Why did their god reject the prayers of those favoring the other team? Surely in the fundamentalist and religion-addled South, there were as many Alabama supporters as Clemson supporters. And surely with the tendency of the religious in particular and of humans in general to jump on bandwagons, the recent success of Alabama football would almost assuredly have given any edge in total numbers to Alabama supporters. Were these Alabama supporters less worthy? Less prayerful or earnest? Did they not pray hard enough or long enough, or use the wrong words? Did they not offer enough burnt offerings? Or maybe, as with so many, many other things in life, a 50-50 binary outcome occurred in the complete absence of any god. One team won because it made more plays than the other team. What an utterly ungodly and cogent thought.

3. Why would a football game be more important than people's lives, or their well-being? In South Carolina, where Clemson University is located, a quarter[4] of the state's children live in poverty. About three people per day die in the state in traffic accidents. This rate is among the very highest in the country.[5] I therefore assume this football-lovin' god is OK with all those people dying in car crashes since he lets so many of them happen. At the exact same time that surgeons and their particular teams all over the world were repairing broken bodies, while teams of firefighters and police and EMTs were rescuing people from danger and death, this god is said to have chosen to reach down to affect the outcome of a particular football game. Oh, but you say, this god also helps in surgery, helps heal the sick and wounded, saves people who would have otherwise died and much more we can't know about.

Oh, but I say back to you, why did that god let them get hurt in the first place? Why does this god let some people become permanently paralyzed or even die, without communicating his reasons for any of these choices, and without any deviation from the normal patterns of physics, biology and chemistry? How many more people would die without the recently human-developed science that brought us even more recently human-developed medicine and vehicle safety technology? Do the lights or the action on football fields in the South draw this god away from hungry children and accident victims like a squirrel distracts a dog? Or do things actually happen or not happen completely due to the action and inaction of actual humans in the real world? And what lesson are we to learn when this god lets little children and babies die horribly and in agony but chooses to help a group of teenage boys win a game? Those who believe or support any of this divine neglect of human well-being are suffering from a disgusting and contemptible delusion which endangers us all.

4. Why would this god care in the slightest about who won or lost that particular game? Does this god intervene in *every* American college football national championship game? How about every American college national championship game in baseball, basketball, volleyball, soccer, lacrosse or indeed any other sport? What about practices or scrimmages for any of these sports? Does it also intervene in conference championships? The regular season? Preseason tournaments? What about professional sports? High school? Little League? Grade school gym class kickball games? Surely if a god wanted to make an impact on someone, in its own mysterious way, it would do it during those younger years when humans are most malleable and subject to influence, not during the college years.

   Or was it indeed for some particular reason that the god who allegedly created everything there ever is and ever will be specifically cared for the outcome of this particular game on this particular day between these particular teams? Or perhaps it is that religion has so profoundly addled the perception and worldview of too many people who don't or won't understand that their miniscule little lives and hopes and dreams are only their own, affected by no supernatural agents of any kind, and that there is no god pulling strings to enable them to win a football game, find a parking space or roll up to the quickest line at the grocery store.

5. Why can't they accept that it is only humans that talk and act in reality? That all the plays on that field and every other were carried out by players, unaffected by any god of any kind?

6. Why does their god care so much more about football than war, crime, disease, poverty or child abuse? To bring things down to a more individual and personal level, the relatively recent ancestors of many of the players on the field during the game didn't get to play many games, because they were otherwise occupied in Christian-supported involuntary incarceration as slaves. Why didn't the god who supposedly affects the outcomes of football games do more about the absolutely inhuman and eternally unforgivable practice of enslaving innocent men, women and children who were fully aware of the evil that they were being made to endure?

   Why did he not intervene in a civil war over that very issue of slavery, a war in which far more than 600,000[6] people died? The god can intervene in something like the outcome of a child's game in which no one will die, but sits on the sidelines eating popcorn and hot dogs while two thirds of a million people died, many of them in utter agony and desperation, not too far from the site of where this football game was played in Florida?

7. Where was this god during the Crusades, during the Inquisition, during the Black Death, during the global Spanish Flu epidemic or during the World Wars of the 20th century? Why was he not intervening in the Holocaust, protecting people in the Killing Fields, or helping the many millions of people who died in Stalin's Russia and Mao's China? Why did he allow the Middle Passage and the slave trade to go on for so interminably long? Why did he allow so many in the Christian South to be so successful at blocking civil rights for people for so long? Why did he ignore the many billions of the most heartfelt and desperate prayers offered to him by brutalized people, when thousands of people were dying and being murdered by sadists on a daily basis?

This unbelievable failure of rational thinking and short-circuiting of actual moral thought is why we nonbelievers thoroughly distrust the religious, and the more faithful you are the more we distrust you. If you can worship a god who would ignore the deaths of tens of millions of people but jumps at the chance to invisibly affect the outcome of a child's game, you are choosing to worship a psychopathic sadist, and your belief is only possible due to your ignorant sadomasochistic infantilism.

There was no god here this week, and there is no god here any week. No Greek gods, no Roman gods, no Hindu gods, no Buddhist gods, no Taoist deities, no monotheistic gods, none of any kind. This is because there are no gods, and such superstitious beliefs are a danger to humanity.

Things happen or do not happen in our lives due to natural processes and the acts of humans, free of supernatural piffle. The only active sources of good and evil in the world are our fellow human beings, and it is only in this life that we will be able to enjoy the fruits of mutual support and positive coexistence.

Relying on or believing in a sky fairy is childish and irresponsible, and leads to failures to take responsibility to reward and support good actions and to punish and stop evil actions.

The sooner we all accept this, and the sooner we all accept the reality that there are no gods and no miracles, the sooner we will achieve a truly moral society.

That would be a true miracle.

Chapter Thirty-Two

Hey, god...Wasn't Here Last Week?

"Who is more humble? The scientist who looks at the universe with an open mind and accepts whatever the universe has to teach us, or somebody who says everything in this book must be considered the literal truth and never mind the fallibility of all the human beings involved?"
— Carl Sagan

This chapter is a followup to the previous one titled "Hey, god Was Here Last Week!" I wrote this one in early 2018 almost exactly one year after I wrote the previous one, for reasons which will become clear.

For those who don't know about college football and those who could not care less, the same two teams who faced off in last year's college football championship game in 2017 met up again this year in the college football playoff. The University of Alabama and Clemson University had another football game. This time it was in the semifinals, but the same teams and the same head coaches were there.

The same Dabo Swinney, yes that is his name and that is what they call him, who coached Clemson last year was back. The same Swinney who tearfully praised his particular god to the heavenly rafters for allegedly helping his team win that game was on the sidelines, and one would assume he was still equipped with the same gridiron-based hotline to heaven.

There was something very different about this year's game, though. Clemson got its ass flat-out kicked in the game, hard enough that there will still be Alabama Crimson Tide cleat marks in the seats of the Clemson players' pants when spring practice starts...in 2023. There was no "hail mary" or "miracle drive" in this game to help this time. No football or any other kind of god or gods helped the team come from behind and save the day, no sir, no ma'am, no way and no how.

After the game, the media of course descended on both teams to ask "What Happened?", as if they hadn't watched the damn game themselves and already knew the answer, but that is the customary practice.

Quick digression - as with all televised sports tragicomedies these days, the postgame audience has to partake in both the jubilation of the victors (Hail!) and the sorrow of the vanquished (Hell!) to get the full experience of the spectacle. The winner has to say "aw, shucks, they're a really great team but we played well enough to win" and the loser has to try not to cry while not saying his team stunk up the joint or that he and his coaches blew the game, while the audience knows that if they said what they really wanted to, the after-game interview would be far more entertaining than the game itself.

So of course good ole' god-boy Dabo, being so assured that his god was the reason his team won last year, would have some similar words about the guy in the sky this year, right?

Wrong.

The website seccountry.com actually posted an article on that subject titled, as you might expect, "What Dabo Swinney said after Clemson's loss to Alabama in Sugar Bowl." Yes indeed. What did he specifically have to say about his god's influence on this game?

Here it is: "……………….………….."

What's that you say, you didn't see anything? Well, that's because he didn't say a damn thing about his god.

He also gave a press conference after the game that is available on YouTube.[1] What did he have to say in that press conference about his god's influence on this game?

He said "……………..…………..…..."

Oh, sorry, you didn't see anything again? Nope, you sure didn't. Because during that entire press conference he didn't say a (god)damn thing about his god.

Where was his god this year? AWOL? Busy elsewhere? Didn't want to get his feet or fingers dirty by spending any time down in the sin-filled Big Easy city of New Orleans? Did he not want to revisit the scene of the crime and get a first-hand look at all the damage he is said to have caused there via hurricane-induced punishment?

Well, I'll ask him. Hold on a moment, I will of course first need to get down onto my hands and knees and kiss the floor in a properly spineless and penitent wormlike praying position.

"Ohhh, Mr. god….. hellooooooooooo…….? Why didn't you help Dabo and the Tigers this year? I'm waiting… Hello? Anybody there? Why won't you talk to me?"

I've now waited ten full minutes, and nothing. No response. My phone didn't ring, I didn't hear any voices in my head and didn't see any patterns in my computer screen spelling out a message. No messages in the clouds, no beams of light, nothing. Nothing.

Which is exactly what this is all about – nothing. No god, no answered prayers, no divine football or other interventions, just nothing at all except one team which played better than the other team did. No magic, ectoplasmic or supernatural tendrils reached down to affect the outcome of the game. It was just flesh and blood young men playing a game. Which means it happened just like it did last year between these teams, and it happened just like every other game that has ever been played in the history of the planet.

But we need to make one thing clear about Swinney's Janus impression. He is actually the epitome of the deeply and sincerely devout believer, doing what they do and saying what they say. When things go well, they thank their particular god for it to its particular high heaven. When things go poorly, they forget their god even exists, or they apologize for the god and attribute their failure or misfortune to themselves or to some ineffable whim of their precious but invisible deity.

Psychologists call this human foolishness 'counting the hits and ignoring the misses[2]', meaning that people ignore the 99 times their prayers didn't result in anything, but that one time they think it did, it became worth a thousand times its weight and blew away the 99% actual failure rate.

The professionals also call it confirmation bias[3] and motivated reasoning[4], and like cockroaches and uninformed opinions, these characteristics are essentially impossible to exterminate despite Herculean efforts to educate people and equally strenuous efforts to instill critical thinking and self-reliance.

People like me call this mental fault by those technical terms as well, but we add four more words, which are much easier for most people to understand. We call it "completely ignorant" and "embarrassingly illogical". Or to be a little southern in the vernacular, such broken logic stinks as bad as a dead skunk's ass in July.

Well, Dabo, there's always next year. Better start prayin' now, to get a good balance in the prayer bank. Never know when the gods might be listening.

Or not.

Chapter Thirty-Three

Resurrections, Another Shipwreck on the Shoals of Evidence

"Come on, you sons of bitches, do you want to live forever?"
- Question allegedly yelled by US Marine Dan Daly[1] at his fellow Marines right before an assault on enemy forces during the Battle of Bellau Wood in France in June 1918. Daly was also one of only 19 people to have twice earned the Medal of Honor for exceptional valor and bravery in combat.

"Uh, yeah, that'd be great."
- What at least some of those Marines must have thought, along with almost everyone else in history when considering this question.

Nobody in their right mind or body wants to die. Many people do kill themselves after much preparation and forethought, and a few do kill themselves in a snap decision on short notice, but this is not normal. There have to be exceptionally harsh and unfortunate circumstances for people in their right minds to want to die, and it normally involves massive damage to and/or the thought of a future of unending pain in their bodies or minds. Such circumstances include people who:

- Are in the process of being burned alive, drowned or murdered in some other horribly painful manner, for instance by the local ruling religious authorities for heresy, blasphemy, witchcraft or some other imaginary offense.

- Are in the process of dying a painful and guaranteed death due to a different source than devout religious stormtroopers and thought police. These include people who have certain forms of cancer or other diseases and know they will suffer constant and sometimes agonizing pain until they die.

- Have had their entire families killed in an accident or murdered by thoroughly evil people, or are suffering from extreme PTSD or other conditions. The constant mental torture a person experiences in such circumstances can be as damaging and hurtful to a person as physical torture caused by disease or other humans.

- Believe they and their families will be raped, tortured and murdered by invading enemy troops, raiders or others. If you think this unlikely and if you have a strong stomach, there is video footage of Japanese suicides on Marpi Point[2] on Saipan. These people killed themselves rather than fall victim to imagined American atrocities during World War II. Beliefs are real in the minds of the believers, no matter how wrong the beliefs are in the real world, and humans will rush to kill themselves even on a false belief if their fear is strong enough.

The entire process of assisted suicide exists for people who have medical problems such as systemic pain or crushing depression and who have decided to end their lives. These people would actually much rather feel better than die, as anyone with a grain of intelligence and anyone who knows such persons can attest. However, medical science has not yet discovered how to cure them or solve their problems.

Also, none of the tens of thousands of unsympathetic, uncommunicative and completely unhelpful gods ever worshiped have ever taken the picosecond it would take to fax, email, carve in stone or otherwise explain to human doctors how to cure agonizing systemic pain and suffering, cancer, Parkinson's or indeed any other affliction whatsoever.

Therefore real people have to make an extremely difficult and unpleasant decision. This decision is never easy, as dying goes against every fiber of a normal person's being, and dying will almost always leave families and friends behind who will miss the dead person very much. The decision is much harder for those adhering to the Christian cult and the Catholic sub-cult.

I have to make the point extremely clear that in all of these cases Christianity's teaching has and still does make these problems far worse that they need to be. The abject hypocrisy and utter sadism of this religion is obvious throughout its teachings and practice, and its manmade origin is revealed through its incremental and involuntary evolution on these positions.

Catholicism had it as dogma for centuries that suicide was a grave sin. Most Christians today believe that suicide is a sin or at least immoral. There has never been a single bit of evidence or proof for these beliefs, but they nevertheless were and in many cases still are held both deeply and sincerely. The catechism of the Catholic cult now states that[3]:

> *"Grave psychological disturbances, anguish, or grave fear of hardship, suffering, or torture can diminish the responsibility of*

> *the one committing suicide. We should not despair of the eternal salvation of persons who have taken their own lives. By ways known to him alone, God can provide the opportunity for salutary repentance. The Church prays for persons who have taken their own lives."*

Well now, isn't that nice of them. After torturing, killing and burning people alive for centuries, often on the direct instructions of or clearance from the allegedly infallible popes, this cult now says that wanting to die or killing oneself while being tortured by Inquisitors means that the "sin" in this act is "diminished" and that we should not flail about in existential angst about the suicide's treatment in the afterlife.

From their perspective, of course, the guilty party in this situation is the one having the suicidal thoughts or actions, not the officially sanctioned torturer and murderer.

This dirty rotten suicidal sinner is said to be able to do "salutary repentance" for the crime of wanting to accelerate the process of death while, for instance, being burned alive by that god's own sadists in action, or to put an end to agony which their god itself could end in an instant but refuses to do so. This is all at one time a horrific collection of textbook sadism and masochism wrapped in a nice little toxic package of utter religious evil and hypocrisy.

We could also ask the faithful to explain why an allegedly good, loving and moral god would invent horrible diseases and tortures then inflict them on innocent humans in the first place. Clearly every god ever invented is of an equal mind on this, in that they all quite strongly desire the continued and uninterrupted existence of the torturer as well as the killer amoeba, virus and bacterium. But I have addressed that elsewhere so will not revisit it here.

Note that all that is necessary for people to commit horrible acts, such as killing their own children and then themselves, is a belief. There need be no actual proof whatsoever of the truth of the belief, and the faithful care not a whit about the mountains of proof that disprove their belief. The belief is all they need or want. When the Japanese leaped off those cliffs, they had a deeply and sincerely held conviction which was 100% wrong but was nevertheless the cause of many suicides and many murders of family members.

The attacks of September 11, 2001 were wholly faith-based acts of mass murder and suicide, committed by extremely deeply and sincerely devout believers. Most of the crimes against humans and humanity committed by Christians have been faith-based. So let's not ever quibble about mere beliefs in completely insane and evil religions and theologies not having the power to enable people to justify completely insane and evil actions.

So people sometimes take their own lives. People also die even after doing everything humanly possible to avoid death. No matter what we do, until medical science discovers how to make our personalities eternal through some combination of organic or inorganic media, our brains will one day die, and with it so too will die all that makes us who we are. If our brain were kept alive outside our body, we would in the most important sense still be who we are, because our personality is the product of processes within the brain. But once our brains die, we die forever, and no part of us magically drifts off to any sort of magical supernatural fantasyland afterlife.

I will get into questions involving inorganic personality storage at the end of this chapter but for now I will deal with current medical science.

So what about death? Is it the end? Once the brain scan shows flatlines, are we truly gone forever? The nontheist will roll his eyes and say "Of course that's the end. I thought we just went over that again for the trillionth time." The monotheist will normally say some variation of "Oh, no, when we die, the physical brain dies but a magical unseen, unseeable and unprovable essence of us which contains our personality and all our memories is whisked away by magical, unseen, unseeable and unprovable forces to either a magical, unseen, unseeable and unprovable heavenly kingdom someplace or to a really bad magical, unseen, unseeable and unprovable dungeon of fire."

The deist, pantheist, Hindu, Buddhist and all other folk on the planet who think there might just be something after death will believe whatever they personally pick to be behind that particular door based on what they've been told or on whatever makes them feel good to think.

All that said, it is the rare person who, if asked if she would like to live on in some supernatural form or realm after her earthly death, particularly when she would be relieved of all pressures and anxieties and able to cavort in an eternity of happiness with all her favorite people, animals and whatever worldly hobbies could be possible in this Wonderland, would say no to such a 'pie in the sky when you die' eternal vacation.

Even dedicated and hardcore scientists, empiricists and antitheists like me would jump at the chance to have such an outcome. But first you have to believe it exists, and that nagging little missing thing called "any evidence at all" prevents me and people like me from giving the idea more than an idle review.

But what if there were an alternative to permanent death, or what if we could actually get a return trip from the beyond? The resurrection idea has been a staple of fantasy and science fiction from its first invention thousands of years ago up to the present day. The resurrection is an idea born of the fear and angst at the loss of loved ones in the real world and the agonizing wish that they could return. It is also due to the uncertainty about one's own death and one's own desire to return to the world.

There are actually religions which have resurrection fables at the core of their systems. If you are a monotheist you may be familiar with at least one of these.

So what is a resurrection? What does it mean to resurrect something or someone?

- The online Cambridge dictionary says[4] it is "to bring someone back to life, or bring something back into use or existence after it disappeared..."

- Merriam-Webster online calls[5] it "1: to raise from the dead" and "2: to bring to view, attention, or use again."

- The Oxford Dictionary online says[6] it is to "1 Restore (a dead person) to life" and to "1.1 Revive or revitalize (something that is inactive, disused, or forgotten)."

So to resurrect someone is to bring that person back to life or raise him from the dead. There are many problems with these definitions, even if one just does the science fiction-fantasy unconstrained analysis on them. Some of these include the following:

1. Does the "someone" or "person" being referred to mean the physical body or the personality of the dead person? One must have both to be the same combination of body and mind that was once present. If just the person's body is brought back to life but the personality is not provided, is it still a resurrection? Or if the full original personality is brought back but in a body that is different from the dead person, say in another gender or race or even another species, is it still a resurrection? Is it now a reincarnation? Or is it some other action?

2. If the original body is brought back, is it actually the original body that has been reconstructed from its now-scattered atoms and molecules, or is a new but genetically identical body supplied or created? The biblical account of Eve's creation from Adam's rib makes this an uncertainty. If there is no body to

donate tissue from which the new body can be grown, will that tissue have to come from a different body? Will that donor body have to be living? Though if so one hopes for the donor's sake that he or she will be medicated or sleeping, as per the Old Testament? Or can the donor body be freshly dead? Would the freshly dead person have to have died of natural causes or would being murdered be a problem?

3. At what chronological age will the resurrected person be formed? At the age when he died or earlier? If the person died on his hundredth birthday, this would not seem to do much good. It would be quite a dirty trick to place a centenarian back into a 100-year old body, because he would not have much life left to go before departing once again. If a person who died later in life were resurrected into an infant body, would the resurrectee have the same memories as the recently departed, or would the new baby have the usual blank mental slate? Would we again have to split hairs on reincarnation vs resurrection?

4. If a person died at age 75, and was resurrected with a 25 year old body, will all the memories the long-dead person accrued be put into the person's brain, or just those that were experienced until age 25? If there are multiple resurrections covering multiple lives on earth, will these memories all be cumulative and all jammed into memory? What would happen to the age 25 through 75 memories if they are not installed into the resurrectee? Will they at some point be placed into the resurrectee's "eternal consciousness", say after he or she dies again? How will those be reconciled? How is this information management process being worked?

5. If a person is resurrected after being dead for ten years, will all of the memories experienced by the disembodied spirit of the dearly departed during those ten years also go into the resurrected meat body, or just those that were generated by the living meat brain up and until the moment of death? Are there two sets of memories with a one-way valve blocking the post-death memories accrued in some supernatural realm from going into the meat brain of a resurrectee in the real world? If so, how is this supernatural information management process done?

6. If, while in some spirit realm, the resurrected person saw any gods, angels, divinities or other supernatural creatures, or even just ancients like Socrates, Democritus, Epicurus, or either of the fictional biblical First Couple of Adam and Eve, would the person be allowed to talk about any of that after coming back to an earthbound meat body? One would think that in the Christian afterlife everyone would be lining up to talk with at least Adam, Eve and Jesus shortly on waking up in the big party in the sky and would have a great deal to say about them and the great time they would ostensibly be having while dead. Indeed, if such tales were believable and congruent they might actually strengthen a religion's claims in an evidence-demanding world.

7. Will a resurrected body age at the same rate as the original? Will a resurrected body retain any scars the original had at the time of death? Will a resurrected body have the same congenital problems, birth defects, susceptibility to disease and pollutants, mental processing flaws, low IQ, missing digits or limbs, speech impediments and other problems as the original person did?

8. Will the resurrectee be placed naked into the world, like the cybernetic killers popping into the past as depicted[7] in the Terminator movies, or will he or she be outfitted with clothing? And will the clothing be period-appropriate from the year of death or upgraded to modern styles?

9. Where will a resurrectee be expected to appear? One option is clawing her way out of the ground as is so freakishly described in the zombie gospel of Matthew (KJV), which says that "{27:52} And the graves were opened; and many bodies of the saints which slept arose, {27:53} And came out of the graves after his resurrection, and went into the holy city, and appeared unto many." Will returnees be magicked into existence standing on top of their graves, decked out in newly washed and pressed robes and spiffy new sandals? Popped into a meadow somewhere like the fictional First Couple and made to find their way to civilization? Materialized onto a Fox News or televangelist's stage ready for their fifteen minutes, or more, of fame?

10. Will the resurrectee be gifted with the ability to speak the current language? Who will sponsor the returnee and help him

acclimate to a culture which may have nothing in common with the one he left? The person will need to have food, clothing, shelter and a way to communicate. Given the current levels of bigotry, xenophobia, distrust of science and general dumbassery in today's hyperchristian America, however, a person being resurrected in a conservoreligiofascist Republican red state would be unable to produce an ID card, papers or other identification. He would therefore be treated as an illegal immigrant and potential perpetrator of voter fraud, and would be jailed pending deportation to Guantanamo for torture and interrogation or outright shot or killed with pitchforks as a modern day witch or warlock.

To avoid that, would the resurrectee need to have a divine passport or green card or other authorization document signed by Jesus or some lesser figure, say Saint Peter, Saint Paul or Martin Luther? Who will tell the local authorities that this person claiming to be resurrected is not in fact totally insane and need not be in a mental institution? Or would the old "free will" special pleading come into play and the resurrectee would damn well need to fend for himself?

11. When and where has there ever been, in the entire history of the world, any actual evidence at all of any kind of resurrection?

All of the above questions are of course just as productive as imagining a bunch of angels circle-jerking each other on the head of a pin. Angels, afterlives, resurrections and all the rest are wholly imaginary, and having to deal with the sorts of people who believe these sorts of infantile fantasies is beyond annoying. The religious will protest that I am mocking their deeply and sincerely held beliefs in resurrection. They will, in a rare moment of concurrence with we nonbelievers, be absolutely correct.

Such beliefs deserve not just mockery and derision but a good dose of the strappado, which was an instrument of torture used during the Inquisition on innocent people by the most dangerous people on earth, that being deeply and sincerely faithful men utterly committed to their religion.

There's more evidence that my big toe has magical powers than does the claim that a dead Jewish preacher came back to life 2000 years ago, or that anyone else ever did. My toe is real. I can see and touch my big toe. I can even wiggle it and move it around without tugging on it with a string. I can actually prove my big toe exists. My family has seen that toe and there are multiple medical records such as X-rays, scans and examination reports that attest to its existence, along with some family photos which clearly show not just one but two big toes in my personal possession. Also, more than 99+% of our species who have ever lived have had big toes. Having big toes is to be expected in our species, as is the ability to wiggle it on demand.

Resurrections are all the way across the universe in terms of the levels of believability and available proof than big toes. The only things that can be offered up in support of the alleged resurrection of Jesus or any other character in history are some ancient and terribly garbled science fiction/fantasy stories, all enabled by fear-driven wishful thinking and ignorant superstition.

As with all the other supernatural claims of Christianity, its claims of resurrection share both a total lack of evidence and a vast number of similar claims found in other religious mythologies. Resurrection stories are not unusual in the Christian and Jewish canon, and they are as common as house dust in many of the fantasies and fairy tales of the ancient Mediterranean region. What would we expect but that the copycatting writers of the Christian bible would include resurrection stories in their own fables and imbue one or more of their lead characters with this ability.

They borrowed the idea from other people's mythologies, just as they borrowed a very large number of other things from those stories, because people wouldn't be nearly as impressed with the stories or as likely to join a belief system without there being some allegedly miraculous power on offer.

Belief in supernaturally driven resurrections of formerly dead preachers or anyone else does nothing but prove humans can believe the most insane things without a shred of proof. Such beliefs and the people who hold them need to be treated like a belief that fire hydrants talk to aliens during the full moon.

Now that we have properly mocked and dismissed the idiotic idea of supernatural resurrection, we can now discuss those questions about organic and inorganic personality storage. Such a technique could come about due to advances in computer and medical science and technology.

One day we may be able to remove a person's living brain from the skull, with the eyes still connected and the rest of the brain linked into mechanical sensors and appendages that enable it to hear, communicate and manipulate physical objects. Would this brain, if we could reinsert it into a different body, be the centerpiece of a virtual resurrection? The body is just an ambulatory tool-equipped shell the brain wears, after all, and the personality is wholly resident inside the brain.

Granted hormonal and other differences would result after such an operation if not corrected for, such as a difference in testosterone levels in males, but in general if such a procedure could result in a stand-alone functioning brain, a virtual resurrection would be possible with no god needed. Would this brain have the same identity the once-whole person did? Are the brain and its personality the essence of the person? Do the retinal pattern and fingerprints truly identify a person?

The legal system will one day need to opine on such questions, and must do so without any counterproductive interference from any imaginary god's very unhelpful fans.

What happens if we are eventually able to embed a person's personality into a future hypercomputer such that it is indistinguishable from the organic brain's personality? If the human source of the personality died, would the computer containing the personality be a vessel for a virtual resurrection? If that personality were to be played back into a different brain in a different body, or into the brain of a clone made from the personality donor, would this count as a resurrection? If we copied that personality from one computer to a hundred more, would we now have multiple resurrections?

If part of our future involves activities in a virtual world, with some of these activities including commerce and actual compensated productive work, the avatar of this computer-based personality would be indistinguishable from that of a meat-based personality.

Could any organic or electronic vessels containing the original personality have the same Constitutional and other rights as a live meat human?

What if we could only imperfectly replicate or duplicate the personality and brain? How close does the copy have to come to the original in body and personality to be considered a "resurrected" person?

The faithful, Catholics in particular, will work extremely hard to inject their dogma into such discussions. They will try to say that we have not considered the soul. I respond by telling Catholics when they can show me a soul or prove its existence, we can engage on that topic. Until that time, any discussion on the subject of souls is just another counterproductive theological circle jerk on imaginary dogma and should be remanded to the custody of safely isolated theologians, with an impenetrable wall of separation between them and the real world.

The wall of separation between mythology and reality is constructed from blocks of solid logic and reason. We must be building that wall as high and as long as possible, not weakening it by removing whichever blocks some superstitious people want to take out because it makes them feel better or because a 2000 year old book of mythology says they should.

Whether the faithful talk about souls, miracles or resurrections, they always fail to acknowledge how we were all born too late to see any actual supporting evidence. We all missed the boat on being eyewitnesses to the stories of near-daily godly intervention in our ignorant and recording device-free past, and the only so-called evidence offered today is stories from ancient mythology rather than image, video and forensic evidence.

Apparently you had to be living in a microscopic part of the earth during a microscopic period of time in fantastically ignorant cultures thoroughly soaked in superstition and completely ignorant of science to be able to view or experience a miracle or resurrection in person.

And once those fleeting non-miraculous events were done, apparently the god or Jewish preacher responsible for all those alleged signs and miracles and resurrections decided that the world had seen enough, and there would be no more of these divinely sparked resurrections, ever.

So the only resurrections we see today are those on the pages of fanciful ghost stories written down by our long-dead storytelling ancestors. Resurrection stories are just ghost stories and zombie tales fit for scaring little children around a campfire or as plots for low-budget movies.

The world we live in and its history are exactly as we would expect if people lie, if they make up and spread fairy tales, and if they believe things not due to any evidence but due to an overpoweringly strong desire to live beyond death.

And an apparently uncontrollable desire to return to Mother Earth for an Act II.

Part Eight

The Proper Path

Chapter Thirty-Four

What to Put In Religion's Place, And How

"Almost any sect, cult, or religion will legislate its creed into law if it acquires the political power to do so."
- Robert A. Heinlein

"The further the spiritual evolution of mankind advances, the more certain it seems to me that the path to genuine religiosity does not lie through the fear of life, and the fear of death, and blind faith, but through striving after rational knowledge. In this sense I believe that the priest must become a teacher if he wishes to do justice to his lofty educational mission."
- Albert Einstein

I sincerely hope that if you have read this far, you will agree that religion absolutely needs to completely go before it kills us all. It has already killed far too many people for far too long, and has damaged the bodies, minds and lives of more billions of people than we will ever know.

If any single human being acted on his own like the masses of the faith-based mobs and their gods do, he would rightly be treated as psychopathic, antisocial, antihuman, hyperbigoted, immoral and evil. But groupthink, mindless in-group loyalty, xenophobia, the mob mentality and a stunning inability to think rationally and critically have enabled religion to live like a life-draining parasite in the collective human body for millennia.

The three monotheisms are an unholy trinity of antihuman savagery. They and the whole body of religion live on in humanity like the parasites they are due to the weakness and fears of fragile people, the social inertia of the herd and the flawed psychology of our species. We are many years past the time when we should have thrown off the mental and in all too many cases the physical shackles of religion.

The vast majority of the mysteries that frightened, confused, damaged and literally plagued our ancestors are now solved. Every day brings new and wonderful discoveries and explanations. Yet there are billions of people who go about their days like brainwashed drones, believing in imaginary gods and unable and/or unwilling to accept a godless reality.

In the 21st century far more of us know how immoral and dangerous religion is than ever before, but the infection has gone extremely deep into the brains and societies of too many people. Most of them will not be able to go cold turkey and abruptly live without their addiction. The religious parasite must be removed from its host, but the process of separating humanity from its parasitic religions will have to be somewhat deliberate. Nevertheless, this separation is required in order to enable one of the pair to survive.

Humans can survive quite well without religion, indeed they tend to flourish when it is absent, but religion cannot survive without people. Thus this badly needed separation will result in humanity living an independent life completely free of its collection of life-sucking parasites, and will stand on its own, unshackled and finally free to think clearly and independently, for the first time in far too many millennia.

Because the rot has gone so deep, and because social inertia is so powerful, the process will not be easy. But as with all extremely good things, hard work and a bit of a struggle to gain the prize is well worth it. But what do we put in religion's place, and how do we do that?

That is actually both easy and hard to answer. The easy answer is to go back to the seven reasons people believe in gods, and work across a wide range of populations and actions to address those problems. As a reminder, these problems are weakness, ignorance, inertia, masochism, mental challenges, power and persecution, and programming.

The methods for addressing these problems range from the draconian, such as forcibly closing all houses of worship, forbidding all public displays of worship, deporting all clergy and banning religious activities, to more peaceful options such as ensuring well-being for all citizens, ensuring each child receives a solid fact and reason-based education, removing tax subsidies for religions and the religious, and requiring empirically-derived facts and reason as the basis for all decisions, particularly all governmental, educational and moral decisions.

Accomplishing the Nirvana of a marvelously religion-free America and world will probably require another civil rights movement, although I fervently hope it does not end in yet another civil war. We do not need an armed conflict in which the armies of the ignorant battle the forces of reason to a bloody end, in a senseless waste of humanity simply to reach a badly needed and completely obvious moral improvement.

As violence should always be the last resort, the best option at least in America is to simply enforce our secular Constitution and ensure the church is really, truly, completely and eternally separate from the state, at all levels and in all ways, and to treat all religion as the infantile mythology it is.

The following are a few simple ways we can address the seven human shortcomings that lead to religion and thus ease the transition to a safe and sensible secular society:

1. **Weakness.**
    - We have to teach children to be mentally strong and courageous, and to put their faith in themselves and not in some imaginary sky fairy. Helping children develop an internal locus of control, to teach them that they have responsibility for their lives and to show them how to handle the stresses and challenges of life are some of the most important things parents and society can instill in children.

    - We need to teach children that there will be uncertainty in their lives, sometimes extreme and long-lasting uncertainty. We therefore have to teach them how to deal with this uncertainty like every nonbeliever does, which is with self-confidence, resilience and mental courage.

    - The faithful will have to work through their stages of grief at losing whichever particular sky daddy and afterlife they believe in, and we will have to help them through this.

    - We must all laugh at the silliness and ignorance of all of the religious mythologies, and treat them as no more real than any other fairy tales, in order to keep religion from accreting any sort of respect or acceptance. We need to shape our culture so that belief in any religious mythology is as socially unacceptable as openly defecating on stage during a graduation ceremony.

2. **Ignorance.**
    - We have to eliminate religion in the same way that our species has slowly, painfully and unfortunately not yet completely eliminated ignorant beliefs that the earth is flat, that it is the center of the universe, and that angry gods send diseases and bad weather. We must use science and facts as the basis for our lives, not ignorant fiction written thousands of years ago.

    - We must teach our children the real truth about the real world, which is that all religions are just local fables people used to tell each other to amuse, to entertain, to try to explain and to control people. We need to ensure they know that religions have no more factual basis than the superhero stories and characters found in modern comic books.

- We have to teach children that human ignorance, unsavory human characteristics and weakness enabled the ancient fables to turn into formal religions, and that inertia and the demonic grip on power by the religious establishments have enabled the religions to carry on even into our scientifically enlightened societies. We have to inform them that religion is a time-tested method for manipulating other people through fear or guilt, and it works both at close range and over extended distances and times. We then have to reinforce this education in our schools, institutions and culture.

- We must ensure a properly empirical fact and science-based education for every citizen to eliminate accidental ignorance. We must also reject deliberate ignorance, no matter how apparently small, at every turn. Both of these processes must occur simultaneously and with extreme vigor, because most people are born with and tend toward ignorance, and this sort of psychological entropy will be a part of our species for many hundreds if not thousands of years.

3. Inertia.

- We need to remove religion from our societies as much as possible in order to move our cultural inertia toward facts and away from fantasy. We have to change our paradigms and our culture. We need to make Sundays, Wednesdays, Fridays, Saturdays and any other days people hold as holy as just another day, or ones which will in our newly-improved godless world be repurposed to community and family activities, sports and games, personal and professional development and improvement, or just doing a whole lot of nothing and recharging our batteries.

- Constitutional violations such as government and military chaplains need to be eliminated immediately. These people should be replaced by secular counselors and advisors, informed by psychology, moral philosophy and fact-based counseling, not the ignorance and immorality of ancient Mediterranean or any other mythology.

- Missionaries need to be discontinued or repurposed to purely secular causes such as healing the sick, providing fresh and clean water and food and other improvements in human well-being. These must be done, by the way, to suit the local culture and what the locals want, not what we arrogant outsiders think they should want. If the natives want fully modern lifestyles or something less, no problem. But it should also be no problem if they want to live in the Stone Age. They must decide whether they want other people's technology or not. We should offer, not order, regardless of how much oil or other natural resource wealth is on or under their land. And no religion of any kind will be allowed in this process.

- We need to eliminate all forms of religious inculcation and iconography. No more religious schools, no more charter schools, no more crosses on billboards and in graveyards and no more churches on every street corner. Repetition breeds familiarity and acceptance, and to be truly moral and empirical all of these religious institutions and symbols need to be repurposed to secular causes or placed in museums next to the displays of trephination, leeching and the reading of bones and entrails.

4. **Masochism.**
   - We need to require that people think critically, empirically and as independent and self-reliant individuals, and that they use these capabilities to grow into or become independent and self-reliant individuals. We need to show people that they will be far better off if they stop believing dehumanizing, degrading and insulting nonsense on no evidence whatsoever. We need to loudly and continually ridicule people who believe ridiculous and self-demeaning things, because constantly teaching people they are evil or sinners or deserve punishment is itself an evil act. Teaching independence and self-worth will help remove the disgusting and degrading masochism which religions preach.

   - As we teach self-respect, self-reliance, the importance of helping others and not persecuting them, and the need to be a part of the community and not trying to bend it to your religious will, we also have to eliminate the teaching of the abject masochism which permeates religion and especially monotheism so thoroughly. Some people do have a tendency toward masochism, and we may never eliminate this as a psychological state, but the worst thing we can do is allow religions or any other group of people or system of thought to teach that humans are a so-called "fallen" species of dirty rotten sinning worms. Humans are amazing creatures who have unbelievable potential and who can live amazing lives when they are not hamstrung by the perverse and sadistically inculcated psychological millstone of institutionalized religious masochism. We can grow a great society populated by great people, but we have to show them the way.

5. **Mental challenges.**
   - We have to treat insanity and mental instability as conditions that should be handled like the breakdown or malfunction in the human brain which they are. We must continue to fund medical research into mental challenges and their cures, and to correct or mitigate them through an appropriate combination of psychological, sociological, chemical and potentially surgical treatments.

   - To be a truly caring and moral society, we must ensure that the less intelligent and most gullible and susceptible of us never fall victim to the lies and deceit of the devout. This can only be done by challenging immoral proselytizing societal vampires at every turn, preventing them from spreading their lies among the vulnerable and by doing what we can to educate, inform and protect those who need it most.

6. **Power and persecution.**
   - When religion loses its power to control others, and immoral and evil people lose the ability to use religion to persecute and make profit off of others, religion becomes as irrelevant and impotent as dryer lint. Because the psychopathic and the sociopathic faithful invoke their gods and the commands of their gods to gain power and to persecute others, they attempt to leverage an assumed ultimate authority for their evil, and far too many people fall victim to this. We must defang, defund, neuter and neutralize religion at every turn. We must show our children that the best societies and lives are those in which people are in control of their own lives and destinies

and that those lives and destinies are maximized through a strong civilization and strong altruistic individuals, devoid of any gods. Religion must be stripped of all of its secular powers and any ability to manipulate, persecute and siphon resources from people.

- The unbelievable amounts of liquid, real property and other wealth held by religious organizations and criminal religious organizations like the Catholic cult must be completely distributed to the poor and the needy, exactly as the Jewish preacher they allegedly worship told them to. We must eliminate theocratic states which serve to enrich and empower the clerical elite. The pastors, preachers and monsignors must vacate their sprawling massive mansions and convert them to houses for the needy. They must sell their private jets and fancy cars and give that cash to the needy, just like one of their gods said they should.

- Every single religious accommodation must be removed from our cultures and our legal systems. We must totally eliminate every single tax exemption or other fiscal reward or loophole which a faith-based organization or person can thread in order to gain access to even a single penny of taxpayer money, avoid paying tax or accrue secular power.

7. **Programming.**
   - As already mentioned, all crosses, Stars of David, the Star and Crescent and all other religious symbols need to be viewed as embarrassing relics of an ignorant and hateful past like the Nazi swastika, not venerated as the symbols of murder and ignorance they are.

   - All of the so-called holy books need to be moved to the fiction, mythology and horror sections in our libraries and bookstores. They need to be removed from hotels, motels, lobbies, bus stations and other places where children and innocent people might accidentally see one.

   - Any other reminders of the ignorant past ought to be removed, meaning religious sayings and mottos should be erased from our money, buildings, police cars and other such placements. The phrase "in god we trust" ought to be as impossible to find on courthouses and police cars in America as it currently is to find ones sporting “allahu akbar”.

- Religious services need to be changed into community and family support activities, neighborhood cleanups, games, entertainment, socializing, listening to public speakers and whatever other such positive and productive activities the people want to do.

And as a catch-all, each and every person on earth will have to be honest and carefully extract what microscopically few good bits we can locate in any of the thousands of religions both past and present. If it makes sense to do so and will truly improve the well-being of our fellow humans, animals and our shared environment, we should incorporate useful and productive bits of advice into truly moral but wholly secular unified codes of behavior and law. Each and every bit will have to be screened extremely thoroughly and codified as a secular tenet, with no tie whatsoever to any religion.

The really hard part is how to put much of this into practice. This will be hard enough in the current mindlessly and aggressively fundamentalist Christian America, but it will be extremely difficult in even more fundamentalist nations such as many Islamic countries. It won’t even be easy in most nominally Christian countries.

Humans are extraordinarily hard-headed and stubborn, and often doggedly refuse to do even those things which would immediately improve their lives. Convincing them to lose their religion and get a grip on reality will be at least as difficult as getting a drunk to finally put down the bottle or a compulsive gambler to permanently stay away from the casino.

Compounding the problem, but making it even more essential that we eliminate all religions, is the fact that religions cannot coexist either with each other or with the happily godless. Like sparks and gasoline, when they get too close to each other there is an explosion, and the fire that follows inevitably burns both people and property.

According to Wikipedia, there are several methods for conflict resolution[1], including negotiation, mediation, arbitration and something called “creative peacebuilding”. These are perfectly usable techniques for rational people, but religion is by definition and in fact not at all rational. Religious people are convinced that their particular god out of all the thousands out there has spoken to them, the god’s words are final, and everybody else therefore damn well needs to follow its orders too.

Once a person believes he has the word of his god in his pocket, or more accurately in his arsenal, the believer is beyond rationality, will not compromise and cannot resolve the conflicts he will inevitably cause. Thus we must actually carry out Solomon's recommendation. The sharp sword of reason must slice clean through the body of religion, and the two dead halves must be buried once and for all. Nontheists can negotiate with theists, but only after the godly give up their gods, because that is a non-negotiable starting point.

As mentioned before in this book and as noted many times in empirical fact, all we need do is look to the Scandinavian countries which are essentially godless but which have very high well-being, extremely high social cohesion and don't fret about sin or being a fallen worm or having to cannibalize or worship the dead or any of the utter nonsense religion teaches.

Not only do we not need religion to be happy or good, history has proven again and again that societies become worse places the more religious they become and better places as they lose their religion.

I again ask the question, if you think religion is so great, name the five theocracies in which you would most like to live, and more pointedly, to which of these manmade hells on earth you would consign your daughters.

So religion will have to go, and the only way it will happen is by the nonreligious taking a long view and a long-range strategy. We must first swallow our disgust and become highly politically active, to fully counter the never-ending treasonous and seditious efforts by the faithful to overtly and covertly destroy our great secular Constitution.

We must come out of the darkness, brave the inevitable attacks and death threats by the faithful and proudly proclaim our lack of belief in myths and fairy tales.

We must loudly and frequently extol the virtues of an educated and informed secular life.

We must get to the point where a person can easily win elections while proclaiming her acceptance of reality, thus her godlessness.

We must wage a constant war on ignorance in our schools, our public and private discourse, and anywhere we encounter ignorance.

We must insist on facts and reality being the drivers of discourse and decisions, and we must be loud and proud about it.

We must work to change the culture so that belief in the supernatural and the mystical is ridiculed rather than revered.

We must speak and act with the tenacious fervor and the truly moral righteousness of the women's and civil rights movements.

And we must understand that as has been so often said about the "war on terrorism" that the war on the mindless perversion of religion will take generations. The fight will have to be taken up by our children and grandchildren, until religion becomes like a once-feared disease which we have wiped out through education and information, in a mental rather than physical campaign of healing and vaccination.

All nonbelievers in every corner of the earth will have to do their part if we are going to win this transgenerational, transnational and transcultural war, and it is definitely a war. It is a war for reason, for civilization and for the freedom of our children, and potentially for the survival of our species.

Whether as an individual, as a small group of like-minded individuals or as members of larger groups such as The Freedom From Religion Foundation, American Atheists and dozens of other brave and tenacious organizations, everyone must work toward the same goals laid out in this chapter. It won't be easy, it won't be quick, and there will be setbacks, but if there is one thing that human history has taught us it is that when smart and dedicated people work toward a goal and they have the facts on their side, they can accomplish anything.

There is no time to waste, so get moving!

Chapter Thirty-Five

What To Do Instead of Going To Church

"Pray, v. To ask that the laws of the universe be annulled in behalf of a single petitioner, confessedly unworthy."

"Faith, n. Belief without evidence in what is told by one who speaks without knowledge, of things without parallel."
— Ambrose Bierce, from The Unabridged Devil's Dictionary

People pray when they go to church, despite the fact that prayer is probably the least productive and most foolish thing humans do. Prayer, in the spirit in which Ambrose Bierce described it in the first quote above, is usually just wishing that an imaginary magical being will magically do something that you ought to damn well do yourself, or that it will magically tell you what to do when you damn well ought to figure it out for yourself.

At other times prayer is just wishing that this magical being will stop something from happening that you yourself damn well ought to stop from happening.

Yet other times, as in the case of healing an extremely sick family member or stopping a religious zealot from carting you off to the rack or the stake, it is wishing that this being will intervene in the real world in a way that it has never actually done before in the entire history of the world.

The rock-rolling Sisyphus actually gets more out of his eternally fruitless endeavor than all the people who pray get out of their wishful thinking, because at least he's getting in a good workout while the prayerful are merely engaged in a mindless and pointless escapist copout.

But since people don't go to church only to pray, we have to replace both the act of praying in public and the other things people get out of going to church. So what do people actually get out of church, whether they admit it or not? There are several things. One is of course the prayer and fire and brimstone which used to be all the rage but which has generally quieted down somewhat. Another is having someone outside your family tell you once or twice a week how screwed up you are, re-tell the cautionary tales of all those who have screwed up before you and how badly they were punished, and how you can avoid this punishment if you do what the preacher says and give him some money.

There is more in this vein to be had, but the main reason many people go to church is the herd instinct. People like to be around other people like them, and they enjoy the groupthink reinforcement which they get from a bunch of people committed to the same wacky ideas like gods and prayers. They also like the social aspect of it, in which they have an excuse to get together outside the home and workplace. For many people church is less a god-and-prayer experience than a social one. They still do the god stuff, of course, and many actually believe it, but the social and herd reinforcement aspects are much of the draw.

With all this in mind, we can quite definitely redirect the faithful from their houses of worship to other venues or even the same venue once it has been repurposed for different and infinitely more practical activities. Since social engagements, groupthink and mutual reinforcement are what the religious need to be happy, let's replace them with far more positive activities. I've listed just a few suggestions below in no particular order, but of course anyone can add others that suit the local people, culture and situations.

1. If people still want to get together, they can meet at their former house of worship after its conversion into a community support center outfitted with tools, games, books and other useful items. Groups can volunteer to take some of these tools, books and useful items to people and places which need them and can get to work. Possible activities could include visiting the elderly and doing difficult chores or tasks for them while giving them some companionship, teaching the elderly about how to use modern technology, setting up and distributing food to the needy, doing highway cleanups, reading to underprivileged and/or poor children, volunteering to teach classes on useful subjects to anyone who wants to attend the training, doing neighborhood cleanups, helping people to recover from natural disasters, power-washing homes, doing painting jobs, cleaning gutters and clearing brush and leaves from high roofs, repairing vehicles and/or appliances, delivering food or just spending time with lonely people.

   These are all great activities, very socially productive and they increase the well-being of the performer as well as the recipient of the service. They also do a hell of a lot more for the community in the spirit of the liberal community servant and community organizer aspect of Jesus than does sitting around in a church talking about how great Jesus was and how much better the faithful think things would be if their bible became law.

2. Donating time to perform some pro bono legal or minor medical work, or at least dispensing professionally-informed legal or medical advice.

3. Spending time in large or small groups talking about issues of the day or just things of interest to the people in the community. Talk about what makes people happy, tips on how to be a good mother/father/son/daughter/friend/citizen, how to do a garden, how to cook or create or do art, how to use or apply the latest scientific discovery, or even how to be more moral and ethical.

4. Call or visit your parents, or your children if they are grown and have left the home. Or both! Make sure to tell them you love them, and find a way to prove it.

5. Play with or spend time with your kids if they are still at home. Teach them something. Make sure to tell them you love them, and find a way to prove it.

6. Relook your life to see if you have been doing the right things, and if not, put down a list of goals for doing better. Start making those goals happen.

7. Exercise, preferably with family. Also, take your pets for a walk, preferably with family.

8. See if your elderly neighbors need anything.

9. Put down social media for a day and do constructive things with actual live people.

10. Learn something new and interesting. Feel free to share it with like-minded people.

11. Try to put yourself in other people's shoes and truly see the world from their perspective. This includes people who are not at all like you. Imagine the world from the eyes of those who have been oppressed or are in unfortunate circumstances. Imagine you are a female in Afghanistan under the Taliban.

Imagine you were a slave under the lash in pre-Civil War America. Imagine yourself as a poor black atheist lesbian non-English-speaking immigrant from a Muslim-majority country in the uber-xenophobic and increasingly bigoted America of 2018. Imagine yourself as any of these people and many more, and consider how other people would treat you. Use this exercise to inform how you interact with other people no matter what they look like or sound like, where they come from or any other harmless characteristic. Learn to judge people by their words and deeds, not superficialities.

12. Be thankful and appreciative for the abilities you have, the people you have and for the comforts you enjoy, but don't ever thank any god for it.

Chapter Thirty-Six

New Jobs For Displaced Clergy

Some professions that no longer exist: switchboard operator, Inquisitionist, headsman, knockerupper, trephiner, phrenologist, horse cavalryman, musket infantryman, bombardier, town crier, slave, alchemist, court jester, professional mourner, body collector, mummifier, Vestal virgin.

Everybody, including women even though this is contrary to much Islamic and Christian belief, should have the opportunity to work for a living. Working brings in the coinage one needs to purchase food, shelter, clothing and education in this modern world. We need those meals, dwellings, clothes and educations to be a participating and indeed a surviving member of a modern society.

Families must make their own internal decisions about how they will live, which members of the family will work, which ones will care for children or who will do both. In practice many women work and many do not work, but women should have at least an equal say in the decision.

Also, many people are born into different cultures or they eschew modernity, from the hunter-gatherers who still populate some of our nations to groups like the Amish who only accept those parts of modernity which their group decides are acceptable. Some families own farms, others travel for a living and do not have a permanent residence, while others have other arrangements which the modern human would call nontraditional, or “off the grid”.

No matter what the individual circumstances are, working outside the home brings in an income and it also brings the satisfaction of accomplishing something and being a part of something bigger than ourselves. All across the world many people in most places have jobs outside the home, or they did before they became disabled or retired. As of this writing, the vast majority of active clergy have jobs outside the home, or technically inside the home for those sects who happen to live in the same building in which they operate.

When humanity eventually grows up and throws off its superstitions, as we nonbelievers fervently wish would happen before next weekend, all of these religiously employed types will need to find new jobs. As I listed in the epigraph to this chapter, there are many, many professions which were once key jobs in former times and places, but which have gone quite extinct, just as most species and religions have.

And guess what? People found new things to do and the world didn't end. And it won't end when we are finally able to add "clergy" to that growing list of dead professions. So we will need to help all these preachers, pastors, priests and other clerical folk finally find something tangible and useful to do. Any reader of this book will be able to come up with dozens of possible professions, but I will list a few here.

Before I do so, though, we have to be brutally honest and state quite clearly that many people who entered the clergy did so for reasons which will disqualify them for certain professions.

As an example, just as many people with psychopathic tendencies tend to enter politics and cut-throat business, and those with violent tendencies tend to enter the armed forces and law enforcement, many men have apparently become Catholic priests so that they could have access to children from a position of power and thus sexually assault and rape them.

These individuals may or may not be homosexual, a characteristic which should not automatically disqualify them for any job, but they are pedophiles and sexual sadists, which does absolutely and permanently disqualify them for any position where they can have any access to people under the age of 21. They should also be kept separate from people over 21 who are mentally challenged and thus vulnerable to physical and mental predation.

Not all clergy are like this, of course, and not even all priests are pedophiles or pedophile protectors, but anyone who self-selected to be a cleric and especially a priest is abnormal enough to require very careful screening before placement into subsequent employment. Those who need to be jailed for crimes already committed, from child rape to the conspiracies to cover up the rapes to the actual coverups and lies, will need to be put away for a very long time, preferably forever. With that said, some possible professions for non-sadistic and non-pedophilic clergy are listed below, in no particular order.

1. **Salesman.** This is a superb option for most clergy, since they've been selling an invisible, imaginary, intangible and undeliverable imaginary product for millennia, and apparently fairly successfully. Selling the humanistic benefits of social service and a secular godless pro-human morality could underlie whatever other sales position the person can hold.

2. **Volunteer organizer, community organizer, food bank or charity operator.** Those who truly entered the clergy partly out of a desire to help others would be well suited for this

option, following proper legal and moral screening. This need not only be in their country of origin; there are plenty of secular nonprofits and other agencies which would be outstanding organizations for former clergy interested in helping others with human and secular needs.

3. **Social worker or counselor.** As with the above professions, these will work well for those who entered the clergy for honorable reasons and who have a talent for helping with emotional needs. Former male clergy will have to undergo extremely rigorous screening before being allowed around children, though. Also all counseling and work would have to be 100% secular at all times, no violations allowed.

4. **High school art teacher.** Art will be the best choice, since unless they have an almost unimaginable and complete change of heart there's no way they will be able to teach the rational and logical disciplines required for science. The same screening for male clergy as mentioned above applies here.

5. **Merchant seaman or Navy sailor.** Clerics are often used to living a somewhat isolated life with a rigid chain of command up to a Supreme High Command. They are also used to being around other men for a long period of time, so these professions could work out well.

6. **Gay rights activist**. This is of course generally for gay clergy. There's no longer a need to hide in the closet so they might as well go "all in" and help the cause. And they have done so much to oppress and demean homosexuals for so long that spending a few solid years undoing that damage would be appropriate and a sign of true repentance.

7. **Women's rights activist.** This would be a viable option for a truly repentant cleric for the same reasons they would do well to serve as gay rights activists. Religions and their clergy have been so evil to women for so long that this should be one of the top professions for a former god-botherer.

8. **Librarian or archivist.** They've lived their lives focused on a book, so being around lots of books or collections of data would seem to be a natural fit. Plus many clergy don't want

much to do with society, so the more introspective and introverted ones might really enjoy this kind of work.

9. **Banker or fundraiser.** Most of them have spent years asking for donations so that will help ease the transition to one of these professions.

10. **Physician, physician's assistant or nurse.** These options would definitely require significant retraining, and some time, but the disbanding religious organizations could provide the time and certainly the funding for the retraining process. Clergy would have to completely renounce any belief in, practice in or recommendations for prayer or any other faith healing nonsense in order to qualify for this profession. But it is an option for those who really want to help others. Former Catholic priests and nuns in particular would be well served by doing volunteer work in Africa taking care of all the people dying of AIDS, the relatives of the dying and the dead, and handling the dead themselves, as they and their religion had said so clearly and often that condoms were much worse than AIDS itself. They can help fix the damage they caused.

11. **Handyman or construction.** Repairing homes and appliances or even doing construction would be an outstanding positive contribution to society and homeowners, and the former clergy would be helping people with real needs. Plus it is the rare person who does not get personal satisfaction from creating a tangible product, be it for commercial sale or for personal enjoyment. It would be a massively improved personal reward to leave a customer with a tangible and real benefit than the imaginary benefits clergy and religions have on offer.

12. **Humanist celebrant.** If the person were really, truly and substantially recovered from and repentant for his time as a cleric, and wanted to truly perform moral acts and do good for others, this would be a great option. They can also still continue to do things like weddings, rites of passage and even serve as moral philosophers and guidance counselors, if they never go back to their bad old religious ways and only do these functions as a full secularist. The need for rigorous screening applies here as well.

13. **Police officer.** They've spent their lives telling other people what not to do or else they will face punishment. Now they can do something about it in the real world. Plus they can try to prevent violence and misbehavior from happening, and be a public servant and a steward of public safety at the same time. Part of being a police officer actually includes things like talking to people, knowing the people on the beat, counseling, crisis resolution and other positive interventions, so hopefully some of the cleric's training could be used in a secular manner as a law enforcement professional. Rigorous screening applies here again, of course.

14. **Firefighter.** As with being a police officer, this job enables the person to be a public servant and a steward of public safety. And they would often work in an isolated workplace with mostly other men, so this job environment would not be unfamiliar to many of them.

15. **Magician.** They will have left a profession in which their main business is actually trying to convince people that magic is real, so this profession would be a very small leap for them to make. It even lets them tickle their compulsion for supernaturalism without actually preaching it.

16. **Conservationist.** Preservation of our shared environment is one of the most important human responsibilities. The anti-environmental conservo-religious right in America has enabled legions of dogmatic environmental destroyers to enter our government and do massive damage to the environment and our economy. Making up for the damage these clergy have supported and enabled would make this a great option.

17. **Secular leader.** This may seem counterintuitive, but when a person really, truly and substantially realizes just how many lies and untruths there are in religion, which is a position a person can reach after being on the inside as a cleric, the rejection of religion can enable a transition to a hugely and positively productive life as a secular leader.

    While few people who were once so devoted to a faith position do voluntarily leave the clergy, some of those who do leave become strongly independent nonbelievers, focused on actual good and actual morality with one could almost call a "religious" fervor.

People like Dan Barker of the Freedom From Religion Foundation and many others like him who have truly rejected the whole religious kit and caboodle have become strong logical, Constitutional and moral opponents of religion and strong leaders for their communities and for former clergy. Also, like a magician who tells the world how the tricks of the trade are done, former clergy can expose religions and their fakery to the world and have a better understanding of the human weaknesses that make people fall victim to religion's spells.

Those are all good options, and as I noted there are dozens more that would be good fits for former clergy. All readers are free to recommend these and other appropriate professions to their local congregations and any and all clergy. That would greatly help the cause of increased global secularism.

So there are plenty of things for former clergy to do. They need not be thrown out on the street or made homeless, they need not be rejected or jailed unless they have abused or violated a child, lied about it, protected or facilitated a child abuser or rapist in any manner whatsoever. Those clergy who are not criminals can assuredly be added to the workforce, as long as they are properly screened.

I encourage everyone to begin coaxing the clergy in their communities to give up the nonsense, reject the supernatural and seek productive employment in a field based on reality and morality. Many of them already very much want to do this, as many former clergy have admitted, and as you can see in The Clergy Project. Found on the Internet at http://clergyproject.org/, the Project's website notes that its mission is "…to provide support, community, and hope to current and former religious professionals who no longer hold supernatural beliefs." If you want true inspiration, put your holy books down, go to this website and read some of the stories of people who threw off the shackles of ignorance and now live far happier and more rewarding lives outside of the churches. Know that there are thousands of them even now living a lie they would love to escape.

Whether you do it overtly, through conversing with a current cleric and trying to convince him or her to jump off of the sinking ship of religion, or covertly by dropping hints or by leaving Clergy Project information and job announcements in strategic places, the more we tell these people they can and should join the community of the rational, the more likely it is that they will do so. Do it with compassion, but do it with strong conviction, and be prepared to help with the transition.

With enough people doing this, we can move people one at a time from the pulpits to the real working world. Just as every vote is important, every single clerical departure from the pulpit is important and will eventually cleanse the world of its peddlers of bigotry, evil, immorality and superstition.

What are you waiting for? Go do it!

Part Nine

In Conclusion, Plus Quotable Quotes

Chapter Thirty-Seven

In Conclusion

"No man ever believes that the Bible means what it says: He is always convinced that it says what he means."
- George Bernard Shaw

"If you could reason with religious people, there would be no religious people."
- Dr. House

I sincerely hope that you have used the experience of reading this book to gain an understanding of how religion came to be, why people still believe in gods and the supernatural, and why such beliefs are not just illogical but evil, highly immoral and often murderous in practice.

I hope you have a much better understanding of how religion sprang from utter human ignorance and how it has been allowed to metastasize across the world as a result of our continuing ignorance, our childish fears and weaknesses, the inherent flaws in human psychology and the sheer evil and complicity of far too many people.

I hope you now see very clearly that religion was an accident of circumstance, that it is the illegitimate bastard offspring of fear and ignorance and that it is a huge mistake that must be immediately corrected if our species is ever to become truly mature and moral. Our very survival is likely at stake.

Many of the topics and perspectives in this book, and the undeniable facts of nature, physical science, psychology, religion and history will have been new to many people, particularly the faithful. Religions work extremely hard to insulate their flocks from facts, reality and differing perspectives. Many current Catholics, even given that cult's evil history and the proliferation of rapists in its priesthood, will not have read such direct attacks on every element of their faith and its arrogantly and obviously incorrectly asserted moral superiority.

Many Christians will never have heard of the anti-religious perspectives presented in this book and will have been as unaware of the history and words of their faith as they were unaware of the deep scars it has left on American history and on Americans themselves. Many of these scars were actual physical scars, caused by implements such as whips and chains, in addition to the billions of mental and societal scars from the deep psychological and legal wounds these faiths inflict.

I doubt that very many so-called moderate believers will have understood before reading this book just how egregiously their support for religion provides cover for fundamentalists of all stripes, and that every person of any faith is an enabler of destructive fundamentalism.

I hope that if you started reading this book as a believer, and if you have read all the way through the book, that you now see the world quite differently than you saw it before you began reading. I fervently hope that you now have a burning desire to support, protect and defend all people against the scourge of religion.

I hope every believer also now understands, as nonbelievers do, that faith can never be trusted and that wherever it is found it will always cause needless suffering, conflict, pain and death. We nonbelievers would love to welcome you into the moral sphere as a new nonbeliever.

I hope that if you were an atheist or agnostic when you started reading that you are permanently and even more strongly recommitted to your original position. I trust that you are now well-armed with facts and have an extremely strong motivation to get out there and fight against religion however you can in arenas both great and small.

I hope that if you were a religiously uninterested or neutral person that you have chosen to permanently and actively join the forces of secular godless reason, morality and human decency and that you will fight with us for those goals. We look forward to joining arms with you and preserving our world and all that is in it for our children and all future generations.

Given all the above, and given the contents of the book, I sincerely doubt that very many people who began reading as strong believers in religion will have had the mental and moral fortitude to keep reading to this point. But if you have persevered and are reading this I want to sincerely thank and congratulate you for doing so, for it cannot have been easy. I sincerely hope that you have gained a very wide and somewhat deep understanding of why faith in the supernatural is not merely illogical, it is also immoral, inhumane and very dangerous.

I hope that you have begun to see why we secularists have no choice but to distrust the faithful, including you yourself, and that we must always be on guard against religious aggression.

I hope you understand why you need to leave your faith immediately and bring as many other people as you can with you out of the darkness. I hope that you have had the same experience which Dorothy had in the Land of Oz, and that you will respond appropriately to the curtain being pulled back on the façade of the supernatural to reveal not magic or splendor but fakery and trickery and with no one but weak, untruthful and fallible humans pulling the strings. I hope you have realized that the little human behind your particular religion's curtain deserves your contempt and pity, not your worship.

I hope that you have decided that you would rather live a life of realities and not fables, a life of truths and not lies, and a life of actual morality and not bigotry and immorality. If so, we heartily welcome you with open arms and we know you will enjoy the rest of your life as a free person, no longer a slave to the depravities of religion's ignorant and hateful history.

If you were a believer and have somehow read every page of this book and still think religion is where you want to be, I encourage you to do two things. One is to re-read the book, or at least Chapters Three, Six, Twelve and Fourteen, with emphasis on the logical fallacies in Chapter Twelve and the reasons people actually believe in religion detailed in Chapter Fourteen. You might want to read Chapter Six a couple of times just to get the point. The other thing I encourage you to do is take a few weeks to think about things. There is no need to make a decision right now, or even next month.

The wonderful thing about truth and reason is that they will always be there for you, not in the imaginary sense that religions pretend to be, but in the actual way that truth and reality are always there. Think about what you have read in this book. Think about everything you have been told by religion, but now think about things through an empirical lens, using critical thinking, reason and facts. Think, reason, assess, think some more, and work things out at your own pace. Talk to religious people and atheists. Do your own research, but use independent and opposition sources outside your religion which do not have a vested interest in fantasy or make-believe.

Ask extremely tough questions, and they need to be the kind you have been afraid to ask because deep down you didn't really want to know the answers. If you are honest in this, your research will have to include every single one of the world's past and present religions, since each of them has an equal chance of being the One True Faith. Balance the words, the perspectives and the actual evidence and decide if you want to live in the real world or the fantasy world of religion.

If you started this book as a Catholic and you are still a Catholic, I honestly have nothing good to say to you. Your logic and morality are broken beyond repair and/or you are too weak to actually leave that criminal and immoral cult despite the tsunami of pain and suffering you and it have been inflicting on innocent people across the world for centuries. I only hope the forces of moral secular law treat your evil, immoral and antihuman child-raping church and its rapists and rapist enablers as the criminals they are, and that mass imprisonments, massive fines and other long-overdue legal actions dissolve that immoral cult despite your deliberate complicity in its utter evil.

For all readers, except those already well-steeped in its contents, it can take a fair bit of time and effort to take in all of the information in a book like this, and will take even longer to synthesize it, because it contains large amounts of information which the faithful and even many nonbelievers will probably have never seen before, particularly in such concentrated form.

The book provides facts to show what really happened in history, how humanity is very susceptible to a wide range of biases and weaknesses, just how ignorant most people are about science and religion even today, and that religions and societies have lied to people for a very long time. It is hard to accept that one has been so severely misled for one's entire life, and even worse that the deception was perpetrated by people in positions of authority and respect. Thus believers will face much dissonance and mental inertia as they process this book.

People have a very hard time changing their minds, and the pernicious ways which religion impacts the psyche make changing a believer's perspective on faith far more difficult than it should be. It can be hard to relook paradigms one took for granted and trusted implicitly. It can be much harder to realize that one's worldview has been built on sand and that an honest acceptance of even a small part of reality is sufficient to bring the whole illogical fabricated religious contraption tumbling down.

But large numbers of people do these things every single day, and they are far better off for it. Every new morning brings a truly new dawn into every person's life, and every day is another opportunity to throw off the blinders of religion, to decide that light is far better than darkness, to accept that love is actually better than hate and that truth really is better than lies, even comforting lies.

Humans are a mix of the savage murderous ape and the self-sacrificing protective mother. There are many critically-thinking geniuses and many people who are totally unable to talk or think, much less use reason. We can be compassionate, gentle, kind and loving and we can commit acts of unspeakable evil. For seven million years we have evolved to be both helpful and savage to others of our kind, and over time we adapted evolutionarily and societally successful strategies which evolved into informal then formal codes of morality. Then humans invented religion.

Religion is the only self-described system of conduct on the planet which by its own words and through its own constructs deliberately orders and enables acts of ultimate immorality and evil which are mandated by an ultimate godhead.

Since the religious holy books contain these orders, since the holy books are the foundation of at least the monotheisms, and since the holy books cannot be changed, they must all go, and the religions must go with them. Humanity can and one day will collectively do a far better job of codifying and achieving morality once it has rejected all of its holy books.

Without religion, we would still have criminals, murderers, psychopaths and other stains on our societies. These are products of damaged minds and damaged societies. But without religion these people and societies will not be able to use the allegedly divine words of an imaginary god's holy book to justify their evil. And in far too many cases it is religion itself that damages these minds and these societies. Religion has never and will never be able to stop itself from repeating these crimes.

Religions have never given a single actual truthful answer to any question humans have ever asked. Since humans first learned to chip rocks into tools, science has proven that it is the only path to true understanding and true human progress and well-being. Religion kills, and science heals. Religion obscures, and science enlightens.

We must reconfigure ourselves as a species, and we must teach our children the truth about the world and about life. We need to instill in them the ability to think critically and morally, with a universal core of morality that does actually achieve the highest levels of well-being for the most people. We need to remove the antihuman and evil disease of religion from our world. If we were to do this, our world would take an immediate quantum leap forward in intellectual prowess, ethical and moral accomplishment and safety, security and well-being for the most people.

Religion is a clear and present danger to humanity and our world, and we must stop at nothing to eliminate it. We need no more inspiration for this than the admittedly fictional Eve. She is an inspirational example of a person who chose to think critically, to be independent, to set aside blind faith and to rely on reason. The biblical Garden story was written as a threatening fable intended to intimidate and to reinforce the orders to both young children and adult children that not blindly following orders would be brutally punished. We need look no further than the cowardly Nazi Nuremberg Defense and the entire history of religion for the best historical examples of how such behavior destroys humans and destroys humanity.

All of us need to be much more like Eve. She and her like-minded sisters and brothers need to be our inspirations, not the pathetically weak and spineless Adam or enablers like him. People like Adam give us the Inquisition, the Crusades, the Holocaust and the Nuremberg Defense. People like Eve give us moral secular law, the Nuremberg Trials and properly informed advances in morality, science, medicine and human well-being.

Think how bad life would be if we had not had a long succession of Eves. Women like her who have told the establishment to damn well kiss their asses and to accept that it was time to be enlightened and moral once and for all have done much to improve, enhance and educate our world. From scientific discoveries to voting and civil rights to demanding equal treatment both in society and under the law, strong women have made our world much stronger.

Far from discouraging women and girls from being like Eve, what this world needs is an Eve in every household, in every courtroom, in every boardroom and laboratory, in every level of government and in every group where important decisions are made.

Thank god For Eve!

Chapter Thirty-Eight

Quotable Quotes

This chapter is a collection of my words from the book plus others which I did not include in it but will in a later work or have already done so in other works. I know it might seem a bit self-serving to collect my own quotes into a single chapter, but there are good reasons for this, in light of our fast-moving societies and increasingly sound-bite and meme-happy cultures. Also, this chapter is intended for several kinds of people.

If you were intrigued by the cover or the title, opened the book, got intimidated by the page count and flipped to the back looking for the summary or conclusion, this chapter is for you. It collects many of the quotable quotes from the book which you would have to spend quite a few hours reading into a place where in just a few minutes you can get the flavor of the book and also put some handy verbal ripostes into your mental armory, ready for use on the next believer who tries to god-talk you.

If you are a committed antitheist, secularist, atheist or other like-minded person, the totality of the book will, I hope, greatly inform you, but you are already on the Reason Team. You don't need to see the light because you're already happily basking in it. You can take all the time you want to go through the book at your leisure, but you can use these quotes immediately, maybe even tonight.

If you are a student in high school or college, especially if you are in a religion, philosophy, history or psychology class, you can use some of these quotes to start some really interesting discussions, and get the class or your peer group to veer off into a much more engaging session than you were going to have. Have fun!

If you are a person in a hurry who would take the time to read the book if you had the time, or if you will have the time sometime in the future but don't have it right now, the few minutes you need to run through the next few pages will give you much to think about and much to discuss with others. Plus you will have a hell of a conversation-starter on your coffee table during your next social event, and you can cite some of the quotes and really blow some people's minds.

If you are a highly religious person who would never commit the venial or mortal sin of actually reading through such a book as this, you can peruse these quotes and use them to think about the world from the perspective of the two billion-plus people on the planet (and growing!) who reject all superstition.

With that, here they are…enjoy!

1. “The hurdles one must cross today to be both a religious person and a scientist require multiple simultaneous catastrophic breakdowns of reason and also leaps of faith long enough to hurdle the Grand Canyon in a single evidence-free bound.”

2. “Your mythology doesn’t get a vote in civil society.”

3. "When a person feels his god is on his side, nothing is sacred."

4. "Absolute power corrupts absolutely, which is why god is absolutely corrupt."

5. "Psychopaths have kept armies in brisk and bloody business for millennia."

6. "Humans can screw up even the best ideas."

7. "Humans are not yet genetically ready for a truly civil society. This is why we need government."

8. "Religion makes people dumb, and it makes societies dumb. The more religious a person or society is, the worse off it is."

9. "Name five of the theocracies in the history of the world in which you would most want to live, and give your reasons."

10. "Sincerely Held Religious Belief = My god Says You're A Piece of Shit!"

11. "If you burn a paper Quran, you will cause riots. If you delete a digital Quran, you will not cause riots. This is religion in a nutshell.”

12. "Bacteria are smarter than creationists."

13. "The human brain at birth is essentially an organic computer with an operating system and computing power. It will either work or

fail to work based on what is input into it and how many viruses it downloads."

14. "The second I hear someone say he 'believes' something rather than saying he knows something due to peer-reviewed scientific and empirical proof, I immediately know he doesn't have a clue what he is talking about, he is deliberately lying, or both."

15. "Denial of science such as evolutionary theory is as contemptible as when a person knows all of the benefits of modern medicine, but only takes the decision whether to use them or not after tossing animal bones to decide if he should go to the doctor or just pray longer and harder."

16. "The faithful can tell a woman what to do with her uterus and vagina when women and atheists can tell men what to do with their penises and testicles."

17. "Science binds us. Religion divides us."

18. "Science lights the hall of knowledge. Religion snuffs out the candles."

19. "Science provides the joy of understanding and enlightenment. Religion provides the angst of guilt and depression."

20. "Science takes us to the future. Religion takes us to the past."

21. "Science lets us expand our vision. Religion keeps our blinders firmly in place."

22. "Science welcomes everyone. Religion welcomes only who it chooses, and excludes many."

23. "Science shows us the path to true morality. Religion shows us the path to true bigotry."

24. "Religion doesn't have the courage of its convictions. Like an insecure bully, it has a compulsive need to plaster itself all over, to unconstitutionally place its statues and signs and teachings in public property and in schools. Secularists wouldn't have to go around putting up statues of freethinkers or signs saying "Thank

Science There Is No God!" if religion would just leave people alone."

25. "We humans are nothing more than extremely evolved microorganisms, but at the same time so much more. We are animated collections of dead stars, shining with an internal light which expresses itself in the form of wondrous creations of art and architecture, prose and poetry, the exultation of the pursuit of knowledge, the love we feel for our families and the acts of kindness bestowed on others."

26. "There is no such thing as 'beyond science'. That infantile 'get god out of jail free' tactic is as obvious and childish an attempt as obfuscation as is a two year old trying to lie about stealing cookies while standing there with one in each hand. It is an attempt to apply the god of the gaps argument wherever and whenever you think your god needs an alibi or an excuse. Such nonsense is, more accurately, 'beyond childish'."

27. "The difference between mythology and religion is, of course, that all religion is mythology except to the true believer, who thinks that everybody else's religions are mythology."

28. "The brain does not have some kind of sump pump or filter which turns on when the level of sewage and misinformation has risen above some threshold and the garbage gets sucked out. For many people, garbage in is always garbage in, and it is garbage out in terms of what the person then does and says based on that garbage."

29. "Most people will believe whatever makes them feel good. And freethinkers know how true it is that trying to use logic on the faithful is like pissing on an electric fence. You'll be quite shocked at the result."

30. "The devout adherents of the world's religions have had their pious knees on humanity's throats for far too many centuries. Every single act of evil committed in the name of these religions, every act of bigotry and persecution perpetrated in accordance with their teachings, has been another cruel and inhuman lash across the bare and bleeding back of our collective humanity."

31. "The lack of knowledge of their own and of other faiths is why the devotees of religion are called "believers" and not "knowers". They *believe* quite a lot, but don't actually *know* very much."

32. "If you are a believer and you say a god is necessary for universe formation, you should also say a god is needed for snowflake formation. If you don't, you're being inconsistent and you have fallen victim to the trap of a self-serving and special pleading bias where every conclusion derives from what you personally think is possible and if it ends up with your solution incorporating the god you personally like."

33. "For some reason, religious people and many politicians believe they are entitled to the arrogant hypocrisy of trusting pilots, doctors, dentists, or engineers, but say they need not trust scientists if the findings of those scientists damage or disprove their religion."

34. "Science has led to more human well-being and more actual knowledge on every measurable subject than any other human endeavor. Yet the faithful feel that what is preached in their religion and what is in their churches and holy books should somehow, on no justification whatsoever except tribal allegiance and fear, supersede science."

35. "One day we ourselves may be able to create universes after we discover that it really wasn't that hard after all."

36. "You are an organic robot, part water and part other matter, composed of elements forged in the hearts of long-dead stars. Your body is an organic machine driven by chemicals, chemical processes and electricity. Inside that organic machine lies the brain, which is itself yet another organic machine which controls the body and also functions using chemicals, chemical processes and electricity."

37. "Your Jesus said to render unto Caesar what was Caesar's. Your Jesus strove to convert people through preaching and teaching, not packing courts and violating secular law. Your Jesus preached a sermon of peace and service on a mountain; he did not tell his disciples or followers to find ways to order other people around, insult and degrade them and try to take away their human dignity

and civil rights. What would your Jesus do if he caught you lying, cheating, bending the truth, not telling the truth, the whole truth and nothing but the truth so help you Jesus, and subverting the rule of law to mandate his religion? I tell you three times, he would take that whip and beat so many American Christians with it so hard and for so long that even his supernatural self would get tired."

38. "Those who actually do value facts and reality must respond with passion and a loud voice to keep the queen of Reason firmly seated on her throne and to boot the court jester Belief outside into the cold rain."

39. "We unfortunately cannot see into the past to observe how these religions were created and shaped solely by humans for human purposes, or all of them would evaporate overnight."

40. "Religion is merely philosophical alchemy."

41. "Religion minus bigotry, weird rituals and ghost stories equals nothing. Simpler is better."

42. "Occam's Razor cuts religion to shreds."

43. "Take your pick – get rid of religion or get rid of science. I know which one I'd choose."

44. "Priests are even worse than politicians. They're both liars but at least the politician usually doesn't try to rape your kids."

45. "The coin of their realm is that of Janus, in which heads Christianity wins, tails Christianity still wins, and nothing you say or demonstrate to them will ever change or open their closed minds."

46. "Christians in America are doggedly perverse in their compulsion to deliberately pick out a narrative in which a god in the image of a human but with supernatural powers created the universe, and either does or does not intervene thereafter, according to the conclusion they want to reach."

47. "People who think they have to force their religion on other people do not mix well with those who don't use fiction as personal or public policy."

48. "If we could pile up the bodies of all of the men, women, children and the unborn who have been murdered in the name of the world's religions in one place, we would have a pile of human flesh the size of a small moon."

49. "Ignorance dissolves like a crystal of salt in the hot clarifying solution of science."

50. "Science eats away the accumulated accretions of ignorance and stupefaction which, like hardened deposits in arteries and water pipes, continue to clog and foul our minds and are always ready to completely cut off the life-giving sustenance of rationality our species has worked so hard to achieve."

51. "Religion is a mind-numbing, critical thinking-killing mental opiate which is antithetical to a truly humanistic society."

52. "If you are a Catholic, you believe that Jesus and his father are glued together by some kind of supernatural ectoplasm called the Holy Ghost, and together these three somehow form both one collective and three separate entities at the same time. There has never in the history of the world been a coherent explanation of this bizarre dogma, and it is just one of many extremely bizarre dogmas dreamed up in that weird cult over the centuries."

53. "So the god of the bible knew as much about science as a person who lived in the Iron and Bronze Ages. What an unremarkable and damning surprise."

54. "People who don't have a clue what they are talking about are the most likely to be ignorant stubborn horse's asses and to hold completely uninformed opinions on any subject whatsoever which wouldn't survive an elementary-school level critical analysis."

55. "Monotheism is masochism masquerading as religion."

56. "Religions rely on fables built from faith which was itself derived from ignorance and wishful thinking."

57. “We are hagridden by the primitive superstitions of our ignorant ancestors.”

58. “Facts only matter to those who choose to accept them.”

59. “Faith: where the insecure go.”

60. “Churches are asylums for the incurably unreasonable.”

61. “God the father. Who’d you expect, god the mother?”

62. “Monotheism: All men, all the time...nothing weird about that, right?”

63. “‘God Had No Mommy’: the original Disney cartoon.”

64. “You say Jesus, we nonbelievers hear Poseidon and Mithra and Amun-Ra, blah blah blah; they’re all the same.”

65. “Faith is a diaper of the mind.”

66. “Faith is mind candy…too much makes you sick.”

67. “Faith means a person will believe absolutely anything no matter how ridiculous."

68. “Religion - no, pigs can't fly, but horses can!"

69. “Religion - women are the root of all evil. It's their damn fault I lust for them!"

70. “Religion - my faith says I can do and say anything I want no matter how bigoted or foolish."

71. “Religion – when the irrational believe the invisible.”

72. “Hate the belief; not the believer.”

73. Religion: "Oh Daddy, please please *please* love me forever, and tell me exactly what to do for the rest of eternity because I can't think for myself. Oh, and what were all those rules about sex again? You have a really weird hangup about that."

74. “The bible: 837,000 words and it still can't figure out what it’s trying to say.”

75. “The bible: A collection of Very Grim Fairy Tales hand-picked by religious bigoted misogynistic self-hating, sex-hating zealots, given to you as the Final Solution.”

76. “God wrote the books on it - genocide, slavery, murder, rape, incest, mass abortion, misogyny, bigotry, persecution and child abuse.”

77. “Religion - magic is real, I promise!”

78. “Religion is the solace of small minds.”

79. “Religion is just 1001 bullshit excuses to fuck with people.”

80. “The bible is proof of perpetual (e)motion. It is an eternal source of bigotry and persecution.”

81. “Everlasting Godwhoppers: False beliefs which the immature and insecure suck on for comfort.”

82. “The bible. Rules to live by, if you're a psychotic sadomasochist.”

83. “Jesus is love. He loves us so much, he made Hell just for us!”

84. “The bible. Witches and warlocks are real! And you must kill them all! Now!”

85. “The bible. More pure hatin' and killin' in one book than you can get through in one sittin'!”

86. “Who killed more unborn babies - Hitler, Stalin or Jehovah? (Hint...it starts with a J).”

87. "Mommy, that boy's sitting in my spot." "Well, Johnny, the bible says if someone is occupying your land, you must surely put him to death, so let's grab some stones and do as the Lord commands."

88. “The Vatican is a fully theocratic dictatorship under the control of a person who is elected but who is also nonetheless a theocratic

maximum leader whose word is not just law, but infallible when he feels like having an Immaculate Communication."

89. "Far from being either true or helpful, religion is merely astrology on steroids, alchemy claiming divine writ and human ignorance codified in highly immoral and thoroughly corrupt human tribal institutions."

90. "In the eyes of the faithful, no lie is too big, no deception is too immoral and no campaign of misinformation is too inappropriate, because they believe they are doing their god's will."

91. "Religion is not a fact-based belief system; it is a belief-based belief system."

92. "Religion's original justifications became brain dead long ago, and the entire current corpus of religion is being kept alive on forced life support by the believers who cannot bear to be without the crutch of their faith."

93. "Do you believe in Mithra, Baal, Shiva, Manitou, Thor, Amun-Ra, Marduk, Ishtar or Mena? No? Then congratulations! You're an atheist, just like me."

94. "Your god doesn't fill any gaps at all except those in your own mind."

95. "Science is an objective method of implementing critical thinking in order to advance human understanding and our store of objective facts and information. Religion's perspective on this is to accuse the scientist of arrogance because the scientist refuses to allow the religionist to claim dogma and opinion as fact."

96. "Arrogance is the spawn of ignorance."

97. "We again return to the astronomical arrogance of the believer who says that he personally knows the mind of his god so unbelievably well that he can personally say which words from the holy book are to be taken as literal, which ones are metaphor or allegory and which ones can be safely ignored."

98. "Science is no more a contributor to arrogance than claiming to know that two plus two equals four demonstrates some sort of arrogance."

99. "Mass destruction of rationality invariably leads to the mass destruction of humanity."

100. "Religion is like a machine gun, in that both are extraordinarily dangerous man-made weapons that can kill a great many people in a very short time. When an irresponsible and/or immoral person or state picks up a machine gun or a religion and points it at other people and starts firing, only horrible things result."

101. "The faithful scream at the top of their lungs that they deserve the right to discriminate against people they don't like because they say that preventing them from discriminating against certain people infringes on their religious freedom."

102. "Persecution and oppression, rather than kindness and forgiveness, seem to invariably be the *sine qua non* of nearly all organized religions, regardless of the size of the sect or the obfuscations of its apologists."

103. "Compared to the level of knowledge we have today, the average person living in the ancient Middle East during biblical times knew about as much about physics, chemistry, psychology and other sciences as did our flint-knapping ancestors of two million years ago. And we should treat their books accordingly."

104. "There is actually far more evidence for Santa Claus than for any god, because on Christmas morning there really are actual presents under the tree. No god has ever delivered anything to humanity."

105. "People often strongly hold their beliefs, no matter how irrational or unjustified, for no reason they can explain. They just "feel that way", which is enough for them to hold that opinion, and that attitude extends to religion. This is summed up in that trite expression "if it feels good, do it", but repurposed to "if it feels good, think it". They then reject evidence that disproves their belief."

106. "Religion makes people dumb and immoral, and it makes societies and nations dumb and immoral. Any system of thought that turns to ignorance as its first problem-solving option is intellectually and morally bankrupt and will always generate far more human misery than human well-being."

107. "Unjustified beliefs have never resulted in good things for humanity."

108. "Part of being a responsible human being is acknowledging that opinions can start a debate but actual facts and actual evidence must always win the debate. This never happens in religion."

109. "If you took the 100 most innovative, intelligent and impactful scientists throughout history and erased them and their contributions, we would almost certainly still be reading scripture by candlelight and many people would still be trying to avoid being burned for witchcraft or heresy."

110. "If you took the 100 most powerful and influential clerics throughout history and removed them and their effects on the world, we might have landed on the moon by the year 1600, cured cancer and most all other afflictions by 1700 and eliminated poverty and human suffering by 1800, not to mention avoiding the uncountable millions of murders that have been carried out by the faithful."

111. "Dogma destroys debate, it destroys truth and it never leads to positive outcomes."

112. "Religion is choosing to answer any as-yet unanswered question and indeed in extreme cases literally everything else with "god did it", a response which is the weakest and most pathetic copout of all time."

113. "Religion is a manmade invention with dogma at its core whose tenets can be tested by science, and found to either be as advertised or not. Science is a manmade invention, but it is wholly free of dogma and its methods and tenets are that of empirical testing. Religion cannot study science or anything else, but science can study religion. That is the core of the irreconcilable difference

between the two, and why humans selectively ignore science when it disagrees with or disproves their closely held beliefs."

114. "Even the slowest mind can see the idiotic hypocrisy so prevalent today, in that we seem to be happy to let people believe what they want to in religion but we don't let them believe they can break laws or make up their own science. Why don't we allow people to pass math and science tests by happily giving them credit for whatever they want to believe are the answers? People aren't questioned about their belief in a "personal god", so why don't we let them believe in a "personal science" or a "personal math", consisting of whatever they feel like believing?"

115. "Religion is a system of magical thinking which scurries about like a cockroach just like its gods do, frightened of the light and hiding in the ever-shrinking shadows science has not yet penetrated and in the highly unfortunate cracks in human psychology."

116. "Science holds the key to human happiness and prosperity. Religion lays the path to division, conflict and sorrow."

117. "Like Bertrand Russel's orbiting teapot, like Carl Sagan's garage-bound dragon, like the Christian heaven, like the Islamic afterlife or like my imaginary adult playhouse for imaginary beings, there is no end to the amount of nonsense a person can make up and pretend to be real simply by saying there is no proof it doesn't exist. Absence of evidence indeed does not always equal evidence of absence, but this only applies when people use reason and are reasonable, which does not occur in religion."

118. "Far too many people are too weak to manage their fear and uncertainty, and choose to respond to their stresses by opening up a bottle of highly distilled religious dogma and drinking themselves into existential oblivion."

119. "Christianity has never been a religion of turning cheeks; it has always been a religion of bloody swords."

120. "The disease of religion multiplies the misery of humanity."

121. "We discovered everything we know solely on our own, we partly hairless semi-evolved godless primates, and we figured them out despite the religions and their holy books, not because of them."

122. "Inconsistency and hypocrisy, both of which are often immoral, form the rope from which the noose of religion is woven."

123. "Ideas can be challenged, but in the mind of the believer the god and the faith cannot be challenged, therefore his imaginary borrowed power will last for as long as he believes in the god, and far more importantly, as long as other people do."

124. "In the spirit of circumcision, it is clear that we should slice away and discard some of the utterly ignorant and immoral practices and beliefs which far too many people share. It is time to cut off the ancient clawed hands which still have a death grip on billions of people, and throw the entire immoral body of religion into the pits of perdition from whence it came."

125. "The book which billions of people claim to use as a moral guide for their lives is an instructional manual on immorality."

126. "The Christian holy book was written by madly bipolar angry schizophrenics, whether or not they were divine or human."

127. "To the secularist, nothing is holy or sacred, particularly religious superstition, and for a bunch of woman-hating sexless semi-evolved mostly hairless aged male primates to gather around ancient mythology and use it to come up with the story that a magic cracker is a reincarnated piece of a two-millennia-dead Israeli demigod preacher who'd already been resurrected once is an invitation in thousand-cubit high flaming letters in the sky for scorn of the highest degree."

128. "Voltaire's timelessly wise words about those who can make a person believe absurdities can make them commit atrocities will echo forever through the tombs of the dead Jews and all the uncountable millions of other people killed in the faithful service of religion."

129. "Any sufficiently credulous or uninformed mind sees the uncommon as miraculous."

130. "The spool of our lives spins very fast, and we know it must end all too soon. Is it any wonder that we fear death and want to live forever after our physical death?"

131. "The biblical Jesus, the one actually described in the book, was a penniless, sandal-wearing, long-haired, turn-the-other-cheek flaming liberal socialist hippie, not the gun-totin' religiously intolerant aggressive Republican bigot that American conservatives have turned him into."

132. "Discussing theology is as pointless and unproductive as imagining a bunch of angels circle-jerking each other on the head of a pin."

133. "The wall of separation between mythology and reality is constructed from blocks of solid logic and reason. We must be building that wall as high and as long as possible, not weakening it by removing whichever blocks some superstitious people want to take out because it makes them feel better or because a 2000 year old book of mythology says they should."

134. "The three monotheisms are an unholy trinity of antihuman savagery."

135. "No matter where science looks, they say, no matter with what instruments or with what discipline, they say, their god can scurry away from the oncoming spotlight like a supernatural supercockroach, never to be glimpsed or found."

136. "We must teach our children the real truth about the real world, which is that all religions are just local fables people used to tell each other to amuse, to entertain, to try to explain and to try to control people. We need to ensure they know that religions have no more factual basis than the superhero stories and characters found in comic books."

137. "Humans are amazing creatures who have unbelievable potential and who can live amazing lives when they are not hamstrung by the perverse and sadistically inculcated psychological millstone of institutionalized religious masochism."

138. "Humanity should not rest until religion becomes like a once-feared disease which we have wiped out through education and

information, in a mental rather than physical campaign of healing and vaccination."

139. "Prayer is wishing that an imaginary magical being will magically do something that you ought to damn well do yourself, or that it will tell you what to do when you damn well ought to figure it out for yourself. At other times prayer is just wishing that this magical being will stop something from happening that you yourself damn well ought to stop from happening."

140. "The wonderful thing about truth and reason is that they will always be there for you, not in the imaginary sense that religions pretend to be there, but in the actual way that reality is always there."

141. "Religion is the only self-described system of conduct on the planet which by its own words and through its own constructs deliberately orders and enables acts of ultimate immorality and evil which are mandated by an ultimate godhead."

142. "No pope is any more infallible than a comparably sized hot air balloon."

143. "The most dangerous people on earth are deeply and sincerely faithful men utterly committed to their religion."

144. "The sharp sword of reason must slice clean through the body of religion, and the two dead halves must be buried once and for all."

145. "Your religious freedom is not an open invitation to be a bigoted fundamentalist asshole. Your Constitutional right to religious freedom does not give you any right whatsoever to tell other people that they are pieces of shit."

146. "And while there is no heaven and no hell, and there will be no Second Coming, if there were I would happily trade my eternal soul for the opportunity to watch the big show when you religious bigots desperately try to explain your hate and bigotry to Jesus while he's picking up that whip of his to beat some human decency into you and asking you "*what the hell goddamn part of 'be nice to people' didn't you understand?*"

Part Ten

References and Further Reading

Notes, Credits and Permissions

I used the King James Version of the Christian bible throughout the book except where specifically noted. I also used the Ahmed Ali translation of the Quran.

Cover Imagery. My extremely talented painter and general creative artist wife translated the vision and tone of the "Thank god for Eve" story in Chapter 24 into the main image on the front cover and its carryover onto the back cover.

Chapter One.

1. Article at History.com titled "*Women Who Fought for the Vote*", written by History.com authors, last updated August 21, 2018. Web resource at https://www.history.com/topics/womens-history/women-who-fought-for-the-vote-1
2. Senator Mitch McConnell states why Republicans voted to silence Senator Elizabeth Warren on February 7, 2017. YouTube video at https://www.youtube.com/watch?v=m2YY9_NfW8w
3. Creationist Ray Comfort attempts to use a banana to justify his preferred god's existence. YouTube video at https://www.youtube.com/watch?v=nfv-Qn1M58I
4. Ouroboros is a mythological snake which eats its own tail. Religions are terminally circular in their logic, thus the reference. Web resource at https://en.wikipedia.org/wiki/Ouroboros

Chapter Two.

1. Pew Research Center study, released May 12, 2015, titled "America's Changing Religious Landscape." Web resource at http://assets.pewresearch.org/wp-content/uploads/sites/11/2015/05/RLS-08-26-full-report.pdf
2. Apollonius of Tyana; web resource at https://en.wikipedia.org/wiki/Apollonius_of_Tyana
3. Herring Amy H, Attard Samantha M, Gordon-Larsen Penny, Joyner William H, Halpern Carolyn T. Like a virgin (mother): analysis of data from a longitudinal, US population representative sample survey BMJ 2013; 347 :f7102. Web resource at https://www.bmj.com/content/347/bmj.f7102
4. Pew Research Center publication titled "*U.S. Religious Knowledge Survey*", dated September 28, 2010. Web resource at http://www.pewforum.org/2010/09/28/u-s-religious-knowledge-survey/
5. National Science Foundation study on American knowledge of science and engineering. National Science Board. 2018. *Science and Engineering Indicators 2018*. Alexandria, VA: National Science Foundation (NSB-2018-1). Web resource at https://www.nsf.gov/statistics/2018/nsb20181/report
6. Quote is alleged to be from Thomas Jefferson but whether he used the actual words is unproven. Web resource at https://www.monticello.org/site/jefferson/educated-citizenry-vital-requisite-our-survival-free-people-spurious-quotation
7. Web resource at https://www.merriam-webster.com/dictionary/non%20sequitur

Chapter Three.

1. Web resource at https://www.merriam-webster.com/dictionary/logic

2. Web resource at https://www.merriam-webster.com/dictionary/inference
3. Web resource at https://www.merriam-webster.com/dictionary/rational
4. Web resource at https://www.merriam-webster.com/dictionary/reason
5. Web resource at https://www.merriam-webster.com/dictionary/statistics
6. Web resource at https://www.merriam-webster.com/dictionary/probable
7. Web resource at https://www.merriam-webster.com/dictionary/probability
8. Web resource at https://www.merriam-webster.com/dictionary/proof
9. Web resource at https://www.merriam-webster.com/dictionary/evidence
10. Dartmouth handout on the Scientific Method. Web resource at https://www.dartmouth.edu/~bio31/01a.SciMethod_Handout.pdf
11. Web resource at https://en.wikipedia.org/wiki/Statistical_significance
12. Web resource at https://www.nolo.com/dictionary/reasonable-doubt-term.html
13. Web resource at https://www.law.cornell.edu/wex/clear_and_convincing_evidence
14. Web resource at https://www.law.cornell.edu/wex/preponderance_of_the_evidence
15. Web resource at https://www.merriam-webster.com/dictionary/evidence - legalDictionary
16. Web resource at https://www.law.cornell.edu/wex/burden_of_proof

Chapter Four. From the January 2002 Pennsylvania Department of Education's Academic Standards for Science and Technology. Web resource at https://www.stateboard.education.pa.gov/Documents/Regulations and Statements/State Academic Standards/ScienceandTechnologyStandards.pdf

Chapter Five. No external references.

Chapter Six.

1. Smithsonian.com article on "*Pilgrims' Progress*". Web resource at https://www.smithsonianmag.com/history/pilgrims-progress-135067108/
2. Web resource at https://www.britannica.com/event/Enlightenment-European-history
3. Web resource at https://www.britannica.com/event/Salem-witch-trials
4. Web resource at https://en.wikipedia.org/wiki/First_Great_Awakening
5. Transcript of the Declaration of Independence. Web resource at https://www.archives.gov/founding-docs/declaration-transcript
6. Transcript of the US Constitution. Web resource at https://www.archives.gov/founding-docs/constitution-transcript
7. Web resource at https://en.wikipedia.org/wiki/Second_Great_Awakening
8. Constitution of the United States. Web resource at https://www.senate.gov/civics/constitution_item/constitution.htm
9. Yale Law School Avalon Project, "*The Barbary Treaties 1786-1816; Treaty of Peace and Friendship, Signed at Tripoli November 4, 1796*". Web resource at http://avalon.law.yale.edu/18th_century/bar1796t.asp
10. Web resource at https://en.wikipedia.org/wiki/Burned-over_district
11. Web resource at https://en.wikipedia.org/wiki/Dispensationalism
12. Web resource at https://en.wikipedia.org/wiki/Joseph_Smith_and_the_criminal_justice_system
13. Web resource at https://en.wikipedia.org/wiki/Mormonism
14. Web resource at https://en.wikipedia.org/wiki/Millerism
15. Web resource at https://en.wikipedia.org/wiki/Great_Disappointment

16. Web resource at https://en.wikipedia.org/wiki/Spiritualism
17. Web resource at https://en.wikipedia.org/wiki/Third_Great_Awakening
18. National Park Service information on casualties in the Civil War. Web resource at https://www.nps.gov/nr/travel/national_cemeteries/death.html
19. Web resource at https://en.wikipedia.org/wiki/Seventh-day_Adventist_Church
20. The 13th Amendment to the U.S. Constitution: Abolition of Slavery. Web resource at https://www.ourdocuments.gov/doc.php?flash=false&doc=40
21. Article on the Ku Klux Klan. Web resource at http://www.pbs.org/wgbh/americanexperience/features/klansville-faq/
22. The Constitution of the United States. Web resource at https://www.senate.gov/civics/constitution_item/constitution.htm
23. Web resource at https://www.britannica.com/event/Comstock-Act
24. Web resource at https://www.britannica.com/event/Jim-Crow-law
25. Cornell Law School's Legal Information Institute information on the 1896 Plessy v. Ferguson Supreme Court case. Web resource at https://www.law.cornell.edu/supremecourt/text/163/537
26. Public Broadcasting Service information on Margaret Sanger. Web resource at https://www.pbs.org/wgbh/americanexperience/features/pill-margaret-sanger-1879-1966/
27. Centers for Disease Control and prevention information on the 1918 flu pandemic. Web resource at https://www.cdc.gov/features/1918-flu-pandemic/index.html
28. Public Broadcasting Service information on Prohibition. Web resource at http://www.pbs.org/kenburns/prohibition/
29. The Constitution of the United States. Web resource at https://www.senate.gov/civics/constitution_item/constitution.htm
30. National Center for Science Education information on the Butler Act; site shows the Butler act in image form. Web resource at https://ncse.com/files/pub/legal/Scopes/Butler_Act.pdf
31. Web resource at https://www.britannica.com/event/Scopes-Trial
32. Article titled "*The Ban on the Teaching of Evolution Reaches the U.S. Supreme Court*", the third in a series on "Creationism in the United States". Article written by Randy Moore, found in *The American Biology Teacher*, Volume 60, No. 9, November/December 1998. Web resource at Mississippi and Arkansas enact Butler Act-like anti-evolution laws
33. Web resource at https://en.wikipedia.org/wiki/Christian_nationalism - United_States
34. Web resource at https://en.wikipedia.org/wiki/Operation_Paperclip
35. Article from the University of Virginia's Miller Center titled "*McCarthyism and the Red Scare*". Web resource at https://millercenter.org/the-presidency/educational-resources/age-of-eisenhower/mcarthyism-red-scare
36. University of Southern California article titled "*The 1950s – Powerful Years for Religion*", written by Carol Tucker, dated June 16, 1997. Web resource at https://news.usc.edu/25835/The-1950s-Powerful-Years-for-Religion/
37. Web resource at https://en.wikipedia.org/wiki/Scientology
38. Cornell Law School's Legal Information Institute information on the 1654 Brown v. Board of Education Supreme Court case. Web resource at https://www.law.cornell.edu/supremecourt/text/347/483
39. Web resource at https://www.britannica.com/event/Pledge-of-Allegiance-to-the-Flag-of-the-United-States-of-America
40. Web resource at https://en.wikipedia.org/wiki/Timeline_of_the_civil_rights_movement

41. Article on Smithsonian.com titled "*The Space Race*", written by Michael Kernan for Smithsonian Magazine, dated August 1997. Web resource at https://www.smithsonianmag.com/history/the-space-race-141404095/
42. Web resource at https://en.wikipedia.org/wiki/In_God_We_Trust
43. Web resource at https://en.wikipedia.org/wiki/In_God_We_Trust
44. Article on NASA.gov titled "*Sputnik and The Dawn of the Space Age*", updated October 10, 2007 by Steve Garber, NASA History Web Curator. Web resource at https://history.nasa.gov/sputnik/
45. Web resource at https://en.wikipedia.org/wiki/Sputnik_1
46. Web resource at https://en.wikipedia.org/wiki/Timeline_of_the_civil_rights_movement
47. Web resource at https://en.wikipedia.org/wiki/Fourth_Great_Awakening
48. Web resource at https://en.wikipedia.org/wiki/Timeline_of_the_civil_rights_movement
49. Web resource at https://en.wikipedia.org/wiki/Timeline_of_the_civil_rights_movement
50. Web resource at https://en.wikipedia.org/wiki/Christian_reconstructionism
51. Web resource at https://en.wikipedia.org/wiki/Timeline_of_the_civil_rights_movement
52. Web resource at https://en.wikipedia.org/wiki/Epperson_v._Arkansas
53. Web resource at https://en.wikipedia.org/wiki/McLean_v._Arkansas
54. Cornell Law School's Legal Information Institute information on the 1987 Edwards v. Aguillard Supreme Court case. Web resource at https://www.law.cornell.edu/supremecourt/text/482/578
55. University of California, Berkeley article titled "*Intelligent Design: Is it scientific*?" Web resource at https://undsci.berkeley.edu/article/id_checklist
56. Article in *The Nation* online magazine, titled "*Bush's War on Science*", written by Katrina vanden Heuvel, dated July 20, 2004. Web resource at https://www.thenation.com/article/bushs-war-science/
57. Council on Foreign Relations article, titled "*The Iraq Invasion Ten Years Later: A Wrong War*", an interview. Interviewer Bernard Gwertzman, interviewee Richard N. Haass, dated March 13, 2013. Web resource at https://www.cfr.org/interview/iraq-invasion-ten-years-later-wrong-war
58. Article in *MIT Technology Review* online, titled "*The Anti-Science President*", by David Ewing Duncan, dated July 12, 2007. Web resource at https://www.technologyreview.com/s/408236/the-anti-science-president/
59. National Center for Science Education article titled "*Kitzmiller v. Dover: Intelligent Design on Trial*". Web resource at https://ncse.com/library-resource/kitzmiller-v-dover-intelligent-design-trial
60. White House statement titled "*IMPACT REPORT: 100 Examples of President Obama's Leadership in Science, Technology, and Innovation*", from the Office of the Press Secretary, dated June 21, 2016. Web resource at https://obamawhitehouse.archives.gov/the-press-office/2016/06/21/impact-report-100-examples-president-obamas-leadership-science
61. Article in *The Scientist*, titled "*TS Picks: Obama's Science Legacy*", by Joshua A. Krisch, dated January 17, 2017. Web resource at https://www.the-scientist.com/the-nutshell/ts-picks-obamas-science-legacy-32194
62. The 2012 Republican Party of Texas Report of Platform Committee and Rules Committee. Web resource at https://www.empowertexans.com/wp-content/uploads/2012/07/2012-GOP-Platform-Final.pdf
63. Article in *Scientific American* titled "*The Trump Administration's War on Science Agencies Threatens the Nation's Health and Safety*", by Andrew A. Rosenberg

and Kathleen Rest, dated January 1, 2018. Web resource at https://www.scientificamerican.com/article/the-trump-administration-rsquo-s-war-on-science-agencies-threatens-the-nation-rsquo-s-health-and-safety/

64. British Broadcasting Corporation article titled "*Jerusalem embassy: Why Trump's move was not about peace*", by Barbara Plett Usher, dated May 15, 2018. Web resource at https://www.bbc.com/news/world-us-canada-44120428

Chapter Seven.

1. The Executive Summary of the Irish Commission to Inquire into Child Abuse (CICA) is available at http://www.childabusecommission.ie/rpt/pdfs/CICA-Executive Summary.pdf
2. News reference to the Pennsylvania grand jury report on Catholic clergy sex abuse from "The Inquirer" article titled "*Catholic Church clergy sex abuse: Read the full grand jury report*", by Nathaniel Lash, updated August 14, 2018. Web resource at http://www2.philly.com/philly/news/catholic-church-clergy-sex-abuse-read-the-full-grand-jury-report-20180814.html

Chapter Eight.

1. News reference to Donald Trump's admission of making things up to suit his narrative. Article from The Guardian titled "*Donald Trump admits making up 'facts' in trade meeting with Justin Trudea*u", by David Smith, dated March 15, 2018. Web resource at https://www.theguardian.com/us-news/2018/mar/15/donald-trump-admits-made-up-facts-justin-trudeau
2. Report of Donald Trump's perspective on facts and reality. Article titled "*Billy Bush on Trump's lies about 'Apprentice' ratings: 'You just tell them and they believe*", at usatoday.com, by Jayme Deerweister, published March 17, 2018. Web resource at https://www.usatoday.com/story/life/tv/2018/03/17/billy-bush-continues-his-redemption-tour-talks-trump-real-time-bill-maher/434927002/
3. Reference to a cosmic or world egg. Web resource at https://en.wikipedia.org/wiki/World_egg
4. Reference to a World Turtle. Web resource at https://en.wikipedia.org/wiki/Turtles_all_the_way_down

Chapter Nine.

1. NASA.gov article on "*How Do We Weigh Planets?*", NASA Official Kristen Erickson, Program Manager Heather Doyle, updated November 16, 2018. Web resource at https://spaceplace.nasa.gov/planets-weight/en/
2. Article in *American Scientist* titled "*Why Is Human Childbirth So Painful?*" by Pat Lee Shipman, from the November-December 2013 issue, Volume 101, number 6, page 426. Web resource at https://www.americanscientist.org/article/why-is-human-childbirth-so-painful
3. Article at Palomar.edu titled "*Humans*", by Dennis O'Neil, copyright 1998-2012. Web resource at https://www2.palomar.edu/anthro/primate/prim_8.htm
4. Pregnancy and childbirth are dangerous to many women, more so in developing nations. World Health Organization article titled "*Maternal mortality*", dated 16 February 2018. Web resource at http://www.who.int/en/news-room/fact-sheets/detail/maternal-mortality
5. The famous "Vast Wasteland" speech from 1961. Web resource at https://www.youtube.com/watch?v=9dGRgLfaGwo

Chapter Ten.

1. Several misused or misinterpreted terms in science. Article in *Scientific American* titled "*"Just a Theory": 7 Misused Science Words*", by Tia Ghose, LiveScience, dated April 2, 2013. Web resource at https://www.scientificamerican.com/article/just-a-theory-7-misused-science-words/
2. Article in *MIT News* on animal dreams titled "*Animals have complex dreams, MIT researcher proves*", by the MIT News Office, dated January 24, 2001. Web resource at http://news.mit.edu/2001/dreaming
3. Article in *Psychology Today* titled "*Do Animals Dream? Science Shows Of Course They Do, Rats Too*", by Marc Bekoff, Ph.D., posted December 4, 2012. Web resource at https://www.psychologytoday.com/us/blog/animal-emotions/201212/do-animals-dream-science-shows-course-they-do-rats-too
4. Article at Reducing Suffering.org titled "*How Many Wild Animals Are There?*" by Brian Tomasik, first published 2009 and last edited 3 June 2018. Web resource at https://reducing-suffering.org/how-many-wild-animals-are-there/ - Mammals
5. Charter schools often deliberately teach religious nonsense rather than actual science. Article at Slate.com titled "*Texas Public Schools Are Teaching Creationism*", by Zack Kopplin, date January 16, 2014. Web resource at https://slate.com/technology/2014/01/creationism-in-texas-public-schools-undermining-the-charter-movement.html
6. Voucher schools teach creationism. Article on PBS.com titled "*Hundreds Of Voucher Schools Teach Creationism In Science Classes*", by Maria Goodavage, dated January 29, 2013. Web resource at http://www.pbs.org/independentlens/blog/hundreds-of-voucher-schools-teach-creationism-in-science-classes/
7. Even those who partially accept science are challenged when trying to find nonfundamentalist home schooling materials. Article in *The Atlantic* titled "*Old Earth, Young Minds: Evangelical Homeschoolers Embrace Evolution*", by David R. Wheeler, dated March 8, 2013. Web resource at https://www.theatlantic.com/national/archive/2013/03/old-earth-young-minds-evangelical-homeschoolers-embrace-evolution/273844/

Chapter Eleven.

1. Hájek, Alan, "Pascal's Wager", *The Stanford Encyclopedia of Philosophy* (Summer 2018 Edition), Edward N. Zalta (ed.). First published May 2, 1998; substantive revision September 1, 2017. Web resource at https://plato.stanford.edu/entries/pascal-wager/
2. Web resource at http://www.adherents.com/
3. Article at religioustolerance.org titled "*Numbers of adherents of major religions, their geographical distribution, date founded, and sacred texts*", by B.A. Robinson, copyright 1997 to 2015 by Ontario Consultants on Religious Tolerance, last update December 2, 2015. Web resource at http://www.religioustolerance.org/worldrel.htm
4. Web resource at https://www.britannica.com/topic/defense-mechanism#ref195659
5. Web resource at https://www.britannica.com/topic/defense-mechanism - ref195659
6. Web resource at https://www.britannica.com/science/eugenics-genetics
7. Web resource at https://www.britannica.com/topic/inquisition
8. Web resource at https://en.wikipedia.org/wiki/Inquisition

9. Pew Research Center release titled "*The Global Religious Landscape*", December 18, 2012. Web resource at http://www.pewforum.org/2012/12/18/global-religious-landscape-exec/
10. Catechism of the Catholic Church. Web resource at http://www.vatican.va/archive/ccc_css/archive/catechism/p2s2c1a3.htm

Chapter Twelve.

1. Video clip from *The Sixth Sense*. Web resource at https://www.youtube.com/watch?v=QUYKSWQmkrg
2. Pew Research Center publication titled "*Christians remain world's largest religious group, but they are declining in Europe*", by Conrad Hackett and David McClendon, published April 5, 2017. Web resource at http://www.pewresearch.org/fact-tank/2017/04/05/christians-remain-worlds-largest-religious-group-but-they-are-declining-in-europe/
3. Pew Research Center publication titled "*The Global Catholic Population*", dated February 13, 2013. Web resource at http://www.pewforum.org/2013/02/13/the-global-catholic-population/ - how-has-the-population-changed-in-heavily-catholic-countries-in-the-past-decade
4. Time Labs publication titled "*See The Change In Catholic Population Around The World*", dated September 21, 2015. Web resource at http://labs.time.com/story/catholic-population/
5. Sloman, Steven and Fernbach, Philip. *The Knowledge Illusion: Why We Never Think Alone*. New York: Riverhead Books, 2018.
6. *Time Magazine* Interview with Clint Eastwood, article contacts Ty Trippet and Nicole Willner, dated February 20, 2005. Web resource at http://content.time.com/time/press_releases/article/0,8599,1029815,00.html
7. Web resource at https://www.britannica.com/science/DNA
8. Oklahoma State University's Department of Agricultural Economics Food Demand Survey. Web resource at http://www.agecon.okstate.edu/agecon_research.asp
9. Oklahoma State University's Department of Agricultural Economics Food Demand Survey, Volume 2, Issue 9: January 16, 2015. Web resource at http://agecon.okstate.edu/faculty/publications/4975.pdf
10. YouTube video clip from the cartoon series *Futurama*, depicting the refusal of the religious community to accept science, in this case the facts of evolution. Web resource at https://www.youtube.com/watch?v=q-RUHhCzgxI
11. YouTube video clip from the cartoon series *Futurama*, depicting the extreme irritation with the reality-denial of the religious community that is shared by many in the empiricist and scientist communities Web resource at https://www.youtube.com/watch?v=RaY9k6tuLog
12. Video of Oklahoma Senator James Inhofe demonstrating his complete scientific ignorance and dogmatism on the floor of the US Senate. Web resource at https://www.youtube.com/watch?v=4J4JtMxCJ94
13. Dartmouth Medical School *Review of Clinical and Functional Neuroscience*, Chapter 12, copyright/site editor Rand Swenson, DC, MD, PhD, 2006. Web resource at https://www.dartmouth.edu/~rswenson/NeuroSci/chapter_12.html
14. Hladky, Stephen B. and Margery A. Barrand. "Mechanisms of fluid movement into, through and out of the brain: evaluation of the evidence" *Fluids and barriers of the CNS* vol. 11,1 26. 2 Dec. 2014, doi:10.1186/2045-8118-11-26. Web resource at https://www.ncbi.nlm.nih.gov/pmc/articles/PMC4326185/

15. Laterra J, Keep R., Betz LA, et al. "Blood—Cerebrospinal Fluid Barrier". In: Siegel GJ, Agranoff BW, Albers RW, et al., editors. *Basic Neurochemistry*: Molecular, Cellular and Medical Aspects. 6th edition. Philadelphia: Lippincott-Raven; 1999. Web resource at https://www.ncbi.nlm.nih.gov/books/NBK27998/
16. Stiles, Joan and Terry L. Jernigan. "The basics of brain development" *Neuropsychology review* vol. 20,4 (2010): 327-48. Web resource at https://www.ncbi.nlm.nih.gov/pmc/articles/PMC2989000/
17. Emory University School of Medicine article titled "*Understanding How the Brain Develops*", by Maureen Mulligan LaRossa, R.N. and Sheena L. Carter, Ph.D. Web resource at http://www.pediatrics.emory.edu/divisions/neonatology/dpc/brain.html
18. Wehrwein, Hakan S., Orer, HS and Barman, Susan M. "Overview of the Anatomy, Physiology, and Pharmacology of the Autonomic Nervous System". In *Compr Physiol.* 2016; 6(3): 1239–1278. Web resource at https://www.ncbi.nlm.nih.gov/pubmed/27347892
19. Web resource at https://www.merriam-webster.com/dictionary/reflex
20. Stanford Children's Health article titled *Newborn-Reflexes*. Web resource at https://www.stanfordchildrens.org/en/topic/default?id=newborn-reflexes-90-P02630
21. Web resource at https://www.merriam-webster.com/dictionary/instinct
22. Link to National Public Radio's *Hidden Brain* main web page. Web resource at https://www.npr.org/series/423302056/hidden-brain
23. Web resource at https://www.amazon.com/Thinking-Fast-Slow-Daniel-Kahneman/dp/0374533555
24. Web resource at https://www.amazon.com/Righteous-Mind-Divided-Politics-Religion-ebook/dp/B0052FF7YM
25. Kahneman, Daniel. *Thinking, Fast and Slow*. New York: Farrar, Straus and Giroux, 2013, 20.
26. Web resource at https://en.wikipedia.org/wiki/M%C3%BCller-Lyer_illusion
27. Kahneman, Daniel. *Thinking, Fast and Slow*, 52.
28. Ibid., 62.
29. Ibid., 114.
30. Ibid., 158.
31. Ibid., 240.
32. Ibid., 284.
33. National Aeronautics and Space Administration article titled "*Will The Universe Expand Forever?*" by NASA/WMAP Science Team, dated January 24, 2014. Web resource at https://map.gsfc.nasa.gov/universe/uni_shape.html
34. Haidt, Jonathan. *The Righteous Mind: Why Good People Are Divided By Politics And Religion*. New York: Vintage Books, 2012, 3.
35. Ibid., 4.
36. Ibid., 22.
37. Ibid., 45.
38. Ibid., 45.
39. Web resource at https://en.wikipedia.org/wiki/Anchoring
40. Web resource at https://en.wikipedia.org/wiki/Anthropocentrism - Cognitive_psychology
41. Web resource at https://www.britannica.com/topic/anthropomorphism
42. Web resource at https://en.wikipedia.org/wiki/Attentional_bias
43. Web resource at https://en.wikipedia.org/wiki/Authority_bias
44. Web resource at https://en.wikipedia.org/wiki/Availability_cascade

45. Web resource at https://www.goodreads.com/quotes/7676946-history-is-strewn-thick-with-evidence-that-a-truth-is
46. *Scientific American* article titled "*How to Convince Someone When Facts Fail*", by Michael Shermer, date January 1, 2017. Web resource at https://www.scientificamerican.com/article/how-to-convince-someone-when-facts-fail/
47. Web resource at https://en.wikipedia.org/wiki/Bandwagon_effect
48. Web resource at https://en.wikipedia.org/wiki/Belief_bias
49. Web resource at https://en.wikipedia.org/wiki/Bias_blind_spot
50. Web resource at https://en.wikipedia.org/wiki/Choice-supportive_bias
51. Web resource at https://en.wikipedia.org/wiki/Clustering_illusion
52. Web resource at https://en.wikipedia.org/wiki/Confirmation_bias
53. Web resource at https://en.wikipedia.org/wiki/Curse_of_knowledge
54. Web resource at https://en.wikipedia.org/wiki/Dunning%E2%80%93Kruger_effect
55. Web resource at https://en.wikipedia.org/wiki/Egocentric_bias
56. Web resource at https://en.wikipedia.org/wiki/False_consensus_effect
57. Web resource at https://www.britannica.com/science/Barnum-Effect
58. Web resource at https://en.wikipedia.org/wiki/Framing_effect_(psychology)
59. The University of Texas at Austin McCombs School of Business article titled "*Fundamental Attribution Error*", with video, copyright 2018. Web resource at https://ethicsunwrapped.utexas.edu/glossary/fundamental-attribution-error
60. Web resource at https://www.britannica.com/topic/hindsight-bias
61. Gilbert, E.T., Brown, R.P., Pinel, E.C. and Wilson, T.D. "The illusion of external agency". *J Pers Soc Psychol*. 2000 Nov; 79(5): 690–700. Web resource at https://www.ncbi.nlm.nih.gov/pubmed/11079235
62. Web resource at https://en.wikipedia.org/wiki/Illusory_truth_effect
63. Web resource at https://en.wikipedia.org/wiki/Illusory_correlation
64. Article in *Psychology Today* titled "*In-groups, out-groups, and the psychology of crowds*", by Susan Krauss Whitbourne, Ph.D., dated December 7, 2010. Web resource at https://www.psychologytoday.com/us/blog/fulfillment-any-age/201012/in-groups-out-groups-and-the-psychology-crowds
65. Santa Clara University's Markkula Center for Applied Ethics article titled "*The Just World Theory*", by Claire Andre and Manuel Velasquez, posted November 13, 2015; originally published in *Issues in Ethics* - V. 3, N. 2 Spring 1990. Web resource at https://www.scu.edu/ethics/ethics-resources/ethical-decision-making/the-just-world-theory/
66. Web resource at https://en.wikipedia.org/wiki/Law_of_the_instrument
67. Article in *Psychology Today* titled "*To know me is to like me I: Mere exposure*", by Art Markman, Pd.D., dated November 20, 2008. Web resource at https://www.psychologytoday.com/us/blog/ulterior-motives/200811/know-me-is-me-i-mere-exposure
68. Web resource at https://en.wikipedia.org/wiki/Na%C3%AFve_realism_(psychology)
69. Web resource at https://en.wikipedia.org/wiki/Ostrich_effect
70. Web resource at https://en.wikipedia.org/wiki/Overconfidence_effect
71. Web resource at https://www.merriam-webster.com/dictionary/pareidolia
72. Article in *Scientific American* titled "*The Believing Brain: Why Science Is the Only Way Out of Belief-Dependent Realism*", by Michael Shermer, dated July 1, 2011. Web resource at https://www.scientificamerican.com/article/the-believing-brain/

73. Web resource at https://www.merriam-webster.com/dictionary/post hoc, ergo propter hoc
74. Web resource at https://en.wikipedia.org/wiki/Choice-supportive_bias
75. Web resource at https://en.wikipedia.org/wiki/Selective_perception
76. Article in *Psychology Today* titled "*Selective Perception and Attention at the World Cup*", by Alan Castel, Ph.D., dated June 26, 2014. Web resource at https://www.psychologytoday.com/us/blog/metacognition-and-the-mind/201406/selective-perception-and-attention-the-world-cup
77. Article in *Psychology Today* titled "*The Self-Serving Bias - Definition, Research, and Antidotes*", by Alice Boyes, Ph.D., dated January 9, 2013. Web resource at https://www.psychologytoday.com/us/blog/in-practice/201301/the-self-serving-bias-definition-research-and-antidotes
78. Web resource at https://en.wikipedia.org/wiki/Semmelweis_reflex
79. Web resource at https://en.wikipedia.org/wiki/Shared_information_bias
80. Web resource at https://en.wikipedia.org/wiki/Status_quo_bias
81. Web resource at https://en.wikipedia.org/wiki/Subjective_validation
82. Web resource at https://dictionary.cambridge.org/us/dictionary/english/sunk-cost-fallacy
83. Article in *Psychology Today* titled "*Letting Go of Sunk Costs*", by Robert L. Leahy, Ph. D., dated September 24, 2014. Web resource at https://www.psychologytoday.com/us/blog/anxiety-files/201409/letting-go-sunk-costs
84. Web resource at https://en.wikipedia.org/wiki/Survivorship_bias
85. Article at PLOS One titled "*Shared Brain Lateralization Patterns in Language and Acheulean Stone Tool Production: A Functional Transcranial Doppler Ultrasound Study*", by Natalie Thais Uomini and Georg Friedrich Meyer, dated August 30, 2013. Web resource at https://journals.plos.org/plosone/article?id=10.1371/journal.pone.0072693
86. Smithsonian Institution National Museum of Natural History article titled "*What does it mean to be human*?", last updated August 24, 2018. Web resource at http://humanorigins.si.edu/evidence/human-fossils/species/homo-sapiens
87. National Center for Biotechnology Information. *PubChem Compound Database*; CID=681, https://pubchem.ncbi.nlm.nih.gov/compound/681 (accessed Nov. 23, 2018). Web resource at https://pubchem.ncbi.nlm.nih.gov/compound/dopamine - section=Top
88. Sathyanarayana Rao, T S et al. "The biochemistry of belief" *Indian journal of psychiatry* vol. 51,4 (2009): 239-41. Web resource at https://www.ncbi.nlm.nih.gov/pmc/articles/PMC2802367/
89. Schmack, Katharina et al. "Linking unfounded beliefs to genetic dopamine availability" *Frontiers in human neuroscience* vol. 9 521. 30 Sep. 2015, doi:10.3389/fnhum.2015.00521. Web resource at https://www.ncbi.nlm.nih.gov/pmc/articles/PMC4588007/
90. Krummenacher P, Mohr C, Haker H, Brugger P., "Dopamine, paranormal belief, and the detection of meaningful stimuli". *J Cogn Neurosci.* 2010 Aug;22(8):1670-81. doi: 10.1162/jocn.2009.21313. Web resource at https://www.ncbi.nlm.nih.gov/pubmed/19642883/
91. Article in *Psychology Today* titled "*Your Sixth Sense*", by Matthew Hutson, published July 3, 2012. Web resource at https://www.psychologytoday.com/us/articles/201207/your-sixth-sense
92. Web resource at https://en.wikipedia.org/wiki/Days_of_Prayer_for_Rain_in_the_State_of_Texas

93. Article in *Scientific American* titled "*Fossil Reveals What Last Common Ancestor of Humans and Apes Looked Like*", by Charles Q. Choi, LiveScience, dated August 10, 2017. Web resource at https://www.scientificamerican.com/article/fossil-reveals-what-last-common-ancestor-of-humans-and-apes-looked-liked/
94. Web resource at https://en.wikipedia.org/wiki/Moore%27s_law
95. Article in *Popular Science* titled "*The incredible evolution of supercomputers' powers, from 1946 to today*", by Sara Chodosh, dated April 22, 2017. Web resource at https://www.popsci.com/supercomputers-then-and-now
96. Morgane PJ1, Austin-LaFrance R, Bronzino J, Tonkiss J, Díaz-Cintra S, Cintra L, Kemper T, Galler JR., "Prenatal malnutrition and development of the brain". *Neurosci Biobehav Rev.* 1993 Spring;17(1):91-128. Web resource at https://www.ncbi.nlm.nih.gov/pubmed/8455820
97. From the Emory University School of Medicine, Department of Psychiatry and Behavioral Sciences, article by the Maternal Substance Abuse and Child Development Project titled "*The effects of maternal stress and anxiety during pregnancy*", point of contact is Katrina C. Johnson, Ph.D. Web resource at http://www.psychiatry.emory.edu/PROGRAMS/GADrug/Feature Articles/Mothers/The effects of maternal stress and anxiety during pregnancy (mot07).pdf
98. Kinsella, Michael T and Catherine Monk. "Impact of maternal stress, depression and anxiety on fetal neurobehavioral development" *Clinical obstetrics and gynecology* vol. 52,3 (2009): 425-40. Web resource at https://www.ncbi.nlm.nih.gov/pmc/articles/PMC3710585/
99. Bublitz, Margaret H and Laura R Stroud. "Maternal smoking during pregnancy and offspring brain structure and function: review and agenda for future research" *Nicotine & tobacco research: official journal of the Society for Research on Nicotine and Tobacco* vol. 14,4 (2011): 388-97. Web resource at https://www.ncbi.nlm.nih.gov/pmc/articles/PMC3313781/
100. Posecion, N et al. "Detection of Exposure to Environmental Pesticides During Pregnancy by the Analysis of Maternal Hair Using GC-MS" *Chromatographia* vol. 64,11-12 (2006): 681-687. Web resource at https://www.ncbi.nlm.nih.gov/pmc/articles/PMC1934610/
101. Mayo Clinic article titled "*Fetal alcohol syndrome*", by the Mayo Clinic Staff, dated January 10, 2018. Web resource at https://www.mayoclinic.org/diseases-conditions/fetal-alcohol-syndrome/symptoms-causes/syc-20352901
102. Article by the Division of Birth Defects and Developmental Disabilities, NCBDDD, Centers for Disease Control and Prevention titled "*Facts about Anencephaly*", page last updated on August 2, 2017. Web resource at https://www.cdc.gov/ncbddd/birthdefects/anencephaly.html
103. Article by the National Center for Emerging and Zoonotic Infectious Diseases (NCEZID), Division of Vector-Borne Diseases (DVBD), Centers for Disease Control and Prevention, titled "*Pregnancy*", page last updated November 6, 2017. Web resource at https://www.cdc.gov/zika/pregnancy/index.html
104. Article by the National Center for Emerging and Zoonotic Infectious Diseases (NCEZID), Division of Vector-Borne Diseases (DVBD), Centers for Disease Control and Prevention, titled "*Microcephaly & Other Birth Defects*", page last updated March 1, 2018. Web resource at https://www.cdc.gov/zika/healtheffects/birth_defects.html
105. Article by the University of Akron, from the Phineas Gage information page, titled "*The Phineas Gage story*", created by Malcolm Macmillan of the School of

Psychology, Deakin University; School of Psychological Sciences, University of Melbourne, Victoria. Australia and now maintained by The Center for the History of Psychology at The University of Akron, Ohio. Web resource at https://www.uakron.edu/gage/story.dot

106. Article from *Smithsonian Magazine* titled "*Phineas Gage: Neuroscience's Most Famous Patient*", by Steve Twomey, January 2010. Web resource at https://www.smithsonianmag.com/history/phineas-gage-neurosciences-most-famous-patient-11390067/
107. Article from National Public Radio titled "*Why Brain Scientists Are Still Obsessed With The Curious Case Of Phineas Gage*", by Jon Hamilton, dated May 21, 2017. Web resource at https://www.npr.org/sections/health-shots/2017/05/21/528966102/why-brain-scientists-are-still-obsessed-with-the-curious-case-of-phineas-gage
108. Web resource at https://www.merriam-webster.com/dictionary/hallucination
109. McGrath JJ, Saha S, Al-Hamzawi A, et al. Psychotic Experiences in the General Population A Cross-National Analysis Based on 31 261 Respondents From 18 Countries. *JAMA Psychiatry*. 2015;72(7):697–705. doi:10.1001/jamapsychiatry.2015.0575. Web resource at https://jamanetwork.com/journals/jamapsychiatry/fullarticle/2298236
110. Teeple, Ryan C et al. "Visual hallucinations: differential diagnosis and treatment" *Primary care companion to the Journal of clinical psychiatry* vol. 11,1 (2009): 26-32. Web resource at https://www.ncbi.nlm.nih.gov/pmc/articles/PMC2660156/
111. Ibid.
112. Mayo Clinic article titled "*Lewy body dementia*", by the Mayo Clinic Staff. Web resource at https://www.mayoclinic.org/diseases-conditions/lewy-body-dementia/symptoms-causes/syc-20352025
113. Article from the United Kingdom's National Health Service, titled "*Charles Bonnet syndrome*", page last reviewed February 15, 2016. Web resource at https://www.nhs.uk/conditions/charles-bonnet-syndrome/
114. ScienceDirect resource "Anton-Babinski syndrome". Web resource at https://www.sciencedirect.com/topics/medicine-and-dentistry/antonbabinski-syndrome
115. Web resource at https://en.wikipedia.org/wiki/Peduncular_hallucinosis
116. Sahoo, Swapnajeet et al. "Acute Hyperglycemia Associated with Psychotic Symptoms in a Patient with Type 1 Diabetes Mellitus: A Case Report" *Innovations in clinical neuroscience* vol. 13,11-12 25-27. 1 Dec. 2016. Web resource at https://www.ncbi.nlm.nih.gov/pmc/articles/PMC5300708/
117. Article at Huffpost.com titled "*Not Quite Enough: The Consequences of Sleep Deprivation*", by Brandon R. Peters, M.D., posted July 11, 2013 and updated September 10, 2013. Web resource at https://www.huffpost.com/entry/sleep-deprivation_b_3536674
118. Article in *Scientific American* titled "*The Sensed-Presence Effect*", by Michael Shermer, dated April 1, 2010. Web resource at https://www.scientificamerican.com/article/the-sensed-presence-effect/
119. Article from National Public Radio titled "Guardian Angels Or The 'Third Man Factor'?", from the program *All Things Considered*, dated September 13, 2009. Web resource at https://www.npr.org/templates/story/story.php?storyId=112746464.
120. Web resource at https://www.amazon.com/Third-Man-Factor-Surviving-Impossible/dp/1602861293

121. Article at abc.net.au titled "*Finding God in a seizure: the link between temporal lobe epilepsy and mysticism*", by Jess Hill, dated December 10, 2014. Web resource at https://www.abc.net.au/radionational/programs/archived/encounter/the-link-between-temporal-lobe-epilepsy-and-mysticism/5956982
122. Landtblom AM, Lindehammar H, Karlsson H, Craig AD, "Insular cortex activation in a patient with "sensed presence"/ecstatic seizures". *Epilepsy Behav.* 2011 Apr;20(4):714-8. doi: 10.1016/j.yebeh.2011.01.031. Epub 2011 Mar 25. Web resource at https://www.ncbi.nlm.nih.gov/pubmed/21440512
123. Article at LiveScience.com, titled "*Near-Death Experiences Explained by Science*", by Charles Q. Choi, September 12, 2011. Web resource at https://www.livescience.com/16019-death-experiences-explained.html
124. Article by the National Institute on Drug Abuse; National Institutes of Health; U.S. Department of Health and Human Services titled "*What are hallucinogens?*", revised January 2016. Web resource at https://www.drugabuse.gov/publications/drugfacts/hallucinogens
125. Web resource at https://en.wikipedia.org/wiki/Dopamine_agonist
126. Article by the National Institute on Drug Abuse; National Institutes of Health; U.S. Department of Health and Human Services titled "*What are marijuana effects?*" Web resource at https://www.drugabuse.gov/publications/research-reports/marijuana/what-are-marijuana-effects
127. Web resource at https://en.wikipedia.org/wiki/MDMA
128. YouTube video clip of Beavis's absolutely hilarious hallucinogenic trip from the movie *Beavis and Butt-Head Do America*. Web resource at https://www.youtube.com/watch?v=xo61f8aXrbg
129. Reference from the U.S. Department of Veterans Affairs and the National Center for PTSD. Web resource at https://www.ptsd.va.gov/
130. Sherin, Jonathan E and Charles B Nemeroff. "Post-traumatic stress disorder: the neurobiological impact of psychological trauma" *Dialogues in clinical neuroscience* vol. 13,3 (2011): 263-78. Web resource at https://www.ncbi.nlm.nih.gov/pmc/articles/PMC3182008/
131. Article by the National Institute of Mental Health; National Institutes of Health; U.S. Department of Health and Human Services titled "*Post-Traumatic Stress Disorder*". Web resource at https://www.nimh.nih.gov/health/topics/post-traumatic-stress-disorder-ptsd/index.shtml
132. Web resource at https://www.britannica.com/topic/imprinting-learning-behaviour
133. Quote from Steven Weinberg. Web resource at https://en.wikiquote.org/wiki/Steven_Weinberg

Chapter Thirteen.

1. Web resource at https://www.merriam-webster.com/dictionary/religion
2. Web resource at https://en.oxforddictionaries.com/definition/religion
3. Web resource at https://www.merriam-webster.com/dictionary/science
4. Web resource at https://en.oxforddictionaries.com/definition/science
5. YouTube video clip of George W. Bush demonstrating his absolute ignorance of history and common sense, amplified by his unthinking religiosity, in his statement about the "Crusade" to wage war in the Middle East. Web resource at https://www.youtube.com/watch?v=7TRVcnX8Vsw
6. YouTube video referencing George W. Bush's belief that his god wanted him to be elected president. Clip also shows footage of Bush being blatantly pandering

to the religious right and proving how totally and immorally addled by religion he was. Web resource at https://www.youtube.com/watch?v=et2QWsQUTvM

7. Article at the University of Arizona web site The Unofficial Stephen Jay Gould Archive, with Gould's article titled "*Nonoverlapping Magisteria*". The site cited the article as: Stephen Jay Gould, "Nonoverlapping Magisteria," *Natural History* 106 (March 1997): pp. 16-22; Also reprinted in *Leonardo's Mountain of Clams and the Diet of Worms*, New York: Harmony Books, 1998, pp. 269-283.] Web resource at http://www.blc.arizona.edu/courses/schaffer/449/Gould Nonoverlapping Magisteria.htm
8. Pew Research Center publication titled "*Which countries still outlaw apostasy and blasphemy?*", by Angelina E. Theodorou, dated July 29, 2016. Web resource at http://www.pewresearch.org/fact-tank/2016/07/29/which-countries-still-outlaw-apostasy-and-blasphemy/
9. YouTube video clip of the short-tenured and much-ridiculed White House Press Secretary Sean Spicer giving his infamous First Big Lie on January 21, 2017, a lie which will indeed forever live in infamy and was the first official indication of how Orwellian that administration would be, about the size of the presidential inauguration. Web resource at https://www.youtube.com/watch?v=PKzHXelQi_A
10. YouTube video of Colonel Nathan Jessup's famous "You can't handle the truth!" zinger back to Tom Cruise's character in the movie *A Few Good Men*. Web resource at https://www.youtube.com/watch?v=UXoNE14U_zM
11. Gallup article titled "*In U.S., Belief in Creationist View of Humans at New Low*", by Art Swift, dated May 22, 2017. Web resource at https://news.gallup.com/poll/210956/belief-creationist-view-humans-new-low.aspx
12. Gallup article titled "*Most Americans Still Believe in God*", by Frank Newport, dated June 29, 2016. Web resource at https://news.gallup.com/poll/193271/americans-believe-god.aspx
13. Gallup article titled "*Record Few Americans Believe Bible Is Literal Word of God*", by Lydia Saad, dated May 15, 2017. Web resource at https://news.gallup.com/poll/210704/record-few-americans-believe-bible-literal-word-god.aspx
14. Stork, NE. "How Many Species of Insects and Other Terrestrial Arthropods Are There on Earth?", *Annu Rev Entomol*. 2018 Jan 7;63:31-45. doi: 10.1146/annurev-ento-020117-043348. Epub 2017 Sep 22. Web resource at https://www.ncbi.nlm.nih.gov/pubmed/28938083
15. Article from Business Insider titled "*New York's commuter nightmare is bad, but the world's most crowded cities are even worse*", by Chris Weller, dated July 11, 2018. Image shows the unbelievably crowded conditions people endure in the 21st century. Web resource at https://www.businessinsider.com/inside-worlds-most-crowded-cities-2017-7/ - kolkata-is-far-from-indias-densest-city-mumbais-73000-people-per-square-mile-forces-homes-to-shrink-to-unthinkably-small-sizes-6
16. Web resource at https://www.britannica.com/science/sedimentary-facies
17. European Geosciences Union (EGU). (2013, November 5). Oldest ice core: Finding a 1.5 million-year record of Earth's climate. *ScienceDaily*. Web resource at https://www.sciencedaily.com/releases/2013/11/131105081228.htm
18. Article from Answers in Genesis website titled "Geologic Evidences for the Genesis Flood Part I: An Overview", by Dr. Andrew A. Snelling, dated September 18, 2007. Web resource at https://answersingenesis.org/the-flood/geologic-evidences-for-the-genesis-flood/

Chapter Fourteen.

1. Web resource at https://www.merriam-webster.com/dictionary/atheist
2. Web resource at https://www.merriam-webster.com/dictionary/agnostic
3. Web resource at https://www.merriam-webster.com/dictionary/freethinker
4. Web resource at https://www.merriam-webster.com/dictionary/pantheism
5. Web resource at https://www.merriam-webster.com/dictionary/deism
6. Web resource at https://www.merriam-webster.com/dictionary/theism
7. Web resource at https://www.merriam-webster.com/dictionary/animism
8. Web resource at https://www.merriam-webster.com/dictionary/faith
9. Web resource at https://www.merriam-webster.com/dictionary/reason
10. Web resource at https://www.merriam-webster.com/dictionary/proof
11. Web resource at https://www.merriam-webster.com/dictionary/evidence
12. Web resource at https://www.merriam-webster.com/dictionary/delusion
13. YouTube video titled "*Bertrand Russell's Teapot: The Skeptic's View on the Supernatural*", referring to Russell's imaginary teapot. Web resource at https://www.youtube.com/watch?v=scC7gY227Cc
14. YouTube video titled "*The Dragon in My Garage*", referring to Sagan's imaginary dragon. Web resource at https://www.youtube.com/watch?v=jJRy3Kl_z5E
15. Smithsonian Institution article titled "*What does it mean to be human?*" Site has an interactive human evolution timeline and was last updated November 21, 2018. Web resource at http://humanorigins.si.edu/evidence/human-evolution-timeline-interactive
16. Article from *National Geographic* titled "*World's Oldest Cave Art Found—And Neanderthals Made It*", by Michael Greshko, published February 22, 2018. The article notes that the art is more than 64,000 years old, which is many tens of thousands of years older than biblical literalists think the universe has been around. Web resource at https://news.nationalgeographic.com/2018/02/neanderthals-cave-art-humans-evolution-science/
17. National Aeronautics and Space Administration article titled "*How Old is the Universe?*" by the NASA/WMAP Science Team, dated December 10, 2010. Web resource at https://map.gsfc.nasa.gov/universe/uni_age.html
18. National Aeronautics and Space Administration article titled "*Precocious Earth*", NASA Official Dr. Mamta Patel Nagaraja, published January 16, 2001 and last updated November 23, 2018. Web resource at https://science.nasa.gov/science-news/science-at-nasa/2001/ast17jan_1
19. Article from *National Geographic* titled "*World's Oldest Fossils Discovered Due to Climate Change*", by Brian Clark Howard, published September 1, 2016. Web resource at https://news.nationalgeographic.com/2016/09/worlds-oldest-fossils-stromatolites-discovered-climate-change/
20. Article from *National Geographic* titled "*This May Be the Oldest Known Sign of Life on Earth*", by Nadia Drake, published March 1, 2017. Web resource at https://news.nationalgeographic.com/2017/03/oldest-life-earth-iron-fossils-canada-vents-science/
21. YouTube video clip of the dancing magical brooms from *The Sorcerer's Apprentice* section of the Disney film *Fantasia*. Web resource at https://www.youtube.com/watch?v=Rrm8usaH0sM
22. Article from the University of California at Berkeley's *Understanding Evolution* web reference, titled "*Early Concepts of Evolution: Jean Baptiste Lamarck*".

Credits at https://evolution.berkeley.edu/evolibrary/credits.php. Web resource at https://evolution.berkeley.edu/evolibrary/article/history_09
23. Web resource at https://en.wikipedia.org/wiki/Inquisition
24. Web resource at https://www.britannica.com/topic/inquisition
25. Web resource at https://en.wikipedia.org/wiki/Edict_of_Milan
26. Web resource at https://en.wikipedia.org/wiki/Constantine_the_Great_and_Christianity - Patronage_of_the_Church
27. Web resource at https://www.britannica.com/event/Council-of-Nicaea-Christianity-325
28. The Nicene Creed, from the US Conference of Catholic Bishops web page. Web resource at http://www.usccb.org/beliefs-and-teachings/what-we-believe/
29. Web resource at https://en.wikipedia.org/wiki/Edict_of_Thessalonica
30. Web resource at https://en.wikipedia.org/wiki/Locus_of_control
31. Actual article from The Guardian titled "*Evangelical Trump adviser tells people to skip flu shots in favor of prayer*", by Jamiles Lartey, dated February 7, 2018. Web resource at https://www.theguardian.com/society/2018/feb/07/evangelical-flu-shot-pray-donald-trump-gloria-copeland
32. Web resource at https://www.britannica.com/topic/flagellation
33. YouTube video clip from the movie "*Monty Python and The Holy Grail*", showing the monks engaging in an unusual form of self-punishment. It isn't quite flagellation but it is as pointless and unproductive a compulsion. Web resource at https://www.youtube.com/watch?v=e4q6eaLn2mY
34. Article from *Psychology Today* titled "*Religious Self Harm*", by Mark D. Griffiths Ph.D., posted may 4, 2017. Web resource at https://www.psychologytoday.com/us/blog/in-excess/201705/religious-self-harm
35. Web resource at https://en.wikipedia.org/wiki/Jerusalem_syndrome
36. Article from *Psychology Today* titled "*Your Sixth Sense*", by Matthew Hutson, published July 3, 2012 and last reviewed June 9, 2016. Web resource at https://www.psychologytoday.com/us/articles/201207/your-sixth-sense

Chapter Fifteen.

1. Article in NewScientist titled "*How life evolved: 10 steps to the first cells*", by Nick Lane and Michael Le Page, dated October 14, 2009. Web resource at https://www.newscientist.com/article/dn17987-how-life-evolved-10-steps-to-the-first-cells/
2. Cooper GM. *The Cell: A Molecular Approach*. 2nd edition. Sunderland (MA): Sinauer Associates; 2000. "The Origin and Evolution of Cells". Web resource at https://www.ncbi.nlm.nih.gov/books/NBK9841/
3. Article from National Public Radio titled "*When Did Human Speech Evolve?*" by Barbara J. King, September 5, 2013. Web resource at https://www.npr.org/sections/13.7/2013/09/05/219236801/when-did-human-speech-evolve
4. Article from PLOSOne titled "*Shared Brain Lateralization Patterns in Language and Acheulean Stone Tool Production: A Functional Transcranial Doppler Ultrasound Study*", by Natalie Thaïs Uomini and Georg Friedrich Meyer, dated August 30, 2013. Web resource at https://journals.plos.org/plosone/article?id=10.1371/journal.pone.0072693
5. Pew Research Center publication titled "*The Global Religious Landscape*", dated December 18, 2012. Web resource at http://www.pewforum.org/2012/12/18/global-religious-landscape-exec/

6. Web resource at https://www.britannica.com/topic/Middle-Eastern-religion
7. YouTube video clip titled "*The Burned Over District - Religions of Central NY by Dr. Mike McKenzie*". Web resource at https://www.youtube.com/watch?v=S5ToJ9TQ7q8
8. Web resource at https://www.britannica.com/topic/Canaanite-religion
9. Web resource at http://www.cslewis.com/us/
10. Web resource at https://en.wikipedia.org/wiki/C._S._Lewis
11. Web reference from the US Conference of Catholic Bishops, titled "*The Real Presence of Jesus Christ in the Sacrament of the Eucharist: Basic Questions and Answers*". Web resource at http://www.usccb.org/prayer-and-worship/the-mass/order-of-mass/liturgy-of-the-eucharist/the-real-presence-of-jesus-christ-in-the-sacrament-of-the-eucharist-basic-questions-and-answers.cfm
12. Article from the Public Broadcasting Service's Global Connections section on the Middle East, titled "*Religion: Three Religions, One God*". Web resource at http://www.pbs.org/wgbh/globalconnections/mideast/themes/religion/
13. References for Table 2.
    a. Is Religion Important To You?
        i. Gallup article titled "*Religiosity Highest in World's Poorest Nations*", by Steve Crabtree, dated August 31, 2010. Web resource at http://news.gallup.com/poll/142727/religiosity-highest-world-poorest-nations.aspx
        ii. Web resource at https://en.wikipedia.org/wiki/Importance_of_religion_by_country
        iii. Pew Research Center article titled "*Americans are in the middle of the pack globally when it comes to the importance of religion*", by Angelina E. Theodorou, dated December 23, 2015. Web resource at http://www.pewresearch.org/fact-tank/2015/12/23/americans-are-in-the-middle-of-the-pack-globally-when-it-comes-to-importance-of-religion/
        iv. Web resource at https://ipfs.io/ipfs/QmXoypizjW3WknFiJnKLwHCnL72vedxjQkDDP1mXWo6uco/wiki/Importance_of_religion_by_country.html#cite_note-gallup-religiosity-poor-nations-1https://www.telegraph.co.uk/travel/maps-and-graphics/most-religious-countries-in-the-world/
        v. Web resource at https://wowair.us/magazine/religion-in-iceland/
        vi. Reference at sidmennt.is website; is a speech by Sigurdur Holm Gunnarson from the International Atheist Conference in Reykjavik, Iceland held on June 24 and 25, 2006. from the Web resource at http://sidmennt.is/2006/08/13/the-status-of-religion-and-religious-life-in-iceland-today/
    b. Score of 15 Year Olds on 2015 Program for International Student Assessment
        i. The Programme for International Student Assessment (PISA) 2015 Results in Focus. Web resource at https://www.oecd.org/pisa/pisa-2015-results-in-focus.pdf
        ii. Results from the 2015 PISA for Beijing-Shanghai-Jiangsu-Guangdong (China). Web resource at https://www.oecd.org/pisa/PISA-2105-Financial-Literacy-Beijing-Shanghai-Jiangsu-Guangdong-China.pdf
    c. Official State Religion
        i. Pew Research Center article titled "*Many Countries Favor Specific Religions, Officially or Unofficially*", dated October 3, 2017. Web resource at http://www.pewforum.org/2017/10/03/many-countries-favor-specific-religions-officially-or-unofficially/

ii. Pew Research Center full report on "*Many Countries Favor Specific Religions, Officially or Unofficially*", dated October 3, 2017. Web resource at http://assets.pewresearch.org/wp-content/uploads/sites/11/2017/09/29162847/FULL-REPORT-FOR-WEB.pdf
iii. The CIA World Factbook. Web resource at https://www.cia.gov/library/publications/the-world-factbook/
iv. Iceland "Quick Facts". Web resource at https://www.iceland.is/the-big-picture/quick-facts

d. World Economic Forum's Global Gender Gap Report 2014. Web resource at http://reports.weforum.org/global-gender-gap-report-2014/rankings/
e. United Nations Office on Drugs and Crime (UNODC) Global Study on Homicide 2013. Web resource at https://www.unodc.org/documents/gsh/pdfs/2014_GLOBAL_HOMICIDE_BOOK_web.pdf
f. Year of Female Suffrage.
    i. Web resource at https://en.wikipedia.org/wiki/Timeline_of_women%27s_suffrage
    ii. Women's suffrage timeline, from the University of British Columbia in Vancouver, Canada, led by Dr. Veronica Strong-Boag. Web resource at http://womensuffrage.org/?page_id=69
    iii. Article from the Public Broadcasting Service titled "*Timeline of Women's Rights in Afghanistan*", dated October 11, 2011. Web resource at http://www.pbs.org/wnet/women-war-and-peace/uncategorized/timeline-of-womens-rights-in-afghanistan/
    iv. Article from Elections Canada titled "*A History of the Vote in Canada; Chapter 2 From a Privilege to a Right 1867-1919*", dated August 27, 2018. Web resource at http://www.elections.ca/content.aspx?section=res&dir=his&document=chap2&lang=e
    v. Article from the United Nations Development Programme titled "*70th Anniversary since women gained voting rights in Montenegro*", dated June 28, 2016. Web resource at http://www.me.undp.org/content/montenegro/en/home/presscenter/pressreleases/2016/06/28/70th-anniversary-since-women-gained-voting-rights-in-montenegro.html
    vi. Article from the Library of Congress titled "*Women in History: Voting Rights*", by Kelly Buchanan, dated March 3, 2015. Web resource at https://blogs.loc.gov/law/2015/03/women-in-history-voting-rights/
g. Life Expectancy At Birth (Yrs).
    i. Article from the Organisation for Economic Co-Operation and Development titled "*Life expectancy at birth*", OECD (2018), Life expectancy at birth (indicator). doi: 10.1787/27e0fc9d-en. Web resource at https://data.oecd.org/healthstat/life-expectancy-at-birth.htm
    ii. World Health Organization data from the Global Health Observatory data repository on "*Healthy life expectancy (HALE) Data by country*", last updated April 6, 2018. Web resource at http://apps.who.int/gho/data/node.main.HALE?lang=en
    iii. The World Bank information on "*Life expectancy at birth, total (years)*", Web resource at https://data.worldbank.org/indicator/SP.DYN.LE00.IN
h. Under-Five Mortality Rate, Both Sexes, per 1000.

i. United Nations Development Programme Human Development Report titled "*Under-five mortality rate (per 1,000 live births)*", from the Inter-agency Group for Child Mortality Estimation (2013). Web resource at http://hdr.undp.org/en/content/under-five-mortality-rate-1000-live-births
ii. Information from The World Bank on "*Mortality Rate, under-5 (per 1,000 live births)*". Web resource at https://data.worldbank.org/indicator/SH.DYN.MORT?locations=EC
iii. United Nations data from UNdata on the Under-five mortality rate for China and Hong Kong SAR. Web resource at http://data.un.org/Data.aspx?q=China%2C+Hong+Kong+SAR&d=GenderStat&f=inID%3A42%3BcrID%3A94%2C97

i. Halliwell, J., Layard, R. & Sachs, J. (2017). World Happiness Report 2017, New York: Sustainable Development Solutions Network. Web resource at http://worldhappiness.report/ed/2017/
j. Information from The World Bank on Gross Domestic Product (GDP) per capita (current US$). Web resource at https://data.worldbank.org/indicator/NY.GDP.PCAP.CD?year_high_desc=true

14. Information on the Program for International Student Assessment (PISA). Web resource at https://nces.ed.gov/surveys/pisa/
15. Web resource at https://en.wikipedia.org/wiki/Opium_of_the_people

Chapter Sixteen.

1. Web resource at https://en.oxforddictionaries.com/definition/morality
2. Gert, Bernard and Gert, Joshua, "The Definition of Morality", *The Stanford Encyclopedia of Philosophy* (Fall 2017 Edition), Edward N. Zalta (ed.), First published Wed Apr 17, 2002; substantive revision Mon Feb 8, 2016. Web resource at https://plato.stanford.edu/entries/morality-definition/ - DescDefiMora

Chapter Seventeen.

1. YouTube video clip from the movie *Schindler's List*, in which the psychopathic Nazi camp commander played by Ralph Fiennes shoots and terrorizes people from his balcony simply out of cruelty. Web resource at https://www.youtube.com/watch?v=FcyLRTB9fRY
2. Web resource at https://en.wikipedia.org/wiki/List_of_English_Bible_translations
3. Article from Reuters titled "*Pope blames devil for Church divisions, scandals, seeks angel's help*", by Philip Pullella, dated October 8, 2018. Web resource at https://www.reuters.com/article/us-pope-devil/pope-blames-devil-for-church-divisions-scandals-seeks-angels-help-idUSKCN1MI10M
4. Article in The Washington Post titled "*A modern pope gets old school on the Devil*", by Anthony Faiola, dated May 10, 2014. Web resource at https://www.washingtonpost.com/world/a-modern-pope-gets-old-school-on-the-devil/2014/05/10/f56a9354-1b93-4662-abbb-d877e49f15ea_story.html?utm_term=.ad98cf8d5dc2
5. Article from Slate titled "*Witness to an Exorcism*", by Laura Helmuth, dated May 12, 2014. Web resource at https://slate.com/human-interest/2014/05/exorcism-in-catholic-church-pope-francis-says-satan-the-devil-is-real.html
6. "St Paul and temporal lobe epilepsy" *Journal of neurology, neurosurgery, and psychiatry* vol. 50,6 (1987): 659-64. Web resource at https://www.ncbi.nlm.nih.gov/pmc/articles/PMC1032067/
7. YouTube video clip of emasculated bootlicking federal cabinet members groveling to the president, who accepts the hosannas like a king accepting tribute

from the unwashed masses. Web resource at https://www.youtube.com/watch?v=-8GkzsUaQVU

Chapter Eighteen.

1. Article at History.com titled "*Crusades*", by History.com authors, originally published June 7, 2010 and last updated September 14, 2018. Web resource at https://www.history.com/topics/middle-ages/crusades
2. Web resource at https://www.britannica.com/event/Crusades
3. Article from National Public Radio titled "*The Origins Of The Shiite-Sunni Split*", by Mike Shuster, dated February 12, 2007. Web resource at https://www.npr.org/sections/parallels/2007/02/12/7332087/the-origins-of-the-shiite-sunni-split
4. "The Roman Catholic Church, the Holocaust, and the demonization of the Jews: Response to "Benjamin and us: Christianity, its Jews, and history" by Jeanne Favret-Saada" HAU : *Journal of Ethnographic Theory* vol. 4,8 (2014): 329-333. Web resource at https://www.ncbi.nlm.nih.gov/pmc/articles/PMC4800995/
5. Article from The Independent titled "*Salman Rushdie: Iranian state media renew fatwa on Satanic Verses author with $600,000 bounty*", by Samuel Osborne, dated February 21, 2016. Web resource at https://www.independent.co.uk/news/people/salman-rushdie-iranian-state-media-renew-fatwa-on-satanic-verses-author-with-600000-bounty-a6887141.html
6. YouTube video titled "*How the Biblical Canon Was Formed*", from the Museum of the Bible, published on May 8, 2015. Web resource at https://www.youtube.com/watch?v=nFEBwfYZBJY
7. Brief explanation of psychopathy in *Psychology Today*. Web resource at https://www.psychologytoday.com/us/basics/psychopathy
8. Web resource at https://www.britannica.com/biography/Hirohito
9. Web resource at https://www.britannica.com/topic/blood-libel
10. Web resource at https://en.wikipedia.org/wiki/Host_desecration
11. Web resource at https://www.theonion.com/
12. Catholic church catechism. Web resource at http://www.vatican.va/archive/ccc_css/archive/catechism/p2s2c1a3.htm
13. Web resource at https://en.wikipedia.org/wiki/PZ_Myers
14. YouTube video clip titled "*Bill on Desecration*" of Bill Donahue performing desecration on decency, common sense, humanity and reason in a rant about desecration of an allegedly magic cracker. Posted by The Catholic League on September 20, 2011. Web resource at https://www.youtube.com/watch?v=-ZzjtTE0KMI
15. Web resource at https://en.wikipedia.org/wiki/Cum_nimis_absurdum
16. Article from Haaretz titled "*This Day in Jewish History 1555: Pope Paul IV Orders Jews to Live in a Ghetto*", by David B. Green, dated July 14, 2013. Web resource at https://www.haaretz.com/jewish/1555-roman-jews-banished-to-ghetto-1.5295240
17. Catholic church catechism. Web resource at http://www.vatican.va/archive/ccc_css/archive/catechism/p123a9p4.htm
18. Web resource at https://www.merriam-webster.com/dictionary/fascism
19. *Summi Pontificatus* - Encyclical Of Pope Pius XII On The Unity Of Human Society To Our Venerable Brethren: The Patriarchs, Primates, Archbishops, Bishiops, And Other Ordinaries In Peace And Communion With The Apostolic See. Web resource at http://w2.vatican.va/content/pius-xii/en/encyclicals/documents/hf_p-xii_enc_20101939_summi-pontificatus.html

20. *Mystici Corporis Christi* - Encyclical Of Pope Pius XII On The Mistical Body Of Christ To Our Venerable Brethren, Patriarchs, Primates, Archbishops, Bishiops, And Other Local Ordinaries Enjoying Peace And Communion With The Apostolic See. Web resource at http://w2.vatican.va/content/pius-xii/en/encyclicals/documents/hf_p-xii_enc_29061943_mystici-corporis-christi.html
21. Article from National Public Radio titled "*From Laundering To Profiteering, A Multitude Of Sins At The Vatican Bank*", by the NPR Staff, dated January 30, 2015. Web resource at https://www.npr.org/2015/01/30/382374060/from-laundering-to-profiteering-a-multitude-of-sins-at-the-vatican-bank
22. Article in Business Insider titled "*The craziest financial schemes that the Vatican Bank tried to cover up*", by Stephanie Yang, dated February 27, 2015. Web resource at https://www.businessinsider.com/gods-bankers-financial-scandals-at-the-vatican-2015-2
23. Article from National Public Radio titled "*Document Renews Debate On Vatican Role in WWII*", by Sylvia Poggioli, dated January 21, 2005. Web resource at https://www.npr.org/templates/story/story.php?storyId=4461958
24. Article from The Guardian titled "*Red Cross and Vatican helped thousands of Nazis to escape*", by Dalya Alberge, dated May 25, 2011. Web resource at https://www.theguardian.com/world/2011/may/25/nazis-escaped-on-red-cross-documents
25. YouTube video titled "*The Vatican Ratline - Primetime Live! With Sam Donaldson*", published on September 28, 2013. Web resource at https://www.youtube.com/watch?v=s5h6re0bV_c
26. Partial listing of papal documents on the Vatican website. Web resource at http://www.vatican.va/offices/papal_docs_list.html
27. Article at CNN titled "*Jewish leaders condemn Vatican's plan to venerate Pope Pius XII*", by Kevin Flower, dated December 22, 2009. Web resource at http://edition.cnn.com/2009/WORLD/meast/12/22/vatican.israel.pope.pius/
28. Web resource at https://www.britannica.com/topic/German-Christian

Chapter Nineteen. Smithsonian Institution article titled "*What does it mean to be human?*" subtitled "*Human Characteristics: Humans Change the World*", last updated November 21, 2018. Web resource at http://humanorigins.si.edu/human-characteristics/humans-change-world

Chapter Twenty. No external references.

Chapter Twenty-One.

1. US Conference of Catholic Bishops article titled "*Rationale for Catholics Reading the Old Testament*", by Michael Steier, D.MIN. Web resource at http://www.usccb.org/bible/understanding-the-bible/study-materials/articles/rationale-for-catholics-reading-the-old-testament.cfm
2. US Conference of Catholic Bishops article titled "*Bible at Core of Catholic Beliefs*" by Monsignor J. Brian Bransfield. Web resource at http://www.usccb.org/bible/understanding-the-bible/study-materials/articles/bible-at-core-of-catholic-beliefs.cfm
3. US Conference of Catholic Bishops article titled "*Morality*", excerpted from the *United States Catholic Catechism for Adults*. Web resource at http://www.usccb.org/beliefs-and-teachings/what-we-believe/morality/

4. Catholic church catechism. Web resource at http://www.vatican.va/archive/ccc_css/archive/catechism/p1s2c1p2.htm
5. US Conference of Catholic Bishops daily reading from March 19, 2018. Web resource at http://www.usccb.org/bible/readings/031918.cfm
6. Catholic church catechism. Web resource at http://www.vatican.va/archive/ccc_css/archive/catechism/p3s2c2a6.htm
7. Article from LiveScience titled "*The World's Catholic Population (Infographic)*", by Ross Toro, dated February 19, 2013. Web resource at https://www.livescience.com/27244-the-world-s-catholic-population-infographic.html
8. Declaration On The Relation Of The Church To Non-Christian Religions, *Nostra Aetate*, Proclaimed By His Holiness Pope Paul VI on October 28, 1965 Web resource at http://www.vatican.va/archive/hist_councils/ii_vatican_council/documents/vat-ii_decl_19651028_nostra-aetate_en.html
9. Web resource at https://en.wikipedia.org/wiki/Spanish_Inquisition - End_of_the_Inquisition
10. Web resource at https://www.britannica.com/event/Lateran-Treaty
11. Web resource at https://www.britannica.com/biography/Adolf-Hitler/Dictator-1933-39

Chapter Twenty-Two. Article from the Christian Research Institute titled "*Should We Control the Number of Children We Have?*" by Hank Hanegraff, dated August 31, 2010. Web resource at https://www.equip.org/hank_speaks_out/should-we-control-the-number-of-children-we-have/

Chapter Twenty-Three.

1. Wynn, Rolf et al. "Psychopathy in women: theoretical and clinical perspectives" *International journal of women's health* vol. 4 (2012): 257-63. Web resource at https://www.ncbi.nlm.nih.gov/pmc/articles/PMC3379858/
2. Ali, Ahmed (translation). *Islam: The Quran.* New York: Princeton University Press, 1988.

Chapter Twenty-Four.

1. Gallup article titled "*Three in Four in U.S. Still See the Bible as Word of God*", by Lydia Saad, posted June 4, 2014. Web resource at https://news.gallup.com/poll/170834/three-four-bible-word-god.aspx
2. Gallup article titled "*Three in Four in U.S. Still See the Bible as Word of God*", by Lydia Saad, posted June 4, 2014. Web resource at https://news.gallup.com/poll/170834/three-four-bible-word-god.aspx

Chapter Twenty-Five.

1. Article at the Jewish Virtual Library titled "*Jewish Prayers: Early Morning Blessings (Birkhot Hashahar)*", by Shira Schoenberg. Web resource at https://www.jewishvirtuallibrary.org/early-morning-blessings-birkhot-hashahar
2. Ali, Ahmed (translation). *Islam: The Quran.* New York: Princeton University Press, 1988.
3. Pew Research Center article titled "*The divide over ordaining women*", by David Masci, posted September 9, 2014. Web resource at

http://www.pewresearch.org/fact-tank/2014/09/09/the-divide-over-ordaining-women/

4. The Church of Jesus Christ of Latter Day Saints *Guide to the Scriptures*, referencing "*Discernment, Gift of*". Web resource at https://www.lds.org/scriptures/gs/discernment-gift-of?lang=eng

Chapter Twenty-Six.

1. Web resource at https://en.wikipedia.org/wiki/Haruspex
2. Article in *Psychology Today* titled "*Motivated Reasoning*". Web resource at https://www.psychologytoday.com/us/basics/motivated-reasoning
3. Article in *The Atlantic* titled "*White Evangelicals Believe They Face More Discrimination Than Muslims*", by Emma Green, dated March 10, 2017. Web resource at https://www.theatlantic.com/politics/archive/2017/03/perceptions-discrimination-muslims-christians/519135/
4. Article in *The Hill* titled "*Roy Moore's five most controversial remarks*", by Ben Kamisar, dated September 24, 2017. Web resource at https://thehill.com/homenews/campaign/351966-gop-senate-hopeful-roy-moores-five-most-controversial-remarks
5. Article at CNN.com titled "*Ten Commandments judge removed from office*", dated November 14, 2003. Web resource at http://www.cnn.com/2003/LAW/11/13/moore.tencommandments/
6. Article at Patheos.com titled "*Senate Confirms Christian Extremist Jeff Sessions Attorney General*", by Michael Stone, dated February 8, 2017. References negative Sessions comments about the ability of secular people to serve in government. Web resource at https://www.patheos.com/blogs/progressivesecularhumanist/2017/02/senate-confirms-christian-extremist-jeff-sessions-attorney-general/
7. Article in *Slate* titled "*A Prayer for the Real Victims*", by Dahlia Lithwick, dated July 19, 2017. Article references several church-state issues and legal battles. Web resource at http://www.slate.com/articles/news_and_politics/jurisprudence/2017/07/jeff_sessions_is_desperate_to_win_the_fight_for_christians_religious_liberty.html
8. Article from National Public Radio titled "*'An Appalling And Detestable Lie': 5 Highlights From Sessions' Senate Testimony*", by Miles Parks, dated June 13, 2017. Web resource at https://www.npr.org/2017/06/13/532827514/an-appalling-and-detestable-lie-5-highlights-from-sessions-senate-testimony
9. Article at Christianity Today titled "*Why Did So Many Christians Support Slavery?*", from Issue 33: Christianity & the Civil War, 1992, by the Editors. Web resource at https://www.christianitytoday.com/history/issues/issue-33/why-christians-supported-slavery.html
10. Article from the Public Broadcasting Service / Thirteen.org titled "*The Slave Experience: Religion*", by Kimberly Sambol-Tosco. Web resource at https://www.thirteen.org/wnet/slavery/experience/religion/history2.html
11. Web resource at https://www.britannica.com/event/Comstock-Act
12. National Center for Science Education site showing image of the Butler Act. Web resource at https://ncse.com/files/pub/legal/Scopes/Butler_Act.pdf
13. Article at History.com titled "*Monkey Trial begins*", by the History.com editors, published November 24, 2009 and last updated August 21, 2018. Web resource at https://www.history.com/this-day-in-history/monkey-trial-begins

14. Article at History.com titled "*Prohibition*", by the History.com editors, published October 29, 2009 and last updated August 21, 2018. Web resource at https://www.history.com/topics/roaring-twenties/prohibition
15. Web resource at https://www.britannica.com/event/Jim-Crow-law
16. Article in *The Economist* titled "*A hot line to heaven*", December 16, 2004. Web resource at https://www.economist.com/united-states/2004/12/16/a-hot-line-to-heaven
17. Web resource at https://abcnews.go.com/Politics/video/pence-introduces-christian-conservative-republican-40758022
18. Article from Public Broadcasting Service *Frontline – The Jesus Factor*, titled "Invoking God and Faith", with many commentators discussing the religious views of George W. Bush, posted April 29, 2004. Web resource at https://www.pbs.org/wgbh/pages/frontline/shows/jesus/president/invoking.html
19. Article from Military.com titled "*US Commander in Turkey Criticized for Religious Statements*", by Oriana Pawlyk. Web resource at https://www.military.com/daily-news/2016/10/07/us-commander-in-turkey-criticized-for-religious-statements.html
20. Article from the Southern Poverty Law Center titled "*Lieutenant General William G. "Jerry" Boykin (Ret.)*". Site contains numerous statements by Boykin reflecting his Christian extremist and fundamentalist views. Web resource at https://www.splcenter.org/fighting-hate/extremist-files/individual/lieutenant-general-william-g-jerry-boykin-ret
21. Web resource at https://www.merriam-webster.com/dictionary/subversion

Chapter Twenty-Seven.

1. The Constitution of the United States: A Transcription. Web resource at https://www.archives.gov/founding-docs/constitution-transcript
2. The Bill of Rights: A Transcription. Web resource at https://www.archives.gov/founding-docs/bill-of-rights-transcript - toc-amendment-i
3. The Constitution: Amendments 11-27. Web resource at https://www.archives.gov/founding-docs/amendments-11-27
4. The Constitution: Amendments 11-27. Web resource at https://www.archives.gov/founding-docs/amendments-11-27
5. Yale Law School Avalon Project, "*The Barbary Treaties 1786-1816; Treaty of Peace and Friendship, Signed at Tripoli November 4, 1796*". Web resource at http://avalon.law.yale.edu/18th_century/bar1796t.asp

Chapter Twenty-Eight.

1. Article from Mott Community College titled "*Spanish Treatment of the Natives*", by Agulyas, dated March 18, 2014. Web resource at https://history.mcc.edu/wordpress/history/2014/03/18/spanish-treatment-of-the-natives/
2. Article from the Library of Congress titled "*Colonial Settlement, 1600s-1763*". Web resource at http://www.loc.gov/teachers/classroommaterials/presentationsandactivities/presentations/timeline/colonial/indians/
3. Article at TheHumanist.com titled "*Unelectable Atheists: U.S. States That Prohibit Godless Americans From Holding Public Office*" by Matthew Bulger, dated May 25, 2012. Web resource at

https://thehumanist.com/news/national/unelectable-atheists-u-s-states-that-prohibit-godless-americans-from-holding-public-office

4. Texas State Constitution. Web resource at https://tlc.texas.gov/docs/legref/TxConst.pdf
5. Arkansas State Constitution. Web resource at http://www.arkleg.state.ar.us/assembly/Summary/ArkansasConstitution1874.pdf
6. Web resource at https://www.britannica.com/event/Jim-Crow-law
7. Web resource at https://en.wikipedia.org/wiki/Cum_nimis_absurdum

Chapter Twenty-Nine. No external references.

Chapter Thirty.

1. Web resource at https://www.merriam-webster.com/dictionary/miracle
2. Web resource at http://www.usccb.org/
3. Article at US Conference of Catholic Bishops titled "*Saints*". Web resource at http://www.usccb.org/prayer-and-worship/prayers-and-devotions/saints/index.cfm
4. Article uploaded to the US Conference of Catholic Bishops website titled "*How the Church Recognizes Saints*", by Jeannine Marino. Web resource at http://www.usccb.org/upload/how-church-recognizes-saints-usccb-media-blog-post.pdf
5. Pew Research Center article titled "*Papal saints: Once a given, now extremely rare*", by Michael Lipka and Tim Townsend, April 24, 2014. Web resource at http://www.pewresearch.org/fact-tank/2014/04/24/papal-saints-once-a-given-now-extremely-rare/
6. Pew Research Center article titled "*Religion Among the Millennials*", dated February 17, 2010. Web resource at http://www.pewforum.org/2010/02/17/religion-among-the-millennials/

Chapter Thirty-One.

1. Web resource at https://ftw.usatoday.com/2017/01/dabo-swinney-emotional-postgame-interview-reaction-clemson-alabama-championship-college-football
2. Article at Patheos.com titled "*Clemson QB On Win: 'It's What God Wanted*'", by Michael Stone, dated January 10, 2017. Article notes not just the QBs words but the team's religiosity. Web resource at https://www.patheos.com/blogs/progressivesecularhumanist/2017/01/clemson-qb-win-god-wanted/
3. YouTube video clip titled "*Wilson Football Factory*", published January 15, 2008. Web resource at https://www.youtube.com/watch?v=K8kZP4dRe3w
4. Data from United States Census. Web resource at https://factfinder.census.gov/faces/tableservices/jsf/pages/productview.xhtml?pid=ACS_15_5YR_DP03&src=pt
5. Data from the Insurance Institute for Highway Safety, Highway loss Data Institute. Web resource at https://www.iihs.org/iihs/topics/t/general-statistics/fatalityfacts/state-by-state-overview
6. Article from the National Public Service titled "*Death and Dying*", by Drew Gilpin Faust. Web resource at https://www.nps.gov/nr/travel/national_cemeteries/death.html

Chapter Thirty-Two.

1. YouTube video clip titled "*Alabama 24 Clemson 6: Dabo Swinney, Kelly Bryant, Clelin Ferrell postgame press conference*", published January 1, 2018. Web resource at https://www.youtube.com/watch?v=xGotlOwbCww
2. Article at *Psychology Today* titled "*Hits, Misses and Delusions*", by Dan Gardner, posted December 29, 2010. Web resource at https://www.psychologytoday.com/us/blog/future-babble/201012/hits-misses-and-delusions
3. Web resource at https://www.sciencedaily.com/terms/confirmation_bias.htm
4. Article at *Psychology Today* defining "*Motivated Reasoning*". Web resource at https://www.psychologytoday.com/us/basics/motivated-reasoning

Chapter Thirty-Three.

1. Web resource at https://en.wikipedia.org/wiki/Daniel_Daly
2. YouTube video clip titled "*Mass Suicides on Marpi Point, Saipan - WW2 in HD and Color*", published April 27, 2010. Web resource at https://www.youtube.com/watch?v=eDUy0uzmaU4
3. Catechism of the Catholic Church. Web resource at http://www.vatican.va/archive/ccc_css/archive/catechism/p3s2c2a5.htm
4. Web resource at https://dictionary.cambridge.org/us/dictionary/english/resurrect
5. Web resource at https://www.merriam-webster.com/dictionary/resurrect
6. Web resource at https://en.oxforddictionaries.com/definition/resurrect
7. YouTube video clip titled "*Terminator 2: Judgment Day – T800 arrival*". Web resource at https://www.youtube.com/watch?v=PqxqxjnPGJs

Chapter Thirty-Four. Web resource at https://en.wikipedia.org/wiki/Conflict_resolution

Chapter Thirty-Five. No external references.

Chapter Thirty-Six. No external references.

Chapter Thirty-Seven. No external references.

Chapter Thirty-Eight. No external references.

Further Recommended Reading

Altemeyer, Bob. *The Authoritarians.* Self-Published; available in hard copy at Lulu.com, in electronic form at https://www.theauthoritarians.org/ and in audio book format on Audible.com or Amazon.com.

Bard, Mitchell G. (Editor). *The Holocaust.* San Diego: Greenhaven Press, Inc., 2001.

Barker, Dan. *Godless: How an Evangelical Preacher Became One of America's Leading Atheists.* Berkeley: Ulysses Press, 2008.

Bierce, Ambrose G. *The Unabridged Devil's Dictionary.* Athens: The University of Georgia Press, 2002.

Boyer, Pascal. *Religion Explained.* New York: Basic Books, 2001.

Bradbury, Ray. *Fahrenheit 451.* New York: Simon and Schuster, 2013.

Cochran, Gregory and Harpending, Henry. *The 10,000 Year Explosion: How Civilization Accelerated Human Evolution.* New York: Basic Books, 2010.

Confucius. *The Analects.* New York: Oxford University Press, 2000.

Coyne, Jerry A. *Faith Versus Fact.* New York: Viking, 2015.

Coyne, Jerry A. *Why Evolution Is True.* New York: Penguin Books, 2010.

Dalley, Stephanie (Editor). *Myths From Mesopotamia: Creation, The Flood, Gilgamesh and Others.* Edited and translated with an introduction and notes. New York: Oxford University Press, Inc., 2008.

Dawkins. Richard. *The Blind Watchmaker: Why The Evidence of Evolution Reveals A Universe Without Design.* New York: W.W. Norton, 1996.

Dawkins. Richard. *The God Delusion.* New York: Houghton Mifflin Harcourt, 2008.

Dawkins. Richard. *Unweaving The Rainbow: Science, Delusion and the Appetite For Wonder.* New York: Houghton Mifflin, 1998.

Diamond, Jared M. *Guns, Germs and Steel: The Fates of Human Societies.* New York: W.W. Norton, 2017.

Diamond, Jared M. *The Third Chimpanzee: The Evolution and Future of The Human Animal.* New York: HarperCollins, 1993.

Erasmus, Desiderius. *In Praise Of Folly.* New York: Penguin Books, 1993.

Fisher, Helen E. *Anatomy of Love: The Mysteries of Mating, Marriage and Why We Stray.* New York: Ballantine Books, 1992.

Freud, Sigmund. *The Future of an Illusion.* New York: W. W. Norton and Company, 1961.

Geiger, John. *The Third Man Factor: Surviving The Impossible.* New York: Weinstein Books, 2009.

Haidt, Jonathan. *The Righteous Mind: Why Good People Are Divided By Politics And Religion.* New York: Vintage Books, 2012.

Hamilton, Virginia. *In The Beginning: Creation Stories From Around The World.* San Diego: Harcourt Brace Jovanovich, Publishers, 1988.

Hare, Robert D. *Without Conscience: The Disturbing World of The Psychopaths Among Us.* New York: The Guilford Press, 1999.
Harris, Sam. *The End of Faith.* New York: W.W. Norton and Company, Inc., 2004.

Harris, Sam. *The Moral Landscape: How Science Can Determine Human Values.* New York: Free Press, 2011.

Hazen, Robert M. *The Story of Earth: the First 4.5 Billion years, From Stardust to Living Planet.* New York: Penguin, 2013.

Heinlein, Robert A. *Starship Troopers.* New York: Ace, 1987.

Hitchens, Christopher E. *god is not Great.* New York: Twelve, 2007.

Hitler, Adolf. *Mein Kampf.* Translated by Ralph Manheim. New York: Houghton Mifflin Mariner, 1999.

Huxley, Aldous. *Brave New World.* New York: Harper Perennial, 2006.

Johnston, Sarah Iles (General Editor). *Religions of the Ancient World; A Guide.* Cambridge: The Belknap Press of Harvard University Press, 2004.

Kahneman, Daniel. *Thinking, Fast and Slow.* New York: Farrar, Straus and Giroux, 2013.

Kertzer, David I. *The Popes Against The Jews.* New York: Vintage Books, 2002.

Leach, Marjorie. *Guide To The Gods.* Santa Barbara: ABC-CLIO, 1992.

Lewis, C.S. *The Problem of Pain.* New York: HarperCollins Publishers, 2001.

Lewis, Sinclair. *It Can't Happen Here.* New York: Penguin Group, 2014.

Maestripieri, Dario (Editor). *Primate Psychology.* Cambridge: Harvard University Press, 2003.

Mencken, H.L. *A Religious Orgy In Tennessee: A Reporter's Account Of The Scopes Monkey Trial.* Hoboken: Melville House Publishing, 2006.

Mencken, H.L. *Treatise On The Gods.* Baltimore: The Johns Hopkins University Press, 1997.

Mills, David. *Atheist Universe; The Thinking Person's Answer to Christian Fundamentalism.* Berkeley: Ulysses Press, 2006.

Minchin, Tim. From the unparalleled master of modern satirical antireligious musical comedy, I recommend starting with his song *Thank You, God*, found on YouTube at https://www.youtube.com/watch?v=IZeWPScnolo, then his hilarious but expletive-filled *Pope Song* at https://www.youtube.com/watch?v=Be5hkPwwjeM, then the rest of his oeuvre.

Pinker, Steven A. *The Blank Slate; The Modern Denial of Human Nature.* New York: Penguin Books, 2003.

Posner, Gerald. *God's Bankers.* New York: Simon & Schuster, 2015.

Russell, Bertrand. *Why I Am Not A Christian, And Other Essays On Religion And Related Subjects.* New York: Simon and Schuster, 1957.

Sagan, Carl. *The Demon-Haunted World: Science As A Candle In The Dark.* New York: Ballantine Books, 1997.

Shermer, Michael B. *The Moral Arc; How Science Makes Us Better People.* New York, St. Martin's Griffin, 2015.

Shermer, Michael B. *The Believing Brain: From Ghosts and Gods to Politics and Conspiracies- How We Construct Beliefs and Reinforce Them As Truths.* St. Martin's Griffin, 2011.

Shubin, Neil. *Your Inner Fish: A Journey Into The 3.5-Billion Year History Of The Human Body.* New York: Vintage Books, 2009.

Sidman, Murray. *Tactics of Scientific Research.* New York: Basic Books, Inc.,

Publishers, 1960.

Sloman, Steven and Fernbach, Philip. *The Knowledge Illusion: Why We Never Think Alone.* New York: Riverhead Books, 2018.

Stringer, Chris. *Lone Survivors: How We Came To Be The Only Humans On Earth.* New York: St. Martin's Griffin, 2012.

Tzu, Sun. *The Art of War.* Translated, With Introductions And Commentary By Ralph D. Sawyer, with The Collaboration of Mei-Chun Sawyer. Boulder: Westview Press, 1994.

Varley, John. *Steel Beach.* New York: Ace Books, 1993.

Varley, John. *The Golden Globe.* New York: Ace Books, 1999.

Vile, John R. *A Companion to the United States Constitution and Its Amendments.* New York: Rowman and Littlefield Publishers, Inc., 2011.

Wade, Nicholas. *The Faith Instinct: How Religion Evolved And Why It Endures.* New York: Penguin Books, 2010.

Wade, Nicholas. *Before the Dawn: Recovering the Lost History of Our Ancestors.* New York: Penguin Books, 2006.

Wagner, Tony *The Global Achievement Gap; Why Even Our Best Schools Don't Teach the New Survival Skills Our Children Need - and What We Can Do About It.* New York: Basic Books, 2008.

Watson, James D. *The Double Helix: A Personal Account of the Discovery of the Structure of DNA.* Edited by Gunther S. Stent. New York: W.W. Norton and Company, 1980.

About The Author

Max Humana is a retired US military servicemember who still supports and defends the Constitution against all enemies both foreign and domestic through writing on the never-safe subjects of religion, politics, morality, humanity, truth and reason.

He is working to advance human knowledge and to rid the world of ignorance, stupidity and evil, and wants you to do the same.

Made in the
USA
Lexington, KY

54406412R00435